# Soil-Structure-Interaction Analysis in Time Domain

*Prentice Hall International Series
in Civil Engineering and Engineering Mechanics*

*William J. Hall, editor*

# Soil-Structure-Interaction Analysis in Time Domain

**John P. Wolf**

*Electrowatt Engineering Services Ltd.*
*Swiss Federal Institute of Technology*

Prentice Hall, Englewood Cliffs, New Jersey 07632

Library of Congress Cataloging-in-Publication Data

Wolf, John P. (date)
  Soil-structure-interaction analysis in time domain.

  Bibliography: p.
  Includes index.
  1. Soil dynamics.  2. Structural dynamics.  I. Title.
  TA710.W594   1987      624.1′5136       87-17493
  ISBN 0-13-822974-0

Editorial/production supervision: Sophie Papanikolaou
Manufacturing buyer: S. Gordon Osbourne

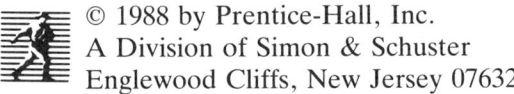

© 1988 by Prentice-Hall, Inc.
A Division of Simon & Schuster
Englewood Cliffs, New Jersey 07632

All rights reserved. No part of this book may be
reproduced, in any form or by any means,
without permission in writing from the publisher.

Printed in the United States of America

10  9  8  7  6  5  4  3  2  1

ISBN 0-13-822974-0  025

PRENTICE-HALL INTERNATIONAL (UK) LIMITED, *London*
PRENTICE-HALL OF AUSTRALIA PTY. LIMITED, *Sydney*
PRENTICE-HALL CANADA INC., *Toronto*
PRENTICE-HALL HISPANOAMERICANA, S.A., *Mexico*
PRENTICE-HALL OF INDIA PRIVATE LIMITED, *New Delhi*
PRENTICE-HALL OF JAPAN, INC., *Tokyo*
SIMON & SCHUSTER ASIA PTE. LTD., *Singapore*
EDITORA PRENTICE-HALL DO BRASIL, LTDA., *Rio de Janeiro*

# Contents

Foreword     xv

Preface     xvii

## 1 Introduction     1

1.1 Statement of Problem    **1**
1.2 Significant Features    **4**
1.3 Substructure and Direct Methods    **4**
1.4 Free-Field Response of Site    **7**
1.5 Organization of Text    **8**
Summary    **9**

## 2 Approximate Dynamic Model of Embedded Foundation     11

2.1 Dynamic-Stiffness Coefficient in Frequency Domain    **13**
2.2 Standard Lumped-Parameter Model    **14**

2.3 Spherical Cavity with Symmetric Waves  16
    2.3.1 Dynamic-Stiffness Coefficient  16
    2.3.2 Special Cases  20
2.4 Basic Discrete Model  20
2.5 Disk Foundation  23
2.6 Embedded Cylindrical Foundation  25
2.7 Embedded Prism Foundation  30
2.8 Strip Foundation  35
2.9 Square Foundation on Layered Half-Space  38
2.10 Material Damping  41
2.11 Disk with Mass on Half-Space  48
2.12 Hammer Foundation with Partial Uplift of Anvil  50
2.13 Complex Modal Analysis  54
2.14 Partial Uplift of Basemat  58
    2.14.1 Disk on Tensionless Half-Space  58
    2.14.2 Structural Model on Tensionless Half-Space  60
    2.14.3 Dynamic-Stiffness Coefficient for Circular Cavity in Tensionless Thin Layer  62
Summary  65
Problems  66

# 3 Direct Method  76

3.1 One-Dimensional Wave Propagation  78
    3.1.1 Prismatic Rod  78
    3.1.2 Conical Rod in Shear  81
    3.1.3 Spherical Cavity with Symmetric Waves  84
    3.1.4 Conical Rod in Torsion  86
3.2 Semi-Infinite Rod on Elastic Foundation  88
    3.2.1 Types of Waves  88
    3.2.2 Influence of Dispersion and Cutoff Frequency on Motion  91
    3.2.3 Propagation of Narrow-Banded Pulse  92
    3.2.4 Dynamic-Stiffness Coefficient in Frequency Domain  94
    3.2.5 Exact Solution of Benchmark Problem  96

        3.2.6  Spatial and Temporal Discretizations of Region
               up to Artificial Boundary  97
3.3   Superposition Boundary     **98**
        3.3.1  Averaging of Final Solutions with Symmetric
               and Antimetric Boundary Conditions  98
        3.3.2  Cancellation of Reflected Waves as They
               Occur  100
               *Concept, 100*
               *Theoretical formulation, 102*
               *Numerical implementation, 105*
               *Benchmark problem, 106*
3.4   Fictitious Material Damping     **106**
3.5   Viscous Damper     **107**
3.6   Doubly Asymptotic Approximation     **109**
3.7   Paraxial Boundary     **111**
        3.7.1  Concept  111
        3.7.2  Benchmark Problem  114
        3.7.3  Generalization  118
3.8   Extrapolation Algorithm     **119**
        3.8.1  Concept Illustrated with Benchmark Problem  119
               *Case 1, 121*
               *Case 2, 122*
        3.8.2  Extension  123
3.9   Location of Artificial Boundary     **125**
        3.9.1  Overview  125
        3.9.2  Body Waves  127
               *Spherical wave, 127*
               *Far field, 129*
        3.9.3  Surface Waves  132
               *Love wave, 132*
               *Rayleigh wave, 142*
3.10  Free-Field Loading     **147**
3.11  Structures with Elasto-Plastic Base Isolation
      Permitting Uplift and Slipping     **148**
        3.11.1  Computational Procedure  148
        3.11.2  Reactor Building Permitting Uplift
                and Slipping  151
        3.11.3  Reactor Building with Elasto-Plastic Base
                Isolation  153
Summary     **157**
Problems     **160**

## 4 Summary of Substructure Method in Frequency Domain — **175**

4.1 Basic Equation of Motion **176**
4.2 Dynamic Stiffness of Site **178**
    4.2.1 Three-Dimensional Wave Equation in Cartesian Coordinates 178
    4.2.2 Out-of-Plane Motion 181
    4.2.3 In-Plane Motion 182
    4.2.4 Three-Dimensional Wave Equation in Cylindrical Coordinates 184
    4.2.5 Assemblage of Dynamic-Stiffness Matrix 187
4.3 Free-Field Response of Site **187**
    4.3.1 Definition of Task 187
    4.3.2 Body Wave 188
    4.3.3 Surface Wave 189
4.4 Dynamic Stiffness of Embedded Foundation **189**
    4.4.1 Green's Function 189
        *Full infinite space, 190*
        *Full infinite plane, 190*
        *Axisymmetric layered half-space, 191*
        *Layered half-plane, 193*
    4.4.2 Boundary-Integral Equation 194
        *Reciprocity Theorem of Maxwell-Betti, 194*
        *Representation Theorem, 195*
    4.4.3 Different Formulations of Boundary-Integral Equations and Their Spatial Discretization 195
        *Weighted-residual method, 196*
        *Indirect boundary-element method, 198*
        *Direct boundary-element method, 199*
        *Dynamic stiffness of free field, 201*
4.5 Scattered Motion **201**
    4.5.1 Introductory Remarks 201
    4.5.2 Boundary-Integral Equation 202
    4.5.3 Different Formulations of Boundary-Integral Equations and Their Spatial Discretization 203
        *Weighted-residual method, 203*
        *Indirect boundary-element method, 204*
        *Direct boundary-element method, 204*
        *Irregular site, 205*
    4.5.4 Load Vector 208
Summary **209**

## 5 Hybrid Frequency—Time-Domain Analysis  **212**

5.1 Formulation  **213**
    5.1.1 Basic Procedure  213
    5.1.2 Convergence Criterion  216
    5.1.3 Comments  216

5.2 Mass Connected by Nonlinear Spring to Semi-Infinite Rod on Elastic Foundation  **217**

5.3 Stability Criterion for Harmonic Excitation  **220**
    5.3.1 One-Degree-of-Freedom System  220
        *Scope of investigation,* 220
        *Criterion,* 222
    5.3.2 Multiple-Degree-of-Freedom System  225

5.4 Properties of Convergence for Iterative Scheme Working with Segments of Time Interval  **227**
    5.4.1 Segmenting Procedure  227
    5.4.2 Nonlinear Two-Degree-of-Freedom System with Frequency-Independent Parameters  228
        *Investigated system,* 228
        *Rate of convergence,* 229
    5.4.3 Rigid Block on Half-Space with Partial Uplift  232
        *Investigated system,* 232
        *Results,* 234
        *Number of iterations,* 235

5.5 Stability Criterion for Transient Excitation  **236**
    5.5.1 Initial-Value Theorem  236
    5.5.2 Criterion  237
    5.5.3 Numerical Validation  238

Summary  **242**

## 6 Substructure Method Using Dynamic Stiffness of Soil  **244**

6.1 Basic Equation of Motion  **245**
    6.1.1 Stiffness Formulation  245
    6.1.2 Flexibility Formulation  248
    6.1.3 Scattered Motion  249

6.2 Dynamic-Stiffness Coefficient in Time Domain  **249**

  6.2.1 Calculation in Time Domain 249
  6.2.2 Transformation from Frequency to Time Domain 253
    *Fourier-Transform pair, 253*
    *Initial characteristics of regular part and corresponding high-frequency behavior, 255*
    *Layered half-space, 256*
    *Vertical dynamic-stiffness coefficient of disk on damped half-space, 257*
6.3 High-Frequency Behavior of Waves from Vibrating Source **258**
  6.3.1 Increased Directionality 258
  6.3.2 Dynamic-Stiffness Coefficient for Infinite Frequency 260
    *Plane vibrating surface, 260*
    *Curved vibrating surface, 263*
    *Material damping, 266*
6.4 Dynamic-Flexibility Coefficient in Time Domain **267**
  6.4.1 Calculation in Time Domain 267
  6.4.2 Transformation from Frequency to Time Domain 270
    *Fourier-Transform pair, 270*
    *Initial characteristics and corresponding high-frequency behavior, 271*
    *Initial value, 272*
    *Layered half-space, 272*
6.5 Hysteretic Damping **276**
  6.5.1 Non-Causal Behavior 276
  6.5.2 Spherical Cavity with Symmetric Waves 277
6.6 Computational Procedure **278**
  6.6.1 Newmark Time-Integration Scheme 278
  6.6.2 Explicit Algorithm 279
    *Predictor-corrector scheme, 279*
    *Stiffness formulation, 280*
    *Flexibility formulation, 281*
  6.6.3 Implicit Algorithm 282
    *Stiffness formulation, 282*
    *Flexibility formulation, 283*
6.7 Structure with Base Isolation **284**
6.8 Recursive Evaluation of Convolution Integral in Frequency Domain **286**
  6.8.1 Introductory Remarks 286

Contents     xi

      6.8.2   Spherical Cavity with Mass with Symmetric Waves   287
      6.8.3   Successive Fourier Transformations   289
            *Fourier series, 289*
            *Discrete Fourier Transform, 291*
            *Spherical cavity, 292*
      6.8.4   Recursive Procedure   293
            *Stiffness formulation with explicit time integration, 293*
            *Flexibility formulation with explicit time integration, 297*
            *Stiffness formulation with implicit time integration, 299*
            *Flexibility formulation with implicit time integration, 300*
      6.8.5   Number of Operations   301
6.9   Recursive Evaluation of Convolution Integral in Time Domain   **302**
      6.9.1   Introductory Remarks   302
      6.9.2   Impulse-Invariant Method   304
            *System of equations for recursive coefficients, 304*
            *Special cases, 306*
            *Optimum agreement of dynamic-stiffness coefficients, 307*
            *Exact recursive equation, 309*
      6.9.3   Segment Approach Based on z-Transform   310
            *Concept, 310*
            *Derivation, 311*
      6.9.4   Recursive Equation Directly from Dynamic-Stiffness Coefficient in Frequency Domain   315
            *Derivation, 315*
            *Least-square method, 317*
            *Example, 318*
      6.9.5   Number of Operations   321
      6.9.6   Semi-Infinite Rod on Elastic Foundation   322
            *Benchmark problem, 322*
            *Nonlinear structure, 324*
Summary   **326**
Problems   **333**

# 7   *Substructure Method Using Green's Function of Soil*   **361**

7.1   Green's Function   **362**
      7.1.1   Spherical Cavity with Symmetric Waves   362
            *Calculation in time domain, 362*
            *Transformation from frequency to time domain, 366*

- 7.1.2 Scalar Wave in Full Infinite Space 368
  - *Three-dimensional case, 368*
  - *Two-dimensional case, 370*
- 7.1.3 Elasto-Dynamic Wave in Full Infinite Space 372
  - *Three-dimensional case, 372*
  - *Two-dimensional case, 373*
- 7.1.4 Properties 374
- 7.1.5 Layered Half-Space 376
  - *General considerations, 376*
  - *Transformation from time-space domain to frequency-wave number domain, 376*
  - *Vertical load on free surface of half-space, 379*
  - *Horizontal load on free surface of half-space, 381*
  - *Comparison with two-dimensional case (half-plane), 383*

7.2 **Time-Dependent Boundary-Integral Equation in Elasto-Dynamics** **383**
- 7.2.1 Fundamentals 383
- 7.2.2 Reciprocity Theorem of Maxwell-Betti 384
- 7.2.3 Representation Theorem 386

7.3 **Spatial and Temporal Discretizations of Boundary-Integral Equation** **388**
- 7.3.1 Basic Procedure 388
- 7.3.2 Weighted-Residual Method 389
- 7.3.3 Indirect Boundary-Element Method 393
- 7.3.4 Direct Boundary-Element Method 400
- 7.3.5 Number of Operations 402

7.4 **Loaded Spherical Cavity with Symmetric Waves** **404**
- 7.4.1 Analytical Solution 404
- 7.4.2 Discretization 407
- 7.4.3 Result 407

7.5 **Flexibility of Rigid Circular Disk** **410**

7.6 **Structure with Partial Basemat Uplift** **411**

7.7 **Embedded Foundation with Separation of Sidewall and Uplift of Basemat** **414**
- 7.7.1 Illustrative Example 414
- 7.7.2 Computational Procedure 415
- 7.7.3 Linear Analysis 416
- 7.7.4 Nonlinear Analysis 418

Summary **421**
Problems **423**

*References* **431**

*Index* **437**

# Foreword

Remarkable advances have been made during the past decades in the development of methods for analysis of the dynamic behavior of civil engineering structures. One important reason for this recent rapid progress is that in the previous era, structural dynamics was not considered an important subject for civil engineers. Almost by definition, civil engineering structures were intended to resist only static loads, and effects of any possible dynamic variation of such loads were dealt with by use of an impact factor. Dynamic loads were recognized to be important in the design of structures intended to move—such as ships or airplanes—but these were not in the province of civil engineering. In rare instances, internally-applied harmonic loads were considered in the design of building or factory structures, but the essential concept of a civil engineering structure was that its supports were fixed in position hence the potential for its dynamic excitation was quite limited.

It was recognition of the importance of earthquake hazard that effectively brought structural dynamics into the field of civil engineering, and it was the concurrent development of digital computers that made it possible to carry out the more complicated calculations required for a structural dynamic analysis. In the United States, the Tehachapi, California earthquake of 1952 served as a very strong motivating factor; it clearly demonstrated the seismic vulnerability of many types of traditional civil engineering structures and also it occurred at the time when the first automatic digital computer systems were becoming available. This evidence of the potential for disaster if a severe earthquake should strike a major population center produced the important result that earthquake loads received increased emphasis in the teaching and practice of structural engineering in California. However, in view of the traditional concept that civil engineering structures were supported by a rigid foundation, it was natural to assume that the earthquake had the effect merely of introducing a specified motion at the base of the structure. In fact, for the tall slender buildings that were the principal earthquake hazard concern of structural engineers at that time, the fixed base seismic input was probably an adequate assumption in most cases. These tall buildings tended to be light and flexible, and typically they were built in very firm foundations; hence the dynamic response analysis was focused on evaluation of the building deformation relative to a rigid base.

The analytical procedures that were formulated to deal with this type of problem concentrated on the idealization of the structure, typically by the finite element method using direct stiffness assembly, and very rapid progress was made in development of general purpose structural analysis programs that were efficient in calculating the static and dynamic behavior of buildings. But when these same computer programs began to be applied to the earthquake response analysis of nuclear power generating facilities, it soon became evident that the basic assumption of a flexible structure fixed at the base was no longer valid. The nuclear power sturctures tended to be very massive and stiff, and frequently they had to be located on relatively soft soils; in such cases it was apparent that the response of the structure would induce significant dynamic deformations of the foundation system. Recognition of this fact led the field of structural dynamics to acquire an important new phase—dynamic soil-structure interaction—which was observed to play a major and often controlling role in the earthquake performance of nuclear power plants and other massive, stiff structures.

With his previous book, *Dynamic Soil-Structure Interaction* Dr. John P. Wolf made a great contribution to the engineering profession in setting forth the entire subject of dynamic soil-structure interaction. Material which originally was developed by many different researchers (including himself) was assembled and organized so as to be accessible to all dynamics analysts. However, the formulation presented in that text has two disadvantages—both resulting from performing the analysis in the frequency domain. First and most important is the fact that a frequency domain analysis can deal only with linear response, in which the response is linearly related to the applied loads. This means that the calculation does not serve one major purpose of earthquake engineering analysis—to predict the degree of damage to be expected from the design earthquake—because "damage" implies nonlinearity. The other disadvantage is that the typical structural analyst is not accustomed to working in the frequency domain; his natural approach is to consider the sequence of developments from one time to the next—that is, to apply the time domain concept. This mental obstacle has greatly limited the general acceptance of analytical procedures using frequency domain formulations.

Now in this new book, Dr. Wolf has brought the subject of dynamic soil-structure interaction into the time domain, making it accessible to engineers who have been prepared to deal with the dynamic response analysis of the structure but not of the foundation material. This is not to say that the analysis suddenly is made easy; developing time domain expressions for the foundation medium and accounting for realistic nonlinearities in both foundation and structure are major tasks. However, when these are done as well as possible, the dynamic response analysis can be performed by standardized step-by-step procedures, so the soil-structure interaction no longer requires a totally new way of thinking about the analysis. This, I think, is the important contribution made by Dr. Wolf in this new book, and I expect it to lead to much broader acceptance of soil-structure interaction as a phenomenon that should be given consideration in the dynamic analysis of all civil engineering structures.

Ray W. Clough<br>
Berkeley, California

# *Preface*

Linear soil-structure-interaction analysis based on the substructure method is generally performed in the frequency domain. These procedures are well established and are described in a textbook written by the same author (*Dynamic Soil-Structure Interaction*, Prentice-Hall, 1985). It is, however, more natural when solving a dynamic problem not to transform the equations of motion to the frequency domain but to remain in the time domain. It is then also possible to solve nonlinear systems, which play an increasing role in practical applications. This book describes the various methods of performing soil-structure-interaction analysis directly in the time domain.

The book is organized as follows. The substructure method working in the frequency domain is summarized in Chapter 4. The frequency-dependent dynamic stiffness of the unbounded soil is calculated using the boundary-element method, which allows the radiation condition to be taken into account rigorously. In Chapter 5, the hybrid frequency-time domain procedure is addressed, in which a series of linear analyses is performed in the frequency domain with pseudo-loads recalculated in the time domain after each linear analysis. Nonlinearities thus affect only the right-hand side of the equations of motion in this iterative procedure. The time-domain substructure method, involving convolution integrals of the dynamic stiffnesses or of the Green's functions, is discussed in Chapters 6 and 7. The convolution integrals arise because the boundary conditions are global in space and time. Their recursive evaluation makes this method computationally competitive. A simple discrete model with frequency-independent springs, dampers, and masses to represent the unbounded soil is examined in Chapter 2. The coefficients of this lumped-parameter model with only one additional degree of freedom are determined in such a way as to achieve an optimum fit of the dynamic stiffness in the frequency domain. The direct method of analyzing soil-structure interaction is addressed in Chapter 3. The various types of transmitting boundaries, which impose conditions local in both space and time, are examined. These enable only approximate models to be made of wave propagation to infinity.

The book serves both as a tutorial and a state-of-the-art compilation of these time-domain procedures, which now have reached a reasonable stage of development. Rigorous procedures based on the boundary-element method in the time domain and approximate models that still capture the essential features of the unbounded soil are discussed. All concepts are developed and illustrated using the same two simple one-dimensional examples that the reader can follow step by step. The same nonlinear applications from actual practice, such as the partial uplift of the basemat, are solved for different levels of sophistication by applying the various methods. The book can be regarded as volume 2, with only a minimal overlap, of *Dynamic Soil-Structure*

*Interaction* or as an independent text describing the time-domain analysis of (nonlinear) soil-structure interaction.

A summary is included at the end of each chapter. Then problems which enhance and extend the chapter material are presented, with the solution procedure specified in detail. References are included to help the reader find a more complete treatment of the subject.

The course on which this text is based has been taught for the past few years to graduate students in civil engineering at the Swiss Federal Institute of Technology in Zurich. Essential prerequisites are some knowledge of structural dynamics, as well as an understanding of elementary frequency-domain procedures for soil-structure-interaction analysis.

Without the valuable contributions of colleagues and students, the compilation of this book would not have been possible. In particular, the author is indebted to Messrs. G. Darbre, S. Mohasseb, M. Motosaka, P. Obernhuber, P. Skrikerud, D. Somaini, R. Vogt, and D. Wepf for their dedicated work. The financial support of the Swiss Federal Institute of Technology and of Electrowatt Engineering Services Ltd. is gratefully acknowledged.

John P. Wolf

# 1

# Introduction

## 1.1 STATEMENT OF PROBLEM

The fundamental objective of a dynamic soil–structure–interaction analysis is illustrated in Fig. 1-1. A structure which is subjected to a specified time-varying load is partially embedded in the ground, which extends to infinity. The dynamic response of the structure—and, to a lesser extent, of the ground—is to be determined. Instead of an exterior load applied directly to the structure (arising, for example, from rotating machinery) the excitation can also consist of an incident wave propagating in the ground (for example, caused by an earthquake or an explosion). The ground can be either (soft) soil or rock. In the following, the word *soil* is used in general for *ground*, as in the well-established term *soil–structure interaction*. It is also worth mentioning that the interaction effects are more pronounced if the structure is embedded in soil instead of founded on rock.

Typical cases, which can be analyzed based on the procedures developed in this text, are shown in Fig. 1-2. The three-dimensional nature of the problem can be represented properly. The methods are well established when the semi-infinite soil consists of horizontal layers resting on a half-space, both consisting of isotropic viscoelastic material with hysteretic damping (Fig. 1-2a). The properties vary with depth but remain constant within the individual layers. Today's state-of-the-art approach does not allow nonlinear behavior of this layered half-space to occur. The restriction of enforcing linearity can be justified, in many cases, because the response of that part of the soil located at a sufficient distance from the loaded structure and farther

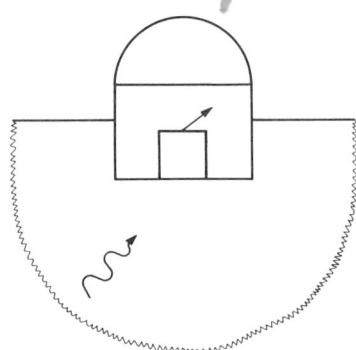

**Figure 1-1** Objective of analysis of soil–structure interaction.

away will be small due to the geometrical spreading. This part of the infinite soil can thus be identified with the layered half-space. In some important cases, the total soil will remain elastic, as, for example, for certain rotating machines for which the permissible displacements are extremely small to guarantee satisfactory performance. Between the layered half-space and the actual structure, an irregular soil region can exist. This finite zone encompasses all geometric irregularities and material inhomogeneities of the soil and can exhibit nonlinear behavior. The structure, which will, in general, be nonlinear, can be partially embedded. The shape of the structure–soil interface is quite general: the basemat and the sidewalls can, for example, be inclined and can vary from flexible to rigid. A buried structure, such as a tunnel, pipe, or underground protective structure, can be examined. More than one structure can be analyzed at the same time, thus taking the through-

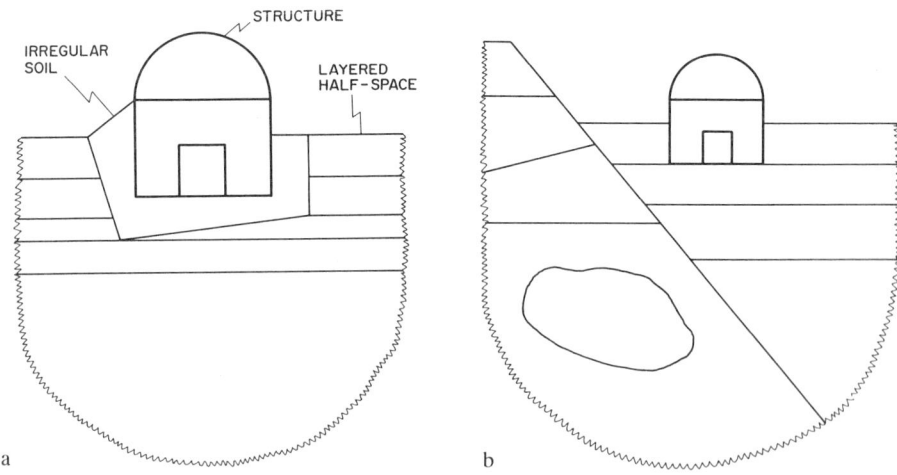

**Figure 1-2** Types of problems encompassed. a. Structure with adjacent irregular soil embedded in layered half-space. b. Structure embedded in system of layered half-spaces with free surfaces on different levels with inclined interface.

soil coupling effects of nearby structures into account. The irregular soil region can be regarded as part of the (generalized) structure exhibiting nonlinear behavior.

Limited experience exists in examining cases such as the one illustrated in Fig. 1-2b, where two (not necessarily horizontally) layered half-spaces are displaced with respect to one another along an inclined interface. The free surfaces of the half-spaces are thus on different levels. Inhomogeneous zones, consisting, for example, of boulders, can occur. In the following, if not stated otherwise, the case shown in Fig. 1-2a, with a horizontally layered half-space for which computationally efficient procedures exist, is addressed.

As already stressed, the layered half-space extending to infinity must remain linear. It is, of course, possible to select its material properties so they are compatible with the (averaged) strains reached during the excitation. This can be achieved by iteration, calculating a series of linear systems, adjusting the material properties at the end of each iteration (quasi-linear analysis).

If the loading does not act directly on the structure, but is introduced into the system through the soil such as for seismic excitation, the so-called free-field analysis has to be performed first to define the loading environment. This is discussed in Section 1.4. In this case, a nonlinear analysis of the layered half-space for the free-field response could, in principle, be performed, although a general wave pattern consisting of inclined body waves and surface waves cannot yet be assumed when a general nonlinear constitutive law is selected. The formulation of soil–structure interaction would still be applicable in this case with a nonlinear free-field analysis, if the additional nonlinearities caused by the actual interaction in the layered half-space extending to infinity can be ignored.

In summary, the procedures derived in this text are applicable to all cases where the nonlinear behavior is restricted to the structure and possibly to an irregular soil region adjacent to the structure, while the infinite soil modeled as a layered half-space is assumed to remain linearly elastic with material damping [W11].* This includes most practical cases. Examples are structures which perform in the nonlinear range for high seismic excitation; base-isolation systems with friction plates exhibiting strong nonlinear characteristics which have to be considered in design; local nonlinearities such as the partial uplift of the basemat and the separation occurring between the sidewalls of the base and the neighboring soil in the case of embedded structures; and the highly nonlinear soil behavior arising adjacent to the basemat.

The soil–structure-interaction analysis should only accommodate its own uncertainties and not, for example, those associated with the definition of the seismic input. The less sophisticated the methods are, the greater the degree of conservatism that has to be introduced. To avoid compounding

---

*The references are listed at the end of the book in alphabetical order.

conservatism upon conservatism, a best-estimate analysis should be performed, adding a factor of safety at the end of the total analysis.

## 1.2 SIGNIFICANT FEATURES

In a standard dynamic structural analysis either the load or the displacement is specified in every point of the structure. In a soil–structure-interaction analysis this is not the case, because the load acting on the structure or its displacement is only known after the response has been determined. In particular, the loading applied to the structure cannot be calculated from the stresses in the soil caused by the incident wave (by multiplying these by a factor of 2, as occurs when a wave is reflected at a fixed boundary). This procedure does not take the (mostly favorable) change of the load caused by the motion of the structure relative to that of the soil arising from the incident wave into account and also ignores the fact that the dynamic properties of the structure are influenced by those of the soil. In reality, dynamic soil–structure interaction involves very complex wave mechanisms [W22]. When the wave front of the incident wave propagating in the soil encounters the structure, scattering of the wave occurs. This leads to a load acting on the structure which will cause a motion of the structure, accompanied by the generation of a radiation wave in the soil and a relief of the loading on the structure. All these mechanisms are coupled: The motion of the structure depends on the loading acting on it, and the loading, in turn, is affected by the structural motion. The same applies to the soil.

The soil is in most cases a semi-infinite medium, an unbounded domain, or so large in extent that the simultaneous modeling together with the structure is impractical. In a dynamic problem it is insufficient to prescribe a zero displacement at a large distance from the structure, as is routinely done in statics. It is assumed, however, that once the waves leave the zone of interest, they will not return during the time of the analysis. It is thus ensured that only outgoing waves are present in the actual interaction analysis. This avoids an infinite energy buildup and will result in damping (called radiation damping) occurring even in an elastic unbounded system. The radiation condition will also lead to a boundary-value problem with a unique solution in an unbounded domain.

## 1.3 SUBSTRUCTURE AND DIRECT METHODS

To analyze the semi-infinite domain of the soil numerically, a surface is chosen which encloses the structure. The properties associated with the nodes on this so-called interaction horizon represent the significant features of the unbounded domain located on the exterior of this surface [S1]. As an example, for an analysis in the frequency domain, the dynamic-stiffness coefficients corresponding to the nodes on the interaction horizon describe the behavior of the unbounded soil. It should be remembered that the latter is always a linear system (Section 1.1). The numerical dynamic model encompasses those nodes which all lie within or on the horizon. A certain arbitrariness exists when selecting the location of the interaction horizon, which actually has no physical significance. Two extremes are possible. The interaction horizon can coincide with the (generalized) structure–soil interface leading to the substructure method; or it can be identical to an artificial boundary up to which the soil is modeled with, for example, finite elements, which results in the direct method.

In the substructure method, illustrated in Fig. 1-3a, the finite region located on the interior of the interaction horizon encompasses, besides the structure, all the geometric irregularities, the material inhomogeneities, and the nonlinear behavior of the adjacent soil, but no part of the linear horizontally layered half-space. If no irregular soil region is present—which is quite common—the interaction horizon will coincide with the structure–soil interface. The finite region made up of the structure and the irregular soil, which forms one of the two substructures, is normally modeled with finite elements. The properties of the other substructure—the unbounded soil on the exterior of the interaction horizon—are represented by a boundary condition linking the degrees of freedom associated with the nodes on the in-

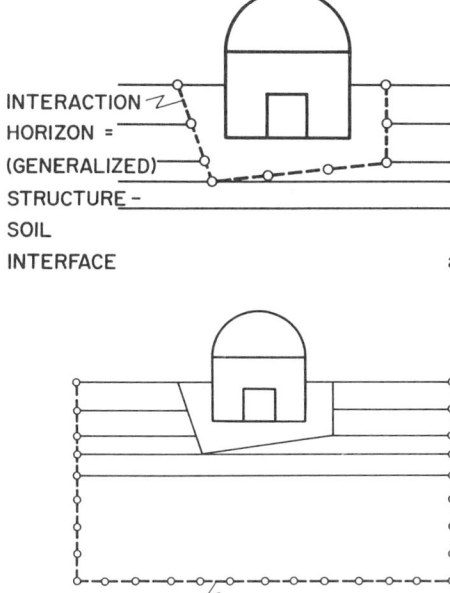

**Figure 1-3** Dynamic models for a. Substructure method. b. Direct method.

teraction horizon.  One can conceptionally view this boundary condition as being determined from the elimination of all dynamic degrees of freedom not lying on the interaction horizon of a mesh extending to infinity.  There are two main possibilities in performing a dynamic analysis: either in the frequency domain, where the excitation is decomposed into a Fourier series and the response is determined independently for each Fourier term corresponding to a specified frequency, or directly in the time domain.  For a calculation in the frequency domain, the boundary condition will be formulated using the frequency-dependent dynamic-stiffness coefficients relating the displacement amplitudes of the nodes on the interaction horizon to the corresponding force amplitudes.  This relationship will be fully coupled (spatial discretization).  For an analysis in the time domain, it will be shown that a total coupling of the time dimension also arises.  Actually, convolution integrals of the dynamic-stiffness coefficients in the time domain and the corresponding displacements have to be evaluated to calculate the forces.  The boundary condition representing the unbounded soil is thus global in space and time in an analysis in the time domain based on the substructure method; that is, all degrees of freedom of the nodes on the interaction horizon from the start of the excitation contribute to the forces.  The dynamic-stiffness coefficients can be determined using the boundary integral-equation procedure, also called the boundary-element method in discretized form.  The substructure method thus permits each substructure to be analyzed by the best-suited computational technique.  Most three-dimensional general purpose programs for the linear analysis of soil–structure interaction, which work in the frequency domain, are based on the substructure method [L9, L8].  The Fast Fourier Transform, coupled with an interpolation scheme in the frequency domain, leads to an efficient procedure for this case.  For an analysis in the time domain, however, the computational effort is substantial, as convolution integrals must be evaluated.  Their recursive evaluation makes this method computationally competitive.

In the direct method, illustrated in Fig. 1-3b, the region of the (linear) soil adjacent to the (generalized) structure–soil interface is also explicitly modeled up to the artificial boundary.  The latter must be introduced, as it is impossible to cover the unbounded soil domain with a finite number of elements with bounded dimensions.  Appropriate boundary conditions must be formulated which must represent the missing soil located on the exterior of the interaction horizon.  Besides modeling the soil's stiffness up to infinity, reflections of the outwardly propagating waves on the interaction horizon must be avoided, which thus acts as a transmitting boundary.  In principle, the same procedure as described for the substructure method should be used.  The boundary conditions must guarantee a unique, well-posed, and stable solution to the problem.  However, as intuitively expected, results of sufficient accuracy can be determined in many cases using only approximate boundary conditions on the interaction horizon, which is located at a significant distance

## Sec. 1.4 Free-Field Response of Site

from the structure. The resulting solution must closely resemble the response of the unbounded domain in the absence of these boundary conditions. Local boundary conditions in space and time can thus be selected. They can easily be chosen to be frequency-independent, as, for example, in the case of a viscous damper. For a time-domain analysis, the direct method can be attractive, although the total dynamic system which must be calculated is larger than in the substructure method.

The substructure and the direct methods are equivalent; that is, if they are implemented consistently, identical results will be obtained. With both methods the important aspects of soil–structure interaction can be captured, which is essential.

## 1.4 FREE-FIELD RESPONSE OF SITE

When the loads do not act directly on the structure, the loading environment must be specified first; that is, the response of the free field of the site must be determined before the actual soil–structure-interaction analysis can be performed. This consists of the spatial and temporal variations of the motion before excavating the soil and superimposing the structure, but of the interaction horizon only. This applies for the substructure and the direct methods. For a calculation based on the substructure method, the procedure is illustrated in Fig. 1-4. Depending on the details of the applied formulation (see Sections 3.10 and 6.1.1), the surface tractions must be determined on the same surface in addition to the displacements. This allows the right-hand side of the basic equation of motion of the soil–structure-interaction analysis to be determined, which, when the interaction horizon is kept fixed, is equal to the forces exerted on the interaction horizon by the motion.

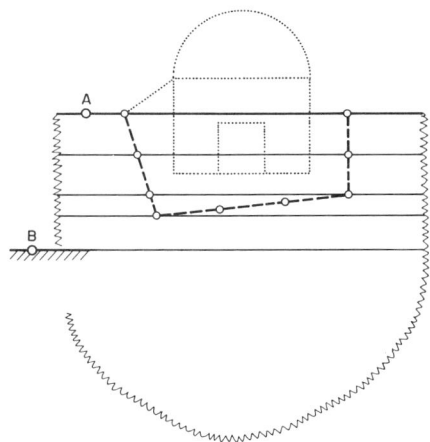

**Figure 1-4** Free field with control point.

The following procedure is applied for seismic excitation. At first, the location of the control point is chosen, either at the free surface (Point A in Fig. 1-4) or at an assumed rock outcrop—that is, on the level of the rock or competent soil—but assuming there is no soil on top (Point B). The latter choice is preferable for an identifiable shallow layer on firm soil. Then a control motion—for example, in form of a seismogram—is selected, and, finally, a wave pattern is assumed, whereby the control motion is rather arbitrarily decomposed into different types of waves. These three aspects are interrelated. The actual free-field analysis of the horizontally layered site is straightforward [S2, W11].

In principle, instead of examining the free field, other dynamic systems could be addressed. Actually, any system which is identical in the region external to the interaction horizon could be used to determine the loading environment. In practical terms, however, the system must be simple to calculate analytically or numerically. If, for instance, a soil–structure-interaction analysis has been performed using a simplified structural model, a calculation using a more detailed structure could be performed using the existing results of the interaction horizon to calculate the right-hand side of the basic equation of motion of the new analysis.

Finally, it is worth mentioning that the free-field analysis is often only one- or two-dimensional, whereas the soil–structure analysis is three-dimensional.

## 1.5 ORGANIZATION OF TEXT

The content of the following chapters is illustrated in Fig. 1-5. Linear soil–structure-interaction analysis is routinely performed in the frequency domain

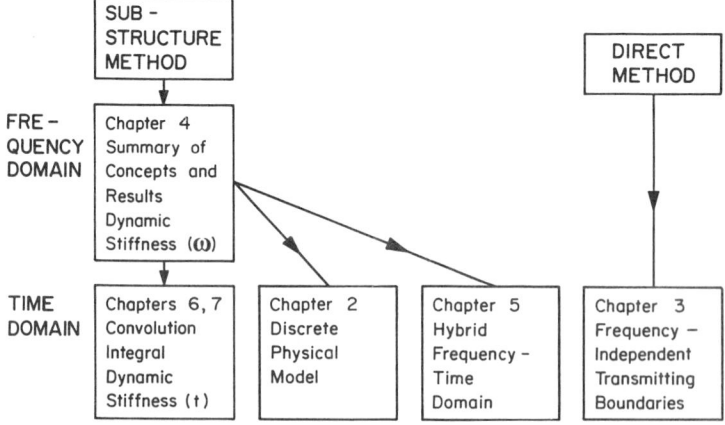

**Figure 1-5** Content of chapters as related to the substructure and direct methods in frequency and time domains.

based on the substructure method, whereby the (unbounded) soil's frequency-dependent dynamic-stiffness coefficients, determined based on the boundary-element method, take the radiation condition into account. The important concepts and results are summarized for easy reference in Chapter 4. For rigid surface and embedded foundations of various shapes and types of soil, the dynamic-stiffness coefficients as a function of the frequency $\omega$ have been published for the various degrees of freedom using the concepts of Chapter 4. Through selecting a simple physical model—actually a semi-infinite truncated cone or a spherically-symmetric full space with a cavity—and then modifying the parameters to achieve an optimum fit of the dynamic-stiffness coefficient in the frequency-domain with the corresponding published exact value, the properties of a discrete physical model are determined. It consists of a spring, dampers, and masses with frequency-independent coefficients. In this discrete model, one of the masses is not directly attached to the foundation node but connected to it through a damper. This concept can be used for analyses in the time domain—the natural choice for a dynamic analysis. The derivation is contained in Chapter 2. The computational procedures and corresponding computer programs which work in the frequency domain cannot be used for a nonlinear analysis, because the latter must be performed in the time domain. An iterative method where the nonlinearities only affect the right-hand sides of the equations of motion through the loads calculated from the so-called pseudo-forces, is, however, an attractive possibility. In this so-called hybrid frequency-time domain analysis described in Chapter 5, a series of linear analyses is performed in the frequency domain with the pseudo-loads recalculated in the time domain after each linear analysis. The time-domain substructure method, involving convolution integrals of the dynamic-stiffness coefficients or of the Green's functions, is discussed in Chapters 6 and 7, respectively. The basic equation of motion in the time domain is derived. The different formulations of the boundary-integral equations and their spatial and temporal discretizations are addressed. Recursive evaluations of the convolution integrals, which lead to a dramatic reduction in the computational effort, are also introduced. In Chapter 3, the direct method is examined, whereby the various approximate frequency-independent boundary conditions which model the wave propagation towards infinity are discussed. Nonlinear soil–structure interaction analyses from actual practice—such as the partial uplift of the basemat and the separation of the wall of the base from the adjacent soil—are presented as examples in the various chapters.

## SUMMARY

1. The computational procedures to analyze soil–structure interaction are applicable to an arbitrary nonlinear structure with an adjacent irregular

nonlinear soil region embedded in a horizontally layered half-space (with material damping), which is assumed to remain linearly elastic.
2. The loads acting on the structure can be determined only after the response of the structure–soil system is known; in particular, they cannot be calculated from the stresses in the soil caused by the incident wave.
3. In the substructure method the interaction horizon, on which the properties of the unbounded soil region are represented, coincides with the (generalized) structure–soil interface. The boundary conditions enforced on the interaction horizon are global in space and time. Convolution integrals occur.
4. In the direct method, part of the linear regular soil is explicitly modeled up to an artificial boundary where the interaction horizon is placed. Approximate boundary conditions, which are local in space and time and which are frequency-independent, are sufficient.
5. To define the loading environment, the free-field response of the interaction horizon is calculated as the first step of any soil–structure-interaction analysis, if the loading does not act directly on the structure.

# 2
# Approximate Dynamic Model of Embedded Foundation

In certain cases, the effect of the actual interaction of the soil and the structure on the response of the latter will be negligible and should thus not be considered. This applies, for example, to a flexible high structure with small mass where the influence of the higher modes (which are actually affected significantly by soil–structure interaction) on the seismic response remains small. It is then permissible to excite the base of the structure with the prescribed earthquake motion. In other cases, which include many everyday building structures, ignoring the interaction analysis can lead to an overly conservative design. It should be remembered that seismic-design provisions [N1] allow for a significant reduction of the equivalent static-lateral force (up to 30%) for soil–structure-interaction effects. To be able to calculate these cases and those where it is impossible to estimate the importance of the interaction before performing the analysis, a simple (discrete) model—which can be approximate—of the soil is needed. Such an approximate engineering-type method is very important for obtaining results of sufficient accuracy—which is limited because of the many uncertainties. Use of these simple models leads to some loss of precision, which in many cases is more than compensated for by their ease of use in parametric studies and by their role in developing the engineers' insight into the problem. As discussed in Section 1.3, dynamic soil–structure interaction is conveniently performed using the substructure method, in which the structure (whose approximations are well established) and the soil are modeled independently. To remain within this concept's framework, the properties of the discrete model of the soil should thus not depend on those of the structure or of the total structure–

soil system. Such a discrete model of the soil will consist of springs, dampers, and masses with frequency-independent coefficients. It can be straightforwardly incorporated into standard dynamic programs working directly in the time domain, which is the natural choice for a dynamic analysis. A nonlinear behavior of the structure can also be taken into account.

As expected, various choices exist for the soil's discrete model—often also called the lumped-parameter system. The simplest consists, for each degree of freedom of the rigid basemat, of a spring and a damper connected to the basemat's node—also associated with a mass of the soil—which is then added to the structure's mass. The properties of this single-degree-of-freedom system are frequency independent. This standard lumped-parameter model is discussed in Section 2.2. Before this critical evaluation, it is appropriate to summarize the important features of a dynamic-stiffness coefficient in the frequency domain, which is used to choose the optimal properties of a discrete system (Section 2.1). The performance of the single-degree-of-freedom system in reproducing the soil's actual response is obviously limited. A better agreement can be obtained by introducing another mass which is not directly attached to the node of the basemat but connected to it through a damper, whereby an additional dynamic degree of freedom is introduced. The dynamic-stiffness coefficient of this discrete model is the same as that of certain one-dimensional continuous models, such as the spherical cavity in a full space with spherically symmetric dilatational waves. This system is addressed in Section 2.3. The basic discrete model is examined in Section 2.4; in particular, how the masses, spring, and damping coefficients affect its dynamic-stiffness coefficient. After enforcing the static stiffness, the remaining coefficients of the discrete model are selected to achieve an optimum fit between the dynamic-stiffness coefficient in the frequency domain and the corresponding exact value published in the literature. The result of this curve-fitting procedure—consisting of the dimensionless coefficients characterizing the dampers and masses and of the formula for the spring coefficient of the discrete model—is presented for various types of foundations in the following sections: for a disk on the surface of a homogeneous half-space in Section 2.5; for an embedded cylinder in Section 2.6; for an embedded prism in Section 2.7; for a strip on the surface of a homogeneous half-plane in Section 2.8; and for a square on the surface of a layered half-space in Section 2.9. The incorporation of material damping is discussed in Section 2.10. A simple illustrative example demonstrates the superiority of the discrete model over the single-degree-of-freedom lumped-parameter model, when the results are compared to the exact solution (Section 2.11). A practical nonlinear case, consisting of a hammer foundation with an eccentrically mounted anvil for which partial uplift occurs, is examined, using a direct time integration procedure, in Section 2.12. For a linear system, a modal analysis can also be performed for the total system—consisting of the structure and the discrete model of the soil—using complex modeshapes (Section 2.13). An approx-

## Sec. 2.1 Dynamic-Stiffness Coefficient in Frequency Domain

imate procedure to model the partial uplift of the basemat using equivalent radii of a disk is described in Section 2.14.

## 2.1 DYNAMIC-STIFFNESS COEFFICIENT IN FREQUENCY DOMAIN

The concept of the dynamic-stiffness coefficient is illustrated through examining a single-degree-of-freedom system consisting of a spring with coefficient $K$, a dashpot with viscous damping coefficient $C$, and a mass $M$ [R4]. For harmonic excitation with frequency $\omega$, the applied load with amplitude $P(\omega)$ will cause a displacement with amplitude $u(\omega)$; these are related by the dynamic-stiffness coefficient in the frequency domain $S(\omega)$, a complex value, as

$$P(\omega) = S(\omega)u(\omega) \tag{2.1}$$

with

$$S(\omega) = K - \omega^2 M + i\omega C \tag{2.2}$$

The undamped natural frequency $\omega_o$ and the damping ratio $\zeta_v$ are equal to

$$\omega_o = \sqrt{\frac{K}{M}} \tag{2.3}$$

$$\zeta_v = \frac{C}{2\sqrt{KM}} \tag{2.4}$$

Nondimensionalizing $S(\omega)$ with its static value $S(\omega = 0) = K$ leads to

$$S(\omega) = K[k(\omega) + i\omega c(\omega)] \tag{2.5}$$

where the dimensionless spring coefficient $k(\omega)$ and the damping coefficient $c(\omega)$, both describing the dynamic part of the stiffness, are specified as

$$k(\omega) = 1 - \omega^2 \frac{M}{K} = 1 - \frac{\omega^2}{\omega_o^2} \tag{2.6a}$$

$$c(\omega) = \frac{C}{K} = 2\frac{\zeta_v}{\omega_o} \tag{2.6b}$$

The real part of $S(\omega)$ characterized by $k(\omega)$ varies with frequency as a second degree parabola, as shown in Fig. 2-1. The inertial effect of the mass $M$ leads to a decrease of $S(\omega)$, resulting in negative values for $\omega > \omega_o$. For a viscous damper, $c(\omega)$ is actually independent of $\omega$. The imaginary part of $S(\omega)$ specified by $\omega c(\omega)$, representing the energy loss in the system, increases linearly with $\omega$. This viscous damping mechanism represents the radiation of energy away from the foundation.

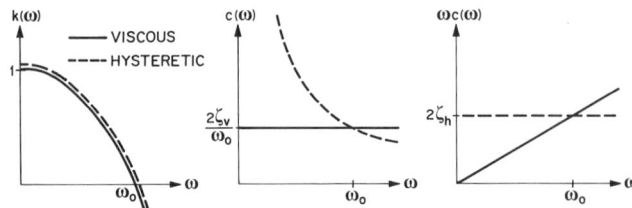

**Figure 2-1** Dynamic-stiffness coefficient of one-degree-of-freedom system.

For material damping which involves frictional losses, in principle, the differential equation of motion is nonlinear. Introducing an equivalent linearization and postulating that the energy loss in the system is independent of $\omega$, which is the case for (linear) hysteretic damping, leads to a complex static stiffness coefficient $K(1 + 2\zeta_h i)$. With $\zeta_h$ denoting the damping ratio, $S(\omega)$ is formulated as

$$S(\omega) = K(1 + 2\zeta_h i) - \omega^2 M \tag{2.7}$$

resulting in

$$k(\omega) = 1 - \frac{\omega^2}{\omega_o^2} \tag{2.8a}$$

$$c(\omega) = 2\frac{\zeta_h}{\omega} \tag{2.8b}$$

whereby Eq. 2.5 again applies. $k(\omega)$ is still a second-degree parabola. $c(\omega)$ varies inversely proportional to $\omega$. The imaginary part of $S(\omega)$ is constant.

## 2.2 STANDARD LUMPED-PARAMETER MODEL

The relationship between the amplitude of the displacement of a rigid massless base embedded in soil and the corresponding amplitude of the applied load is specified for each degree of freedom by the dynamic-stiffness coefficient $S(\omega)$, as in the case of the single-degree-of-freedom system addressed in Section 2.1. Equation 2.5 still applies. The static-stiffness coefficient $K$ can be used to nondimensionalize $S(\omega)$, leading to the dynamic part, which is again a complex number $k(\omega) + i\omega c(\omega)$. The latter is a function of the frequency of excitation $\omega$. The real component $k(\omega)$ mainly represents the stiffness and inertia of the soil, and the imaginary component $c(\omega)$ reflects the radiation of waves propagating away from the foundation and the material damping. For the soil, $k(\omega)$ and $c(\omega)$ will be more complicated functions of $\omega$ than those shown in Fig. 2-1.

For a rigid circular basemat (disk) with mass resting on the surface of an undamped homogeneous half-space, it is possible to model the soil approximately as a single-degree-of-freedom system for each of the 4 degrees of freedom (horizontal and vertical translations, and rocking and torsional rotations). The (frequency-independent) coefficients of the spring, viscous damper, and mass, corresponding to each degree of freedom of this standard lumped-parameter model, are selected to reproduce as closely as possible the actual response of the total dynamic system in the low- and medium-frequency ranges. Contrary to the concept of the substructure method, the curve-fitting technique is thus applied to the total system's dynamic-stiffness coefficient and not to that of the soil alone. The mass or mass moment of the soil's inertia is added to that of the rigid disk, leading to a reduction of the dynamic spring coefficient $k(\omega)$, as shown in Fig. 2-1. This added mass does not mean that an identifiable mass of the soil really exists which moves with the same amplitude and in phase with the disk over the whole range of frequency. It is a totally fictitious quantity which serves only to obtain a better fit between the dynamic-stiffness coefficient of the lumped-parameter model and that of the actual soil (in both cases taking the influence of the disk's mass into consideration). In the higher frequency range this added mass leads to a significant error in $k(\omega)$.

As expected, slightly different coefficients of the lumped-parameter model are derived depending on the details of the curve-fitting procedure [W3, R1]. The following nomenclature is introduced. $m$ denotes the mass for the two translational degrees of freedom, the mass moment of inertia for the rocking motion, and the polar mass moment of inertia for the torsional motion with respect to the center of the disk of radius $a$. $G$, $v$, and $\rho$ are the shear modulus, Poisson's ratio, and the mass density of the soil. The spring coefficient $K$ is equal to the static stiffness. The dimensionless frequency $a_o$ is introduced as

$$a_o = \frac{\omega a}{c_s} \qquad (2.9)$$

where $c_s$ denotes the shear-wave velocity

$$c_s = \sqrt{\frac{G}{\rho}} \qquad (2.10)$$

The damping coefficient $C$ and the mass $M$ (which is actually a mass moment of inertia for the two rotational degrees of freedom) are specified based on dimensionless coefficients $\gamma$ and $\mu$ as

$$C = \frac{a}{c_s} K\gamma \qquad (2.11a)$$

$$M = \frac{a^2}{c_s^2} K\mu \qquad (2.11b)$$

**TABLE 2-1** Static Stiffness and Dimensionless Coefficients of Standard Lumped-Parameter Model for Disk with Mass (Homogeneous Half-Space)

| | Static Stiffness $K$ | Dimensionless Coefficients for | |
|---|---|---|---|
| | | Damper $\gamma$ | Mass $\mu$ |
| Horizontal | $\dfrac{8Ga}{2-\nu}$ | 0.58 | 0.095 |
| Vertical | $\dfrac{4Ga}{1-\nu}$ | 0.85 | 0.27 |
| Rocking | $\dfrac{8Ga^3}{3(1-\nu)}$ | $\dfrac{0.3}{1+\dfrac{3(1-\nu)m}{8a^5\rho}}$ | 0.24 |
| Torsional | $\dfrac{16Ga^3}{3}$ | $\dfrac{0.433}{1+\dfrac{2m}{a^5\rho}}\sqrt{\dfrac{m}{a^5\rho}}$ | 0.045 |

$K$, $\gamma$, and $\mu$ are listed in Table 2-1. For the 2 rotational degrees of freedom, the ratio of the mass moments of inertia of the disk and of the soil arises in the formulas.

As this simplest possible lumped-parameter model for the soil consists only of a single-degree-of-freedom system, the agreement beteween its dynamic-stiffness coefficient as a function of frequency and those of the actual half-space is limited. $c(\omega)$ is constant; $k(\omega)$ is represented by a parabola (Fig. 2-1) which restricts the use of this lumped-parameter model to the low- and intermediate-frequency ranges ($a_o < 1.5$). It is also inappropriate to try generalizing the model to represent a layered half-space or an embedded foundation. This is possible, however, when an additional degree of freedom is introduced, which then leads to the so-called discrete model addressed in Section 2.4. Another possibility is examined in Problem 2.1.

## 2.3 SPHERICAL CAVITY WITH SYMMETRIC WAVES

### 2.3.1 Dynamic-Stiffness Coefficient

A vibrating base embedded in a layered half-space generates a very complicated wave field made up of $P$-, $S$-, and surface waves. To determine the general form one could expect the dynamic-stiffness coefficient to have, a very simple one-dimensional physical model is examined: a spherical cavity embedded in a full undamped space with only spherically symmetric radial $P$-waves occurring. In Fig. 2-2, the spherical cavity of radius $a$ is shown. A uniform pressure with amplitude $p(a_o)$ acts normal to the cavity's wall, re-

## Sec. 2.3  Spherical Cavity with Symmetric Waves

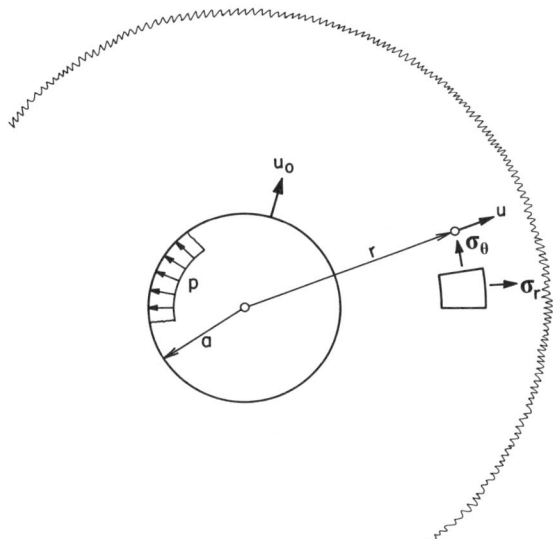

**Figure 2-2** Spherical cavity with uniform pressure (section).

sulting in an amplitude of the radial displacement $u_o(a_o)$ at the same location. $a_o$ denotes the dimensionless frequency of the harmonic excitation

$$a_o = \frac{\omega a}{c_p} \qquad (2.12)$$

defined using the dilatational wave velocity $c_p$

$$c_p = \sqrt{\frac{\lambda + 2G}{\rho}} \qquad (2.13)$$

and the two Lamé constants $\lambda$ and $G$, expressed as

$$\lambda = \frac{\nu E}{(1 + \nu)(1 - 2\nu)} \qquad (2.14a)$$

$$G = \frac{E}{2(1 + \nu)} \qquad (2.14b)$$

$E$ and $\nu$ are the modulus of elasticity and Poisson's ratio. The dynamic-stiffness coefficient $S(a_o)$ relates $u_o(a_o)$ to $p(a_o)$:

$$p(a_o) = S(a_o)u_o(a_o) \qquad (2.15)$$

For spherical symmetry, all variables depend on the radial coordinate $r$ only. The amplitude of the only displacement, which arises in the radial direction, is denoted as $u(a_o)$. The radial strain with amplitude $\varepsilon_r(a_o)$ and the normal strain in any direction perpendicular to $r$ with amplitude $\varepsilon_\theta(a_o)$ are specified as

$$\varepsilon_r(a_o) = u(a_o)_{,r} \qquad (2.16a)$$

$$\varepsilon_\theta(a_o) = \frac{u(a_o)}{r} \qquad (2.16b)$$

The dynamic-equilibrium equation for harmonic excitation is formulated as

$$\sigma_r(a_o)_{,r} + \frac{2}{r}[\sigma_r(a_o) - \sigma_\theta(a_o)] = -\rho\omega^2 u(a_o) \qquad (2.17)$$

where $\sigma_r(a_o)$ and $\sigma_\theta(a_o)$ are the amplitudes of the stresses in the corresponding directions. Substituting Eq. 2.16 into Hooke's law and then introducing the stresses in Eq. 2.17 leads to the equation of motion

$$u(a_o)_{,rr} + \frac{2}{r}u(a_o)_{,r} - \frac{2u(a_o)}{r^2} + \frac{a_o^2}{a^2}u(a_o) = 0 \qquad (2.18)$$

The displacement amplitude can be expressed as a function of the potential $\varphi(a_o)$ as

$$u(a_o) = \varphi(a_o)_{,r} \qquad (2.19)$$

Substituting Eq. 2.19 into Eq. 2.18 leads to the wave equation expressed in the potential. This equation is satisfied if the following equation applies for the product $r\varphi(a_o)$:

$$[r\,\varphi(a_o)]_{,rr} = -\frac{a_o^2}{a^2} r\,\varphi(a_o) \qquad (2.20)$$

The solution of this one-dimensional wave equation is

$$\varphi(a_o) = c_1 \frac{\exp\left(-i\frac{a_o}{a}r\right)}{r} + c_2 \frac{\exp\left(i\frac{a_o}{a}r\right)}{r} \qquad (2.21)$$

with the two integration constants $c_1$ and $c_2$. The displacement amplitude follows from Equation 2.19 as

$$u(a_o) = c_1\left(-\frac{1}{r^2} - i\frac{a_o}{ar}\right)\exp\left(-i\frac{a_o}{a}r\right) + c_2\left(-\frac{1}{r^2} + i\frac{a_o}{ar}\right)\exp\left(i\frac{a_o}{a}r\right)$$

$$(2.22)$$

To interpret Eq. 2.22 physically, recall that the displacement equals the product of the corresponding amplitude and the factor $\exp(+i\omega t)$. As the expression $\exp[i\omega(t - r/c_p)]$ describes a wave propagating in the positive radial direction with the velocity $c_p$, the first term in Eq. 2.22 corresponds to an

## Sec. 2.3 Spherical Cavity with Symmetric Waves

outgoing spherical wave. Analogously, the second is associated with an incoming wave. As only outgoing waves will exist, this radiation condition sets $c_2 = 0$.

For a prescribed displacement amplitude $u_o(a_o) = u(r = a, a_o)$

$$c_1 = -\frac{a^2}{1 + i a_o} \exp(i a_o) u_o \qquad (2.23)$$

follows. The pressure amplitude $p(a_o) = -\sigma_r(r = a, a_o)$ follows from Hooke's law, substituting Eq. 2.16 and using Eqs. 2.22 and 2.23. The dynamic-stiffness coefficient defined in Eq. 2.15 then follows as

$$S(a_o) = 4\frac{G}{a}\left(1 + \frac{-\frac{1}{4}\frac{c_p^2}{c_s^2}a_o^2 + i a_o \frac{1}{4}\frac{c_p^2}{c_s^2}a_o^2}{1 + a_o^2}\right) \qquad (2.24)$$

The static-stiffness coefficient $K$ equals $4G/a$. Nondimensionalizing $S(a_o)$ as

$$S(a_o) = K[k(a_o) + i a_o c(a_o)] \qquad (2.25)$$

defines the dimensionless dynamic-spring and -damping coefficients as

$$k(a_o) = 1 - \frac{\frac{1}{4}\frac{c_p^2}{c_s^2}a_o^2}{1 + a_o^2} \qquad (2.26a)$$

$$c(a_o) = \frac{1}{4}\frac{c_p^2}{c_s^2}\frac{a_o^2}{1 + a_o^2} \qquad (2.26b)$$

$k(a_o)$ and $c(a_o)$ are plotted versus the dimensionless frequency for $\nu = \frac{1}{3}(c_p = 2c_s)$ in Fig. 2-3. The curves corresponding to this case of the spherically symmetric P-waves are quite different from those associated with the single-degree-of-freedom lumped-parameter model shown in Fig. 2-1. However, as demonstrated in Section 2.4, one can select the parameters of a specific

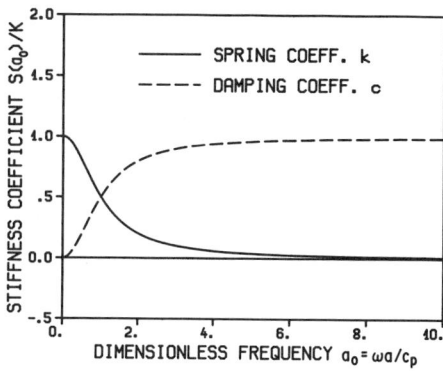

**Figure 2-3** Dynamic-stiffness coefficient of spherical cavity.

### 2.3.2 Special Cases

It is instructive to study two special cases. For an excitation in the very high frequency range ($a_o \to \infty$), $k(a_o)$ converges to $1 - \frac{1}{4} c_p^2/c_s^2$, $c(a_o)$ to $\frac{1}{4} c_p^2/c_s^2$ (Eq. 2.26), and $S(a_o)$ to $i\omega\rho c_p$ (Eq. 2.25); that is, the force–displacement relationship of Eq. 2.15 is formulated for the limit directly in the time domain as

$$p = \rho c_p \dot{u} \qquad (2.27)$$

which represents a dashpot with a viscous damping coefficient $\rho c_p$.

For an incompressible medium ($\nu = 0.5 \to \lambda = \infty \to c_p = \infty$), Eq. 2.24 is transformed to a real quantity

$$S(\omega) = 4 \frac{G}{a} \left[ 1 - \frac{1}{4} \left( \frac{\omega a}{c_s} \right)^2 \right] \qquad (2.28)$$

The corresponding force–displacement relationship of Eq. 2.15 can again be expressed in the time domain as

$$p = 4 \frac{G}{a} u_o + a\rho \ddot{u}_o \qquad (2.29)$$

The first term corresponds to a spring with a coefficient $4G/a$ and the second to a mass $a\rho$. No radiation damping occurs in the incompressible case.

In principle, in this text the same letter is used to designate a variable in the time domain (for example, $u$ for the displacement) and the corresponding amplitude in the frequency domain—whereby as a rule the frequency is added as an argument [$u(\omega)$ or $u(a_o)$]. If there is no possibility of confusion arising, the argument can also be omitted to simplify the nomenclature. To improve clarity, the time can be included as an argument in a variable in the time domain [$u(t)$ or $u(\bar{t})$; $\bar{t}$ being a dimensionless time]. The arguments can be deleted in figures.

## 2.4 BASIC DISCRETE MODEL

The one-dimensional discrete model shown in Fig. 2-4, employed to represent one component of the motion, has two dynamic degrees of freedom: at the basemat (to which the structure is connected), the foundation node $O$ with the mass $M_o$, which is attached to a rigid support with a spring having a coefficient $K$ and with a damper having a coefficient $C_o$; and the node 1 with the mass $M_1$, which is connected to the node $O$ through a damper having a

## Sec. 2.4 Basic Discrete Model

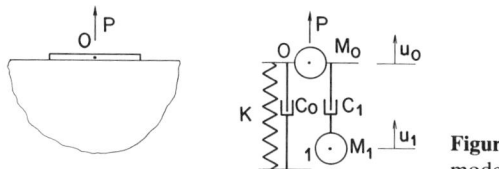

Figure 2-4 One-dimensional discrete model.

coefficient $C_1$. In certain circumstances, some of the elements will be missing. This model is essentially introduced in Ref. [M1].

The dynamic-stiffness coefficient in the frequency domain at the foundation node relating the displacement amplitude $u_o(\omega)$ to the applied load amplitude $P(\omega)$ is calculated as follows [W16]. Formulating the equilibrium equations in the two nodes for harmonic motion of frequency $\omega$ leads to

$$-\omega^2 M_1 u_1(\omega) + i\omega C_1(u_1(\omega) - u_o(\omega)) = 0 \tag{2.30a}$$

$$-\omega^2 M_o u_o(\omega) + i\omega(C_o + C_1)u_o(\omega) - i\omega C_1 u_1(\omega) + K u_o(\omega) = P(\omega) \tag{2.30b}$$

Eliminating $u_1(\omega)$ from Eq. 2.30 results in

$$P(\omega) = K\left[1 - \frac{\dfrac{\omega^2 M_1}{K}}{1 + \dfrac{\omega^2 M_1^2}{C_1^2}} - \frac{\omega^2 M_o}{K} + i\omega\left(\frac{M_1}{C_1}\frac{\dfrac{\omega^2 M_1}{K}}{1 + \dfrac{\omega^2 M_1^2}{C_1^2}} + \frac{C_o}{K}\right)\right]u_o(\omega)$$

(2.31)

The dimensionless frequency $a_o$ and the dimensionless coefficients of the dampers $\gamma_o$, $\gamma_1$ and of the masses $\mu_o$, $\mu_1$ are introduced as

$$a_o = \frac{\omega a}{c_s} \tag{2.32}$$

$$C_o = \frac{a}{c_s} K\gamma_o \tag{2.33a}$$

$$C_1 = \frac{a}{c_s} K\gamma_1 \tag{2.33b}$$

$$M_o = \frac{a^2}{c_s^2} K\mu_o \tag{2.33c}$$

$$M_1 = \frac{a^2}{c_s^2} K\mu_1 \tag{2.33d}$$

where $a$ represents a characteristic length of the foundation (for example, the radius for a disk) and $c_s$ denotes the shear-wave velocity. Equation 2.31 is then transformed to

$$P(a_o) = K\left[1 - \frac{\mu_1 a_o^2}{1 + \frac{\mu_1^2}{\gamma_1^2} a_o^2} - \mu_o a_o^2 + ia_o\left(\frac{\mu_1}{\gamma_1}\frac{\mu_1 a_o^2}{1 + \frac{\mu_1^2}{\gamma_1^2} a_o^2} + \gamma_o\right)\right]u_o(a_o)$$

(2.34)

With $K$ representing the static-stiffness coefficient, the force–displacement relationship is conveniently formulated as

$$P(a_o) = K[k(a_o) + ia_o c(a_o)]u_o(a_o) \qquad (2.35)$$

where $k(a_o)$ and $c(a_o)$ denote the dimensionless dynamic coefficients of the spring and the damper. Comparing the Eqs. 2.34 and 2.35 leads to

$$k(a_o) = 1 - \frac{\mu_1 a_o^2}{1 + \frac{\mu_1^2}{\gamma_1^2} a_o^2} - \mu_o a_o^2 \qquad (2.36a)$$

$$c(a_o) = \frac{\mu_1}{\gamma_1}\frac{\mu_1 a_o^2}{1 + \frac{\mu_1^2}{\gamma_1^2} a_o^2} + \gamma_o \qquad (2.36b)$$

The coefficients $\gamma_1$ and $\mu_1$ appear in both coefficients.

Comparing the dynamic-stiffness of the spherical cavity embedded in a full space (Eqs. 2.25, 2.26) with that of the discrete model (Eqs. 2.35, 2.36), it follows that full agreement for all frequencies can be reached by selecting $K = 4G/a$, $\gamma_o = 0$, $\gamma_1 = c_p/(4c_s)$, $\mu_o = 0$, and $\mu_1 = 1/4$ (observe that $a_o$ is defined with different wave velocities in the two cases, Eqs. 2.12 and 2.32). This means that the discrete model with only 2 degrees of freedom exactly represents the continuous physical model with waves propagating towards infinity. Other physical models exist which lead to the same discrete model: the semi-infinite truncated cone in translation perpendicular to its axis and in rocking [M1] and in torsion [V4]. These are discussed further in Sections 3.1.2 and 3.1.4.

The discrete model shown in Fig. 2-4 forms the starting point from which one can approximately analyze quite general foundations directly in the time domain. The spring with the coefficient $K$ represents the static stiffness. The remaining dimensionless coefficients $\gamma_o$, $\gamma_1$ of the dampers and $\mu_o$, $\mu_1$ of the masses are selected so as to achieve an optimum fit between the dynamic-stiffness coefficient of the discrete model and the corresponding exact value. It turns out that the coefficients $\gamma_o$ to $\mu_1$, which can easily be determined (individually or in groups of two), are real and even positive values (in contrast to the model addressed in Problem 2.1). The approximate procedure is thus semi-empirical. Reviews of the dynamic-stiffness coefficients' exact values can be found in Ref. [G1] and in the references contained in the next sections. Most likely, as more exact results of the dynamic-stiffness coefficients become available in the future, the scope of the discrete model's application can be expanded considerably. However, restrictions do exist. For instance, the discrete model cannot represent the layer built-in at its base, whose damping coefficient of the dynamic stiffness vanishes for an excitation below the fundamental frequency of the layer (cutoff frequency).

The rest of this chapter deals first with various undamped elastic foundations specifying $K$, $\gamma_o$, $\gamma_1$, $\mu_o$, $\mu_1$ in easily accessible tables, including comparisons between the dynamic-stiffness coefficients of the discrete model and those of more rigorous procedures; then it incorporates material damping; and finally it presents linear and nonlinear applications.

## 2.5 DISK FOUNDATION

The static-stiffness coefficients $K$ and the dimensionless coefficients of the dampers $\gamma_o$, $\gamma_1$ and of the masses $\mu_o$, $\mu_1$ for a massless rigid disk resting on the surface of an elastic undamped homogeneous half-space are specified in Table 2-2 [W16]. For each degree of freedom (horizontal, vertical, rocking, torsional), a discrete model independent of the others and acting in the corresponding direction is introduced. The radius of the disk is denoted by $a$, $G$ is the shear modulus, $\nu$ represents Poisson's ratio, and $\rho$ is the mass density. The shear-wave velocity $c_s$ equals $\sqrt{G/\rho}$. The coefficients $\gamma_o$, $\gamma_1$, $\mu_o$, $\mu_1$ depend—with the exception of the torsional motion—on $\nu$. The equations are derived making use of the values for discrete values of $\nu$ specified in Ref. [V3]. The disk is supported by the static spring for all degrees of freedom. For the horizontal motion, a damper with the coefficient $C_o$ in parallel to the spring is also present (Fig. 2-4). For rocking, the dimensionless spring and damping coefficients $k_r$ and $c_r$, determined from Eq. 2.36 using the values in Table 2-2, are plotted as lines as a function of $a_o$ in Fig. 2-5. The more rigorous values of Ref. [V2] are also shown as distinct points. For all three Poisson's ratios, the agreement is satisfactory. For $\nu = 0$ and $= 1/3$, the discrete model consists, in addition to the static spring, only of the mass moment of inertia $M_1$ with its own degree of freedom and the connecting rotational damper with the coefficient $C_1$. These two elements are well suited to model the decrease of $k_r$ and the increase of $c_r$ with increasing $a_o$. As $c_r$ is zero for $a_o \rightarrow 0$, the damper with the coefficient $C_o$ is not present. For $\nu > 1/3$, the mass $M_o$ is also present; it is attached to the disk and thus vibrates in phase. Such a mass also occurs for the case of the incompressible spherical cavity (Section 2.3.2). This element is responsible for the strong decrease of $k_r$, which even becomes negative for sufficiently large $a_o$. The corresponding comparison for the vertical degree of freedom is shown in Fig. 2-6.

**TABLE 2-2 Static Stiffness and Dimensionless Coefficients of Discrete Model for Disk Foundation (Homogeneous Half-Space)**

| | Static Stiffness $K$ | Dampers | | Masses | |
|---|---|---|---|---|---|
| | | $\gamma_o$ | $\gamma_1$ | $\mu_o$ | $\mu_1$ |
| Horizontal | $\dfrac{8Ga}{2-\nu}$ | $0.78 - 0.4\nu$ | — | — | — |
| Vertical | $\dfrac{4Ga}{1-\nu}$ | $0.8$ | $0.34 - 4.3\nu^4$ | $\nu < \dfrac{1}{3}$ : $0$ <br> $\nu > \dfrac{1}{3}$ : $0.9\left(\nu - \dfrac{1}{3}\right)$ | $0.4 - 4\nu^4$ |
| Rocking | $\dfrac{8Ga^3}{3(1-\nu)}$ | — | $0.42 - 0.3\nu^2$ | $\nu < \dfrac{1}{3}$ : $0$ <br> $\nu > \dfrac{1}{3}$ : $0.16\left(\nu - \dfrac{1}{3}\right)$ | $0.34 - 0.2\nu^2$ |
| Torsional | $\dfrac{16Ga^3}{3}$ | — | $0.29$ | — | $0.2$ |

Sec. 2.6   Embedded Cylindrical Foundation   25

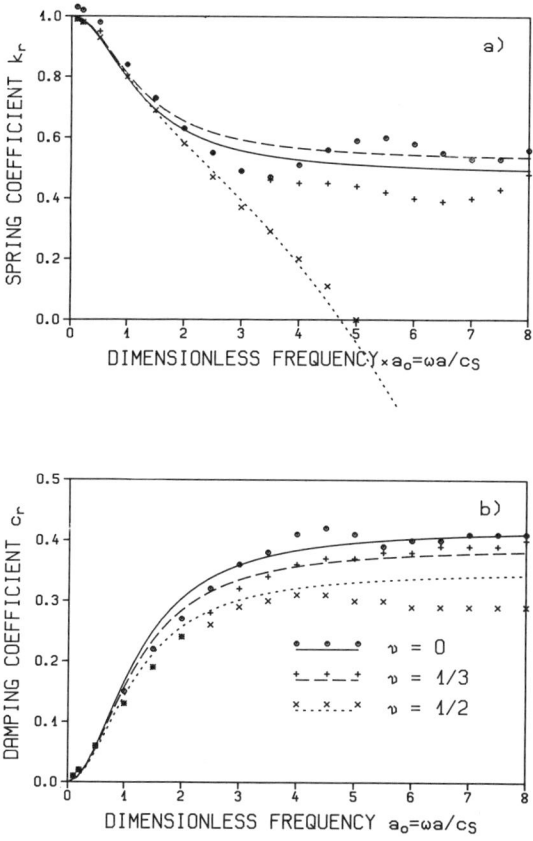

**Figure 2-5**  Dynamic-stiffness coefficient of disk for rocking motion.

The agreement is better for the damping coefficient than for the spring coefficient. In this case, for $\nu > 1/3$, the discrete model consists of all elements shown in Fig. 2-4.

## 2.6 EMBEDDED CYLINDRICAL FOUNDATION

The discrete model of a massless rigid cylindrical foundation with a vertical axis embedded in an elastic undamped homogeneous half-space is addressed (Fig. 2-7). The embedment of the cylinder of radius $a$ is denoted as $e$. $G$, $\nu$, and $\rho$ are the shear modulus, Poisson's ratio, and the mass density of the elastic half-space.

For the vertical and the torsional degrees of freedom, the discrete models acting in the corresponding directions can be used directly. For an embedded foundation, a (nonnegligible) dynamic-stiffness coefficient, which couples the

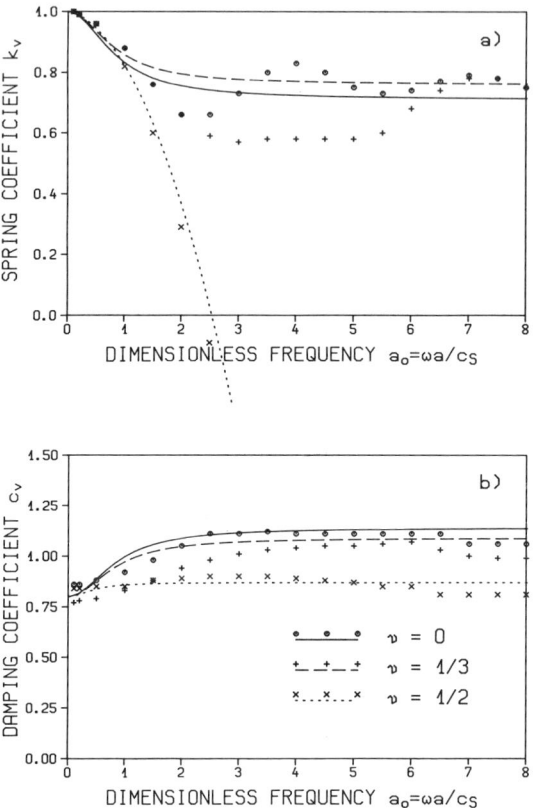

**Figure 2-6** Dynamic-stiffness coefficient of disk for vertical motion.

horizontal and rocking degrees of freedom referred to the center 0 of the circular basemat, arises. To take this effect into account, the discrete model corresponding to the horizontal degree of freedom is connected eccentrically to point 0 (Fig. 2-8). The eccentricities of the horizontal static spring with a coefficient $K_h$ and of the damper with a coefficient $C_{oh}$ are denoted as $f_K$

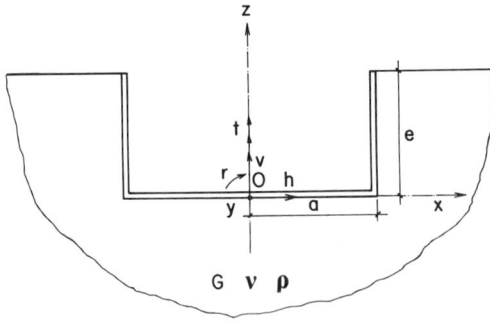

**Figure 2-7** Embedded cylindrical foundation.

## Sec. 2.6  Embedded Cylindrical Foundation

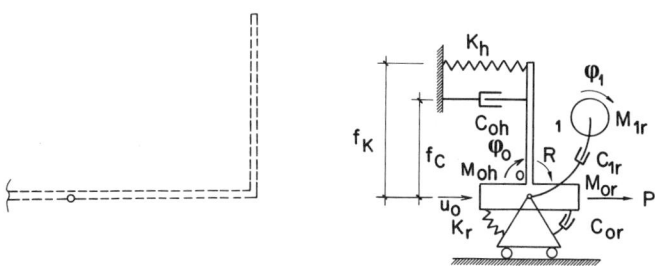

**Figure 2-8** Discrete model to represent coupling of horizontal and rocking motions for embedded foundation.

and $f_C$, respectively. The vertical bar connecting these two elements to point 0 is rigid. A mass $M_{oh}$ at point 0 is also present for the horizontal degree of freedom. For the rocking degree of freedom, all elements shown in Fig. 2-4 arise. A (second) subscript $r$ is used to denote all coefficients corresponding to this degree of freedom. For Poisson's ratio $\nu = 1/4$, for which the dimensionless coefficients are specified in the following, the coefficients $M_{oh}$, $C_{or}$, and $M_{or}$ vanish. They become non-zero when material damping is introduced (Section 2.10).

To study the behavior of the coupling term of the horizontal and rocking degrees of freedom, the force–displacement relationships at point 0 are established. The amplitudes of the horizontal load and of the moment are denoted as $P(\omega)$ and $R(\omega)$; the amplitudes of the horizontal displacement and rotation are denoted as $u_o(\omega)$ and $\varphi_o(\omega)$. Formulating the horizontal- and rotational-equilibrium equations in point 0 and the rotational-equilibrium equation in point 1 and eliminating $\varphi_1(\omega)$ from these relations leads to

$$P(\omega) = K_h\left[1 - \frac{\omega^2 M_{oh}}{K_h} + i\omega \frac{C_{oh}}{K_h}\right]u_o(\omega) + K_h f_K\left[1 + i\omega \frac{f_C}{f_K}\frac{C_{oh}}{K_h}\right]\varphi_o(\omega)$$

(2.37a)

$$R(\omega) = K_r f_K\left[1 + i\omega \frac{f_C}{f_K}\frac{C_{oh}}{K_h}\right]u_o(\omega)$$

$$+ K_r\left[1 - \frac{\frac{\omega^2 M_{1r}}{K_r}}{1 + \frac{\omega^2 M_{1r}^2}{C_{1r}^2}} - \frac{\omega^2 M_{or}}{K_r} + i\omega\left(\frac{M_{1r}}{C_{1r}}\frac{\frac{\omega^2 M_{1r}}{K_r}}{1 + \frac{\omega^2 M_{1r}^2}{C_{1r}^2}} + \frac{C_{or}}{K_r}\right)\right]\varphi_o(\omega)$$

(2.37b)

The dynamic-stiffness coefficients on the main diagonal—that is, the coefficient of $u_o(\omega)$ in Eq. 2.37a and that of $\varphi_o(\omega)$ in Eq. 2.37b—are in an identical form as for the one-dimensional discrete model (Eq. 2.31). Thus, introducing the dimensionless frequency $a_o$ (Eq. 2.32) and, separately for the horizontal and rotational motions, the dimensionless coefficients of the dampers $\gamma_{oh}$, respectively $\gamma_{or}$, $\gamma_{1r}$ (Eqs. 2.33a, 2.33b), and of the masses $\mu_{oh}$, respectively $\mu_{or}$, $\mu_{1r}$ (Eqs. 2.33c, 2.33d), transforms the terms on the main diagonal of Eq. 2.37 to the form of Eq. 2.34 formulated for the horizontal and rocking degrees of freedom. Equation 2.36 is also valid. The definitions of the coefficients of the dampers and masses are thus not affected.

The force–displacement relationships can be formulated as

$$P(a_o) = K_h[k_h(a_o) + ia_o c_h(a_o)]u_o(a_o) + K_{hr}[1 + ia_o c_{hr}(a_o)]\varphi_o(a_o) \quad (2.38a)$$

$$R(a_o) = K_{rh}[1 + ia_o c_{rh}(a_o)]u_o(a_o) + K_r[k_r(a_o) + ia_o c_r(a_o)]\varphi_o(a_o) \quad (2.38b)$$

where $k_h(a_o)$, $c_h(a_o)$ and $k_r(a_o)$, $c_r(a_o)$ are specified in Eq. 2.36 for the two degrees of freedom. Comparing the off-diagonal terms of Eq. 2.38 with those of Eq. 2.37 results in

$$K_{hr} = K_{rh} = K_h f_K \quad (2.39)$$

$$c_{hr}(a_o) = c_{rh}(a_o) = \frac{f_C}{f_K} \gamma_{oh} \quad (2.40)$$

The static-stiffness coefficients $K$ and the dimensionless coefficients of the dampers $\gamma_o$, $\gamma_1$ and of the masses $\mu_o$, $\mu_1$ for the cylindrical foundation embedded in an elastic undamped half-space are specified in Table 2-3 [W16]. To calculate the damping coefficients $C_o$, $C_1$ and the masses $M_o$, $M_1$, Eq. 2.33 is applied with $c_s = \sqrt{G/\rho}$. The eccentricity of the horizontal spring $f_K$ and that of the horizontal damper $f_C$ follow from

$$f_K = 0.25\, e \quad (2.41a)$$

$$f_C = 0.32\, e + 0.03\, e\left(\frac{e}{a}\right)^2 \quad (2.41b)$$

The static stiffnesses $K$, which are approximate, are taken from Ref. [P1]. The dimensionless coefficients of the dampers and masses specified in Table 2-3 apply for $\nu = 0.25$. The lack of reliable data for other Poisson's ratios does not allow these coefficients to be specified as a function of $\nu$. For $e/a = 0$, the coefficients for a disk on the surface for $\nu = 0.25$ are recovered, as listed in Table 2-2. For the coordinate system chosen, $\gamma_o$ for rocking differs from zero for the embedded case.

The dimensionless spring and damping coefficients in the frequency domain calculated from Eq. 2.36 are plotted as lines for $e/a = 0.5$ 1, and 2 in each of the Figs. 2-9, 2-10, 2-11, and 2-13 for the horizontal, vertical, rocking, and torsional motions. $k_h$ equals 1 for all embedment ratios. The

**TABLE 2-3 Static Stiffness and Dimensionless Coefficients of Discrete Model for Embedded Cylindrical Foundation (Homogeneous Half-Space)**

| | Static Stiffness $K$ | Dimensionless Coefficients for $\nu = 0.25$ of | | | | |
|---|---|---|---|---|---|---|
| | | Dampers | | | Masses | |
| | | $\gamma_o$ | $\gamma_1$ | | $\mu_o$ | $\mu_1$ |
| Horizontal | $\dfrac{8Ga}{2-\nu}\left(1+\dfrac{e}{a}\right)$ | $0.68 + 0.57\sqrt{e/a}$ | — | | — | — |
| Vertical | $\dfrac{4Ga}{1-\nu}\left(1+0.54\dfrac{e}{a}\right)$ | $0.8 + 0.35\dfrac{e}{a}$ | $0.32 - 0.01\left(\dfrac{e}{a}\right)^4$ | | — | $0.38$ |
| Rocking | $\dfrac{8Ga^3}{3(1-\nu)}\left(1+2.3\dfrac{e}{a}+0.58\left(\dfrac{e}{a}\right)^3\right)$ | $0.16\dfrac{e}{a}$ | $0.40 + 0.03\left(\dfrac{e}{a}\right)^2$ | | — | $0.33 + 0.1\left(\dfrac{e}{a}\right)^2$ |
| Torsional | $\dfrac{16Ga^3}{3}\left(1+2.67\dfrac{e}{a}\right)$ | — | $0.29 + 0.09\sqrt{e/a}$ | | — | $0.20 + 0.25\sqrt{e/a}$ |

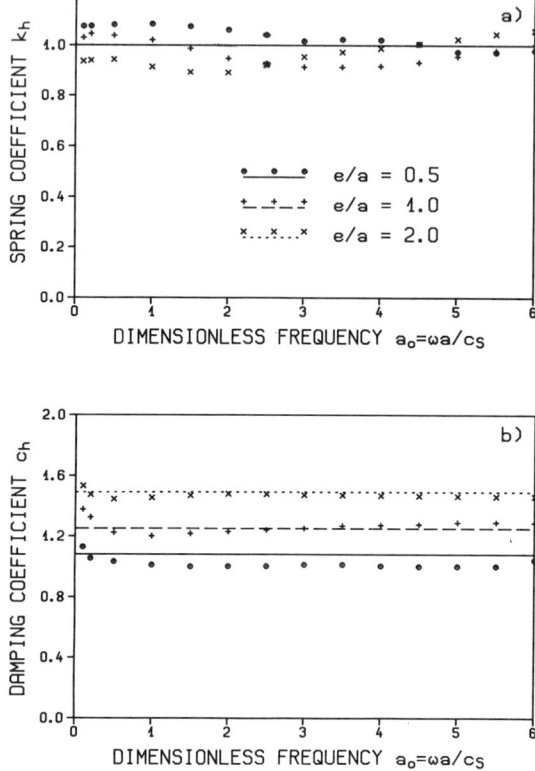

**Figure 2-9** Dynamic-stiffness coefficient of embedded cylinder for horizontal motion.

corresponding "exact" values based on the indirect boundary-element method of Ref. [A2] are shown as discrete points. The values in this reference include some material damping. The undamped values used in the comparison are recalculated approximately from the damped values. In Fig. 2-12, the term for coupling between the horizontal and rocking motions of the dynamic-stiffness matrix is addressed. $k_{hr}$ equals 1 and $c_{hr}(a_o)$ is specified in Eq. 2.40. The agreement is satisfactory.

## 2.7 EMBEDDED PRISM FOUNDATION

An embedded massless rigid rectangular foundation which forms a prism is examined next (Fig. 2-14). The length is denoted as $2l$, the width as $2b$ ($l \geq b$), and the embedment as $e$. There are 6 degrees of freedom at node 0, which coincides with the center of the basemat. $h_x$ denotes the horizontal

Sec. 2.7    Embedded Prism Foundation    31

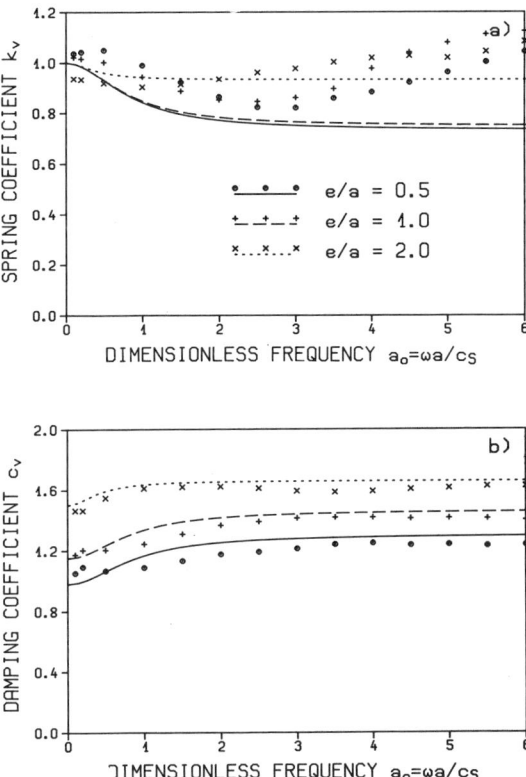

**Figure 2-10**  Dynamic-stiffness coefficient of embedded cylinder for vertical motion.

translation in the $x$-direction and $r_x$ the rocking motion around the $x$-axis. The properties of the undamped elastic homogeneous half-space are defined by $G$, $v$, and $\rho$. The rectangular foundation at the surface is a special case ($e = 0$).

One procedure to treat rectangular foundations consists of determining equivalent circular foundations (which differ, in general, for each degree of freedom) and then using the corresponding equations of the disk. A comparison between the equivalent disk's stiffness coefficients and those found in the literature, which are determined directly for a rectangle using a boundary-element method, indicates that an improved procedure is possible.

Due to the lack of axisymmetry, there are fewer results available for comparison for the rectangle than for the disk. Based on the meagre data published in the literature, the following static-stiffness coefficients are suggested in Ref. [P1]. They assume that the same dependence exists on $v$ as for the disk on the surface (Table 2-2). The increase in the stiffness for

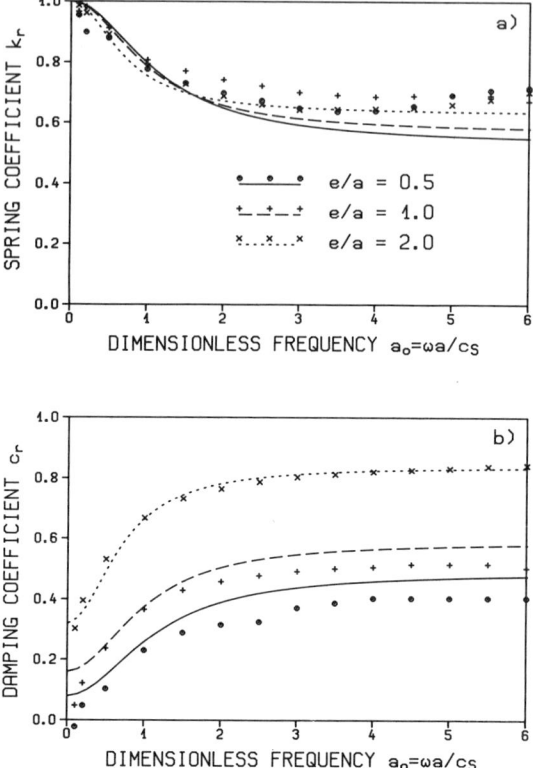

**Figure 2-11** Dynamic-stiffness coefficient of embedded cylinder for rocking motion.

embedment leads to a factor with which the value for the surface foundation is multiplied and which is, for $e = 0$, equal to 1.

$$K_{hx} = \frac{Gb}{2 - \nu}\left[6.8\left(\frac{l}{b}\right)^{0.65} + 2.4\right]\left[1 + \left(0.33 + \frac{1.34}{1 + \dfrac{l}{b}}\right)\left(\frac{e}{b}\right)^{0.8}\right]$$

(2.42a)

$$K_{hy} = \frac{Gb}{2 - \nu}\left[6.8\left(\frac{l}{b}\right)^{0.65} + 0.8\frac{l}{b} + 1.6\right]\left[1 + \left(0.33 + \frac{1.34}{1 + \dfrac{l}{b}}\right)\left(\frac{e}{b}\right)^{0.8}\right]$$

(2.42b)

Sec. 2.7  Embedded Prism Foundation

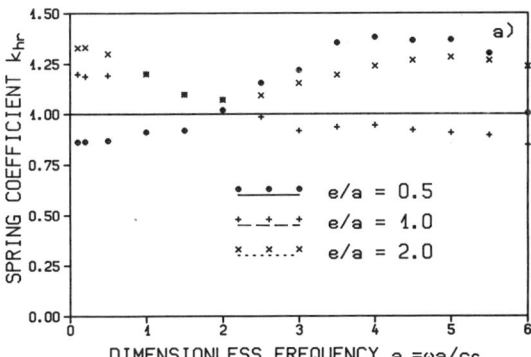

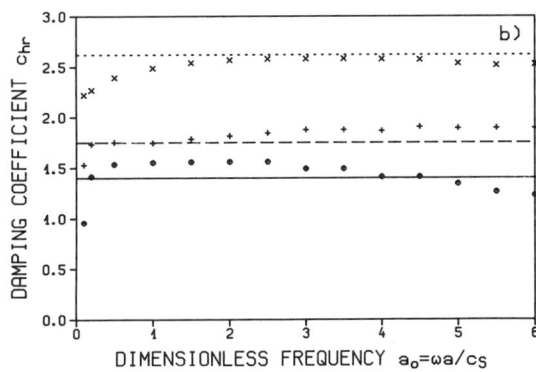

**Figure 2-12** Coupling term of dynamic-stiffness coefficient of embedded cylinder.

$$K_v = \frac{Gb}{1-v}\left[3.1\left(\frac{l}{b}\right)^{0.75} + 1.6\right]\left[1 + \left(0.25 + \frac{0.25b}{l}\right)\left(\frac{e}{b}\right)^{0.8}\right]$$

(2.42c)

$$K_{rx} = \frac{Gb^3}{1-v}\left[3.2\frac{l}{b} + 0.8\right]\left[1 + \frac{e}{b} + \frac{1.6}{0.35 + \frac{l}{b}}\left(\frac{e}{b}\right)^2\right]$$

(2.42d)

$$K_{ry} = \frac{Gb^3}{1-v}\left[3.73\left(\frac{l}{b}\right)^{2.4} + 0.27\right]\left[1 + \frac{e}{b} + \frac{1.6}{0.35 + \left(\frac{l}{b}\right)^4}\left(\frac{e}{b}\right)^2\right]$$

(2.42e)

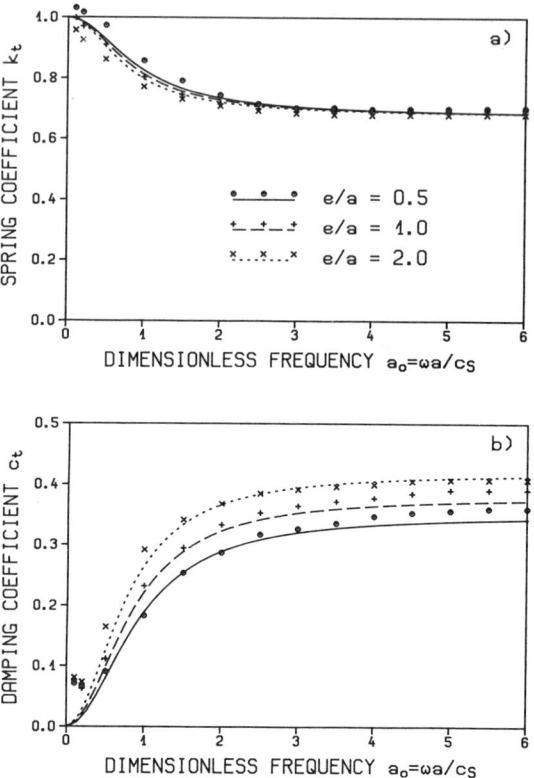

**Figure 2-13** Dynamic-stiffness coefficient of embedded cylinder for torsional motion.

$$K_t = Gb^3 \left[ 4.25 \left( \frac{l}{b} \right)^{2.45} + 4.06 \right] \left[ 1 + \left( 1.3 + \frac{1.32b}{l} \right) \left( \frac{e}{b} \right)^{0.9} \right]$$

(2.42f)

The coupling term is specified as

$$K_{hxry} = \frac{e}{3} K_{hx} \qquad (2.43a)$$

$$K_{hyrx} = \frac{e}{3} K_{hy} \qquad (2.43b)$$

Comparing Eqs. 2.43 and 2.39, one determines the eccentricities of the horizontal springs:

$$f_{Kx} = f_{Ky} = \frac{e}{3} \qquad (2.44)$$

Sec. 2.8  Strip Foundation 35

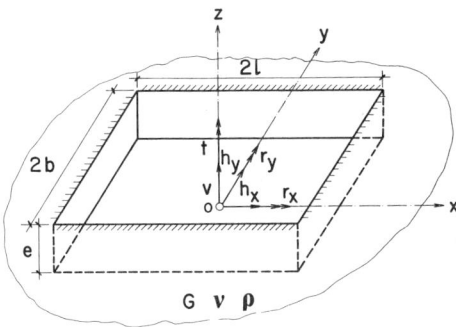

**Figure 2-14** Embedded prism foundation.

For a rectangular foundation at the surface of an elastic half-space with $\nu = 1/3$, applying curve-fitting techniques to the dynamic-stiffness coefficients in the frequency domain and using the results of Ref. [W21] lead to the dimensionless coefficients $\gamma_o$, $\gamma_1$, $\mu_o$, and $\mu_1$ shown in Table 2-4 [W16]. To determine the damping coefficients $C_o$, $C_1$ and the masses $M_o$, $M_1$ of the discrete model, Eq. 2.33 is applied with $b$ replacing $a$ and with $c_s = \sqrt{G/\rho}$.

To check the accuracy, the dimensionless spring and damping coefficients in the frequency domain of the surface foundation calculated with Eq. 2.36 are plotted as lines for $l/b = 1, 2, 3,$ and 4 for the rocking motions around the $x$-axis, the $y$-axis, and for the torsional motion in Figs. 2-15, 2-16, and 2-17. Note that the characteristic length in the dimensionless frequency's definition is $b$.

As can be seen from Table 2-4, the dynamic-stiffness coefficient for rocking around the $x$-axis is assumed to be independent of $l/b$. The "exact" values of Ref. [W21] are shown as discrete points. The agreement is acceptable.

As the variation of the dynamic-stiffness coefficients with frequency has been investigated for an embedded prism only in the low-frequency range, the dependence of $\gamma_o$, $\gamma_1$, $\mu_o$, and $\mu_1$ on $e$ cannot be specified yet. This influence can either be disregarded, or an equivalent radius $a$ can be estimated which then allows some guidance to be gained using the corresponding dimensionless coefficients for the embedded cylinder (Table 2-3). For lack of data and for the sake of simplicity, the eccentricity of the horizontal damper $f_{Cx} = f_{Cy}$ is set equal to that of the spring (Eq. 2.44).

## 2.8 STRIP FOUNDATION

The massless rigid strip resting on the surface of an elastic undamped homogeneous half-plane is examined next. The width of the strip equals $2b$ and node 0 is located in the center of the strip. The half-plane is characterized by $G$, $\nu$, and $\rho$.

**TABLE 2-4 Dimensionless Coefficients of Discrete Model for Rectangular Foundation (Homogeneous Half-Space)**

| | | Dimensionless Coefficients for $\nu = \tfrac{1}{3}$ of | | | |
|---|---|---|---|---|---|
| | | Dampers | | Masses | |
| | | $\gamma_o$ | $\gamma_1$ | $\mu_o$ | $\mu_1$ |
| Horizontal | $h_x h_y$ | $0.75 + 0.2\left(\dfrac{l}{b} - 1\right)$ | — | — | — |
| Vertical | | $0.9 + 0.4\left(\dfrac{l}{b} - 1\right)^{2/3}$ | 0.3 | — | 0.14 |
| Rocking | $r_x$ | — | 0.45 | — | 0.34 |
| | $r_y$ | — | $0.45 + 0.23\left(\dfrac{l}{b} - 1\right)$ | — | $0.34 + 0.55\left(\dfrac{l}{b} - 1\right)$ |
| Torsional | | — | $0.35 + 0.12\left(\dfrac{l}{b} - 1\right)$ | — | $0.28 + 0.63\left(\dfrac{l}{b} - 1\right)$ |

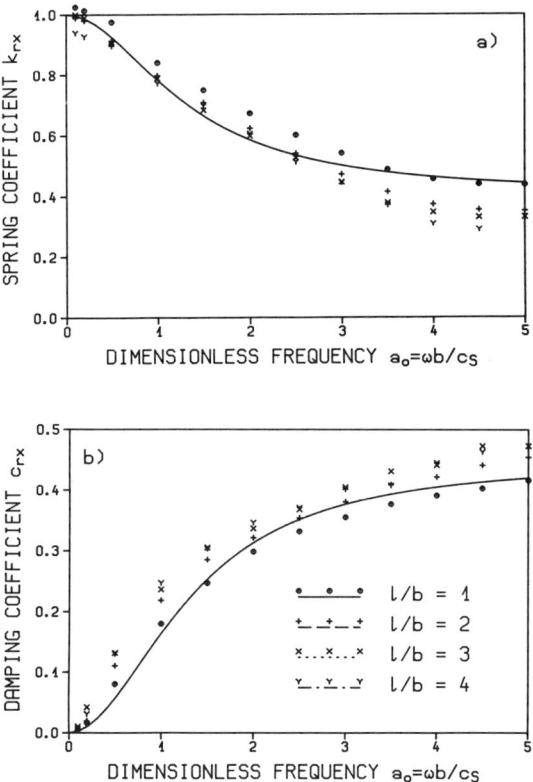

**Figure 2-15** Dynamic-stiffness coefficient of rectangular foundation for rocking motion around $x$-axis.

As the static-stiffness coefficients vanish for the 2 translational degrees of freedom, they cannot be used for the springs connected to the rigid support in the discrete model. The spring constant $K$ and the dimensionless coefficients of the dampers $\gamma_o$, $\gamma_1$ and of the masses $\mu_o$, $\mu_1$ are specified for the three motions in Table 2-5 [W16]. The dampers with the coefficients $C_o$, $C_1$ and the masses $M_o$, $M_1$ of the discrete model follow from Eq. 2.33 using $b$ instead of $a$ and with $c_s = \sqrt{G/\rho}$.

The dynamic-stiffness coefficients determined from Eq. 2.36 for $\nu = 1/3$ and $1/2$ are plotted as lines versus the dimensionless frequency $\omega b/c_s$ for the horizontal and rocking motions in Figs. 2-18 and 2-19. $k_h$ is assumed to be independent of $\nu$. The agreement with the "exact" values of Refs. [W11] for $\nu = 1/3$ and [L3] for $\nu = 1/2$ shown as discrete points is acceptable.

The embedded strip foundation is addressed in Problem 2.3.

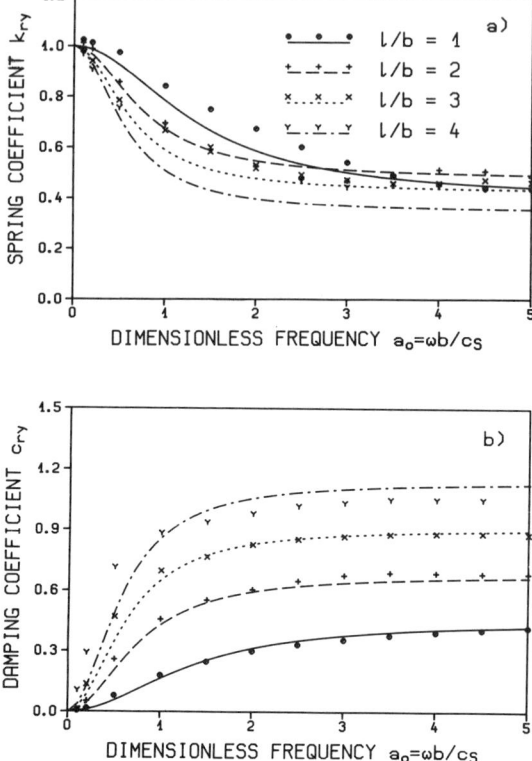

**Figure 2-16** Dynamic-stiffness coefficient of rectangular foundation for rocking motion around $y$-axis.

## 2.9 SQUARE FOUNDATION ON LAYERED HALF-SPACE

The cases discussed in Sections 2.5 to 2.8 apply to a homogeneous half-space. The same curve-fitting technique can also be used for a layered half-space whose properties vary within a certain range. As an example, the site consisting of a layer with a total depth $d$ resting on a homogeneous half-space is addressed (Fig. 2-20). The shear-wave velocity varies linearly with depth across the layer, from $c_s^t$ at the free surface ($t$ stands for top) to $c_s^R$ at its base, which is also the value for the homogeneous half-space ($R$ stands for rock). The mass density of the half-space equals 1.13 $\rho$, where $\rho$ is that of the layer. Poisson's ratio $\nu$ is constant throughout the layered half-space ($= 1/3$). A massless rigid square foundation of length $2b$ rests on the free surface of the layered half-space.

Curve-fitting techniques applied to the "exact" spring and damping

### Sec. 2.9  Square Foundation on Layered Half-Space

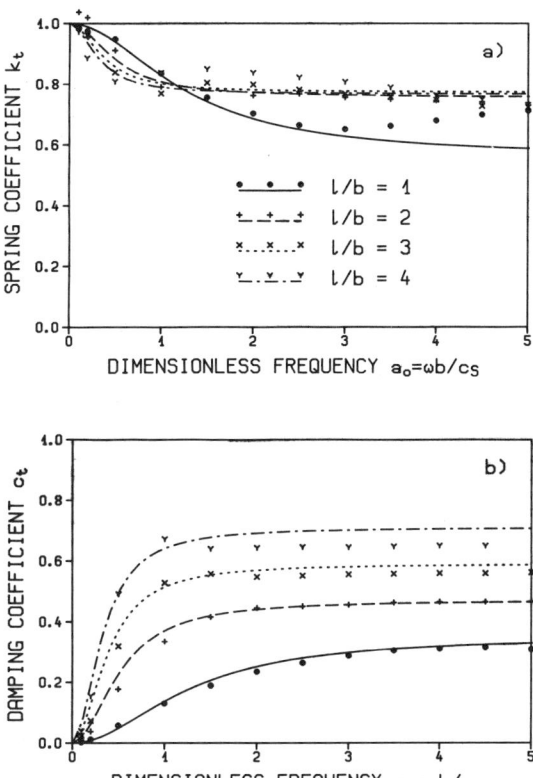

**Figure 2-17**  Dynamic-stiffness coefficient of rectangular foundation for torsional motion.

coefficients of Ref. [W23] (transformed approximately to the corresponding values with zero material damping) lead to the static-stiffness coefficients $K$ and the dimensionless coefficients $\gamma_o$, $\gamma_1$, $\mu_o$, and $\mu_1$ presented for the 4 degrees of freedom in Table 2-6 [W16]. To calculate the damping coefficients $C_o$, $C_1$ and the masses $M_o$, $M_1$ of the discrete model, Eq. 2.33 is applied with $c_s^t$ and with $b$ replacing $a$. $G$ equals $\rho(c_s^t)^2$. The relations in Table 2-6 lead to a good agreement with the exact ones for $3 \leq d/b < \infty$ and $0.5 \leq c_s^t/c_s^R \leq 1$. To estimate the static-stiffness coefficient $K$, the formulas apply for the range starting at $c_s^t/c_s^R = 0.3$. Agreement between Tables 2-4 and 2-6 exists for a square foundation on a homogeneous half-space. The approximation for the torsional motion barely depends on $c_s^t/c_s^R$ and is independent of $d/b$.

The dimensionless spring and damping coefficients in the frequency domain calculated from Eq. 2.36 are, in Fig. 2-21, plotted as lines for the

**TABLE 2-5 Spring Coefficient and Dimensionless Coefficients of Discrete Model for Strip Foundation (Homogeneous Half-Plane)**

| | Spring Coefficient $K$ | Dimensionless Coefficients of | | | | |
|---|---|---|---|---|---|---|
| | | Dampers | | | Masses | |
| | | $\gamma_o$ | $\gamma_1$ | | $\mu_o$ | $\mu_1$ |
| Horizontal | $G(1 + 5v^2)$ | $2 - 2.2v$ | — | | — | — |
| Vertical | $G(1 + 4v^2)$ | $3.5 - 2v$ | — | $v < 1/3$ | $0$ | — |
| | | | | $v > 1/3$ | $4.5(v - 1/3)$ | |
| Rocking | $Gb^2(1.8 + 5.2v^2)$ | $0.14 - 0.24v^2$ | $0.4$ | $v < 1/3$ | $0$ | $0.3$ |
| | | | | $v > 1/3$ | $0.25(v - 1/3)$ | |

Sec. 2.10   Material Damping                                                            41

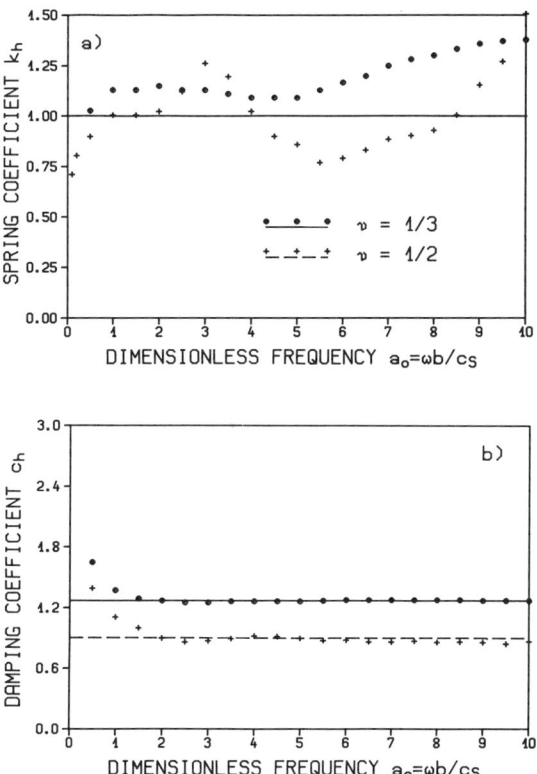

**Figure 2-18**  Dynamic-stiffness coefficient of strip for horizontal motion.

horizontal motion, assuming $d/b = 5$. $c_s^t/c_s^R$ is varied as indicated. The corresponding "exact" values without material damping are shown as discrete points. In Figs. 2-22 and 2-23, the corresponding comparisons are shown for the vertical and rocking motions, respectively, assuming $c_s^t/c_s^R = 0.6$ and varying $d/b$. The curves of $k_v$ for $d/b = 3, 5$, and 10 coincide (Fig. 2-22a). The agreement is surprisingly good for all cases.

## 2.10 MATERIAL DAMPING

Hysteretic damping, which corresponds to material damping, cannot be modeled with this discrete model. As can be seen from Fig. 2-1, the damping coefficient $c(\omega)$ becomes unbounded for $\omega \to 0$ in the case of hysteretic damping. Such a variation is not contained in the corresponding equation of the discrete model (Eq. 2.36b).

Viscous material damping of the Voigt type can, however, be considered

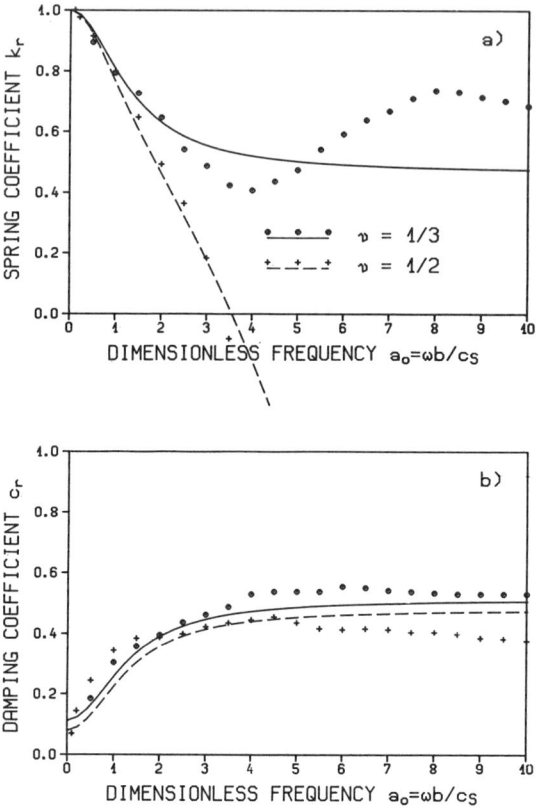

**Figure 2-19** Dynamic-stiffness coefficient of strip for rocking motion.

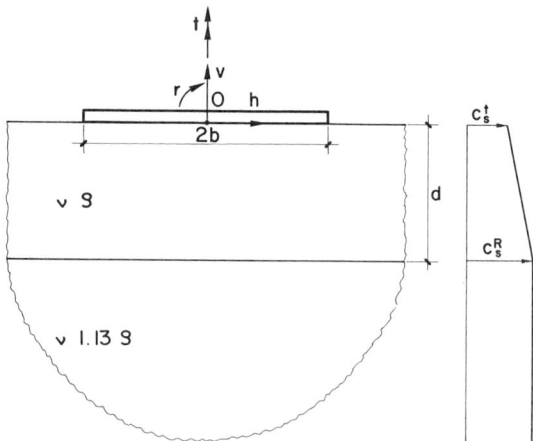

**Figure 2-20** Square foundation on layered half-space.

**TABLE 2-6 Static Stiffness and Dimensionless Coefficients of Discrete Model for Square Foundation (Layered Half-Space)**

| | Static Stiffness $K$ | Dimensionless Coefficients for $\nu = 1/3$ of | | | | |
|---|---|---|---|---|---|---|
| | | Dampers | | | Masses | |
| | | $\gamma_o$ | $\gamma_1$ | $\mu_o$ | | $\mu_1$ |
| Horizontal | $\dfrac{9.2\,Gb}{2-\nu}\left[1 + 3.7\,\dfrac{\left(1 - \dfrac{c_s^t}{c_s^R}\right)^2}{d/b}\right]$ | $0.75 - 1.3\,\dfrac{1 - \dfrac{c_s^t}{c_s^R}}{\sqrt[4]{d/b}}$ | $0.6 - 0.6\,\dfrac{c_s^t}{c_s^R}$ | — | | $0.55 - 0.55\,\dfrac{c_s^t}{c_s^R}$ |
| Vertical | $\dfrac{4.7\,Gb}{1-\nu}\left[1 + 9\,\dfrac{\left(1 - \dfrac{c_s^t}{c_s^R}\right)^2}{d/b}\right]$ | $0.9 - 2.5\,\dfrac{1 - \dfrac{c_s^t}{c_s^R}}{\sqrt{d/b}}$ | $0.7 - 0.4\,\dfrac{c_s^t}{c_s^R}$ | — | | $0.62 - 0.48\,\dfrac{c_s^t}{c_s^R}$ |
| Rocking | $\dfrac{4\,Gb^3}{1-\nu}\left[1 + 3\,\dfrac{\left(1 - \dfrac{c_s^t}{c_s^R}\right)^2}{d/b}\right]$ | — | $0.45 - 0.18\sqrt{\dfrac{1 - \dfrac{c_s^t}{c_s^R}}{d/b}}$ | — | | $0.34 - 0.36\sqrt{\dfrac{1 - \dfrac{c_s^t}{c_s^R}}{d/b}}$ |
| Torsional | $8.3\,Gb^3$ | — | $0.35$ | — | | $0.18 + 0.1\,\dfrac{c_s^t}{c_s^R}$ |

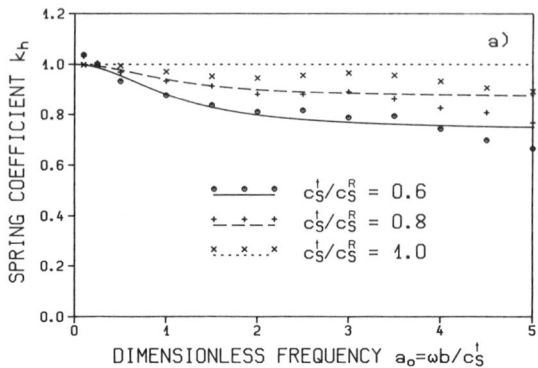
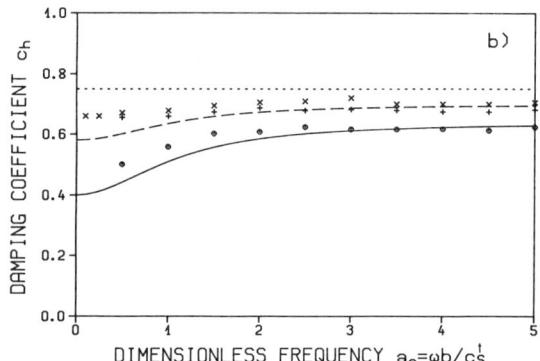

**Figure 2-21** Dynamic-stiffness coefficient of square foundation on layered half-space for horizontal motion.

in the discrete model. The goal is to use the same model, modifying only the coefficients $\gamma_o$ to $\mu_1$ and the eccentricities $f_K$, $f_C$. These values for the damped system are to be expressed as a function of the corresponding undamped ones and the damping ratio. Damping can be introduced into the solution, when working in the frequency domain, by using the correspondence principle. This principle states that the damped solution is obtained from the elastic one by replacing the elastic constants with the corresponding complex ones. A complex modulus $G^*$ is defined as

$$G^* = G(1 + 2\zeta a_o i) \qquad (2.45)$$

$\zeta$ is the nondimensionalized viscous damping ratio. Comparing the imaginary parts of Eqs. 2.45 and 2.5 and using Eq. 2.6b leads to

$$\zeta = \frac{\zeta_v}{\omega_o} \frac{c_s}{a} \qquad (2.46)$$

Sec. 2.10     Material Damping                                                                45

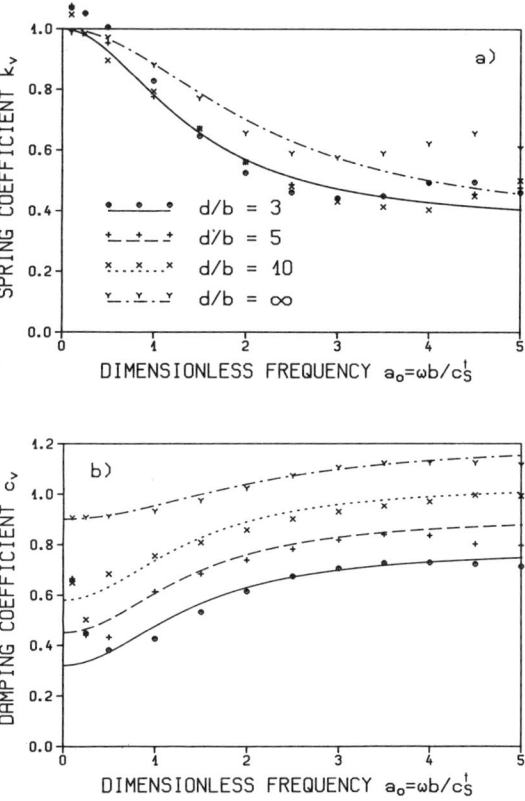

**Figure 2-22** Dynamic-stiffness coefficient of square foundation on layered half-space for vertical motion.

Besides $\zeta_v$, a frequency $\omega_o$ must be selected (or calculated) to determine $\zeta$. The dimensionless frequency

$$a_o^* = \frac{\omega a}{\sqrt{G^*/\rho}} = \frac{a_o}{\sqrt{1 + 2\zeta a_o i}} \qquad (2.47)$$

is also modified by the complex shear modulus. The damping ratio affects the dynamic-stiffness coefficient in three ways. First, the static-stiffness coefficient used to nondimensionalize the expression is multiplied by $(1 + 2\zeta a_o i)$. Second, $a_o^*$ is substituted for $a_o$; and third, $k(a_o)$ and $c(a_o)$ are replaced by their damped (complex) counterparts $k(a_o^*)$ and $c(a_o^*)$.

The dynamic-stiffness coefficients $S$ of the damped basic discrete model thus equals (Eq. 2.35)

$$S = K(1 + 2\zeta a_o i)[k(a_o^*) + ia_o^* c(a_o^*)] \qquad (2.48)$$

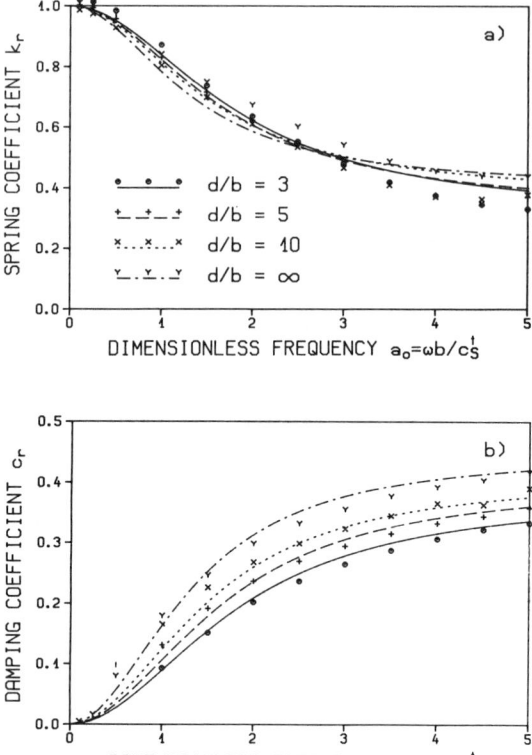

**Figure 2-23** Dynamic-stiffness coefficient of square foundation on layered half-space for rocking motion.

with $k(a_o^*)$ and $c(a_o^*)$ still defined by Eq. 2.36, but with $a_o^*$ replacing $a_o$. Substituting $k(a_o^*)$ and $c(a_o^*)$ after introducing Eq. 2.47 leads (after neglecting terms in $\zeta^2$) to

$$S = K\left[1 - \frac{\mu_1 a_o^2}{1 + \frac{\mu_1^2}{\gamma_1^2}a_o^2} - \mu_o a_o^2 - \frac{\mu_1}{\gamma_1}\frac{\mu_1 a_o^2}{1 + \frac{\mu_1^2}{\gamma_1^2}a_o^2}\zeta a_o^2\left(1 - \frac{2}{1 + \frac{\mu_1^2}{\gamma_1^2}a_o^2}\right) - \gamma_o \zeta a_o^2\right.$$

$$\left. + ia_o\left(\frac{\mu_1}{\gamma_1}\frac{\mu_1 a_o^2}{1 + \frac{\mu_1^2}{\gamma_1^2}a_o^2} + \gamma_o + 2\zeta - \frac{\mu_1 a_o^2}{1 + \frac{\mu_1^2}{\gamma_1^2}a_o^2}2\zeta\left(1 - \frac{1}{1 + \frac{\mu_1^2}{\gamma_1^2}a_o^2}\right)\right)\right] \quad (2.49)$$

The dynamic-stiffness coefficient of the discrete system which shall model the

Sec. 2.10    Material Damping    47

material damping is formulated as (Eq. 2.34)

$$S = K\left[1 - \frac{\mu_1^\zeta a_o^2}{1 + (\mu_1^\zeta/\gamma_1^\zeta)^2 a_o^2} - \mu_o^\zeta a_o^2 + ia_o\left(\frac{\mu_1^\zeta}{\gamma_1^\zeta}\frac{\mu_1^\zeta a_o^2}{1 + (\mu_1^\zeta/\gamma_1^\zeta)^2 a_o^2} + \gamma_o^\zeta\right)\right]$$

(2.50)

where these dimensionless coefficients of the dampers and masses are denoted with the superscript $\zeta$.

Comparing Eqs. 2.49 and 2.50 leads as an approximation to

$$\gamma_o^\zeta = \gamma_o + 2\zeta \tag{2.51a}$$

$$\gamma_1^\zeta = \gamma_1 \tag{2.51b}$$

$$\mu_o^\zeta = \mu_o + (\gamma_1 + \gamma_o)\zeta \tag{2.51c}$$

$$\mu_1^\zeta = \mu_1 \tag{2.51d}$$

For $a_o \to \infty$, these approximations lead to the correct real value of $S$. Introducing material damping thus leads to the damper with the coefficient $C_o$ (or increases $C_o$) which is in parallel to the static-stiffness spring; it also creates a mass $M_o$ attached to the node 0 (or increases $M_o$).

To check the accuracy of the dimensionless coefficients as specified in Equation 2.51, the disk's dynamic-stiffness coefficient for rocking (Table 2-2) is calculated ($\nu = 1/3$, $\zeta = 0.05$). In Fig. 2-24, the value using Eq. 2.36, but determined with the coefficients $\gamma_o^\zeta$, $\gamma_1^\zeta$, $\mu_o^\zeta$, $\mu_1^\zeta$, is compared to the rigorous one based on substituting Eqs. 2.45 and 2.47 in Eq. 2.34. The agreement is good.

Finally, the influence of material damping on the coupling term between the horizontal and rocking motions for an embedded foundation is examined. Applying the correspondence principle to the off-diagonal term $S_{hr}$ in Eq. 2.38 leads to

$$S_{hr} = K_h f_K (1 + 2\zeta a_o i)\left[1 + i\frac{a_o}{\sqrt{1 + 2\zeta a_o i}}\frac{f_C}{f_K}\gamma_{oh}\right] \tag{2.52}$$

whereby Eqs. 2.39 and 2.40 are used. For the discrete system which shall model the material damping, the coupling term is formulated as

$$S_{hr} = K_h f_K^\zeta \left[1 + ia_o \frac{f_C^\zeta}{f_K^\zeta}\gamma_{oh}^\zeta\right] \tag{2.53}$$

Comparing the real parts of Eqs. 2.52 and 2.53 leads to

$$f_K^\zeta = f_K \tag{2.54}$$

whereby the term—that is, the mass term—in $a_o^2$ is suppressed. Comparing

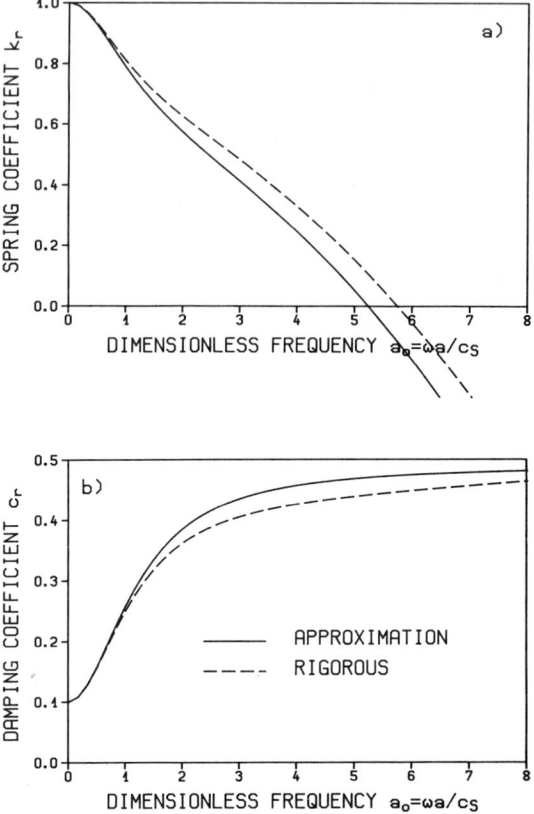

**Figure 2-24** Dynamic-stiffness coefficient of disk for rocking with material damping.

the imaginary terms and substituting Eq. 2.51a results in

$$f_C^\zeta = \frac{f_C \gamma_{oh} + 2 f_K \zeta}{\gamma_{oh} + 2\zeta} \tag{2.55}$$

## 2.11 DISK WITH MASS ON HALF-SPACE

For the sake of illustration, the rigid disk of radius $a$ with a polar mass moment of inertia $m$ resting on an undamped elastic half-space of shear modulus $G$ and mass density $\rho$ is loaded by a torsional moment $T$ (Fig. 2-25a).

$$\begin{aligned} T(t) &= \frac{T_o}{2}\left[1 - \cos\left(2\pi \frac{t}{t_o}\right)\right] & 0 < t < t_o \\ T(t) &= 0 & t > t_o \end{aligned} \tag{2.56}$$

## Sec. 2.11  Disk with Mass on Half-Space

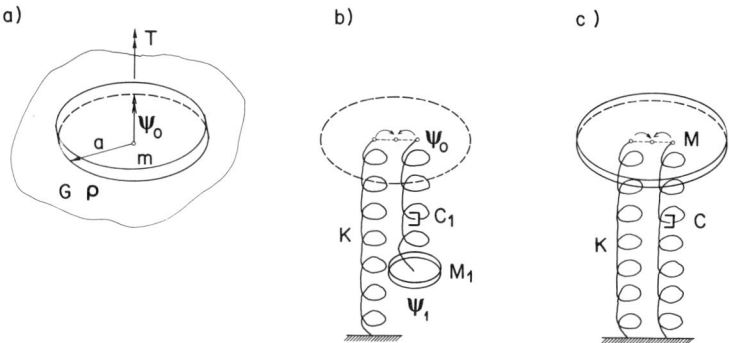

**Figure 2-25**  a. Rigid disk with polar mass moment of inertia loaded by torsional moment.  b. Discrete model with additional torsional degree of freedom.  c. Discrete model with added polar mass moment of inertia (Lumped-parameter model).

This loading is a versed cosine function with a zero value at $t = 0$, the maximum $T_o$ at $t = t_o/2$ and again zero at $t = t_o$. For the actual calculation, $m = 0.3 \, \rho a^5$ and $t_o = a/c_s$ are selected ($c_s$ = shear-wave velocity).

Three calculations are performed. First, for this linear system, the exact solution is determined by working in the frequency domain (Fast Fourier Transform with a period = 41 $t_o$, 2048 points). The accurate dynamic-stiffness coefficients in the frequency domain of Ref. [L2] are used. In Fig. 2-26, the torsional rotation $\psi_o$ (nondimensionalized with the static-flexibility coefficient $T_o/K$) at the disk's center is plotted versus the time (multiplied by $c_s/a$). Second, the soil is represented by the discrete model which introduces an additional torsional dynamic degree of freedom $\psi_1$ with a polar mass moment of inertia $M_1$ connected through a torsional damper with a coefficient $C_1$ to the disk (Fig. 2-25b). The values specified in Table 2-2 are used in Eq. 2.33. The resulting dynamic system is solved by an explicit time-integration scheme, which is discussed in Section 2.12 (Newmark family with $\beta = 0$, $\gamma = 0.5$, $\Delta t = 0.02 \, a/c_s$). The results agree well with the exact solution throughout the time-history. Third, the standard lumped-parameter model shown in Fig. 2-25c is selected for the soil, where a polar mass moment of inertia $M$ is added to the disk. Using the values in Table 2-1 and Eq. 2.11 leads to

$$K = \frac{16 G a^3}{3} \tag{2.57a}$$

$$C = \frac{1}{1 + \dfrac{2m}{\rho a^5}} \sqrt{Km} \tag{2.57b}$$

$$M = 0.24 \, \rho a^5 \tag{2.57c}$$

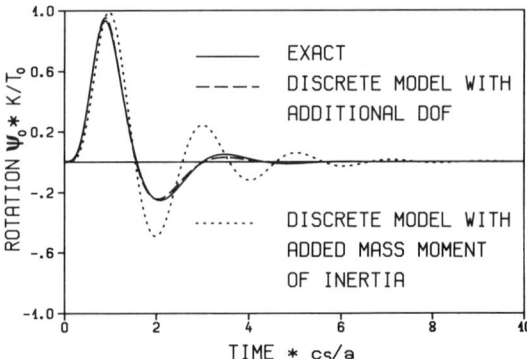

**Figure 2-26** Time-history of torsional rotation of disk.

Applying the same time-integration scheme, the results shown in Fig. 2-26 are calculated. Large deviations exist in the free-vibration phase. The standard lumped-parameter model thus leads to a higher damped frequency and seems to underestimate the radiation damping in this case (see also Section 2.13). The maximum response is quite well predicted, because the static effect is dominant.

## 2.12 HAMMER FOUNDATION WITH PARTIAL UPLIFT OF ANVIL

As an example of a nonlinear soil–structure-interaction analysis, the vibration of a hammer foundation embedded in soil with an eccentrically mounted anvil is examined [H1]. The head impacts against the anvil, which is a massive steel block (Fig. 2-27). This anvil is supported by a viscoelastic suspension (pad) on the foundation block of concrete which is embedded. As a tension-resistant connection is not provided for the pads, the anvil will partially uplift from the block when the dynamic stress (tension) exceeds the static stress (compression). A nonlinear dynamic system thus occurs. Any possible separation of the block from the adjacent soil is disregarded in the following.

More specifically, the head with mass $m_h = 1.5 \cdot 10^3$ kg impacts with a velocity $c_h = 5$ m/s against the anvil, which excites the anvil with an initial velocity. In this conservative simplification, no impact force–time history is used; it is assumed to be infinitely short. The coefficient of restitution (collision) is disregarded. The cylindrical anvil with mass density $7.85 \cdot 10^3$ kg/m³ is mounted with an eccentricity = 0.5 m with respect to the cylindrical block with mass density = $2.5 \cdot 10^3$ kg/m³. The anvil and the block, with the dimensions specified in Fig. 2-27, are modeled as rigid bodies. The soil consists of an undamped elastic half-space with a cylindrical excavation (radius = 2.5 m, depth = embedment of block = 2.5 m) and a Poisson's ratio =

## Sec. 2.12 Hammer Foundation with Partial Uplift of Anvil

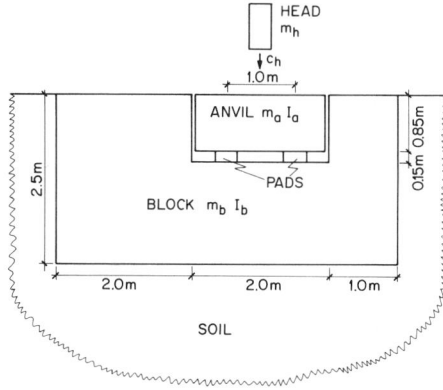

**Figure 2-27** Hammer foundation with inertial block.

0.25, a shear-wave velocity = 150 m/s, and a density = $2 \cdot 10^3$ kg/m³. The pad of the anvil has an area $A = 3$ m², a thickness $d = 0.15$ m, a modulus of elasticity $E = 100$ MPa, a Poisson's ratio $\nu = 0.25$, and a damping ratio $\zeta = 0.05$. The pad is modeled by two discrete elements at a distance of 1.0 m apart (Fig. 2-27). For the analysis with partial anvil uplift, the motion of the anvil, and to a lesser extent of the block, is to be determined, and it is to be compared to the corresponding results of the linear analysis.

The dynamic model, which is shown in Fig. 2-28, has 8 degrees of freedom: the horizontal, vertical, and rotational motions of the anvil $u_a$, $w_a$, $\varphi_a$ and of the block $u_b$, $w_b$, $\varphi_b$ and the additional vertical displacement $w_1$

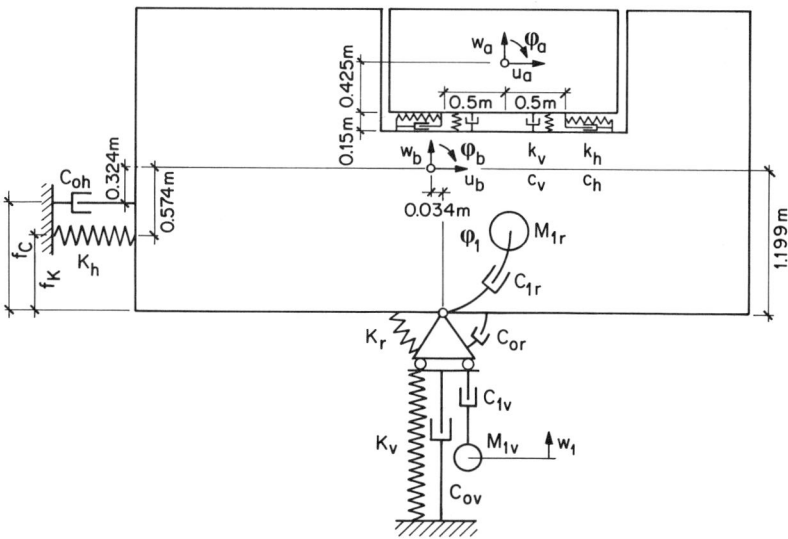

**Figure 2-28** Dynamic model.

and rotation $\varphi_1$ of the discrete model of the soil. The coefficients of the dampers $C_o$, $C_1$ and the mass $M_1$ for the 3 degrees of freedom of the rigid embedded cylindrical foundation follow from Eq. 2.33, with the corresponding dimensionless coefficients $\gamma_o$, $\gamma_1$, $\mu_1$ and the static stiffness $K$ specified in Table 2-3. The eccentricities of the discrete model in the horizontal direction $f_K$, $f_C$ are calculated from Eq. 2.41. The spring and damping coefficients of the two discrete elements of the pad in the horizontal and vertical directions are determined as

$$k_h = \frac{E}{2(1+v)} \frac{A}{2d} \qquad (2.58a)$$

$$c_h = 2\sqrt{k_h m_a}\, \zeta \qquad (2.58b)$$

$$k_v = \frac{EA}{2d} \qquad (2.58c)$$

$$c_v = 2\sqrt{k_v m_a}\, \zeta \qquad (2.58d)$$

where $m_a$ denotes the mass of the anvil. These values apply when no uplift occurs. They vanish in the other case. Uplift starts when the total force in the vertical direction—calculated as the sum of the dead load and the dynamic part in one of the discrete elements of the pad—becomes zero. The corresponding horizontal force is then also set equal to zero. Contact is regained when the displacement of the anvil relative to the block becomes equal to the length of the unloaded spring (penetration).

The dynamic system's response—a free vibration—is triggered by the initial velocity $\dot{w}_a(t=0)$ of the anvil in the vertical direction. Formulating the law of momentum leads to

$$\dot{w}_a(t=0) = -\frac{m_h c_h}{m_a + m_h} \qquad (2.59)$$

The following explicit algorithm, based on the Newmark family of methods with the two parameters $\beta$ and $\gamma$, is used for the time integration. Starting from the known motion at time $(n-1)\Delta t$—that is, $\{u\}_{n-1}$, $\{\dot{u}\}_{n-1}$, $\{\ddot{u}\}_{n-1}$—the displacement and velocity are predicted at time $n\Delta t$ as

$$\{\tilde{u}\}_n = \{u\}_{n-1} + \Delta t\, \{\dot{u}\}_{n-1} + \left(\frac{1}{2} - \beta\right) \Delta t^2\, \{\ddot{u}\}_{n-1} \qquad (2.60a)$$

$$\{\tilde{\dot{u}}\} = \{\dot{u}\}_{n-1} + (1-\gamma)\Delta t\, \{\ddot{u}\}_{n-1} \qquad (2.60b)$$

A tilde (˜) denotes the predicted values. With $\{\tilde{u}\}_n$ and $\{\tilde{\dot{u}}\}_n$, the internal forces—that is, the linear forces in the springs and dashpots of the dynamic model—can be calculated at time $n\Delta t$. Based on the material law of the nonlinear elements, the internal forces in such elements are determined using

## Sec. 2.12  Hammer Foundation with Partial Uplift of Anvil

the distortions. Formulating the equilibrium equations at time $n\Delta t$ results, for a vanishing exterior load, in

$$\{\ddot{u}\}_n = -[M]^{-1}\{F\}_n \tag{2.61}$$

$\{F\}$ is the vector of the resultants of the internal forces. As the mass matrix $[M]$ is a diagonal matrix, the accelerations are calculated on an element basis. The corrected values at time $n\Delta t$ are equal to

$$\{u\}_n = \{\tilde{u}\}_n + \beta \Delta t^2 \{\ddot{u}\}_n \tag{2.62a}$$

$$\{\dot{u}\}_n = \{\tilde{\dot{u}}\}_n + \gamma \Delta t \{\ddot{u}\}_n \tag{2.62b}$$

which concludes the calculations for the time step.

The analysis is performed for $\beta = 0$ and $\gamma = 0.5$, with $\Delta t = 10^{-3}$ s. In Fig. 2-29, the vertical displacement $w_a$ (measured from the static equilibrium position), the horizontal displacement $u_a$ and the rotation $\varphi_a$ at the center of the anvil are plotted as a function of time. For comparison, the linear

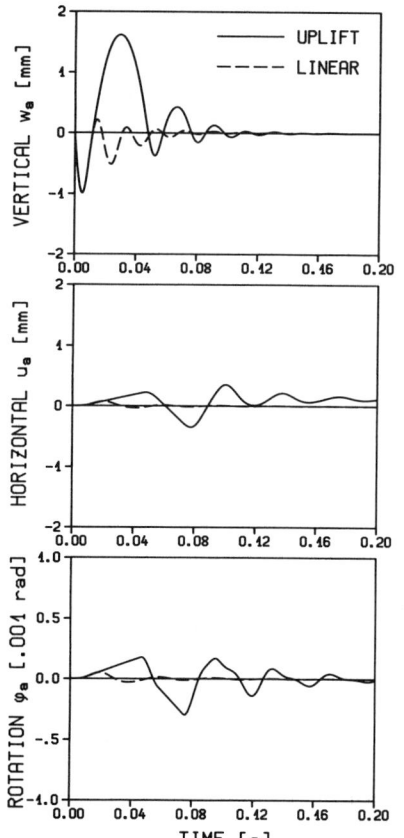

**Figure 2-29** Response at center of anvil.

results are also indicated. As expected, the partial uplift of the anvil increases the motion significantly.

## 2.13 COMPLEX MODAL ANALYSIS

When the structure remains linear, knowledge of the natural frequencies, the mode shapes, and the corresponding modal damping ratios leads to physical insight into the system's dynamic behavior. In general, normal modes, which uncouple the equations of motion, do not exist for the total dynamic system consisting of the structure and the soil. If a transformation to (classical) modal coordinates is performed, the resulting damping matrix will have off-diagonal terms. It is possible, however, to calculate the frequencies of the damped free vibration, the corresponding (complex) mode shapes, and the modal damping ratios using the complex eigenvalue approach. This complex modal analysis—described in detail in a textbook (Chapter 9 of Ref. [H2])—is summarized here and illustrated using the rigid disk with mass in torsional motion discussed in Section 2.11.

Adding to the equations of motion of the structure–soil system with $n$ degrees of freedom

$$[M]\{\ddot{u}\} + [C]\{\dot{u}\} + [K]\{u\} = \{R\} \qquad (2.63)$$

the matrix identity

$$[M]\{\dot{u}\} - [M]\{\dot{u}\} = \{0\} \qquad (2.64)$$

leads to

$$\begin{bmatrix} [O] & [M] \\ [M] & [C] \end{bmatrix} \begin{Bmatrix} \{\ddot{u}\} \\ \{\dot{u}\} \end{Bmatrix} + \begin{bmatrix} -[M] & [O] \\ [O] & [K] \end{bmatrix} \begin{Bmatrix} \{\dot{u}\} \\ \{u\} \end{Bmatrix} = \begin{Bmatrix} \{O\} \\ \{R\} \end{Bmatrix} \qquad (2.65)$$

$\{R\}$ denotes the vector of the applied load. This equation can be written as a set of $2n$ first-order differential equations:

$$[A]\{\dot{y}\} + [B]\{y\} = \{Y\} \qquad (2.66)$$

with

$$[A] = \begin{bmatrix} [O] & [M] \\ [M] & [C] \end{bmatrix} \qquad (2.67a)$$

$$[B] = \begin{bmatrix} -[M] & [O] \\ [O] & [K] \end{bmatrix} \qquad (2.67b)$$

$$\{y\} = \begin{Bmatrix} \{\dot{u}\} \\ \{u\} \end{Bmatrix} \qquad (2.67c)$$

$$\{Y\} = \begin{Bmatrix} \{O\} \\ \{R\} \end{Bmatrix} \qquad (2.67d)$$

Both matrices $[A]$ and $[B]$ are real and symmetric.

## Sec. 2.13  Complex Modal Analysis

For the dynamic system shown in Fig. 2-25b the following relations apply:

$$\{u\} = \begin{Bmatrix} \psi_o \\ \psi_1 \end{Bmatrix} \tag{2.68a}$$

$$\{R\} = \begin{Bmatrix} T \\ O \end{Bmatrix} \tag{2.68b}$$

$$[M] = \begin{bmatrix} m & \\ & M_1 \end{bmatrix} \tag{2.68c}$$

$$[C] = \begin{bmatrix} C_1 & -C_1 \\ -C_1 & C_1 \end{bmatrix} \tag{2.68d}$$

$$[K] = \begin{bmatrix} K & \\ & O \end{bmatrix} \tag{2.68e}$$

$\psi_o$ and $\psi_1$ are the rotations at the center of the disk and at the additional node. Note that the stiffness matrix $[K]$ is singular. As the procedures to calculate the eigenvalues assume $[K]$ can be inverted, it is necessary to eliminate $\psi_1$ (but not $\dot{\psi}_1$) before defining $\{y\}$ and thus Eq. 2.66.

To calculate the eigenvalues $p_i$ and the corresponding mode shapes $\{y_i\}$, the right-hand side of Eq. 2.66 is set equal to zero. As the displacements and the velocities have the form $\exp(p_i t)$,

$$\{\dot{y}_i\} = p_i\{y_i\} \tag{2.69}$$

applies, leading to the eigenvalue problem

$$(p_i[A] + [B])\{y_i\} = \{0\} \tag{2.70}$$

Its solution will yield $2n$ eigenvalues and $2n$ mode shapes. For an underdamped system, $p_i$ is complex with negative real parts and comes in conjugate pairs with corresponding pairs of complex conjugate mode shapes. For an overdamped system, $p_i$ is real and negative, and the associated mode shape is real.

From the complex eigenvalue

$$p_i = \alpha_i + i\beta_i \tag{2.71}$$

the damped frequency $\omega_i$ and the associated damping ratio $\zeta_i$ of the mode shape $\{y_i\}$ can be calculated. Comparing $\exp(p_i t)$ with the damped free-vibration response of a single-degree-of-freedom system

$$\exp(\alpha_i t)\exp(i\beta_i t) = \exp\left[-\frac{\zeta_i \omega_i}{\sqrt{1-\zeta_i^2}} t\right] \exp(i\omega_i t) \tag{2.72}$$

leads to

$$\omega_i = \beta_i \tag{2.73}$$

$$\zeta_i = -\frac{\alpha_i}{\sqrt{\alpha_i^2 + \beta_i^2}} \tag{2.74}$$

Using Eq. 2.68, Eq. 2.70 leads to

$$[mM_1 p_i^3 + C_1(m + M_1)p_i^2 + KM_1 p_i + KC_1]p_i = 0 \tag{2.75}$$

$p_i = 0$ represents a trivial eigenvalue. Substituting Eq. 2.33 and the values from Table 2-2 leads, for $m = 0.3\rho a^5$, to the three eigenvalues

$$p_1 = -3.677 \frac{c_s}{a} \tag{2.76a}$$

$$p_{2,3} = (-1.464 \pm 2.206\, i)\frac{c_s}{a} \tag{2.76b}$$

and to the damped frequency and damping ratio (Eqs. 2.73, 2.74)

$$\omega_{2,3} = 2.206 \frac{c_s}{a} \tag{2.77a}$$

$$\zeta_{2,3} = 0.553 \tag{2.77b}$$

The associated mode shapes corresponding to $\{u\}$ follow from Eq. 2.70

$$\{\varphi_1\} = \begin{Bmatrix} -1.5361 \\ 1 \end{Bmatrix} \tag{2.78a}$$

$$\{\varphi_{2,3}\} = \begin{Bmatrix} -0.0097 \pm 1.5214i \\ 1 \end{Bmatrix} \tag{2.78b}$$

Using the standard lumped-parameter model (Fig. 2-25c) with $K$ and $C$ specified in Eqs. 2.57a and 2.57b and the total mass equal to the sum of $m$ and $M$ (Eq. 2.57c) leads to the undamped frequency $= 3.143\, c_s/a$ and the damping ratio 0.233. The damped frequency equals $3.056\, c_s/a$.

The mode shapes can be used to construct a transformation of coordinates in which the equations of motion are uncoupled. The procedure is analogous to that used for the case of classical modes [H2]. Introducing

$$\{u\} = [\Phi]\{z\} \tag{2.79}$$

where all $\{\varphi_i\}$ are assembled in $[\Phi]$ and $\{z\}$ denotes the vector of the modal amplitudes (coordinates), the equation of motion for each amplitude is formulated as

$$\dot{z}_i - p_i z_i = \frac{\{\varphi_i\}^T \{R\}}{a_i} \tag{2.80}$$

with

$$a_i = 2p_i \{\varphi_i\}^T [M]\{\varphi_i\} + \{\varphi_i\}^T [C]\{\varphi_i\} \tag{2.81}$$

### Sec. 2.13  Complex Modal Analysis

The solution of this first-order differential equation is specified as

$$z_i(t) = \frac{1}{a_i} \int_0^t \{\varphi_i\}^T \{R(\tau)\} \exp[p_i(t - \tau)] \, d\tau \tag{2.82}$$

The amplitudes $z_i(t)$ corresponding to two complex conjugate eigenvalues will also be complex conjugates.

Performing certain transformations, it is possible, in terms of deformations and velocities [V5], to interpret Eq. 2.82 as the response of a single-degree-of-freedom system.

For the loading $T(t)$ described by a versed cosine function (Eq. 2.56), the convolution integral can be evaluated in closed form as

$$\int_0^t T(t) \exp[p_i(t - \tau)] \, d\tau$$

$$= \frac{T_o}{2} \left[ \frac{1}{p_i} (\exp(p_i t) - 1) + \frac{1}{\frac{4\pi^2 c_s^2}{a^2} + p_i^2} \left( p_i \cos\left(\frac{2\pi t}{t_o}\right) \right. \right.$$

$$\left. \left. - \frac{2\pi c_s}{a} \sin\left(\frac{2\pi t}{t_o}\right) - p_i \exp(p_i t) \right) \right] \quad \text{for } t < t_o \tag{2.83a}$$

$$= \frac{T_o}{2} \exp(p_i t)(1 - \exp(-p_i t_o)) \frac{4\pi^2}{4\pi^2 p_i + p_i^3 t_o^2} \quad \text{for } t > t_o \tag{2.83b}$$

In Fig. 2-30, the responses of the first mode and those of the second and third modes combined are plotted. The superposition of the contributions of all three modes leads to the exact result determined in Section 2.11 (Fig. 2-26).

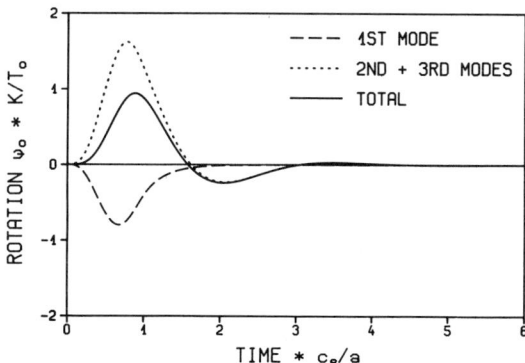

**Figure 2-30** Time-history of torsional rotation of disk determined by complex modal analysis.

## 2.14 PARTIAL UPLIFT OF BASEMAT

### 2.14.1 Disk on Tensionless Half-Space

Large lateral loads acting on a structure—caused, for example, by a severe earthquake—will lead to a substantial overturning moment. This can result in tension occurring in part of the structure's and the soil's basemat area, according to an analysis based on a linear theory. For a circular basemat on the surface of a half-space, if the overturning moment exceeds the product of the vertical force (dead weight minus the effect of the vertical acceleration) and one-third of the radius, then tension will occur in part of the area of contact, assuming a distribution of stress as in the static case. For a strip foundation, the same arises if the eccentricity of the vertical force exceeds a quarter of the total width. As tension is incompatible with the constitutive law of soils, the basemat will become partially separated from the underlying soil. Assuming that only normal stresses in compression and corresponding shear stresses (friction) can occur in the area of contact (tensionless half-space), a method of analyzing soil–structure interaction including partial uplift is derived, which otherwise is based on the elastic behavior of the soil. The area of contact is discretized into (boundary) elements. The rigorous procedure is discussed in Section 7.6, and an approximate one, using springs and dampers with frequency-independent coefficients, is covered in Section 3.11. An even simpler approximate approach, which replaces the actual area of contact by equivalent disks, is examined in this section [W5, W6].

For a sufficiently large overturning moment $M_o$ acting on a rigid basemat, the latter will become partially separated from the soil, as demonstrated for a circular basemat with radius $a$ in Fig. 2-31. The area of contact is

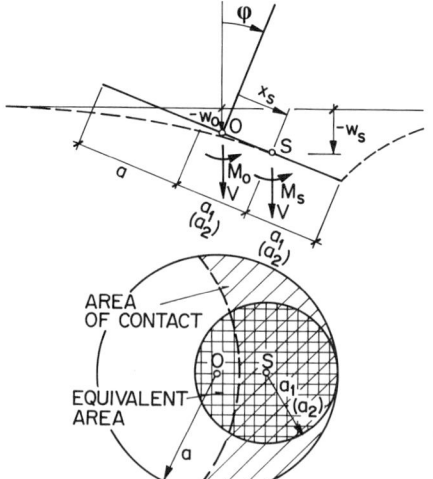

**Figure 2-31** Discretized area of contact between basemat and soil.

## Sec. 2.14 Partial Uplift of Basemat

determined by using a simple iterative procedure based on static influence coefficients of the elastic half-space. It is a function of the ratio $|M_o/Va|$, where $V$ is the vertical load. The actual irregular area of contact is then replaced by an equivalent circle with the same center of gravity $S$, the radius being calculated by equating the areas for the 2 translational degrees of freedom ($a_1$) and the moment of inertia for the rotational degree of freedom ($a_2$). The horizontal interaction force $H$ and corresponding displacement $u$ are not shown in Fig. 2-31. The equivalent radii and the location of $S$, expressed by its coordinate $x_S$ with respect to point $O$, can thus be determined for all ratios of $|M_o/Va|$ a priori, before any dynamic analysis is performed. The approximate expressions for a circular basemat are [W5]

$$\frac{a_1}{a} = -1.074 \left(\frac{M_o}{Va}\right)^2 - 0.068 \left|\frac{M_o}{Va}\right| + 1.142 \quad (2.84a)$$

$$\frac{a_2}{a} = \frac{3}{2}\left(1 - \left|\frac{M_o}{Va}\right|\right) \quad (2.84b)$$

$$\pm \frac{x_S}{a} = \frac{3}{2}\left|\frac{M_o}{Va}\right| - \frac{1}{2} \quad (2.84c)$$

where the upper and lower signs of Eq. 2.84c apply to the rotation $\varphi > 0$ and $\varphi < 0$, respectively. Eqs. 2.84 are valid for

$$\frac{1}{3} \leq \left|\frac{M_o}{Va}\right| \leq 1 \quad (2.85a)$$

and

$$V < 0 \quad (2.85b)$$

To prevent unrealistically small areas of contact, minimum radii are selected as $0.1\,a$. The radii $a_1$ and $a_2$ thus determined are used to calculate the static spring coefficient $K$ and the damping coefficient $C$ of the standard lumped—parameter model in point $S$ using Eq. 2.11 and Table 2-1. The mass $M$ is disregarded in Ref. [W6], on which this section is based.

In this dynamic transient analysis with nonlinearities, the explicit time-integration scheme is appropriate. Using the predicted values of the displacements at 0 at time $n\Delta t$, the vertical displacement $w_S$ at $S$ is calculated based on the compatibility equation

$$w_S = w_o - x_S\varphi \quad (2.86)$$

The tilde and subscript $n$ for the time are dropped. The interaction forces follow from

$$H = K_h u + C_h \dot{u} \quad (2.87a)$$

$$V = K_v w_S + C_v \dot{w}_S \quad (2.87b)$$

$$M_S = K_r \varphi + C_r \dot{\varphi} \quad (2.87c)$$

The moment $M_o$ is equal to

$$M_o = M_S - x_S V \qquad (2.88)$$

Equations 2.84 to 2.88 are used iteratively. After reaching convergence, $H$, $V$, and $M_o$, together with the internal forces of the structure, then allow the accelerations to be determined at time $n\Delta t$.

### 2.14.2 Structural Model on Tensionless Half-Space

To study the global effects of the basemat's partial uplift on the response, the structure is modeled as a one-degree-of-freedom system connected to a rigid disk with mass (Fig. 2-32). The 3 degrees of freedom of the disk with radius $a$, mass $m_o$, and moment of inertia $I_o$ at 0 are the two displacements $u_o$, $w_o$ and the rotation $\varphi$. The lateral displacement (relative to the basemat) of the idealized structure with the mass $m$ located at height $h$, and with the stiffness and damping coefficients $k$ and $c$, respectively, is denoted by $u$. $u_g$ represents the horizontal seismic motion of the free field.

The following four equations of motion are established:

$$m(\ddot{u}_o + h\ddot{\varphi} + \ddot{u}) + c\dot{u} + ku = -m\ddot{u}_g \qquad (2.89a)$$

$$m_o \ddot{u}_o + C_h \dot{u}_o - c\dot{u} + K_h u_o - ku = -m_o \ddot{u}_g \qquad (2.89b)$$

$$(m + m_o)\ddot{w}_o + C_v(\dot{w}_o - x_S\dot{\varphi}) + K_v(w_o - x_S\varphi) = 0 \qquad (2.89c)$$

$$mh(\ddot{u}_o + h\ddot{\varphi} + \ddot{u}) + I_o\ddot{\varphi} - x_S C_v(\dot{w}_o - x_S\dot{\varphi})$$
$$+ C_r\dot{\varphi} - x_S K_v(w_o - x_S\varphi) + K_r\varphi = -mh\ddot{u}_g \qquad (2.89d)$$

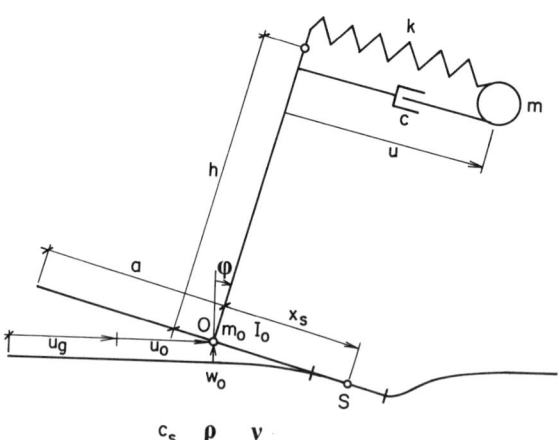

**Figure 2-32** Idealized structural model with rigid basemat, resting on surface of soil, with partial uplift.

$x_S$ and the spring and damping coefficients $K_h$, $C_h$, ... $C_r$ are functions of the area of contact (Eqs. 2.84).

The selected structural parameters correspond to a typical reactor building with $a = 30$ m and $h = 25$ m. The mass $m = 9.5 \cdot 10^7$ kg, the stiffness coefficient $k = 1.35 \cdot 10^{11}$ N/m, and the damping coefficient $c = 3.58 \cdot 10^8$ Ns/m, which correspond to a fixed-base frequency = 6 Hz and a damping ratio = 0.05 of the structure. The values corresponding to the disk are chosen as $m_o = 3 \cdot 10^7$ kg and $I_o = 2.4 \cdot 10^{10}$ kgm². Three different grounds are examined: a soft soil with a shear-wave velocity $c_s = 200$ m/s, a density $\rho = 1.6 \cdot 10^3$ kg/m³, and a Poisson's ratio $\nu = 0.45$; a medium soil with $c_s = 600$ m/s, $\rho = 1.8 \cdot 10^3$ kg/m³, $\nu = 0.4$; and a hard soil (rock) with $c_s = 2000$ m/s, $\rho = 2.5 \cdot 10^3$ kg/m³, $\nu = 0.3$. For all investigations, a 10 s acceleration time history acting in the horizontal direction whose response spectrum follows that of the U.S. Nuclear Regulatory Commission for 7% damping [U2] is used, whereby the maximum acceleration $\ddot{u}_g$ is varied.

To illustrate the phenomenon of uplift, the maximum overturning moment $M_o$ (Fig. 2-33a), the extreme vertical force $V$ (Fig. 2-33b), the minimum equivalent radii $a_1$, $a_2$ (Fig. 2-33c), and the maximum horizontal displacement of the mass point $u_o + h\varphi + u$ (Fig. 2-33d), for increasing ground acceleration

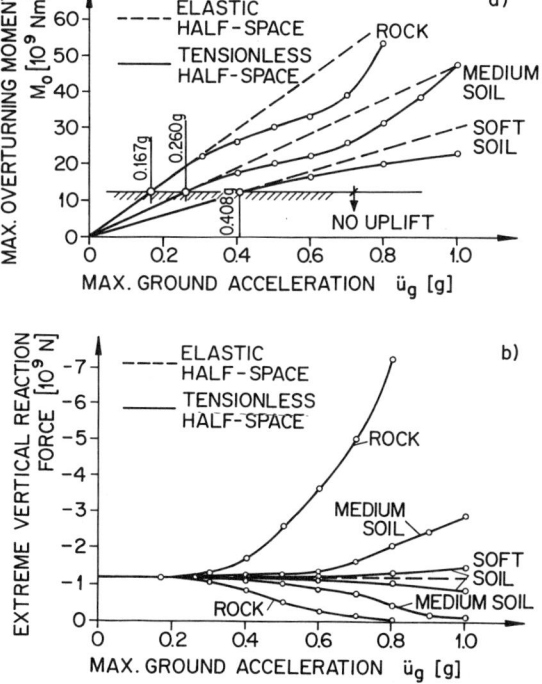

**Figure 2-33** Dynamic response with uplift of basemat.

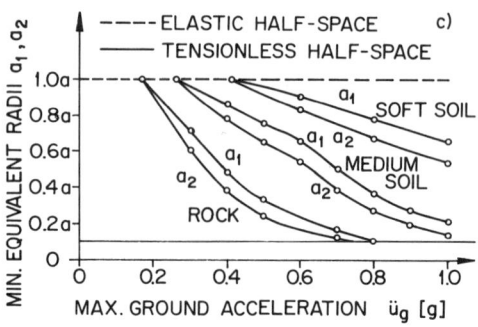

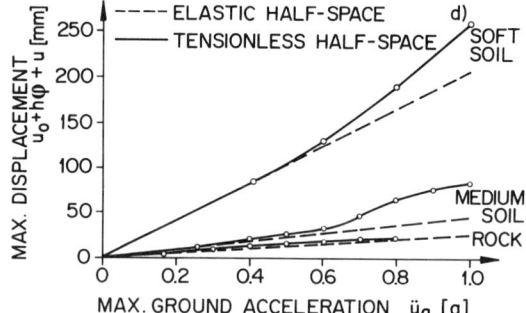

**Figure 2-33** Continued

and for the three soils, are calculated on the basis of the tensionless half-space theory with uplift and of the elastic half-space theory (with tension arising). The overturning moment $M_o$ is representative for the relative lateral displacement $u$, the total acceleration $\ddot{u}_o + h\ddot{\varphi} + \ddot{u} + \ddot{u}_g$, and the shear force in the structure for this simple model. Uplift starts when $M_o = Va/3$ (Fig. 2-33a) and always leads to a reduction of $M_o$ when compared to the elastic solution. For increasing acceleration, this reduction in the overturning moment $M_o$ increases at first. The vertical oscillation, accompanied by a large vertical reaction force $V$ (Fig. 2-33b) tends to decrease this reduction for higher acceleration values. It is important that the analytical model can account for this vertical-force variation, which is caused by the coupling of the vertical to the rocking motion. While the area of contact is significantly reduced for uplift (Fig. 2-33c), the additional lateral displacement of the mass point is small (Fig. 2-33d).

### 2.14.3 Dynamic-Stiffness Coefficient for Circular Cavity in Tensionless Thin Layer

The procedure of analyzing the partial separation of the basemat from the neighboring soil for a surface foundation, described in Section 2.14.1, can

### Sec. 2.14   Partial Uplift of Basemat

be extended to an embedded cylindrical foundation [W19]. The procedure is only outlined in the following, whereby a certain physical insight is gained when separation occurs.

The contributions of the rigid embedded cylindrical foundation's basemat and sidewall to the dynamic stiffness are calculated separately. In addition, the soil adjacent to the sidewall is assumed to consist of independent infinitesimally thin horizontal layers. The formulation is expanded to a no-tension material by using higher-order harmonics in the circumferential direction, because only stresses in compression and, possibly, shear stresses arise on the circumference of the cavity. A gap forms between part of the foundation and the adjacent soil. For a specified frequency, a layer's area of contact is determined from its horizontal displacement. For the basemat, the procedure of Section 2.14.1 is selected, in which an equivalent disk is determined.

The dynamic-stiffness coefficients for a circular cavity of radius $a$ in the unbounded thin layer with shear modulus $G$, density $\rho$, and Poisson's ratio $\nu$ can be calculated as a function of the dimensionless frequency $a_o = \omega a/c_s$ (with the shear-wave velocity $c_s$), based on a linear analysis.

For $a_o = 0.63$ and $\nu = 0.48$, the displacements and tractions along the cavity's circumference corresponding to the dynamic-stiffness coefficient in the horizontal direction $k_h + i\omega c_h$ are plotted on the left-hand side of Fig. 2-34. The corresponding spring and damping coefficients in nondimensionalized form are also specified. Excluding tension—that is, setting the normal and shear stresses on the circumference of the cavity in the zone of separation equal to zero—leads to the values shown in the center of Fig. 2-34. Introducing separation of the soil barely affects $k_h$, but reduces $c_h$ by more than

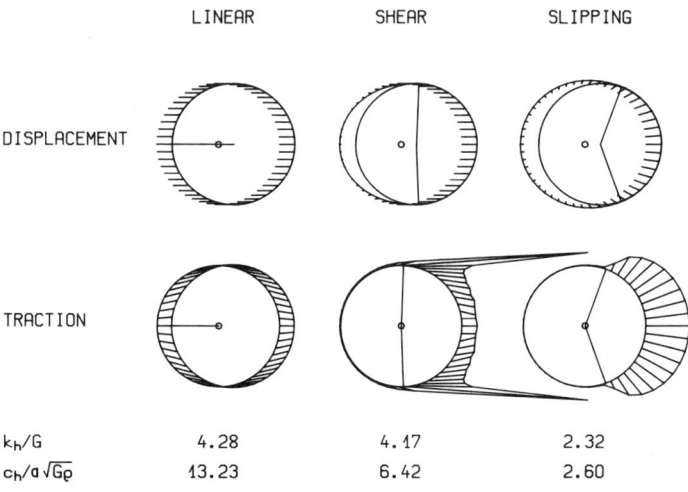

**Figure 2-34**   Dynamic-stiffness coefficient in horizontal direction.

**64**   Approximate Dynamic Model of Embedded Foundation   Chap. 2

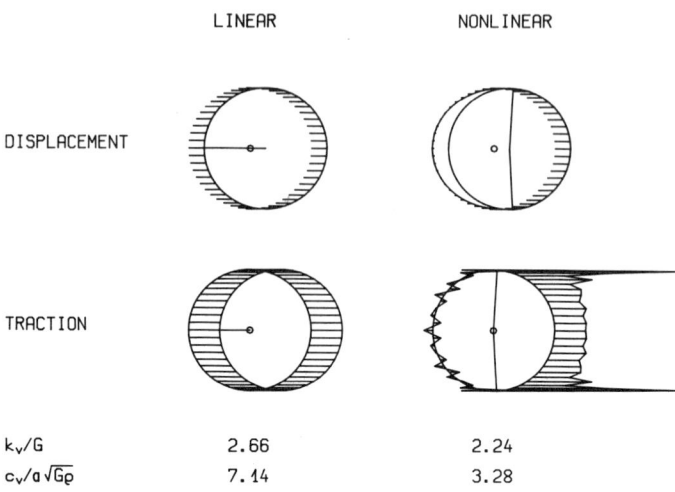

**Figure 2-35** Dynamic-stiffness coefficient in vertical direction.

50% of the linear case results. If, in addition, slipping in the zone of contact is postulated—that is, the shear stress acting on the circumference is zero—$k_h$ as well as $c_h$ are strongly reduced (right-hand side of Fig. 2-34). In this case, the reduction in $c_h$ is caused by the smaller zone of contact through which energy can be radiated—which will predominantly occur only with dilatational waves.

Fig. 2-35 displays the results for the vertical direction using the known zone of contact determined in the analysis in the horizontal direction. The quantities which act in the vertical direction are plotted horizontally. As for

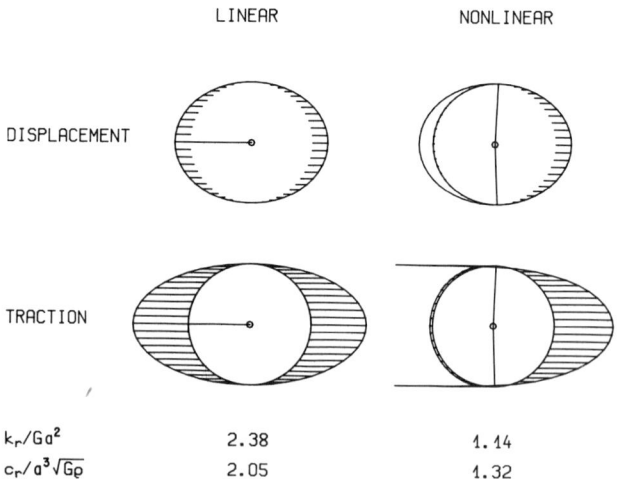

**Figure 2-36** Dynamic-stiffness coefficient in rocking motion.

the horizontal motion, the nonlinearity barely affects the spring constant $k_v$ and halves the damping coefficient $c_v$.

The results for the rocking motion are plotted in Fig. 2-36. The rocking spring coefficient $k_r$ is halved.

## SUMMARY

1. For each dynamic degree of freedom, the standard lumped-parameter system to model the soil consists of a mass $M$ (which is added to that of the structure's base) and of a spring $K$ and a damper $C$ in parallel which are attached to a rigid support. The real part of the corresponding dynamic-stiffness coefficient varies as a second-degree parabola with frequency, and the imaginary part increases linearly. The mass $M$ leads to a negative real part for high-frequency excitation. $K$ is equal to the static stiffness. For simple systems in the low and intermediate ranges, such as the disk with mass on the surface of an elastic half-space, (frequency-independent) expressions for $C$ and $M$ can be developed to achieve an acceptable accuracy (Eq. 2.11, Table 2-1).

2. For more general cases, another discrete model to represent the unbounded soil in a soil–structure-interaction analysis in the time domain is developed. For each dynamic degree of freedom of the foundation node, the discrete model consists of a mass $M_o$ at the foundation node which is attached to a rigid support with a spring $K$ and with a damper $C_o$. In addition, a free node with the mass $M_1$ is introduced, which is connected to the foundation node with a damper $C_1$. All coefficients are frequency-independent and depend only on the properties of the soil.

3. The same discrete model is derived when determining the dynamic-stiffness coefficient of a spherical cavity embedded in full space, assuming spherically symmetric P-waves. Other continuous unbounded systems exist, such as the semi-infinite truncated cone excited in torsion or in rocking, which lead to the same dynamic-stiffness coefficient as for the discrete model.

4. The discrete model used for an actual case is semi-empirical. After enforcing the static stiffness, the remaining parameters are selected to achieve an optimum fit of the dynamic-stiffness coefficients in the frequency domain for the discrete model and for the actual case (curve-fitting).

5. The spring $K$ is equal to the static stiffness. The dampers $C_o$, $C_1$ and the masses $M_o$, $M_1$ follow from the coefficients $\gamma_o$, $\gamma_1$ and $\mu_o$, $\mu_1$, using Eq. 2.33. For the disk on the surface of an elastic half-space, $K$ and $\gamma_o$ to $\mu_1$ are listed as a function of Poisson's ratio in Table 2-2; for the

embedded cylinder as a function of embedment, in Table 2-3; for the rectangular foundation as a function of the length-to-width ratio, in Eq. 2.42 and Table 2-4, including embedment; and, for the strip as a function of Poisson's ratio, in Table 2-5. For an embedded foundation, the discrete model for the horizontal motion is attached with an eccentricity specified in Eq. 2.41. All these specified values apply to a homogeneous half-space. The coefficients can also be determined using the same curve-fitting technique for a layered half-space with the stiffness increasing with depth. The corresponding values $K$ and $\gamma_o$ to $\mu_1$ are listed in Table 2-6 for a square foundation on a layer, whose shear-wave velocity increases linearly with depth, resting on a homogeneous half-space. (Viscous) material damping increases $\gamma_o$ and $\mu_o$.

6. When applying the discrete model of the soil, the resulting equations of motion of the (nonlinear) structure–soil system are conveniently solved using the direct time-integration procedure. For a linear structure, the complex modal analysis method based on complex eigenvalues—which leads to the frequencies of the damped free vibrations, the modal damping values, and complex mode shapes—can also be applied to determine the response for a transient excitation.

7. An approximate method of analyzing soil–structure interaction with partial uplift is discussed, assuming that only normal stresses in compression and corresponding shear stresses can occur in the area of contact between the basemat and the soil. For a given vertical force and moment acting on the rigid disk (basemat), the area of contact is determined for static conditions. The actual irregular area of contact is then replaced by an equivalent disk with the same center of gravity, the radii being calculated by equating the area and the moment of inertia for the translational and rotational degrees of freedom, respectively. This then allows the lumped-parameter model to be applied also for partial uplift. The coupling of the vertical and the rocking motions is properly taken into account. Allowing the structure to uplift always leads to a reduction of the total horizontal acceleration—and hence of the shear force and of the overturning moment—and of the relative lateral displacement within the structure, when compared to the results of the corresponding linear analysis. The latter requires that tensile capacity be provided between the structure and the soil. On the other hand, the total lateral displacement and the vertical response are increased.

# PROBLEMS

**2.1** A generalized lumped-parameter model with an additional degree of freedom $u_1$ is shown in Fig. P 2-1a. Parallel to the spring with the static coefficient $K$ and the damper with the coefficient $C_o$, a component consisting of a spring with the coefficient

Chap. 2    Problems

$K_1$ and a damper with the coefficient $C_1$ in series is present. Determine the coefficients $C_o$, $K_1$, and $C_1$ in dimensionless form so as to achieve an optimal fit between the dynamic-stiffness coefficient of the lumped-parameter model and that of a disk of radius $a$ resting on the surface of an elastic undamped half-space with shear-wave velocity $c_s$ for the torsional motion in the range $0 < a_o < 4$ (exact values presented in Ref. [V4]; see Fig. P 2-1b). Compare the spring and damping coefficients up to $a_o = 8$.

*Solution:*

Equations of motion in frequency domain:

$$i\omega C_1 u_1(\omega) + K_1 u_1(\omega) - K_1 u_o(\omega) = 0$$

$$i\omega C_o u_o(\omega) + (K + K_1) u_o(\omega) - K_1 u_1(\omega) = P(\omega)$$

Elimination of $u_1(\omega)$:

$$P(\omega) = K \frac{1 - \omega^2 \frac{C_o C_1}{K K_1} + i\omega \left( \frac{C_o}{K} + \frac{C_1}{K} + \frac{C_1}{K_1} \right)}{1 + i\omega \frac{C_1}{K_1}} u_o(\omega)$$

$$= K \frac{1 + \frac{\omega^2 C_1^2}{K_1} \left( \frac{1}{K} + \frac{1}{K_1} \right) + i\omega \frac{1}{K} \left( C_o + C_1 + \omega^2 C_o \frac{C_1^2}{K_1^2} \right)}{1 + \frac{\omega^2 C_1^2}{K_1^2}} u_o(\omega)$$

$$= K \overline{S}(\omega) u_o(\omega)$$

with dimensionless dynamic-stiffness coefficient $\overline{S}(\omega)$.
Dimensionless coefficients:

$$a_o = \frac{\omega a}{c_s}$$

$$K_1 = \kappa K$$

$$C_o = \frac{a}{c_s} K \gamma_o$$

$$C_1 = \frac{a}{c_s} K \gamma_1$$

Dimensionless dynamic-stiffness coefficient:

$$\overline{S}(a_o) = \frac{1 - \frac{\gamma_o \gamma_1}{\kappa} a_o^2 + i a_o \left( \gamma_o + \gamma_1 + \frac{\gamma_1}{\kappa} \right)}{1 + i a_o \frac{\gamma_1}{\kappa}} = \frac{1 - a_1 a_o^2 + i a_o a_2}{1 + i a_o a_3}$$

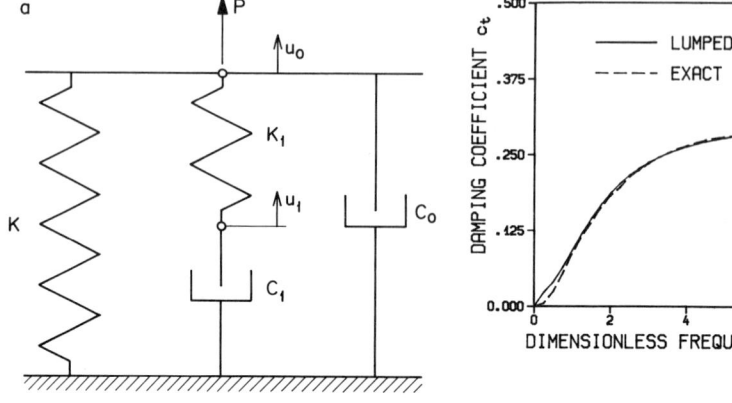

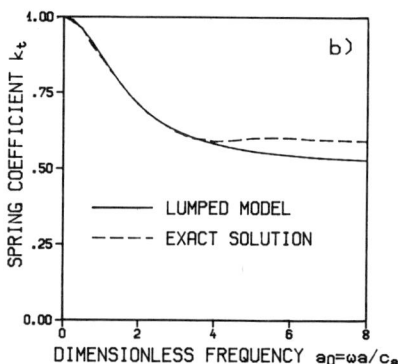

**Figure P 2-1** Generalized lumped-parameter model. a. Arrangement of springs and dampers with additional degree of freedom. b. Dynamic-stiffness coefficient of disk for torsional motion.

with coefficients

$$a_1 = \frac{\gamma_o \gamma_1}{\kappa}$$

$$a_2 = \gamma_o + \gamma_1 + \frac{\gamma_1}{\kappa}$$

$$a_3 = \frac{\gamma_1}{\kappa}$$

Curve fitting of complex dynamic-stiffness coefficient with values at $a_o = 1, 2, 3,$ and 4 based on least square method:

$$a_1 = 0.18123$$

$$a_2 = 0.60574$$

$$a_3 = 0.58882$$

Dimensionless coefficients of spring and dampers:

$$\kappa = -0.49399$$
$$\gamma_o = 0.30779$$
$$\gamma_1 = -0.29087$$

Good agreement of damping coefficient $c_r(a_o)$ not only in range of curve fitting ($0 < a_o < 4$), but also up to $a_o = 8$ (Fig. P 2-1b) and limited deviations of spring coefficient $k_r(a_o)$ in higher frequency range, where the latter is not important.

**2.2** The wave motion corresponding to a circular cavity with radial symmetry embedded in a full plane (two-dimensional case) is examined in Problem 3.7. The dynamic-stiffness coefficient $S_r$, relating the radial displacement at the wall to the force which is equal to minus the radial stress at the same location multiplied by the circumference of the cavity, is calculated as [W11]

$$S_r = \frac{2\pi E}{1 + \nu} (k_r + ia_o c_r)$$

where

$$k_r + ia_o c_r = 1 - \frac{1 - \nu}{1 - 2\nu} a_o \frac{H_0^{(2)}(a_o)}{H_1^{(2)}(a_o)}$$

$$a_o = \frac{\omega a}{c_p} = \sqrt{\frac{(1 + \nu)(1 - 2\nu)}{1 - \nu} \frac{\rho}{E}} \, \omega a$$

$a$ is the radius of the cavity, $E$ the modulus of elasticity, $\nu$ Poisson's ratio, $\rho$ the mass density. $H_0^{(2)}$ and $H_1^{(2)}$ are the zeroth- and first-order Hankel functions of the second kind. $k_r$ and $c_r$ are plotted, for $\nu = 1/3$, as a solid line in Fig. P 2-2.

Select the coefficients $\gamma_o$ to $\mu_1$ of the discrete model shown in Fig. 2-4 to achieve an optimum fit ($\nu = 1/3$).

*Result:*

$$K = \frac{2\pi E}{1 + \nu}$$

$\gamma_o = 0$ $\qquad$ $\gamma_1 = 2.00$

$\mu_o = 0$ $\qquad$ $\mu_1 = 4.19$

whereby, in the definition of $C_1$ and $M_1$ (Eq. 2.33), $c_s$ is replaced by $c_p$. The corresponding dynamic-stiffness coefficient, shown as a dashed line, agrees well with the exact solution (Fig. P 2-2).

**2.3** In Section 2.8, the coefficients of the discrete model for a strip foundation on the surface of a half-plane are established (Table 2-5). Because there is a lack of results available in the literature, the embedded strip foundation in the two-dimensional case cannot be treated now. However, promising possibilities exist. For instance, for the rocking motion referred to the center $O$ of the basemat, $2b$ wide, of a foundation

**70**  Approximate Dynamic Model of Embedded Foundation  Chap. 2

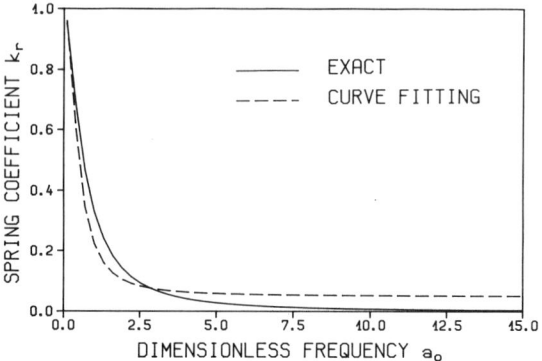

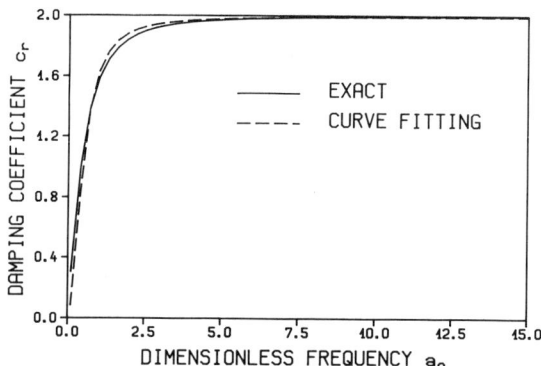

**Figure P 2-2** Dynamic-stiffness coefficient of cylindrical cavity.

with an embedment $e$, the static stiffness $K_r$ can be approximated as

$$K_r = Gb^2 \left( 2.38 + \frac{e}{b} + 1.2 \left(\frac{e}{b}\right)^2 \right)$$

Using the variation of the dynamic-stiffness coefficient for three embedment ratios $e/b$ taken from Ref. [W11] ($\nu = 1/3$) and which are reproduced in Fig. P 2-3, determine the dimensionless coefficients $\gamma_o$ to $\mu_1$ of the discrete model shown in Fig. 2-4.

*Result:*

$$\gamma_{or} = 0.11 + 0.35 \frac{e}{b} + 0.1 \left(\frac{e}{b}\right)^2$$

$$\gamma_{1r} = 0.4$$

$$\mu_{or} \begin{cases} = 0 & \text{for } e \leq b \\ = 0.3 \left(\frac{e}{b} - 1\right) & \text{for } e \geq b \end{cases}$$

$$\mu_{1r} = 0.3$$

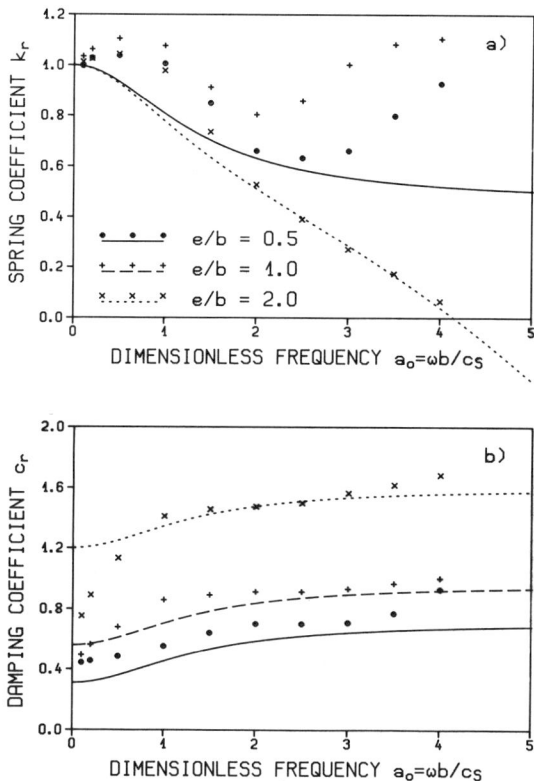

**Figure P 2-3** Dynamic-stiffness coefficient of embedded strip for rocking motion.

The agreement with the exact values is acceptable. The dashed and solid lines coincide in Fig. P 2-3a.

**2.4** A rigid cylindrical block with radius $a$, mass $m$, and moment of inertia $I$ (with respect to the center of mass, which is at a height $h$) rests on the surface of an elastic undamped half-space with Poisson's ratio $v$, shear-wave velocity $c_s$, and density $\rho$ (Fig. P 2-4). For a horizontal excitation, this system has two dynamic degrees of freedom $u_o$ and $\varphi$.

Model the soil with the standard lumped-parameter model (Table 2-1) and with the discrete model of Fig. 2-4. Establish the equations of motion, the undamped natural frequencies and damping ratios corresponding to classical modes (for lumped-parameter model), and then the damped frequencies and damping ratios of the complex modal analysis. Use the following values: $a = 20$ m, $h = 20$ m, $m = 5 \cdot 10^7$ kg, $I = 1.2 \cdot 10^{10}$ kg m², $v = 1/3$, $c_s = 500$ m/s, $\rho = 2 \cdot 10^3$ kg/m³.

*Results:*

**a.** Standard Lumped-Parameter Model (Fig. P 2-4a)

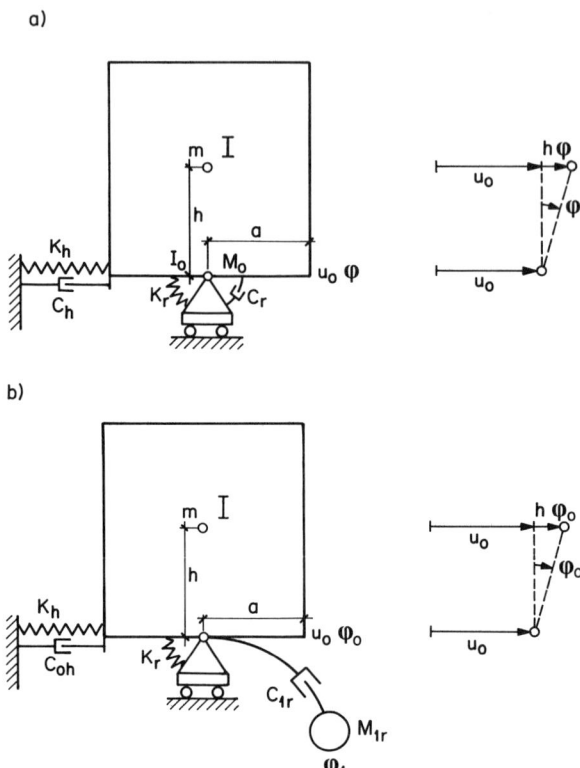

**Figure P 2-4** Rigid body on flexible soil, modeled as a. Lumped-parameter model. b. Discrete model with additional degree of freedom.

Equations of motion:

$$\begin{bmatrix} m + M_o & m h \\ m h & m h^2 + I + I_o \end{bmatrix} \begin{Bmatrix} \ddot{u}_o \\ \ddot{\varphi} \end{Bmatrix} + \begin{bmatrix} C_h & \\ & C_r \end{bmatrix} \begin{Bmatrix} \dot{u}_o \\ \dot{\varphi} \end{Bmatrix} + \begin{bmatrix} K_h & \\ & K_r \end{bmatrix} \begin{Bmatrix} u_o \\ \varphi \end{Bmatrix} = \begin{Bmatrix} 0 \\ 0 \end{Bmatrix}$$

$M_o$, $I_o$, $C_h$, $C_r$, $K_h$, $K_r$ Table 2-1 (a subscript $o$ is added for the mass and the mass moment of inertia of the soil).

Classical modal analysis:
natural frequencies $\omega_i$

$$|[K] - \omega_i^2 [M]| = 0$$

$$\frac{\omega_1}{2\pi} = 2.87 \text{ Hz} \qquad \frac{\omega_2}{2\pi} = 7.13 \text{ Hz}$$

modal damping ratios $\zeta_i$

$$\zeta_i = \frac{\{\varphi_i\}^T [C] \{\varphi_i\}}{2\omega_i \{\varphi_i\}^T [M] \{\varphi_i\}}$$

$$\zeta_1 = 0.111 \qquad \zeta_2 = 0.429$$

Chap. 2 Problems

Complex modal analysis (Eq. 2.70):
eigenvalues $p_i$

$$|p_i[A] + [B]| = 0$$

$$p_{1,2} = -0.00598 \pm 0.05468\,i$$

$$p_{3,4} = -0.00973 \pm 0.02028\,i$$

damped frequencies $\omega_i$ (Eq. 2.73)

$$\frac{\omega_{1,2}}{2\pi} = 2.87 \text{ Hz} \qquad \frac{\omega_{3,4}}{2\pi} = 6.38 \text{ Hz}$$

damping ratios $\zeta_i$ (Eq. 2.74)

$$\zeta_{1,2} = 0.109 \qquad \zeta_{3,4} = 0.433$$

b. Discrete Model (Fig. P 2-4b)
Equations of motion:

$$\begin{bmatrix} m & mh \\ mh & mh^2 + I \\ & & M_{1r} \end{bmatrix} \begin{Bmatrix} \ddot{u}_o \\ \ddot{\varphi}_o \\ \ddot{\varphi}_1 \end{Bmatrix} + \begin{bmatrix} C_{oh} & & \\ & C_{1r} & -C_{1r} \\ & -C_{1r} & C_{1r} \end{bmatrix} \begin{Bmatrix} \dot{u}_o \\ \dot{\varphi}_o \\ \dot{\varphi}_1 \end{Bmatrix}$$

$$+ \begin{bmatrix} K_h & & \\ & K_r & \\ & & 0 \end{bmatrix} \begin{Bmatrix} u_o \\ \varphi_o \\ \varphi_1 \end{Bmatrix} = \begin{Bmatrix} 0 \\ 0 \\ 0 \end{Bmatrix}$$

Complex modal analysis:
stiffness matrix $[K]$ singular
eliminate $\varphi_1$ as follows (3rd equation of motion)

$$\dot{\varphi}_1 = -\frac{M_{1r}}{C_{1r}} \ddot{\varphi}_1 + \dot{\varphi}_o$$

transformation matrix $[T]$ for $\{y\}$ (Eq. 2.67c)

$$\begin{Bmatrix} \ddot{u}_o \\ \ddot{\varphi}_o \\ \ddot{\varphi}_1 \\ \dot{u}_o \\ \dot{\varphi}_o \\ \dot{\varphi}_1 \end{Bmatrix} = \begin{bmatrix} 1 & & & & & \\ & 1 & & & & \\ & & 1 & & & \\ & & & 1 & & \\ & & & & 1 & \\ & & -\dfrac{M_{1r}}{C_{1r}} & & 1 & \end{bmatrix} \begin{Bmatrix} \ddot{u}_o \\ \ddot{\varphi}_o \\ \ddot{\varphi}_1 \\ \dot{u}_o \\ \dot{\varphi}_o \end{Bmatrix}$$

$$\{y\} = [T]\{y_r\}$$

reduced eigenvalue problem (Eq. 2.70)

$$|p_i[T]^T[A][T] + [T]^T[B][T]| = 0$$

eigenvalues $p_i$

$$p_{1,2} = -0.00454 \pm 0.05422\,i$$
$$p_{3,4} = -0.01337 \pm 0.01390\,i$$
$$p_5 = -0.02293$$

damped frequencies $\omega_i$ (Eq. 2.73)

$$\frac{\omega_{1,2}}{2\pi} = 2.91 \text{ Hz} \qquad \frac{\omega_{3,4}}{2\pi} = 5.95 \text{ Hz} \qquad \omega_5 = 0$$

damping ratio $\zeta_i$ (Eq. 2.74)

$$\zeta_{1,2} = 0.083 \qquad \zeta_{3,4} = 0.69 \qquad \zeta_5 = 1$$

**2.5** Discrete models can also be established to calculate the free-field response of a site directly in the time domain. For instance, for an undamped site consisting of soil layers resting on elastic rock, the model shown in Fig. P 2-5 can be used for horizontal earthquake excitation propagating vertically. The spring coefficient is determined as $k_1 = G_1/d_1$, the masses $m_1 = \rho_1 d_1/2$, $m_2 = (\rho_1 d_1 + \rho_2 d_2)/2$, and so on. The rock (half-space) is modeled as a damper with a coefficient $c = \rho_n c_{sn} = \sqrt{\rho_n G_n}$. The equation of motion of node $n$ is formulated in the time domain as

$$\frac{\rho_{n-1} d_{n-1}}{2} \ddot{v}_n + \frac{G_{n-1}}{d_{n-1}} (v_n - v_{n-1}) + \sqrt{\rho_n G_n}\, \dot{v}_n = \sqrt{\rho_n G_n}\, \dot{v}_o$$

where $v_n$ is the free-field displacement (as a function of time) at node $n$ and $\dot{v}_o$ is the velocity of the (prescribed) outcropping control motion. If the control motion is specified at the free surface, the calculation proceeds from this point downwards with zero right-hand sides. The equation of motion of node $n - 1$ then leads to the motion in node $n$.

This discrete model can also be used to calculate inclined SH-waves (out-of-plane motion), if an effective density $\rho^e$ defined as

$$\rho_i^e = \sin^2\psi_i \rho_i \qquad i = 1, 2, \ldots, n - 1, n$$

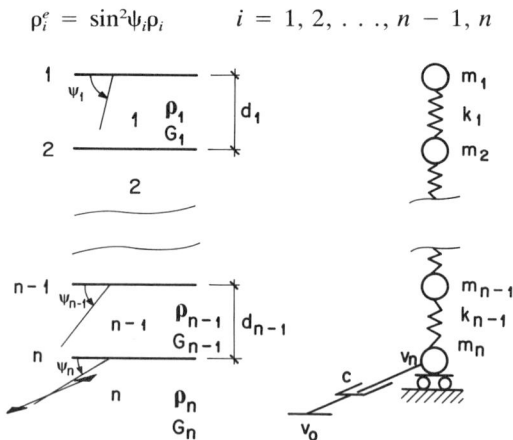

**Figure P 2-5** Discrete soil model for inclined SH-wave.

is used [K3]. The angle of incidence in layer $i$, measured from the horizontal (Fig. P 2-5), is denoted as $\psi_i$. As the apparent velocity $c_a$ is the same for the half-space and for all layers, the angles of incidence satisfy

$$c_a = \frac{c_1}{\cos\psi_1} = \cdots = \frac{c_i}{\cos\psi_i} = \cdots = \frac{c_{n-1}}{\cos\psi_{n-1}} = \frac{c_n}{\cos\psi_n}$$

Derive this formula for the effective density.

*Solution:*

The exact dynamic-stiffness matrix (Eq. 4.22) for the out-of-plane motion of a layer (superscript $L$) is specified as (the index $i$ is dropped for convenience)

$$[S^L] = \frac{ktG}{\sin ktd}\begin{bmatrix} \cos ktd & -1 \\ -1 & \cos ktd \end{bmatrix}$$

and that of the half-space (superscript $R$ for rock) as

$$S^R = iktG$$

where

$$kt = \frac{\omega}{c_a}\sqrt{\frac{1}{\cos^2\psi} - 1} = \frac{\omega}{c_s}\sin\psi = \frac{\omega}{\sqrt{G}}\sqrt{\rho}\sin\psi$$

It is apparent that for $\rho^e = \rho \sin^2\psi$, $kt$ equals $\frac{\omega}{\sqrt{G}}\sqrt{\rho^e}$, which is the expression for vertically propagating waves ($\psi = 90°$) with the mass density $\rho^e$ instead of $\rho$ and the same shear modulus $G$. This is an exact method.

Assuming that the depth of the layer $d$ is small compared to the wave length $\sqrt{G}/(\sin\psi\sqrt{\rho})(2\pi/\omega)$,

$$\sin ktd = \sin(\omega\sqrt{\rho}\sin\psi\, d/\sqrt{G})$$
$$= \sin(\omega\sqrt{\rho^e}\, d/\sqrt{G}) \simeq \omega\sqrt{\rho^e}d/\sqrt{G}$$

and $\cos ktd \simeq 1 - \omega^2\rho^e d^2/(2G)$ can be substituted in $[S^L]$, leading to

$$[S^L] = \frac{G}{d}\begin{bmatrix} 1 & -1 \\ -1 & 1 \end{bmatrix} - \frac{\omega^2\rho^e d}{2}\begin{bmatrix} 1 & \\ & 1 \end{bmatrix}$$

The first term represents the static-stiffness matrix with the spring coefficient $G/d$; the second, the mass matrix with lumped masses $\rho^e d/2$. The dynamic-stiffness coefficient of the half-space equals

$$S^R = i\omega\sqrt{\rho^e_n G_n}$$

$\sqrt{\rho^e_n G_n}$ represents the damping coefficient.

# 3

# DIRECT METHOD

In the direct method of analyzing soil–structure interaction, the linear soil adjacent to the structure (or the irregular soil region) is also modeled up to the artificial boundary with which the interaction horizon coincides (Section 1.3, Fig. 1-3b). The boundary condition, formulated with a finite number of degrees of freedom located on this surface, must simulate the infinite extent of the soil. This transmitting boundary, which will allow energy to propagate only from the interior to the exterior region, is also called a nonreflecting, (energy-) absorbing, or silent boundary.

The rigorous boundary condition will be global in both space and time— that is, to advance one time step in a transient analysis requires information over the entire boundary from all previous time steps. This can be computationally expensive. Several schemes have been developed to formulate highly absorbing local approximations to the perfectly absorbing boundary. These approximations are local in space and time: they use information only from the node or the nearby region of the mesh at the start of the time step or, at most, at a few recent time stations. The boundary condition must be independent of frequency to be able to be used in a transient analysis in the time domain. It should be able to handle (approximately) all types of waves and there should be no restrictions on the geometry and on the material properties, such as Poisson's ratio. Implementation in a standard finite element program should also be straightforward and easy to perform.

In the direct method, modeling of the linear regular soil adjacent to the structure–soil interface leads to many degrees of freedom. It can be advantageous from a computational point of view to introduce generalized coordinates, especially when the total dynamic system behaves linearly and when (frequency-independent) springs, dashpots, and masses are used to model the transmitting boundary. These generalized coordinates (class of Ritz vectors or so-called Lanczos coordinates [W4, B1, R5, N2]) can be generated with a fraction of the numerical effort required for the calculation of classical or complex eigenvectors (mode shapes) and can be used to significantly reduce the number of degrees of freedom of a dynamic system with a constant spatial distribution of the loading.

Certain features of one-dimensional wave propagation in a semi-infinite system are discussed in Section 3.1—in particular, the transmitting boundary condition's general form. The prismatic rod, the truncated cone in shear and in torsion, and the spherical cavity embedded in a full space with symmetric waves are addressed. Various procedures exist to approximately model, on the artificial boundary, the wave propagation towards infinity. In the following, these time-domain transmitting boundaries are examined and compared using a simple one-dimensional system: the semi-infinite undamped rod which is supported elastically [W15]. This system, which is local in space, exhibits the same properties as certain sites which are analyzed in soil–structure interaction. It is a dispersive system and exhibits a cutoff frequency below which no waves propagate. As the apparent velocity is not constant in this system, certain conclusions are also valid for the two- and three-dimensional cases where the apparent velocity in the direction of the normal to the boundary also varies, depending on the angle of incidence of the wave. However, it is recognized that certain other features, such as the presence of various types of waves in a system, cannot be studied using this one-dimensional case. Closed-form solutions, which allow the various procedures' merits to be established, can be derived for this simple system. This semi-infinite elastically supported rod's properties are examined in Section 3.2, which also contains the benchmark problem's definition: At the free end of the semi-infinite rod, the displacement is prescribed as a function of time. Using an explicit integration scheme in the time domain, the various approximate procedures to model the wave propagation towards infinity are described and examined: the superposition boundary (Section 3.3), where the reflected waves in two overlapping boundary zones with complementary boundary conditions are canceled as they occur, the incorporation of fictitious material damping (Section 3.4) to absorb the waves reflected off the artificial boundary; the well-known viscous damper (Section 3.5), which absorbs waves propagating perpendicularly to the boundary; the doubly asymptotic approximation (Section 3.6), which is exact at both high and low frequencies; the paraxial boundary (Section 3.7), which corresponds to a differential equation that transmits waves in only one direction—that is, outwardly; and the extrapolation algorithm (Section 3.8), which uses present and recent past data along a line normal to the artificial boundary to enforce outgoing propagating waves. Besides the comparison of the semi-infinite rod's result, other examples which demonstrate the performance of the transmitting boundaries are included in these sections. In Section 3.9, the artificial boundary location's influence on the result's accuracy is discussed for body and surface waves. The procedures of the transmitting boundaries, described in Sections 3.3 to 3.8, are generalized in Section 3.10 to the case where the load is not applied directly to the structure but is introduced through the soil (free-field response). In Section 3.11, an approximate procedure is addressed that uses frequency-independent far-coupled springs and dampers to model the soil so structures can be analyzed from actual practice, including partial uplift of the basemat.

## 3.1 ONE-DIMENSIONAL WAVE PROPAGATION

### 3.1.1 Prismatic Rod

One-dimensional wave propagation—which by definition is local in space—can be used to develop the basis for frequency-independent transmitting boundaries which are local in time. The semi-infinite prismatic rod is the simplest case and thus is addressed first. Radial effects are disregarded.

The prismatic rod with area $A$, modulus of elasticity $E$, and mass density $\rho$ extending to infinity is shown in Fig. 3-1a. $N$ represents the axial force and $u$ the axial displacement. Formulating equilibrium of the infinitesimal element (Fig. 3-1b)

$$N_{,x} dx - \rho A dx \ddot{u} = 0 \tag{3.1}$$

and substituting the force–displacement relationship

$$N = EA u_{,x} \tag{3.2}$$

leads to the equation of motion

$$u_{,xx} - \frac{\ddot{u}}{c_l^2} = 0 \tag{3.3}$$

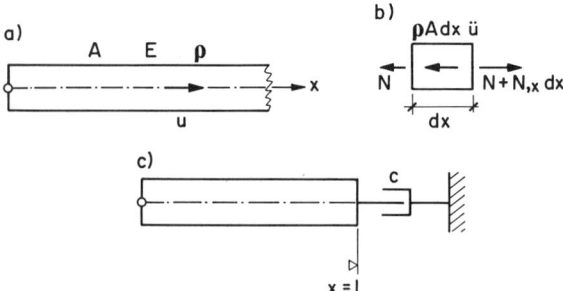

**Figure 3-1**  a. Semi-infinite prismatic rod.  b. Equilibrium of infinitesimal element.  c. Viscous damper modeling truncated rod.

## Sec. 3.1 One-Dimensional Wave Propagation

where $c_l$ denotes the rod velocity

$$c_l = \sqrt{E/\rho} \tag{3.4}$$

The one-dimensional wave equation (Eq. 3.3) can be integrated directly in the time domain after changing the variables $x, t$ to $\xi, \eta$

$$\xi = t - x/c_l \tag{3.5a}$$

$$\eta = t + x/c_l \tag{3.5b}$$

Equation 3.3 is then transformed to

$$u_{,\xi\eta} = 0 \tag{3.6}$$

which can be integrated to give

$$u = f(\xi) + g(\eta) \tag{3.7}$$

or

$$u = f(t - x/c_l) + g(t + x/c_l) \tag{3.8}$$

where $f$ and $g$ are two arbitrary functions of their arguments. Equation 3.8 is the general solution to the one-dimensional homogeneous wave equation. $f(t - x/c_l)$ represents a wave propagating in the positive $x$-direction with a velocity $c_l$, which keeps its original shape; $g(t + x/c_l)$ represents a wave traveling in the negative $x$-direction with the same features.

The property of a transmitting boundary located on the artificial boundary at $x = l$, described by a differential equation, is addressed next (Fig. 3-1c). When the incident wave $f$ encounters the artificial boundary, this wave must pass through it without any modification so it can continue propagating towards $x = +\infty$. No reflected wave $g$, which would propagate in the negative $x$-direction, may arise. The wave $f$ satisfies the boundary condition formulated at $x = l$, but $g$ does not. Because the functions $f(t - x/c_l)$ and $g(t + x/c_l)$ differ by the signs of the $x/c_l$ terms in the arguments, a differentiation with respect to $x$ is appropriate. This involves a differentiation of $f$ and $g$ with respect to the argument—for instance, $f' = df(t - x/c_l)/d(t - x/c_l)$ which is equal to $\dot{f}$. The following differential equations apply:

$$f_{,x} + \frac{\dot{f}}{c_l} = 0 \tag{3.9a}$$

$$g_{,x} - \frac{\dot{g}}{c_l} = 0 \tag{3.9b}$$

Selecting Eq. 3.9a, which is identically satisfied for $f$, as the boundary condition for $u$ at $x = l$

$$u_{,x}(x = l, t) + \frac{\dot{u}}{c_l}(x = l, t) = 0 \tag{3.10}$$

results in $g = 0$. This is easily verified by substituting Eq. 3.8 in Eq. 3.10, which leads to $g' = dg(t + x/c_l)/d(t + x/c_l) = 0$ at $x = l$. Equation 3.10 is also called the radiation condition.

The physical interpretation of the boundary condition at $x = l$ becomes apparent when Eq. 3.10 is multiplied by $EA$

$$EAu_{,x} + \frac{EA}{c_l} \dot{u} = 0 \qquad (3.11a)$$

or, after substituting Eqs. 3.2 and 3.4,

$$N + c\dot{u} = 0 \qquad (3.11b)$$

results with

$$c = A\rho c_l \qquad (3.12)$$

Equation 3.11b expresses equilibrium at the artificial boundary, involving the normal force and the force of a viscous damper with a coefficient $c$, which replaces the part of the rod up to infinity (Fig. 3-1c). $c$ is also called the impedance. Because $c$ is independent of frequency, this transmitting boundary can be used directly for an analysis in the time domain.

The boundary condition is exact for the one-dimensional propagation of longitudinal waves. It can also be applied to the propagation of shear waves with the transverse displacement $w$ in the rod, by replacing $E$ by the shear modulus $G$, and $c_l$ by the shear-wave velocity $c_s = \sqrt{G/\rho}$. The one-dimensional wave equation then equals

$$w_{,xx} - \frac{\ddot{w}}{c_s^2} = 0 \qquad (3.13)$$

Equations 3.10, 3.11b, and 3.12 are transformed to

$$w_{,x} + \frac{\dot{w}}{c_s} = 0 \qquad (3.14)$$

$$Q + c\dot{w} = 0 \qquad (3.15)$$

$$c = A\rho c_s \qquad (3.16)$$

with $Q$ denoting the transverse shear force. No distinction is made between the shear area and the rod's area. These concepts form the basis of the transmitting boundary consisting of such viscous dampers to approximately model two- and three-dimensional cases (Section 3.5).

The energy of the incident wave is continuously being totally dissipated in the damper. This is verified as follows: The wave travels the distance $dx = c_l\, dt$ in the time $dt$. The kinetic energy $dE_k$ of the wave and the potential or strain energy $dE_s$ are equal to

Sec. 3.1   One-Dimensional Wave Propagation

$$dE_k = \frac{1}{2}\int_o^{dx} A\rho \dot{u}^2 dx' \qquad (3.17a)$$

$$dE_s = \frac{1}{2}\int_o^{dx} Eu_{,x}^2 A dx' = \frac{1}{2}\int_o^{dx} A\rho \dot{u}^2 dx' \qquad (3.17b)$$

whereby Eq. 3.10 is substituted. The energy $dE_d$ dissipated in the damper is calculated as the product of the force $A\rho c_l \dot{u}$ and the displacement $\dot{u}dt'$, leading to

$$dE_d = \int_o^{dx} A\rho c_l \dot{u}\dot{u}dt' = \int_o^{dx} A\rho \dot{u}^2 dx' \qquad (3.18)$$

The term $dE_d$ is equal to the total energy of the wave $dE_k + dE_s$.

It is instructive to study the total wave pattern when a pulse $f(t - x/c_l)$ is reflected at a fixed boundary at $x = l$. Formulating $u(x = l, t) = 0$ in Eq. 3.8 leads to $g[t + (x - l)/c_l] = -f[t - (x - l)/c_l]$, resulting in the total wave

$$u(x, t) = f\left(t - \frac{x - l}{c_l}\right) - f\left(t + \frac{x - l}{c_l}\right) \qquad (3.19)$$

At a fixed boundary, the incident wave is reflected back with the same shape but opposite in sign. Analogously, when a pulse $f(t - x/c_l)$ is reflected at a free boundary at $x = l$, $u(x = l, t)_{,x} = 0$ (Eq. 3.2) holds, resulting in $g[t + (x - l)/c_l] = f[t - (x - l)/c_l]$ or, for the total wave,

$$u(x, t) = f\left(t - \frac{x - l}{c_l}\right) + f\left(t + \frac{x - l}{c_l}\right) \qquad (3.20)$$

At a free boundary, the incident wave is reflected back with the same shape and the same sign.

### 3.1.2  Conical Rod in Shear

The semi-infinite rod, with its area varying as in a cone, is addressed next (Fig. 3-2a). As a load applied at the free surface of a half-space leads to stresses acting on an area that increases with depth, such a cone when truncated is a better model for a site than the prismatic rod discussed in Section 3.1.1. This semi-infinite truncated cone has been used to calculate an approximate dynamic-stiffness coefficient in the horizontal direction involving shear waves [M1]. It is thus appropriate to examine the equation of motion for the transverse displacement $w$ in the following.

The equilibrium equation is formulated as (Fig. 3-2b)

$$Q_{,x}dx - \rho A dx \ddot{w} = 0 \qquad (3.21)$$

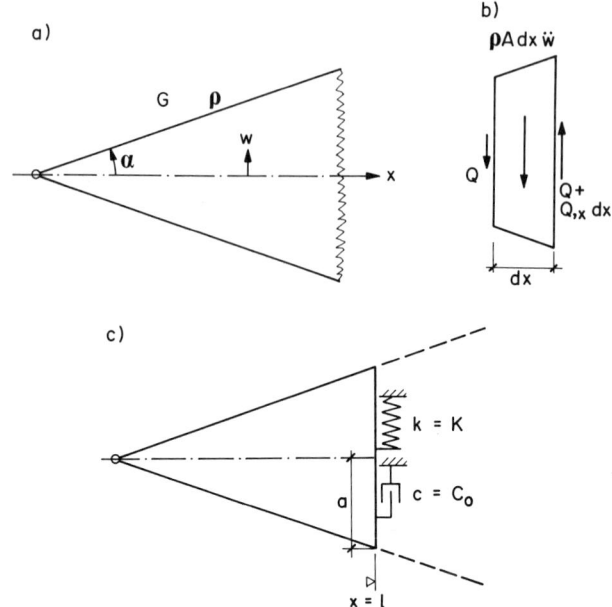

**Figure 3-2** a. Semi-infinite conical rod in shear. b. Equilibrium of infinitesimal element. c. Spring and damper modeling truncating cone.

with the transverse shear force $Q$, the mass density $\rho$, and the area $A(x) = \pi \tan^2\alpha x^2$ ($\alpha$ = angle). Substituting the force–displacement relationship, with $G$ denoting the shear modulus

$$Q = GAw_{,x} \qquad (3.22)$$

results in the equation of motion

$$w_{,xx} + \frac{A_{,x}}{A} w_{,x} - \frac{\ddot{w}}{c_s^2} = 0 \qquad (3.23a)$$

or

$$w_{,xx} + \frac{2}{x} w_{,x} - \frac{\ddot{w}}{c_s^2} = 0 \qquad (3.23b)$$

with the shear-wave velocity $c_s$ defined as

$$c_s = \sqrt{G/\rho} \qquad (3.24)$$

Equation 3.23b is rewritten as

$$(xw)_{,xx} - \frac{(xw)^{\cdot\cdot}}{c_s^2} = 0 \qquad (3.25)$$

Comparing this equation with the wave equation of the prismatic rod (Eq.

Sec. 3.1  One-Dimensional Wave Propagation

3.3) and its solution (Eq. 3.8), $xw$ corresponds to $u$. $w$ is thus formulated as

$$w = \frac{f(t - x/c_s)}{x} + \frac{g(t + x/c_s)}{x} \tag{3.26}$$

The wave pattern is the same as for the prismatic rod, except the amplitude is proportional to $1/x$—that is, for $f$ propagating with the velocity $c_s$ in the positive $x$-direction, the amplitude decreases inversely in proportion to the distance traveled.

To determine the condition enforced at the artificial boundary $x = l$, the procedure is analogous to that for the prismatic rod. The differential equation satisfied for the wave $f/x$ propagating in the positive $x$-direction—but not for $g/x$—can be derived from the radiation condition (Eq. 3.10) by replacing $u$ by $xw$.

$$(xw)_{,x} + \frac{(xw)\dot{}}{c_s} = 0 \tag{3.27}$$

Formulated at $x = l$ this leads to

$$w_{,x}(x = l, t) + \frac{w(x = l, t)}{l} + \frac{\dot{w}(x = l, t)}{c_s} = 0 \tag{3.28}$$

Substituting Eq. 3.26 into this equation results in $g' = dg(t + x/c_s)/d(t + x/c_s) = 0$ at $x = l$, while $f(t - x/c_s)$ cancels. Multiplying Eq. 3.28 by $GA(x = l) = GA$ results in

$$GAw_{,x} + \frac{GA}{l} w + \frac{GA}{c_s} \dot{w} = 0 \tag{3.29a}$$

or, after substituting Eqs. 3.22 and 3.24,

$$Q + kw + c\dot{w} = 0 \tag{3.29b}$$

with the spring and damping coefficients specified as

$$k = \frac{GA}{l} \tag{3.30a}$$

$$c = A\rho c_s \tag{3.30b}$$

Equation 3.29b expresses that the transverse shear force is in equilibrium with the forces in a spring that has coefficient $k$ and in a damper that has coefficient $c$ (Fig. 3-2c). The cone's missing part up to infinity is thus modeled by a spring and a damper that have frequency-independent coefficients.

As already mentioned, this semi-infinite truncated conical shear rod from $x = l$ up to infinity forms the basis for approximately calculating the dynamic-stiffness coefficient in the horizontal direction of a rigid disk of radius

$a = l \tan\alpha$ on the surface of an elastic half-space with shear modulus $G$, Poisson's ratio $v$, and density $\rho$. The angle $\alpha$ of the cone is determined by setting the static-stiffness coefficient $K$ equal to $k$ (Eq. 3.30a):

$$K = \frac{8Ga}{2-v} = \frac{GA}{l} \quad (3.31)$$

leading to

$$\tan\alpha = \frac{8}{(2-v)\pi} \quad (3.32)$$

The dynamic-stiffness coefficient $S$ follows from the second and third terms of Eq. 3.29b. Using $\dot{w} = i\omega w$ and substituting Eqs. 3.30, 3.31, and 3.32 results in

$$S = \frac{8Ga}{2-v}\left[1 + ia_o \frac{\pi(2-v)}{8}\right] \quad (3.33)$$

with the dimensionless frequency

$$a_o = \frac{\omega a}{c_s} \quad (3.34)$$

Comparing with Eqs. 2.35 and 2.36 leads to the dimensionless coefficient $\gamma_o$ of the damper $C_o$ (Eq. 2.33a):

$$\gamma_o = \frac{\pi(2-v)}{8} = 0.79 - 0.39v \quad (3.35)$$

This value is very close to the corresponding value of the discrete model $0.78 - 0.40v$ (Table 2-2), which is selected to obtain an optimal fit between the dynamic-stiffness coefficient of the discrete model (Eq. 3.33) and that of the half-space.

### 3.1.3 Spherical Cavity with Symmetric Waves

This section covers spherically symmetric wave propagation, which, for example, is generated by a symmetrical explosion inside a spherical cavity. This case is discussed for harmonic motion in Section 2.3 (Fig. 2-2). The equation of motion in the radial direction $r$ in the time domain is formulated as (analogous to Eq. 2.18 in the frequency domain)

$$u_{,rr} + \frac{2}{r}u_{,r} - \frac{2u}{r^2} - \frac{\ddot{u}}{c_p^2} = 0 \quad (3.36)$$

where $u$ is the radial displacement and $c_p$ the dilational wave velocity specified in Eq. 2.13. Introducing the potential $\varphi$ defined by

$$u = \varphi_{,r} \quad (3.37)$$

Sec. 3.1    One-Dimensional Wave Propagation

leads to the one-dimensional wave equation for $r\varphi$

$$(r\varphi)_{,rr} - \frac{(r\varphi)^{..}}{c_p^2} = 0 \tag{3.38}$$

Its solution, determined directly in the time domain, is specified in Eq. 3.8 with $r\varphi$ replacing $u$, leading to

$$\varphi = \frac{f(t - r/c_p)}{r} + \frac{g(t + r/c_p)}{r} \tag{3.39}$$

with $f$ and $g$ being two arbitrary functions, as discussed in Section 3.1.1. Substituting in Eq. 3.37 results in

$$u = -\frac{f(t - r/c_p)}{r^2} - \frac{f'(t - r/c_p)}{rc_p} - \frac{g(t + r/c_p)}{r^2} + \frac{g'(t + r/c_p)}{rc_p} \tag{3.40}$$

where $f' = df(t - r/c_p)/d(t - r/c_p)$ and $g' = dg(t + r/c_p)/d(t + r/c_p)$.

For waves propagating in the positive $r$-direction only (outgoing spherical wave), the function $g$ must vanish. The boundary condition to be enforced at the artificial boundary $r = R$ follows from Eq. 3.10, by substituting $r\varphi$ for $u$

$$(r\varphi)_{,r} + \frac{(r\varphi)^{.}}{c_p} = 0 \tag{3.41}$$

To formulate this radiation condition in $u$, Eq. 3.41 is differentiated twice with respect to $r$ leading to

$$u_{,rr} + \frac{3}{r} u_{,r} + \frac{2}{rc_p} \dot{u} + \frac{1}{c_p} \dot{u}_{,r} = 0 \tag{3.42a}$$

or, using the equation of motion (Eq. 3.36), to

$$\frac{2u}{r^2} + \frac{1}{r} u_{,r} + \frac{2}{rc_p} \dot{u} + \frac{1}{c_p} \dot{u}_{,r} + \frac{\ddot{u}}{c_p^2} = 0 \tag{3.42b}$$

These differential equations (Eq. 3.42), which are similar to the equation of motion (Eq. 3.36), lead to solutions which transmit waves only in the positive $r$-direction (see paraxial boundary, Section 3.7).

For a practical application, the missing medium on the exterior of the artificial boundary $r = R$ up to infinity is modeled more appropriately by the discrete model shown in Fig. 2-4. This consists of a spring with the static-stiffness coefficient and a mass with its own (additional) degree of freedom which is connected by a damper to the artificial boundary. This procedure is rigorous when one uses the coefficients for the spherical cavity with symmetric waves specified in Section 2.4. In the formulas, $R$ replaces $a$.

### 3.1.4 Conical Rod in Torsion

Finally, the semi-infinite rod, with its area varying as in a cone loaded by torsion, is discussed (Fig. 3-3a). $T$ is the torsional moment, $\varphi$ is the rotation, $\alpha$ is the angle, $J(x) = \pi \tan^4\alpha x^4/2$ is the polar moment of inertia, $G$ is the shear modulus, and $\rho$ is the mass density. Substituting the force–displacement relationship

$$T = GJ\varphi_{,x} \tag{3.43}$$

into the equilibrium equation (Fig. 3-3b)

$$T_{,x}dx - \rho J dx \ddot{\varphi} = 0 \tag{3.44}$$

leads to the equation of motion

$$\varphi_{,xx} + \frac{J_{,x}}{J}\varphi_{,x} - \frac{\ddot{\varphi}}{c_s^2} = 0 \tag{3.45a}$$

or

$$\varphi_{,xx} + \frac{4}{x}\varphi_{,x} - \frac{\ddot{\varphi}}{c_s^2} = 0 \tag{3.45b}$$

with the shear-wave velocity $c_s$ specified in Eq. 3.24.

Its solution is conveniently determined in the frequency domain. The

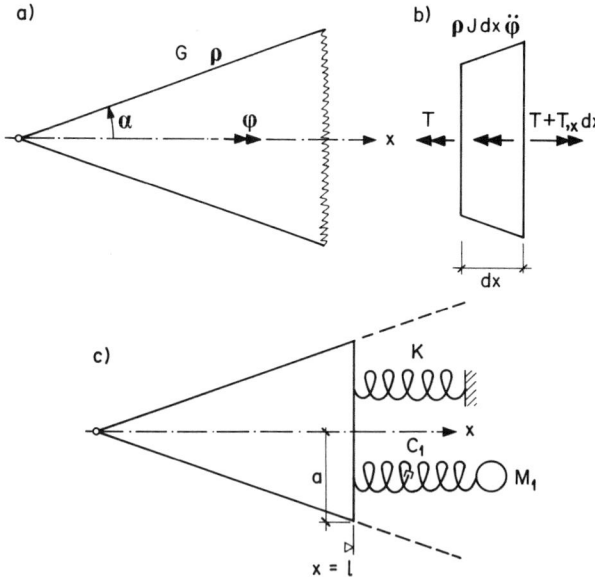

**Figure 3-3** a. Semi-infinite conical rod in torsion. b. Equilibrium of infinitesimal element. c. Discrete model of truncated cone with spring, damper, and mass with additional degree of freedom.

## Sec. 3.1  One-Dimensional Wave Propagation

equation 3.45b is transformed to

$$\varphi(\omega)_{,xx} + \frac{4}{x}\varphi(\omega)_{,x} + \frac{\omega^2}{c_s^2}\varphi(\omega) = 0 \qquad (3.46)$$

The general solution equals [V4, W11]

$$\varphi(\omega) = \frac{c_s^2}{\omega^2 x^2}\left[c_1\left(\frac{ic_s}{\omega x} - 1\right)\exp(-i\omega x/c_s) + c_2\left(-\frac{ic_s}{\omega x} - 1\right)\exp(+i\omega x/c_s)\right] \qquad (3.47)$$

with the integration constants $c_1$ and $c_2$. Remember: the rotation $\varphi(t)$ equals $\varphi(\omega)$ multiplied by $\exp(+i\omega t)$. The first term in Eq. 3.47 with the expression $\exp[i\omega(t - x/c_s)]$ will correspond to an outgoing wave; the second, to an incoming wave.

To derive the condition to be enforced at the artificial boundary $x = l$, which is expressed using the dynamic-stiffness coefficient, the amplitude of the incoming wave is set equal to zero ($c_2 = 0$). Enforcing the boundary condition $\varphi_o(\omega) = \varphi(x = l, \omega)$ leads, with $a = l\tan\alpha$, to

$$\varphi(\omega) = \varphi_o(\omega)\frac{a^3\cot^3\alpha\omega^3}{c_s^3\left(1 + ia\dfrac{\cot\alpha}{c_s}\omega\right)}\frac{1 + i\omega x/c_s}{\omega^3 x^3/c_s^3}\exp\left[-i\omega\left(\frac{x}{c_s} - \frac{a\cot\alpha}{c_s}\right)\right] \qquad (3.48)$$

The amplitude of the torsional moment $T(\omega)$ at the same location is determined by substituting Eq. 3.48 into Eq. 3.43, which, after a change in sign, is denoted as $T_o(a_o)$

$$T_o(a_o) = \frac{3\pi G a^3}{2\cot\alpha}\left[1 - f(a_o) + ia_o\frac{9\pi}{32}f(a_o)\right]\varphi_o(a_o) \qquad (3.49)$$

where the function $f(a_o)$ is specified as

$$f(a_o) = \frac{1}{3}\frac{(9\pi/32\, a_o)^2}{1 + \left(\dfrac{9\pi}{32}a_o\right)^2} \qquad (3.50)$$

and with the dimensionless frequency $a_o$ defined in Eq. 3.34. The coefficient of $\varphi_o(a_o)$ in Eq. 3.49 is the dynamic-stiffness coefficient. It is identical to that of the discrete model shown in Fig. 2-4. Selecting the static-stiffness coefficient $K = 3\pi G a^3/(2\cot\alpha)$ and the dimensionless coefficients $\gamma_o = 0$, $\gamma_1 = 3\pi/32$, $\mu_o = 0$, $\mu_1 = (9\pi/32)^2/3$ and substituting these values into Eqs. 2.35 and 2.36 leads to Eq. 3.49. The discrete model shown in Fig. 3-3c that has the damper with coefficient $C_1$ and the mass with coefficient $M_1$, both determined from Eq. 2.33, thus rigorously models the missing part of the conical rod up to infinity—that is, the semi-infinite truncated cone.

An application is discussed in Problem 3.1

This conical rod in torsion from $x = l$ up to infinity can be used to calculate the approximate dynamic-stiffness coefficient for torsion of a rigid disk of radius $a = l \tan\alpha$ on the surface of an elastic half-space with shear modulus $G$ and density $\rho$ [V4]. The cone's angle $\alpha$ is calculated by equating $K$ (Eq. 3.49) to the static-stiffness coefficient of the elastic half-space:

$$\frac{3\pi G a^3}{2 \cot\alpha} = \frac{16 G a^3}{3} \qquad (3.51)$$

which results in

$$\tan\alpha = \frac{32}{9\pi} \qquad (3.52)$$

As already described, $\gamma_1 = 3\pi/32 = 0.29$ and $\mu_1 = (9\pi/32)^2/3 = 0.26$, which are close to the corresponding respective values 0.29 and 0.20 presented in Table 2-2 for the discrete model with an optimal fit.

## 3.2 SEMI-INFINITE ROD ON ELASTIC FOUNDATION

### 3.2.1 Types of Waves

The various transmitting boundaries are illustrated and their accuracies compared using the semi-infinite rod resting on an elastic foundation shown in Fig. 3-4a [W15]. One could regard this as a model of a pile embedded in soil or as a potential crude model to estimate the dynamic-stiffness coefficient of a rigid foundation on a half-space's surface. In the latter case, Point 0 would lie at the center of the foundation, the rod's area would coincide with the foundation's, and the springs connected to the rod would represent the resistance of that part of the soil not already represented by the rod. This model can indeed be used to discuss the vital aspects of wave propagation towards infinity. However, it should not be used to model an actual site, unlike the conical rods in shear and in torsion (Sections 3.1.2 and 3.1.4).

Before describing the benchmark problem used for the comparison, which is calculated analytically, the wave propagation characteristics are addressed.

The area of the undamped rod is denoted by $A$, the modulus of elasticity by $E$, the mass density by $\rho$, and the static spring stiffness per unit length of the elastic foundation by $k_g$. $N$ represents the normal force and $u$ the axial displacement. Formulating equilibrium (Fig. 3-4b)

$$N_{,x} dx - k_g u \, dx - A \rho \ddot{u} \, dx = 0 \qquad (3.53)$$

and substituting the force–displacement relationship

$$N = EA u_{,x} \qquad (3.54)$$

## Sec. 3.2  Semi-Infinite Rod on Elastic Foundation

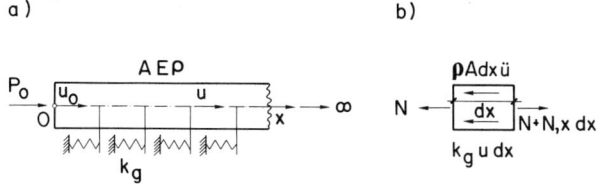

**Figure 3-4** a. Semi-infinite rod resting on elastic foundation. b. Equilibrium of infinitesimal element.

leads to the equation of motion

$$u_{xx} - \kappa^2 u - \frac{\ddot{u}}{c_l^2} = 0 \tag{3.55}$$

where

$$\kappa^2 = \frac{k_g}{EA} \tag{3.56a}$$

$$c_l = \sqrt{E/\rho} \tag{3.56b}$$

$c_l$ denotes the rod velocity.

To solve Eq. 3.55, $u$ of the form $u(\omega) \exp(i\omega t)$ is considered with $u(\omega) = \exp(i\gamma x)$. The equation is thus actually solved in the frequency domain. Introducing the dimensionless frequency $a_o$ and the wave number $k$

$$a_o = \frac{\omega}{c_l \kappa} \tag{3.57}$$

$$k = \frac{\omega}{c} \tag{3.58}$$

where $c$ is the phase velocity, the solution is given by

$$u(x, a_o) = a \exp(-ikx) + b \exp(+ikx) \tag{3.59}$$

with

$$k = \kappa \sqrt{a_o^2 - 1} \tag{3.60}$$

$$c = \frac{a_o}{\sqrt{a_o^2 - 1}} c_l \tag{3.61}$$

Recall that the harmonic motion is represented as $\exp(+i\omega t)$, and $a$ and $b$ are the amplitudes of the waves propagating with the phase velocity in the positive and negative $x$-axes, respectively. Because $c$ depends on $a_o$, the dynamic system is dispersive, which is soon discussed. A cutoff frequency exists at $a_o = 1$, below which the motion does not propagate, but decays exponentially (that is, as $\exp(-\kappa \sqrt{1 - a_o^2}\, x)$). To be able to discuss wave propagation in such dispersive systems, the group velocity $c_g$ is introduced, defined as

$$c_g = \frac{d\omega}{dk} \qquad (3.62)$$

It is equal to

$$c_g = \frac{\sqrt{a_o^2 - 1}}{a_o} c_l \qquad (3.63)$$

The ratios $c/c_l$ and $c_g/c_l$ are plotted in Fig. 3-5. $c$ and $c_g$ are equal to $+\infty$ and 0 at $a_o = 1$, respectively, and both converge to $c_l$ for $a_o \to \infty$. For $a_o < 1$, both $c$ and $c_g$ are imaginary.

Such dispersive systems are the rule and not the exception in soil-structure interaction analysis. Any layered half-space exhibits this property. In a dispersive system, each term of the motion's Fourier series corresponding to a specific frequency propagates with a different velocity, and therefore the shape of the wave will be altered as it travels. Cutoff frequencies also often occur in practice. For instance, for a layer built-in at its base, such a cutoff frequency exists, which is equal to the fundamental frequency (see Section 3.9.3). The response depends strongly on the frequency of excitation. Below the cutoff frequency, the behavior is similar to that in the static case. For instance, for this range of frequencies, the dynamic-stiffness coefficient of a foundation in such an elastic site is real—that is, the damping coefficient vanishes and thus no radiation of energy takes place. To gain some physical insight, the influence of the dispersion and the cutoff frequency on the motion as well as the propagation of a narrow-band pulse are addressed in Sections 3.2.2 and 3.2.3.

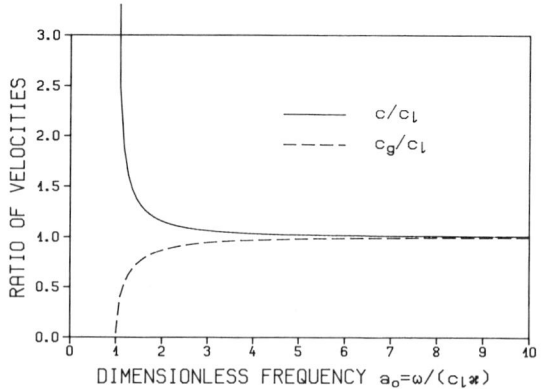

Figure 3-5 Phase and group velocities.

### 3.2.2 Influence of Dispersion and Cutoff Frequency on Motion

To further illustrate wave propagation in a dispersive system with a cutoff frequency, it is useful to examine the displacement along the rod arising from the prescribed support motion $u_o(\bar{t})$ at $x = 0$ (Fig. 3-4a):

$$u_o(\bar{t}) = \frac{u_o}{2}\left[1 - \cos\left(2\pi \frac{\bar{t}}{\bar{t}_o}\right)\right], \quad 0 < \bar{t} < \bar{t}_o$$
$$u_o(\bar{t}) = 0 \quad \bar{t} > \bar{t}_o \quad (3.64)$$

with the dimensionless time $\bar{t}$

$$\bar{t} = tc_l\kappa \quad (3.65)$$

The dimensionless time $\bar{t}_o = t_o c_l \kappa$ is selected as equal to 2. This choice leads to a significant content of the motion in the range below the cutoff frequency. The dimensionless support movement $\bar{u}_o(\bar{t}) = u_o(\bar{t})/u_o$—a rounded triangular pulse—is plotted as a solid line in Fig. 3-6b. Only waves propagating in the positive x-direction will occur ($b = 0$ in Eq. 3.59). To calculate the motion at $x = 5/\kappa$, Fourier transformations in the frequency and wave number domains are performed using Eq. 3.59

$$u_o(a_o) = \frac{1}{c_l\kappa} \int_0^{\bar{t}_o} u_o(\bar{t}) \exp(-ia_o\bar{t}) \, d\bar{t} \quad (3.66a)$$

$$u(x = 5/\kappa, \bar{t}) = \frac{c_l\kappa}{2\pi} \int_{-\infty}^{+\infty} u_o(a_o) \exp(ia_o\bar{t}) \exp\left[-ik(a_o)\frac{5}{\kappa}\right] da_o \quad (3.66b)$$

$\bar{u}(\bar{t}) = u(x = 5/\kappa, \bar{t})/u_o$ is plotted as a solid line in Fig. 3-6a. The essentially triangular pulse applied at $x = 0$ is strongly distorted until it reaches $x = 5/\kappa$; the largest peak even has a different sign. It is instructive to suppress the components of the motion corresponding to frequencies below the cutoff frequency $a_o = 1$. This is performed by deleting the range $-1 < a_o < 1$ from the integral in Eq. 3.66b. While large differences exist at $x = 0$, this no longer applies at $x = 5/\kappa$ (shown as dashed lines), because the components of the motion with a frequency below $a_o = 1$ decay exponentially.

From the time-history of the motion shown in Fig. 3-6a, one can see that the components with high frequencies arrive before those associated with low frequencies. This tendency agrees with the notion that the waves' dominant components travel with the group and not with the phase velocity [G3]. This can be verified as follows. For the (relative) maximum arriving at $\bar{t} = 8.3$ (Fig. 3-6a), the velocity is calculated as $c/c_l = \kappa x/\bar{t} = 5/8.3 = 0.60$. Assuming this is the group velocity, $c_g/c = 0.60$ corresponds to $a_o = 1.25$ (Fig. 3-5) or to the period $\bar{T} = 5$, which agrees well with the value read from Fig. 3-6a.

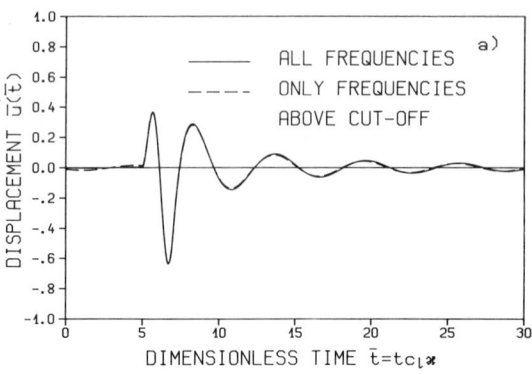

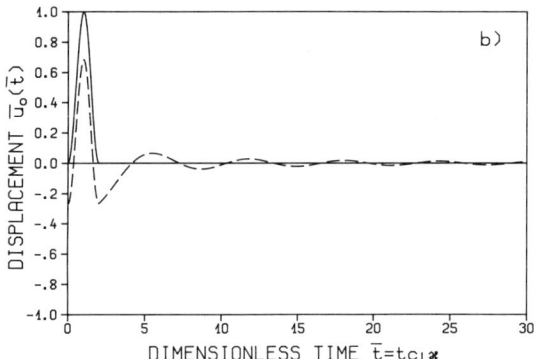

**Figure 3-6** Dispersion and cutoff frequency. a. Displacement at $x = 5/\kappa$. b. Prescribed displacement at $x = 0$.

### 3.2.3 Propagation of Narrow-Banded Pulse

Further insight can be gained by examining the propagation of a finite-duration oscillatory wave whose Fourier amplitudes are clustered around a dominant frequency $\omega_o$ (narrow-banded pulse). The Fourier transform is denoted as $u_o(\omega)$. Applying the inverse Fourier transform to the wave propagating in the positive $x$-direction of Eq. 3.59 leads to

$$u(x, t) = \frac{1}{2\pi} \int_{-\infty}^{+\infty} u_o(\omega) \exp(i\omega t) \exp[-ik(\omega)x] d\omega \qquad (3.67)$$

Expanding $k(\omega)$ in a Taylor series around $\omega_o$

## Sec. 3.2  Semi-Infinite Rod on Elastic Foundation

$$k(\omega) = k_o + \frac{dk}{d\omega}\bigg|_o (\omega - \omega_o) \tag{3.68}$$

with $k_o = \omega_o/c^o$ and introducing the definition of the group velocity $c_g^o$ determined at $\omega_o$ (Eq. 3.62)

$$c_g^o = \frac{d\omega}{dk}\bigg|_o \tag{3.69}$$

transforms Eq. 3.67 to

$$u(x, t) = \exp\left[-i\left(k_o - \frac{\omega_o}{c_g^o}\right)x\right] \frac{1}{2\pi} \int_{-\infty}^{+\infty} u_o(\omega) \exp\left[i\omega\left(t - \frac{x}{c_g^o}\right)\right] \tag{3.70}$$

The integral (with the factor $1/(2\pi)$) represents the original wave form at $x = 0$, which has propagated undistorted up to $x$ with the group velocity $c_g^o$. The exponential factor in Eq. 3.70 contributes nothing more than a phase angle for a specific $x$ [G3].

Equation 3.70 can be examined further. Introducing

$$t' = t - \frac{x}{c_g^o} \tag{3.71}$$

the original wave form is represented as the product of an amplitude $a(t')$ and a wave with the frequency $\omega_o$ and velocity $c_g^o$

$$u_o(t') = \frac{1}{2\pi} \int_{-\infty}^{+\infty} u_o(\omega) \exp(i\omega t')\, d\omega = a(t') \exp(i\omega_o t') \tag{3.72}$$

Substituting Eq. 3.72 into Eq. 3.70 results in

$$u(x, t) = \exp\left[-i\omega_o\left(\frac{x}{c^o} - \frac{x}{c_g^o}\right)\right] a(t') \exp\left[i\omega_o\left(t - \frac{x}{c_g^o}\right)\right] \tag{3.73a}$$

or

$$u(x, t) = a\left(t - \frac{x}{c_g^o}\right) \exp\left[i\omega_o\left(t - \frac{x}{c^o}\right)\right] \tag{3.73b}$$

This equation for the modulated wave represents a harmonic wave with the frequency $\omega_o$ propagating with the phase velocity $c^o$ (the carrier) and an amplitude $a$ which moves undistorted with the group velocity $c_g^o$ (the modulation).

As an example, the wave propagation along the rod is addressed, caused by the following support settlement at $x = 0$ (Fig. 3-7a):

$$\begin{aligned} u_o(\bar{t}) &= u_o \sin\pi \frac{\bar{t}}{t_o} \sin\pi \frac{10\bar{t}}{t_o}, & 0 < \bar{t} < t_o \\ u_o(\bar{t}) &= 0 & \bar{t} > t_o \end{aligned} \tag{3.74}$$

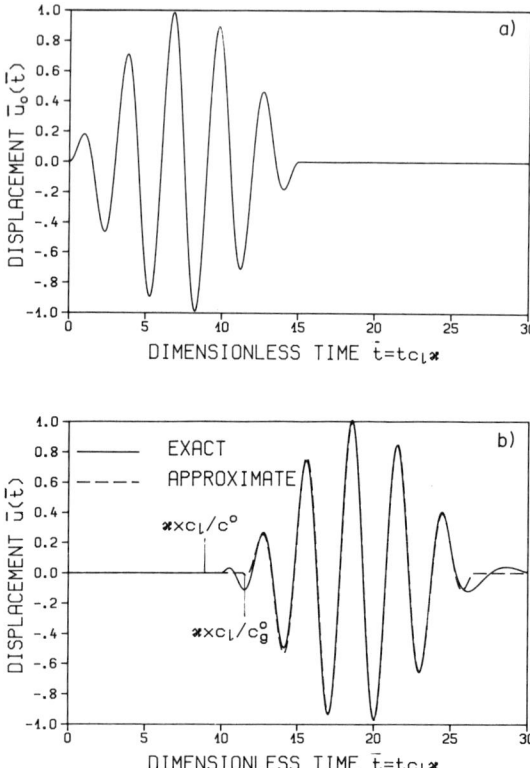

**Figure 3-7** Narrow-banded pulse. a. Prescribed displacement at $x = 0$. b. Displacement at $x = 10/\kappa$.

The dimensionless time $\bar{t}_o = t_o c_l \kappa$ is selected as equal to 15. The dominant frequency $\omega_o$ equals $10\pi/t_o$ or, in dimensionless form, $10\pi/15$. The phase velocity $c^o$ and the group velocity $c_g^o$ follow from Eqs. 3.61 and 3.63. The motion at $x = 10/\kappa$, nondimensionalized as $\bar{u}(\bar{t}) = u(x = 10/\kappa, \bar{t})/u_o$, based on the theory of wave propagation of a narrow-banded pulse, is calculated using the nondimensional forms of Eq. 3.70 or Eq. 3.73b. This approximate value agrees well with the exact one determined from Eq. 3.67 (Fig. 3-7b).

### 3.2.4 Dynamic-Stiffness Coefficient in Frequency Domain

To be able to calculate the exact solution of the benchmark problem used to establish the merits of the various transmitting boundaries, the dy-

## Sec. 3.2 Semi-Infinite Rod on Elastic Foundation

namic-stiffness coefficient in the frequency domain $S(a_o)$ is determined, relating the amplitude of the displacement at the free end $u_o(a_o)$ to the amplitude of the force $P_o(a_o)$ (Fig. 3-4a):

$$P_o(a_o) = S(a_o)u(a_o) \tag{3.75}$$

When determining the dynamic-stiffness coefficient at Point 0 in the semi-infinite system, only outwardly propagating waves exist. Setting $b = 0$ in Eq. 3.59 and enforcing $u(x = 0, a_o) = u_o(a_o)$ leads to

$$u(x, a_o) = u_o(a_o) \exp(-ikx) \tag{3.76}$$

Substituting Eq. 3.76 in

$$P_o(a_o) = -EAu(a_o, x = 0)_{,x} \tag{3.77}$$

results in

$$P_o(a_o) = i\sqrt{EAk_g}\sqrt{a_o^2 - 1}\, u_o(a_o) \tag{3.78}$$

With the static-stiffness coefficient $K$

$$K = \sqrt{EAk_g} \tag{3.79}$$

the dynamic-stiffness coefficient $S(a_o)$ is formulated as

$$S(a_o) = K\overline{S}(a_o) \tag{3.80}$$

where $\overline{S}(a_o)$, which is dimensionless, equals

$$\overline{S}(a_o) = i\sqrt{a_o^2 - 1} \tag{3.81}$$

It is customary to split the generally complex $\overline{S}(a_o)$ into its real and imaginary parts

$$\overline{S}(a_o) = k_o(a_o) + ia_o c_o(a_o) \tag{3.82}$$

$k_o(a_o)$ represents the (frequency-dependent) spring coefficient and $c_o(a_o)$ is proportional to the (frequency-dependent) damping coefficient:

$$a_o < 1: k_o = \sqrt{1 - a_o^2} \tag{3.83a}$$

$$c_o = 0 \tag{3.83b}$$

$$a_o > 1: k_o = 0 \tag{3.83c}$$

$$c_o = \sqrt{1 - 1/a_o^2} \tag{3.83d}$$

As already mentioned, no radiation damping occurs below the cutoff frequency.

The equation 3.78 describes a damper with a coefficient $A\rho c_l \sqrt{1 - 1/a_o^2}$, which is thus frequency dependent.

### 3.2.5 Exact Solution of Benchmark Problem

To examine the performance of the (approximate) local boundary conditions corresponding to the various transmitting boundaries, the undamped semi-infinite rod on an elastic foundation (Fig. 3-4a) is subjected to a prescribed support movement at Point 0.

$$u_o(\bar{t}) = \frac{u_o}{2}\left[1 - \cos\left(2\pi \frac{\bar{t}}{\bar{t}_o}\right)\right], \quad 0 < \bar{t} < \bar{t}_o$$
$$u_o(\bar{t}) = 0 \quad \bar{t} > \bar{t}_o$$
(3.84)

The corresponding dimensionless time $\bar{t}_o = t_o c_l/\kappa$ is selected as equal to 2. As demonstrated in Section 3.2.2, this enforced displacement leads to a significant content of the motion in the range below the cutoff frequency. The dimensionless support movement $\bar{u}_o(\bar{t}) = u_o(\bar{t})/u_o$—a rounded triangular pulse—is plotted in Fig. 3-8a.

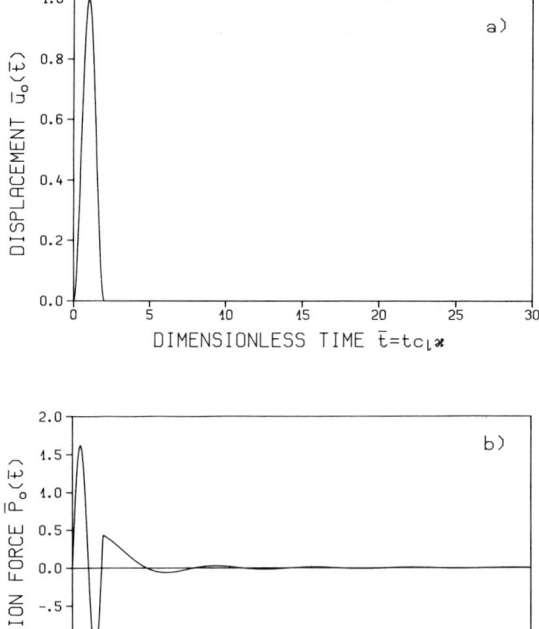

**Figure 3-8** a. Prescribed displacement. b. Reaction force (exact solution).

## Sec. 3.2  Semi-Infinite Rod on Elastic Foundation

The corresponding reaction force $P_o(t)$ at Point 0 for this linear benchmark problem is calculated working in the frequency domain. Using $\overline{S}(a_o)$ specified in Eq. 3.81, the dimensionless force $\overline{P}_o(\bar{t}) = P_o(\bar{t})/(Ku_o)$ follows as

$$\overline{P}_o(\bar{t}) = \frac{1}{2\pi} \int_{-\infty}^{+\infty} \overline{S}(a_o) \left( \int_0^{\bar{t}_o} \bar{u}_o(\bar{t}) \exp(-ia_o\bar{t}) d\bar{t} \right) \exp(ia_o\bar{t}) da_o \quad (3.85)$$

where $\bar{u}_o(\bar{t}) = u_o(\bar{t})/u_o$. The rigorous solution for $\overline{P}_o(\bar{t})$, which is plotted in Fig. 3-8b exhibits a significant value after $u_o(\bar{t})$ is zero again ($\bar{t} > 2$).

The various schemes' abilities to model outwardly propagating waves on the artificial boundary are evaluated by comparing the reaction force $\overline{P}_o(\bar{t})$. In addition, when physical insight can be gained, the ratio of the reflected wave's amplitude to the incident wave's amplitude at the artificial boundary—$b/a$—is also addressed for harmonic motion. $|b/a|$ is examined as a function of $a_o$.

### 3.2.6 Spatial and Temporal Discretizations of Region up to Artificial Boundary

The direct method of analyzing soil–structure interaction models a linear region of soil which is truncated by the artificial boundary. Between Point 0 and the artificial boundary, 10 one-dimensional finite elements of equal length $e$ are chosen (Fig. 3-9). The $x$-coordinate of the artificial boundary $l$ equals $10\ e$. $e$ is selected as $0.1/\kappa$. The static-stiffness matrix of each rod element—which is continuously elastically supported—is based on a linear shape function $[N(x)]$ for the displacement.

$$u(x) = \left(1 - \frac{x}{e}\right) u_1 + \frac{x}{e} u_2 = [N(x)]\{u\} \quad (3.86)$$

where $u_1$ and $u_2$ denote the displacements at the two interfaces of the rod element. The static-stiffness matrix $[K]$ is calculated as

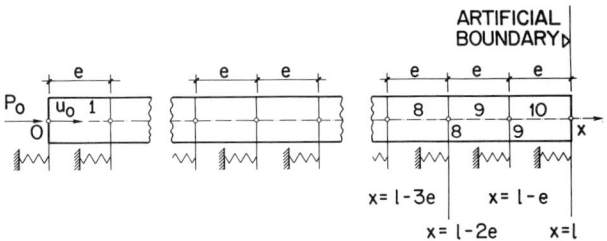

**Figure 3-9**  Finite-element discretization up to artificial boundary used with explicit time integration.

$$\begin{bmatrix} \dfrac{EA}{e} + \dfrac{k_g e}{3} & -\dfrac{EA}{e} + \dfrac{k_g e}{6} \\ -\dfrac{EA}{e} + \dfrac{k_g e}{6} & \dfrac{EA}{e} + \dfrac{k_g e}{3} \end{bmatrix} \tag{3.87}$$

A lumped mass matrix is used. For such a wave-propagation problem, an explicit time-integration scheme is appropriate. The Newmark algorithm, with $\beta = 0$ and $\gamma = 0.5$ (Eqs. 2.60 and 2.62) and with a $\Delta \bar{t} = 0.05$, if not stated otherwise, is used. For the largest group velocity occurring ($= c_l$, Eq. 3.63), the wave partially reflected at the artificial boundary will begin influencing the reaction force $P_o(\bar{t})$ at $\bar{t} = 2$.

In Sections 3.3 to 3.8, the various approximate procedures to model wave propagation towards infinity on the artificial boundary are presented and their accuracies are discussed. These methods lead to boundary conditions which are local in time—that is, they involve the motion at the current time or also at a few steps earlier. Generally, it is assumed that the load acts directly on the structure—that is, that no free-field response must be incorporated into the formulation. The latter is discussed in Section 3.10.

To describe in detail all aspects of these transmitting boundaries is outside this textbook's scope. So the reader can become familiar with specifics, more references than usual are listed in the rest of this chapter.

## 3.3 SUPERPOSITION BOUNDARY

### 3.3.1 Averaging of Final Solutions with Symmetric and Antimetric Boundary Conditions

As is well known, elementary boundary conditions cannot be used to simulate a transmitting boundary. They would consist of prescribed displacements or stresses corresponding to the free field or of zero values when the load is applied directly to the structure. These elementary boundaries are perfect reflectors.

However, correct results using only elementary boundaries can be obtained in certain cases, whereby the dynamic system must be solved at least twice. This superposition-boundary concept decomposes the total solution into a symmetric one and an antimetric one, differing only in the boundary conditions formulated on the line of symmetry, which coincides with the artificial boundary. For the one-dimensional rod, denoting the displacements caused by the original loads in the symmetric and antimetric solutions as $u_s(x, t)$ and $u_a(x, t)$, the two conditions on the artificial boundary at $x = l$ for the symmetric and the antimetric cases are equal to

Sec. 3.3   Superposition Boundary

$$u_s(x = l, t) = 0 \qquad (3.88a)$$

$$u_a(x = l, t)_{,x} = 0 \qquad (3.88b)$$

Equation 3.88a corresponds to a vanishing displacement; Eq. 3.88b to a zero normal force; that is, the two elementary boundary conditions are sequentially enforced. The total solution is equal to the average of the symmetric and antimetric ones $(u(x, t) = [u_s(x, t) + u_a(x, t)]/2)$. Using the rod on the elastic foundation for illustration, the amplitudes of the reflected waves of the solutions in the frequency domain for zero displacement and for zero normal force have the same magnitudes but opposite sign [S3], as is easily verified from Eq. 3.59 using Eq. 3.54:

$$\frac{b}{a} = -\exp(-2i\kappa l \sqrt{a_o^2 - 1}) \qquad (3.89a)$$

$$\frac{b}{a} = +\exp(-2i\kappa l \sqrt{a_o^2 - 1}) \qquad (3.89b)$$

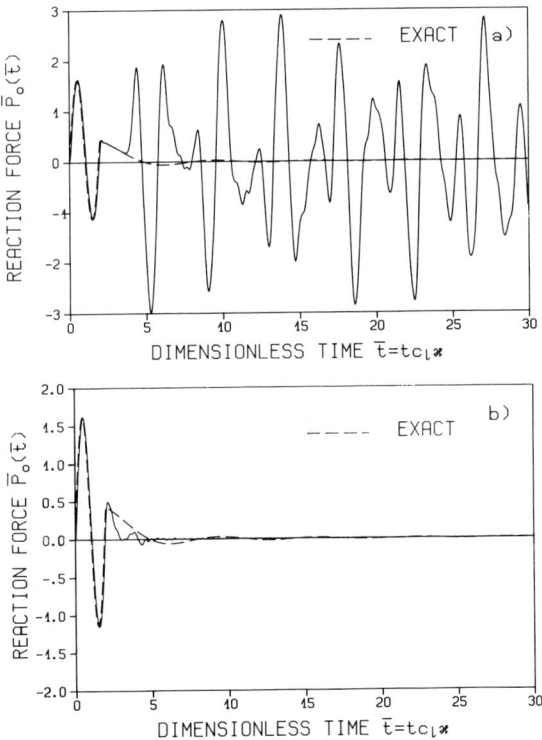

**Figure 3-10** Superposition boundary. a. Averaging of final solutions. b. Averaging of solutions as they occur.

In the frequency domain, averaging the results of the two boundary conditions will lead to the correct solution with only outwardly propagating waves occurring. In the time domain, what is actually being calculated is a system twice the original size, with the addition of a mirror image about the line of symmetry. This is confirmed in Fig. 3-10a, where the reaction force—determined as the average of the final solutions for the two boundary conditions in the explicit time-integration scheme with the mesh of Fig. 3-9—is plotted. The exact solution results (from Fig. 3-8b) for $\bar{t} < 4$, which is the time necessary for the wave to travel from Point 0 to a (fixed) boundary in that system which is twice as large, and back again to Point 0. For $\bar{t} > 4$, the effect of the multiple reflections is clearly visible, leading to an unacceptable solution; reflections involving multiple encounters with the same boundary are thus not eliminated by the superposition of the final solutions and cause the procedure as described so far to fail.

For a single-plane boundary in full space, the procedure also completely eliminates all reflections for two- and three-dimensional cases. For the three-dimensional symmetric and antimetric cases, the conditions to be enforced on the artificial boundary are specified in Eqs. 3.90a and 3.90b, respectively.

$$u_s = 0, \quad v_{s,n} = 0, \quad w_{s,n} = 0 \tag{3.90a}$$

$$u_{a,n} = 0, \quad v_a = 0, \quad w_a = 0 \tag{3.90b}$$

$u$ denotes the component of the displacement in the direction of the normal $n$; $v$ and $w$ are the tangential components on the artificial boundary. From Eq. 3.90a, it follows that, besides $u_s$, the two shear stresses acting on the artificial boundary also vanish; while Eq. 3.90b is equivalent to stating that the normal stress is zero in addition to the two tangential displacements $v_a$, $w_a$, which are elementary boundary conditions. As the procedure is based only on simple considerations of symmetry and antimetry, it applies to all types of waves, such as inclined body waves and (dispersive) surface waves. (See Problem 3.2 for a verification for the in-plane motion.) When more than one boundary needs to be transmitting, more than two solutions must be added to eliminate all reflections. At a two-dimensional corner, four solutions are superimposed, as discussed in Problem 3.3. Of course, the added mirror image (with respect to the artificial boundary) must be compatible with the unbounded domain on the exterior; this is, for instance, not the case when, for a one-dimensional problem, the effective area changes with the coordinate, as for a spherical wave (Section 3.9.2).

### 3.3.2 Cancellation of Reflected Waves as They Occur

**Concept.** The procedure described in Section 3.3.1, which averages the final solutions with symmetric and antimetric boundary conditions, breaks

## Sec. 3.3  Superposition Boundary

down when multiple reflections occur on the same boundary. It can also make more than two complete solutions necessary when reflections need to be eliminated on more than one boundary, as mentioned in the case of a corner. To keep the procedure from failing or becoming computationally unattractive, the following modifications are introduced [C6, K7]. Two overlapping narrow boundary zones—each with half the total stiffness and mass, and with complementary boundary conditions corresponding to the symmetric and antimetric cases—are independently attached to the main mesh. The two-dimensional case is illustrated in Fig. 3-11. These boundary conditions are the same as presented in Eq. 3.90: on the symmetric boundary, the normal displacement $u$ and the shear stresses $\tau$ are prescribed; and on the antimetric boundary, the normal stress $\sigma$ and the tangential displacements $v$ are prescribed. A wave that propagates from the main mesh will enter the two boundary zones simultaneously. After reflection off the artificial boundaries, the amplitudes of the waves in the two zones will have the same magnitudes, but different signs. The superposition of the two solutions in the boundary zones is performed before the reflected waves can reach the main mesh; thus the frequent averaging will cancel the reflections at the source as they occur. The enforced boundary conditions will be zero only up to the first superposition; afterwards they will be equal to the averaged values resulting from the prior superposition. To avoid numerical shocks, it is appropriate to specify prescribed velocities instead of displacements; this can also be substantiated by the theoretical derivation that follows. A bar denotes prescribed values in Fig. 3-11. Typically, the boundary zones are three to four elements wide, and the superposition is performed after every three time steps. This computational procedure is more efficient than when the total solutions are su-

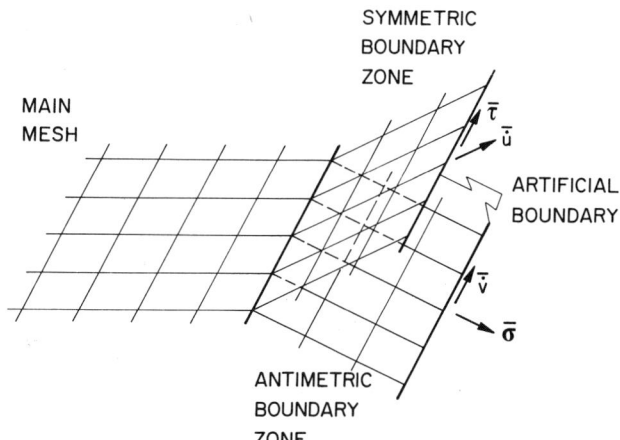

**Figure 3-11** Two independent overlapping boundary zones connected to main mesh.

perimposed, because the calculations are performed only once in the (large) main mesh (and twice in the narrow boundary zones). It is obvious from the derivation that the two boundary zones must behave linearly.

**Theoretical formulation.** As the superposition boundary is based on considerations of symmetry and antimetry, it is sufficient to examine the prismatic rod [D2, K2]. The essential features are captured in this one-dimensional system, and the generalization to two- and three-dimensional cases is straightforward. The one-dimensional wave propagation is addressed in Section 3.1.1. The semi-infinite prismatic rod shown in Fig. 3-12 is subjected to a wave $f(t - x/c)$ which propagates with the velocity $c$ in the positive $x$-direction. For a concise notation, the subscript $l$ used in Section 3.1.1 is dropped, and the origin $x = 0$ is placed on the artificial boundary. The wave splits into two identical waves in the symmetric and antimetric boundary zones. No reflections occur at Point A, as the area of the rod in the boundary zones is half that in the main mesh. At the start of the analysis, the symmetric zone is fixed at the artificial boundary and the antimetric zone has a free boundary. After the waves reach the artificial boundaries, it is desirable to impose a velocity $\bar{u}_s(t)$ equal to the velocity $\dot{u}_a(t)$ of the free boundary at the fixed boundary and also to prescribe a force $\bar{P}_a(t)$ equal to the reaction force $P_s(t)$ of the fixed boundary at the free boundary.

$$\bar{u}_s(t) = \dot{u}_a(t) \tag{3.91a}$$

$$\bar{P}_a(t) = P_s(t) \tag{3.91b}$$

The solutions in the two boundary zones are then averaged through superposition, which eliminates the reflections as they occur. For times before the arrival of the reflected waves at Point A, the displacements in the symmetric $u_s(x, t)$ and antimetric $u_a(x, t)$ boundary zones are formulated as

$$u_s(x, t) = f(t - x/c) - f(t + x/c) + g_s(t + x/c) \tag{3.92a}$$

$$u_a(x, t) = f(t - x/c) + f(t + x/c) - g_a(t + x/c) \tag{3.92b}$$

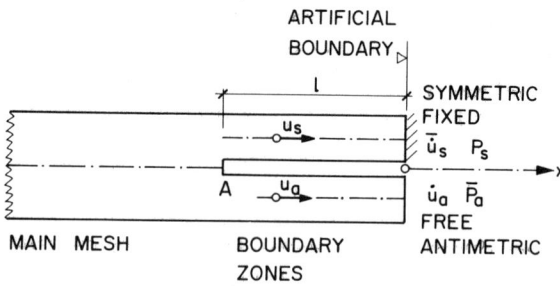

**Figure 3-12** Nomenclature of superposition-boundary theory.

## Sec. 3.3 Superposition Boundary

The first two terms on the right-hand sides of Eqs. 3.92a and 3.92b correspond to the wave patterns generated by the reflections at fixed and free boundaries, respectively (homogeneous boundary conditions, Eqs. 3.19 and 3.20). The functions $g_s(t + x/c)$ and $-g_a(t + x/c)$, representing waves starting at the artificial boundaries and propagating in the direction of the reflected waves, are generated by the enforcement of non-zero values $\bar{u}_s$ and $\bar{P}_a$, respectively. The minus sign of $g_a$ in Eq. 3.92b simplifies the discussion of the derivation's outcome. The velocities and forces at the fixed and free boundaries follow from Eq. 3.92, using Eq. 3.2 with $x = 0$, as

$$\bar{u}_s(t) = g'_s(t) \tag{3.93a}$$

$$P_s(t) = \frac{EA}{c}[-2f'(t) + g'_s(t)] \tag{3.93b}$$

$$\dot{u}_a(t) = 2f'(t) - g'_a(t) \tag{3.94a}$$

$$\bar{P}_a(t) = -\frac{EA}{c} g'_a(t) \tag{3.94b}$$

where the symbol ' denotes a derivative with respect to the argument—for example, $f' = df(t + x/c)/d(t + x/c)$. Enforcing velocities instead of displacements results in a formulation with derivatives instead of the functions themselves, which is compatible with prescribing forces.

To gain some physical insight, it is appropriate to study the behavior of the variables in Eqs. 3.93 and 3.94 when the prescribed values are kept fixed during a few time steps—that is, from $t$ to $t + T$. At the fixed boundary,

$$\bar{u}_s(t + T) - \bar{u}_s(t) = g'_s(t + T) - g'_s(t) = 0 \tag{3.95a}$$

applies, which leads to

$$P_s(t + T) - P_s(t) = -\frac{2EA}{c}[f'(t + T) - f'(t)] \tag{3.95b}$$

Analogously, at the free boundary,

$$\bar{P}_a(t + T) - \bar{P}_a(t) = -\frac{EA}{c}[g'_a(t + T) - g'_a(t)] = 0 \tag{3.96a}$$

holds, which results in

$$\dot{u}_a(t + T) - \dot{u}_a(t) = 2[f'(t + T) - f'(t)] \tag{3.96b}$$

Averaging the change of the velocities and of the forces at the two boundaries leads to

$$\frac{1}{2}[\bar{u}_s(t + T) - \bar{u}_s(t) + \dot{u}_a(t + T) - \dot{u}_a(t)] = f'(t + T) - f'(t) \tag{3.97a}$$

$$\frac{1}{2}[P_s(t+T) - P_s(t) + \overline{P}_a(t+T) - \overline{P}_a(t)] = -\frac{EA}{c}[f'(t+T) - f'(t)]$$

(3.97b)

This is the exact result, which depends on the incident wave only.

Returning to the computational procedure, it has already been stated that ideal conditions would exist for the superposition boundary if Eq. 3.91 could be rigorously enforced. However, this is not practical, as $\dot{u}_a(t)$ and $P_s(t)$ are not known a priori. The following forward-difference scheme with the time step $\Delta t$ can be used as an approximation:

$$\overline{\dot{u}}_s(t) = \frac{1}{2}[\overline{\dot{u}}_s(t - \Delta t) + \dot{u}_a(t - \Delta t)] \qquad (3.98a)$$

$$\overline{P}_a(t) = \frac{1}{2}[\overline{P}_a(t - \Delta t) + P_s(t - \Delta t)] \qquad (3.98b)$$

The averaged values at the two boundaries at the prior time station are thus applied. Substituting Eqs. 3.93 and 3.94 into Eq. 3.98 results in

$$g'_s(t) = \frac{1}{2}[g'_s(t - \Delta t) + 2f'(t - \Delta t) - g'_a(t - \Delta t)] \qquad (3.99a)$$

$$g'_a(t) = \frac{1}{2}[g'_a(t - \Delta t) + 2f'(t - \Delta t) - g'_s(t - \Delta t)] \qquad (3.99b)$$

$g'_s(t)$ can be further transformed as follows. Formulating Eqs. 3.99 at the time station $t - \Delta t$ and substituting $g'_s(t - \Delta t)$ and $g'_a(t - \Delta t)$ on the right-hand side of Eq. 3.99a leads to

$$g'_s(t) = \frac{1}{2}[g'_s(t - 2\Delta t) + 2f'(t - \Delta t) - g'_a(t - 2\Delta t)] \qquad (3.100)$$

Repeating these substitutions $(n - 1)$ times results in

$$g'_s(t) = \frac{1}{2}[g'_s(t - n\Delta t) + 2f'(t - \Delta t) - g'_a(t - n\Delta t)] \qquad (3.101)$$

When the $n$ selected is so large that $t - n\Delta t$ corresponds to a time prior to the arrival of the incident waves at the artificial boundaries, $g'_s(t - n\Delta t)$ and $g'_a(t - n\Delta t)$ will vanish. Thus

$$g'_s(t) = f'(t - \Delta t) \qquad (3.102a)$$

holds and, with zero initial conditions,

$$g_s(t) = f(t - \Delta t) \qquad (3.102b)$$

Analogously,

## Sec. 3.3 Superposition Boundary

$$g_a(t) = f(t - \Delta t) \tag{3.103}$$

applies.

Equation 3.92 is then formulated as

$$u_s(x, t) = f(t - x/c) - f(t + x/c) + f(t - \Delta t + x/c) \tag{3.104a}$$

$$u_a(x, t) = f(t - x/c) + f(t + x/c) - f(t - \Delta t + x/c) \tag{3.104b}$$

After introducing an error function $e(t)$, which can be made arbitrarily small,

$$e(t) = f(t - \Delta t) - f(t) \tag{3.105}$$

Eq. 3.104 is rewritten as

$$u_s(x, t) = f(t - x/c) + e(t + x/c) \tag{3.106a}$$

$$u_a(x, t) = f(t - x/c) - e(t + x/c) \tag{3.106b}$$

Averaging the displacements in the two boundary zones

$$u(x, t) = \frac{1}{2}[u_s(x, t) + u_a(x, t)] = f(t - x/c) \tag{3.107}$$

eliminates the error function, resulting in the incident wave only.

For a discussion of the wave pattern that results when the superposition is not performed as the reflected waves occur, see Problem 3.4.

**Numerical implementation.** For such wave propagation problems, the explicit-time–integration scheme is appropriate, which is applied without modification to all nodes not lying on the artificial boundaries (Section 2.12). For the nodes on the artificial boundaries of the two boundary zones, one needs to distinguish between those degrees of freedom with prescribed velocities and those with prescribed forces, which remain constant during a few consecutive time steps. In a two- or three-dimensional case, both types of degrees of freedom will occur on the boundaries of the symmetric and antimetric zones (Fig. 3-11). For instance, if, for a specific degree of freedom, a velocity is prescribed in the symmetric boundary zone, and then in the antimetric boundary zone a force is enforced (and vice versa). The corresponding computational procedure for the $n$th time step (from $(n - 1)\Delta t$ to $n\Delta t$), based on the explicit algorithm (Newmark family with the two parameters $\beta$, $\gamma$), is summarized in Table 3-1. For the error functions in the symmetric and antimetric boundary zones to have the same amplitudes, all operations must be performed consistently. The superposition of the two solutions in the symmetric and antimetric boundary zones, which is performed every few time steps and leads to new prescribed velocities and forces, is designated with the subscript $j$. $F$ denotes the internal force that can be calculated from the displacements of the boundary node and adjacent ones.

**TABLE 3-1 Computational Procedure for Degrees of Freedom on Artificial Boundaries**

| Time Station $(n-1)\Delta t$ $u_{n-1}, \dot{u}_{n-1}, \ddot{u}_{n-1}$ | |
|---|---|
| Velocity Prescribed | Force Prescribed |
| $\dot{u}_n = \bar{\dot{u}}_j$ $u_n = u_{n-1} + \Delta t \dot{u}_{n-1} + \left(\dfrac{1}{2} - \beta\right)\Delta t^2 \ddot{u}_{n-1}$ $(F_n = P_n)$ $\ddot{u}_n = -\dfrac{1-\gamma}{\gamma}\ddot{u}_{n-1}$ $u_n = \tilde{u}_n + \beta \Delta t^2 \ddot{u}_n$ | $P_n = \bar{P}_j$ $\tilde{u}_n = u_{n-1} + \Delta t \dot{u}_{n-1} + \left(\dfrac{1}{2} - \beta\right)\Delta t^2 \ddot{u}_{n-1}$ $\tilde{\dot{u}}_n = \dot{u}_{n-1} + (1-\gamma)\Delta t \ddot{u}_{n-1}$ $F_n$ $\ddot{u}_n = \dfrac{-F_n + \bar{P}_j}{m}$ $u_n = \tilde{u}_n + \beta \Delta t^2 \ddot{u}_n$ $\dot{u}_n = \tilde{\dot{u}}_n + \gamma \Delta t \ddot{u}_n$ |
| Superposition/Averaging | |
| $\bar{\dot{u}}_{j+1} = \dfrac{1}{2}(\bar{\dot{u}}_j + \dot{u}_n)$ $\bar{P}_{j+1} = \dfrac{1}{2}(P_n + \bar{P}_j)$ | |

For the prescribed velocity, $F$ is calculated only in the time step prior to the superposition, which is indicated by a parenthesis. $m$ is the lumped mass. During the superposition, all displacements, velocities, and accelerations in all nodes of the boundary zones are averaged and used as the starting values for the next time step.

**Benchmark problem.** The benchmark problem of the semi-infinite rod on elastic foundation is addressed (Fig. 3-9). The overlapping boundary zones consist of three elements adjacent to the artificial boundary towards the interior; the main mesh is thus limited to seven elements. The results are averaged every three time steps. The corresponding reaction force shown in Fig. 3-10b demonstrates the dramatic improvement of the result's accuracy compared to when only the total solutions are averaged (Fig. 3-10a).

## 3.4 FICTITIOUS MATERIAL DAMPING

Artificially introduced material damping cannot be used to duplicate radiation damping. The material damping will indeed lead to a decay of the waves' amplitudes. However, the practical implementation is not clear. The amount of material damping and its spatial distribution cannot be specified.

For the sake of illustration, the benchmark problem is analyzed (Section

Sec. 3.5   Viscous Damper

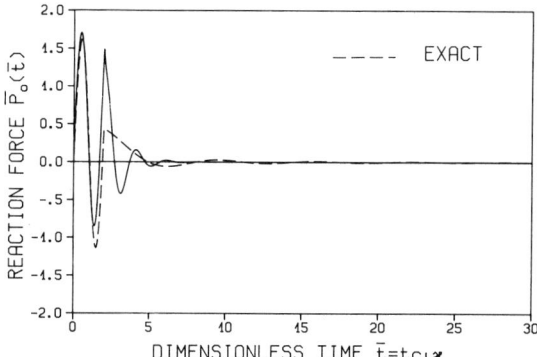

Figure 3-13   Fictitious material damping.

3.2.6). For the finite rods, a damping matrix proportional to the static stiffness matrix with a factor $2\bar{\zeta}/(c_l\kappa)$ is chosen. $\Delta \bar{t}$ is reduced to 0.01 for this case. Selecting the dimensionless damping factor $\bar{\zeta}$ equal to 0.1 and for a zero displacement on the artificial boundary, the reaction force presented in Fig. 3-13 arises. The result is unsatisfactory.

## 3.5 VISCOUS DAMPER

For the one-dimensional wave propagation in a prismatic rod (Section 3.1.1), the exact transmitting boundary consists of a damper with a coefficient equal to the product of the area A, the mass density ρ, and either the rod velocity $c_l$ for longitudinal waves or the shear-wave velocity $c_s$ for shear waves (Eqs. 3.12 and 3.16). Assuming that the waves—for example, in the two-dimensional case (Fig. 3-14)—impinge at a right angle on the artificial boundary, the exact transmitting boundary condition is formulated as

$$\sigma(s) + \rho c_p \dot{u}(s) = 0 \qquad (3.108)$$

$$\tau(s) + \rho c_s \dot{v}(s) = 0 \qquad (3.109)$$

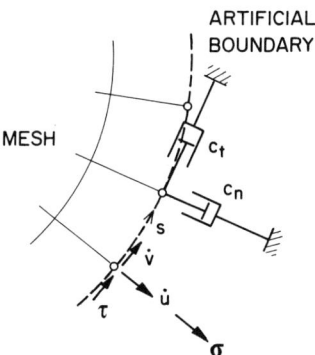

Figure 3-14   Lumped viscous damper.

$\sigma(s)$ and $\tau(s)$ are the normal and shear stresses on the boundary, and $u(s)$ and $v(s)$ are the normal and tangential displacements. $c_p$ represents the dilatational-wave velocity, with $s$ denoting the coordinate on the artificial boundary.

In the three-dimensional case, a second equation like Eq. 3-109 is formulated, applicable to the other tangential direction. In discretized form it is customary to lump the distributed dampers described by Eqs. 3.108 and 3.109, which results in each node in dampers with coefficients $c_n$ and $c_t$ in the normal and the tangential directions

$$c_n = A\rho c_p \tag{3.110}$$

$$c_t = A\rho c_s \tag{3.111}$$

For the three-dimensional case, $A$ represents the applicable area which is replaced in the two-dimensional case by the applicable length. This procedure leads to frequency-independent boundary conditions that are local in space and time. If the shape functions of the neighboring finite elements are used instead of the crude lumping procedure, a narrow-banded damping matrix arises, which is also easy to implement.

Turning to the benchmark problem, a damper with a coefficient $A\rho c_l$ is attached to the artificial boundary. This corresponds, for $x = l$, to

$$N = -A\rho c_l \dot{u} \tag{3.112}$$

or to the differential equation

$$u_{,x} + \frac{\dot{u}}{c_l} = 0 \tag{3.113}$$

This is the exact boundary condition for a "free" prismatic rod but is only an approximation for the elastically supported rod. The rigorous transmitting boundary would consist of a damper with a frequency-dependent coefficient (Section 3.2.4). There is good agreement between the reaction force plotted in Fig. 3-15 and the exact solution.

As just discussed, these local viscous dampers represent the exact solution for $P$- and $S$-waves which impinge at a right angle on the artificial boundary. They are approximate for inclined body waves, whereby the reflected energy is only a small part of the total energy [L6]. In many cases, the farther one chooses the artificial boundary to be from a source which radiates waves, the more the angle of incidence with respect to the artificial boundary will approach 90°, and, thus, the better the viscous dampers will perform (see also Section 3.9.2). The same applies as the frequency of excitation becomes higher. This is illustrated as follows.

For instance, the pressure–displacement relationship for the spherical cavity is described in the limit $\omega \to \infty$ by a damper with the coefficient $\rho c_p$ (Eq. 2.27). For the elastically supported rod, the phase velocity converges

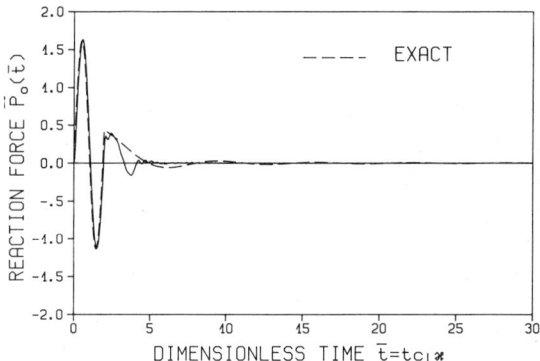

**Figure 3-15** Viscous damper.

for $\omega \to \infty$ to $c_l$ (Eq. 3.61), and the corresponding damping coefficient converges to $A\rho c_l$ (Section 3.2.4). This asymptotic behavior also applies in general to surface waves occurring in a layered half-space [W22]. For $\omega \to \infty$, the phase velocities of all modes of the generalized Love waves converge to the shear-wave velocity of the top layer, and those of the Rayleigh waves converge to the same value (with the exception of the first mode). As proven in Section 6.3.2, the waves propagate perpendicularly to the vibrating source for $\omega \to \infty$, which can be represented by dampers with frequency-independent coefficients. However, for surface waves in the intermediate range of frequencies, the viscous dampers can lead to significant errors for harmonic excitation (Section 3.9.3).

Although a certain caution is appropriate for surface waves, the viscous dampers represent a suitable transmitting boundary for many applications involving both dilatational and shear waves. The accuracy is generally acceptable. The procedure is simple and easily implemented. General-purpose programs developed to calculate structural dynamics can be straightforwardly applied to analyze soil–structure interaction without any additional programming. It is worth mentioning that the viscous dampers lead to permanent displacements at nearly all points of the mesh in an elastic system. The same also applies to the superposition boundary.

## 3.6 DOUBLY ASYMPTOTIC APPROXIMATION

A straightforward procedure to obtain frequency-independent springs and dampers consists of evaluating the soil's dynamic-stiffness coefficients at a specific frequency. The latter can, for example, be equal to the system's fundamental frequency, whereby a different frequency can be chosen for each degree of freedom. The coefficients can also be averaged over the frequency range of interest. In addition, different frequencies can be chosen for calculating the real parts of the dynamic-stiffness coefficients, leading to the springs, and for the imaginary parts, resulting in the dampers. In particular, the former can be determined for the static case ($\omega = 0$) and the latter for frequencies approaching infinity. As for $\omega \to \infty$, the dynamic-stiffness coefficients are dominated by the damping coefficients (because of the factor $\omega$); this procedure, called the doubly asymptotic approximation [U1], is asymptotically exact at both high and low frequencies. The static-stiffness coefficients, conveniently determined for the general case using the boundary-element method, are fully coupled; that is, besides springs that connect each node's degrees of freedom to a rigid support, springs also exist that couple all degrees of freedom of every node on the boundary. As briefly discussed in Section 3.5 and as proven in Section 6.3.2, waves propagate in the perpendicular direction away from a vibrating source for $\omega \to \infty$, which can be rigorously described by distributed dampers, as in Eqs. 3.108 and 3.109. The damping coefficients will thus be the same as those introduced in the transmitting boundary consisting of viscous dampers. In particular, the damping matrix is narrow banded or, if a lumping procedure is applied, local dampers with the coefficients specified in Eqs. 3.110 and 3.111 arise (Fig. 3-14). After the analysis, no permanent displacements exist anymore in an elastic system, as in the case of viscous dampers without springs.

Addressing the benchmark problem, the asymptotic values of the dynamic-stiffness coefficient are calculated from Eq. 3.83. It follows that for $a_o \to 0$, $k_o = 1$ and for $a_o \to \infty$, $c_o = 1$. In the time domain this corresponds to

$$N = -Ku - A\rho c_l \dot{u} \quad (3.114)$$

or the boundary condition at $x = l$ is written as

$$u_{,x} + \kappa u + \frac{\dot{u}}{c_l} = 0 \quad (3.115)$$

From Eq. 3.114 it follows that the doubly asymptotic approximation corre-

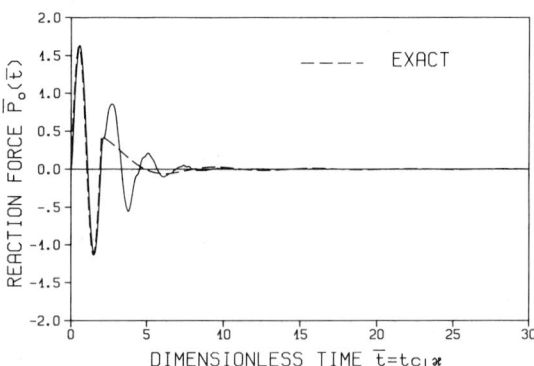

**Figure 3-16** Doubly asymptotic approximation.

Sec. 3.7  Paraxial Boundary                                        111

sponds to a damper–spring system with frequency-independent coefficients placed on the artificial boundary. The same configuration rigorously models the missing domain up to infinity for the conical rod in shear (Section 3.1.2). This approximation's performance is somewhat disappointing (Fig. 3-16) compared to that of the viscous damper (Fig. 3-15), but is still satisfactory.

## 3.7 PARAXIAL BOUNDARY

### 3.7.1 Concept

The solution of the wave's differential equation leads to an outgoing wave pattern and an incoming one, which can be separated from one another. In the paraxial approximation [C3, E1, C4], one constructs a differential equation similar to that of the wave equation, which allows only outgoing waves to propagate. This differential equation is then used as the boundary condition enforced on the artificial boundary. The paraxial boundary, which is only approximate, can thus be used to model waves moving in one general direction and to discriminate against waves moving in the opposite direction.

The concept of the paraxial element can be easily discussed based on the two-dimensional scalar wave equation describing the out-of-plane motion (Fig. 3-17; see also Section 4.2.2). The displacement $v$ is a function of the coordinates $x$ and $z$. Only shear stresses $\tau_{yx}$ and $\tau_{yz}$ occur, acting in the out-of-plane direction $y$. Formulating the equilibrium equation

$$\tau_{yx,x} + \tau_{yz,z} - \rho\ddot{v} = 0 \tag{3.116}$$

and substituting the stress–displacement relationship

$$\tau_{yx} = G v_{,x} \tag{3.117a}$$

$$\tau_{yz} = G v_{,z} \tag{3.117b}$$

results in the wave equation

$$v_{,xx} + v_{,zz} - \frac{\ddot{v}}{c_s^2} = 0 \tag{3.118}$$

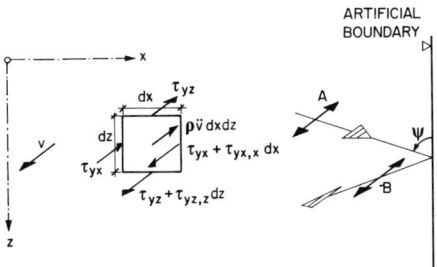

**Figure 3-17**  Nomenclature of out-of-plane motion.

with the shear-wave velocity $c_s = \sqrt{G/\rho}$ ($G$ = shear modulus, $\rho$ = mass density).

For harmonic excitation with frequency $\omega$, the solution can be formulated as

$$v = \exp[i(\omega t - k_x x - k_z z)] \tag{3.119}$$

with $k_x$ and $k_z$ being the wave numbers in the $x$- and $z$-directions. Substituting Eq. 3.119 in Eq. 3.118 leads to

$$k_x^2 + k_z^2 - \frac{\omega^2}{c_s^2} = 0 \tag{3.120}$$

To construct a paraxial boundary condition at $x = l$, Eq. 3.120 is factored as follows

$$\left(k_x - \frac{\omega}{c_s}\sqrt{1 - \left(\frac{c_s k_z}{\omega}\right)^2}\right)\left(k_x + \frac{\omega}{c_s}\sqrt{1 - \left(\frac{c_s k_z}{\omega}\right)^2}\right) = 0 \tag{3.121}$$

For $c_s k_z/\omega < 1$, the first factor leading to

$$k_x = +\frac{\omega}{c_s}\sqrt{1 - \left(\frac{c_s k_z}{\omega}\right)^2} \tag{3.122}$$

corresponds to a wave propagating in the positive $x$-direction, as is seen from Eq. 3.119 (outgoing wave). The second factor is associated with an incoming wave.

To be able to construct a differential equation that models only outgoing waves, the square root in Eq. 3.122 is expanded into the first two terms of a Taylor series producing

$$k_x - \frac{\omega}{c_s} + \frac{c_s}{2\omega} k_z^2 = 0 \tag{3.123a}$$

or

$$\frac{k_x \omega}{c_s} + \frac{k_z^2}{2} - \frac{\omega^2}{c_s^2} = 0 \tag{3.123b}$$

For a solution of the same form as specified in Eq. 3.119, the differential equation leading to Eq. 3.123b is found by inspection as

$$\frac{1}{c_s} \dot{v}_{,x} - \frac{1}{2} v_{,zz} + \frac{\ddot{v}}{c_s^2} = 0 \tag{3.124}$$

Equation 3.124 thus permits (in an approximate manner) waves to occur which propagate only in the positive $x$-direction.

If just the first term of the Taylor series is kept,

## Sec. 3.7  Paraxial Boundary

$$k_x - \frac{\omega}{c_s} = 0 \qquad (3.125)$$

follows, which corresponds to

$$v_{,x} + \frac{\dot{v}}{c_s} = 0 \qquad (3.126)$$

Multiplying by $G$ and using Eq. 3.117a, Eq. 3.126 is transformed to the viscous damper described by Eq. 3.109:

$$\tau_{yx} + \rho c_s \dot{v} = 0 \qquad (3.127)$$

The paraxial boundary can thus be interpreted as a generalization of the viscous damper.

To examine the accuracy of the paraxial approximation, the reflected wave, which is caused by an incident wave impinging upon the artificial boundary, is addressed, varying the angle of incidence $\psi$ (Fig. 3-17). $A$ and $B$ are the amplitudes of the incident and reflected SH-waves. Recalling that the wave number is equal to the ratio of the frequency and the apparent velocity (wave velocity divided by the cosine of the angle of incidence with respect to the corresponding axis),

$$k_x = \pm \frac{\omega \sin\psi}{c_s} \qquad (3.128a)$$

$$k_z = \frac{\omega \cos\psi}{c_s} \qquad (3.128b)$$

the exact solution (Eq. 3.119) is specified as

$$v = [A \exp(-i\omega \sin\psi \, x/c_s)$$
$$+ B \exp(i\omega \sin\psi \, x/c_s)] \exp(-i\omega \cos\psi \, z/c_s) \exp(i\omega t) \qquad (3.129)$$

Substituting Eq. 3.129 into the paraxial approximation (Eq. 3.124) leads, for the artificial boundary at $x = 0$, to

$$\frac{B}{A} = -\left(\frac{1 - \sin\psi}{1 + \sin\psi}\right)^2 \qquad (3.130)$$

The more $\psi$ deviates from 90°, the larger $|B/A|$ becomes and thus the less effective the paraxial boundary is in transmitting outgoing waves. A wave at grazing angle is totally reflected. Analogously, substituting into the boundary condition describing the viscous damper results in

$$\frac{B}{A} = -\frac{1 - \sin\psi}{1 + \sin\psi} \qquad (3.131)$$

As expected, the paraxial boundary (with two terms in the Taylor expansion) performs better than the viscous boundary (which is equal to the paraxial boundary with only one term in the Taylor expansion) for waves which do not impinge at a right angle upon the artificial boundary. This is easily verified by noting that the ratio in Eq. 3.131, which is smaller than 1, is squared in Eq. 3.130. $|B/A|$ are plotted versus $\psi$ in Fig. P 3-6.

### 3.7.2 Benchmark Problem

The benchmark problem consisting of the semi-infinite rod on an elastic foundation (Section 3.2.5) is examined next.

To solve the equation of motion (Eq. 3.55), $u$ of the form $\exp[i(\omega t + \gamma x)]$ is considered. The corresponding quadratic equation can be formulated as

$$(\gamma + k)(\gamma - k) = 0 \tag{3.132}$$

where the roots $\gamma_1$ and $\gamma_2$ equal $-k$ and $+k$, respectively, and with $k$ specified in Eq. 3.60. The first factor $\gamma - \gamma_1 = \gamma + k$ represents a wave propagating in the positive $x$-direction; the second factor, one in the negative direction (see Eq. 3.59). The outwardly propagating wave is thus subjected to (substituting Eq. 3.60)

$$\gamma + \kappa \sqrt{a_o^2 - 1} = 0 \tag{3.133}$$

For the case $a_o < 1$, Eq. 3.133 is reformulated as

$$i\gamma + \kappa \sqrt{1 - a_o^2} = 0 \tag{3.134}$$

The square root can be expanded into a Taylor series. Keeping only the first term (first approximation) leads to

$$i\gamma + \kappa = 0 \tag{3.135}$$

The corresponding equation, which leads to Eq. 3.135 for $u$ of the form $\exp[i(\omega t + \gamma x)]$ is found by inspection as

$$u_{,x} + \kappa u = 0 \tag{3.136}$$

Keeping the first two terms (second approximation) results in

$$i\gamma + \kappa\left(1 - \frac{1}{2}a_o^2\right) = 0 \tag{3.137}$$

with the corresponding differential equation being

$$u_{,x} + \kappa u + \frac{1}{2c_1^2\kappa}\ddot{u} = 0 \tag{3.138}$$

Multiplying this equation by $EA$ and using Eq. 3.54 leads to

## Sec. 3.7 Paraxial Boundary

$$N = -Ku - \frac{\rho A}{2\kappa} \ddot{u} \qquad (3.139)$$

This boundary condition thus corresponds to a spring with a coefficient $K$ and a mass connected to the node on the artificial boundary with the value $\rho A/(2\kappa)$.

Denoting with $a$ the amplitude of the incident wave at the artificial boundary, the following ratios are calculated from Eqs. 3.136 and 3.138:

$$\frac{b}{a} = \frac{\sqrt{1 - a_o^2} - 1}{\sqrt{1 - a_o^2} + 1} \qquad (3.140)$$

$$\frac{b}{a} = \frac{\sqrt{1 - a_o^2} - 1 + a_o^2/2}{\sqrt{1 - a_o^2} + 1 - a_o^2/2} \qquad (3.141)$$

For the case $a_o > 1$, Eq. 3.133 is reformulated as

$$i\gamma + i\kappa a_o \sqrt{1 - \frac{1}{a_o^2}} = 0 \qquad (3.142)$$

Keeping the first term in the Taylor expansion of the square root, the corresponding differential equation equals

$$u_{,x} + \frac{1}{c_l} \dot{u} = 0 \qquad (3.143)$$

This boundary condition is identical to the one arising for the viscous boundary (Eq. 3.113). For other systems—for example, for the rod with an exponentially increasing area, examined in Problem 3.5—this is not the case. Keeping the first two terms leads to

$$\gamma a_o + \kappa a_o^2 - \frac{\kappa}{2} = 0 \qquad (3.144)$$

which corresponds to

$$\dot{u}_{,x} + \frac{\ddot{u}}{c_l} + \frac{\kappa^2 c_l}{2} u = 0 \qquad (3.145)$$

The ratio is determined from Eq. 3.143 as

$$\frac{b}{a} = \frac{\sqrt{a_o^2 - 1} - a_o}{\sqrt{a_o^2 - 1} + a_o} \qquad (3.146)$$

and from Eq. 3.145 as

$$\frac{b}{a} = \frac{\sqrt{a_o^2 - 1} - a_o + 1/(2a_o)}{\sqrt{a_o^2 - 1} + a_o - 1/(2a_o)} \qquad (3.147)$$

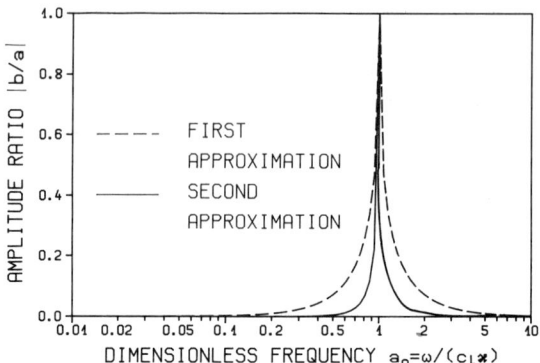

**Figure 3-18** Paraxial boundary.

$\left|\dfrac{b}{a}\right|$ is plotted versus $a_o$ in Fig. 3-18 ( Eqs. 3.140 and 3.141 for $a_o < 1$, Eqs. 3.146 and 3.147 for $a_o > 1$).

In an explicit-time–integration scheme, only one boundary condition valid for the whole frequency range can be used. Using Eq. 3.143, the results are shown in Fig. 3-15.

As a boundary condition, the differential equation of Eq. 3.145 can easily be implemented in a finite difference scheme. In the finite-element procedure it is more difficult. In particular, it can be shown that the first term in Eq. 3.145 will lead to a nonsymmetric damping matrix [C4]. In an explicit algorithm no unsurmountable difficulties arise from this fact. The "boundary" condition is assumed to apply throughout the last finite element of length $e$. With $v(x)$ denoting the weighting function, the corresponding weak form of Eq. 3.145 is formulated as

$$\frac{1}{c_l}\int_o^e \ddot{u}v\,dx + \int_o^e \dot{u}_{,x}v\,dx + \frac{\kappa^2 c_l}{2}\int_o^e uv\,dx = 0 \qquad (3.148\text{a})$$

or

$$\rho A \int_o^e v^T \ddot{u}\,dx + \rho A\, c_l \int_o^e v^T \dot{u}_{,x}\,dx + \frac{k_g}{2}\int_o^e v^T u\,dx = 0 \qquad (3.148\text{b})$$

Substituting Eq. 3.86 and the corresponding relationship for $v$ leads to

$$\rho A \int_o^e [N(x)]^T[N(x)]\,dx\,\{\ddot{u}\} + \rho A c_l \int_o^e [N(x)]^T[dN(x)/dx]\,dx\{\dot{u}\}$$
$$+ \frac{k_g}{2}\int_o^e [N(x)]^T[N(x)]\,dx\,\{u\} = \{0\} \qquad (3.149)$$

which can be rewritten as

Sec. 3.7  Paraxial Boundary

$$[M]\{\ddot{u}\} + [C]\{\dot{u}\} + [K]\{u\} = \{0\} \quad (3.150)$$

whereby the mass matrix $[M]$ (after lumping), the unsymmetric damping matrix $[C]$, and the static-stiffness matrix $[K]$ of the paraxial element are specified as

$$[M] = \rho A e \begin{bmatrix} \frac{1}{2} & \\ & \frac{1}{2} \end{bmatrix} \quad (3.151)$$

$$[C] = \rho A c_l \begin{bmatrix} -\frac{1}{2} & \frac{1}{2} \\ -\frac{1}{2} & \frac{1}{2} \end{bmatrix} \quad (3.152)$$

$$[K] = k_g e \begin{bmatrix} \frac{1}{6} & \frac{1}{12} \\ \frac{1}{12} & \frac{1}{6} \end{bmatrix} \quad (3.153)$$

Applying the explicit algorithm ($\beta = 0$, $\gamma = 0.5$) to the finite-element mesh—consisting of nine standard finite elements representing the standard wave equation (Eq. 3.55) and one finite element at the boundary (element number 10 in Fig. 3-9) governed by the paraxial approximation (Eq. 3.145)—leads to the reaction force plotted in Fig. 3-19a. It is obvious that an interfacing effect exists which inhibits the elastic waves from passing smoothly into the paraxial element, ending in catastrophic results. To circumvent this problem [C4], an interface element is introduced. The system of finite elements thus consists of eight standard finite elements (element numbers 1 to 8), an interface element (number 9), and the paraxial element (number 10). The internal force of the interface element at node 8 (Fig. 3-9) is calculated using the standard static-stiffness matrix (Eq. 3.87); and that at node 9 is calculated using the damping and static-stiffness matrices of the paraxial element (Eqs. 3.152 and 3.153). In other words, the equation of motion for node 8 is formulated as if there were standard finite elements present on both sides, and that for node 9 is formulated as if only paraxial elements contributed. The improvement in accuracy is dramatic (Fig. 3-19b).

Although the rod on an elastic foundation is a one-dimensional system, it can be used to illustrate the paraxial boundary concept. What is decisive is the apparent velocity $c_a$ in the paraxial direction. In the case of the out-of-plane motion with inclined body waves, $c_a$ equals $c_s/\sin\psi$, with $\psi$ measured from the boundary (Fig. 3-17). The square root (Eq. 3.122) represents $c_s/c_a = \sin\psi$ and is expanded into a Taylor series at $c_s/c_a = 1$ ($\psi = 90°$). In the case of the rod on an elastic foundation, dispersion leads to $c_l/c_a =$

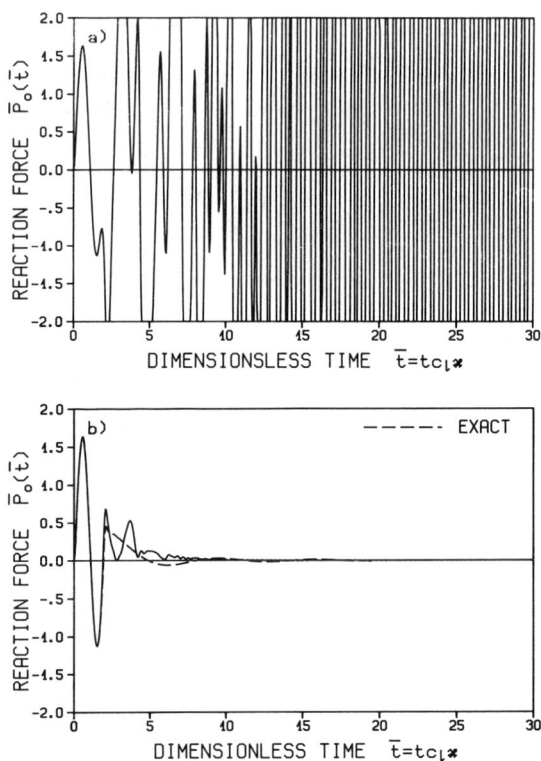

**Figure 3-19** Paraxial element. a. Without interface element. b. With interface element.

$\sqrt{1 - 1/a_o^2}$ (Eq. 3.61). The square root in Eq. 3.142 is expanded at $c_l/c_a = 1$ ($a_o = \infty$).

### 3.7.3 Generalization

The paraxial boundary concept can also be applied to the two-dimensional elasticity wave equations governing the in-plane motion with two displacements. However, the paraxial differential equations are instable for strongly inclined body waves, if Poisson's ratio is larger than 1/3 [C4]. This can cause the reflected waves' amplitudes to be larger than those of the incident waves. The scheme can be made stable by deleting a negative stiffness term in the differential equation. Even for the out-of-plane motion addressed in Section 3.7.1, the paraxial boundary becomes strongly ill-posed when a third term in the Taylor expansion of the square root in Eq. 3.122 is included [E1].

The differential equations of motion are easily implemented in a finite difference scheme. The approximations involve only first-order derivatives in the direction of the preferred wave propagation (that is, normal to the artificial boundary), which will involve only the unknowns one row away from the boundary. The implementation in the finite-element procedure (not discussed in depth in this text) is much more difficult. Interfacing effects inhibit the waves from smoothly proceeding into the paraxial element. Certain modifications in the numerical procedures are necessary to upgrade the boundary's accuracy [C4]. Even with these, in a transient analysis, the paraxial boundary seems to perform only slightly better than the simpler viscous damper. The accuracy for modeling surface waves is discussed in Section 3.9.3.

## 3.8 EXTRAPOLATION ALGORITHM

### 3.8.1 Concept Illustrated with Benchmark Problem

Problems of wave propagation towards infinity can be analyzed using the explicit-time–integration scheme in connection with the extrapolation algorithm [L1]. In the latter, the node on the artificial boundary is processed separately. The corresponding displacement for the next time station is extrapolated from data at earlier times along a line normal to the artificial boundary in the region's interior. Outward wave propagation is thereby enforced. To formulate the wave propagation in one direction only, the velocity along the normal must be introduced. Only simple numerical extrapolations and—if the data points do not coincide with the mesh's nodes—interpolations are performed.

The extrapolation algorithm concept is discussed using the benchmark problem. Although its wave propagation is one-dimensional, the procedure is not trivial for this application, because the apparent velocity $c_a$, which is equal to the phase velocity, depends on the frequency (Eq. 3.61) for the rod on an elastic foundation and thus is not known for a transient wave. For the two-dimensional case, the apparent velocity $c_a$ along the normal to the artificial boundary depends on the angle of incidence $\psi$ of the inclined body wave ($c_a = c_s/\sin\psi$, for $SH$-waves, Fig. 3-17), which varies. (See also the discussion at the end of Section 3.7.2.)

The displacement of the node on the artificial boundary $x = l$ at time $t$ is extrapolated from the corresponding known values at the interior nodes at earlier times (Fig. 3-9). Using fitting of polynomials of various orders results in the following one-sided formula:

$$N = 1 \quad u(l,t) = u(l - e, t - \Delta t) \tag{3.154a}$$

$$N = 2 \quad u(l,t) = 2u(l - e, t - \Delta t) - u(l - 2e, t - 2\Delta t) \tag{3.154b}$$

$$N = 3 \quad u(l,t) = 3u(l - e, t - \Delta t) - 3u(l - 2e, t - 2\Delta t)$$
$$+ u(l - 3e, t - 3\Delta t) \tag{3.154c}$$

The displacements' sampling points are assumed to coincide with the nodes of the finite-element mesh. $N$ denotes the number of terms used in the extrapolation series. For instance, $N = 2$ corresponds to a linear extrapolation. The further removed the interior node is from the artificial boundary, the earlier the corresponding displacement is sampled. For a transient excitation, Eq. 3.154 applies to the total wave, with the displacements at the interior points being calculated using the explicit-time–integration scheme.

For a general $N$, the extrapolation series can be written as

$$u(l,t) = \sum_{j=1}^{N} (-1)^{j+1} C_j^N u(l - je, t - j\Delta t) \quad (3.155)$$

in which $C_j^N$ are the binomial coefficients

$$C_j^N = \frac{N!}{(N-j)!j!} \quad (3.156)$$

For a transient analysis, some weighted average of the apparent velocities is chosen. Using the values specified in Section 3.2.6 [$(e = 0.1/\kappa, \Delta t = 0.05/(c_l\kappa)$], the velocity with which the total wave field propagates towards infinity is chosen as $2c_l$ ($= e/\Delta t$). In Fig. 3-20, the reaction force is plotted for $N = 1, 2,$ and $3$. Increasing the number of terms improves the accuracy of the extrapolation algorithm. Even for a linear extrapolation ($N = 2$), acceptable results are obtained.

Further insight is gained by examining the ratio of the amplitudes of the reflected wave $b$ and the incident wave $a$ at the artificial boundary for harmonic excitation. Using Eq. 3.59, the displacement as a function of time is formulated as

$$u(x, t) = a \exp(-ikx) \exp(i\omega t) + b \exp(+ikx) \exp(i\omega t) \quad (3.157)$$

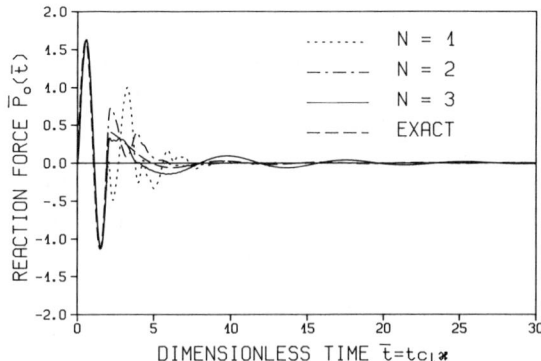

**Figure 3-20** Extrapolation algorithm.

## Sec. 3.8  Extrapolation Algorithm

Two cases exist. In the first one, the displacements at the interior nodes at earlier times are assumed to consist exclusively of waves propagating in the positive $x$-direction—that is, only the amplitude $a$ appears on the right-hand side of Eq. 3.155. This situation occurs when the incident wave starts impinging on the artificial boundary—that is, at the beginning of the transient excitation. In the second case, in addition, waves propagating in the negative $x$-direction occur at the interior nodes, leading to both amplitudes $a$ and $b$ being present on the right-hand side. This arises towards the end of the transient excitation, when significant partial reflection on the artificial boundary has already taken place.

**Case 1.** Substituting Eq. 3.157 for $x = 0$ in Eq. 3.155 leads to

$$(a + b)\exp(i\omega t) = \sum_{j=1}^{N} (-1)^{j+1} C_j^N a \exp(ikje) \exp[i\omega(t - j\Delta t)] \quad (3.158)$$

Using Eq. 3.58, Eq. 3.158 is transformed to

$$\frac{b}{a} = -1 + \sum_{j=1}^{N} (-1)^{j+1} C_j^N \exp\left[i\omega j \left(\frac{e}{c} - \Delta t\right)\right] \quad (3.159)$$

Using the properties of the binomial coefficients, the absolute value of the ratio in Eq. 3.159 follows as

$$\left|\frac{b}{a}\right| = 2^{\frac{N}{2}}\left(1 - \cos\left[\omega\left(\frac{e}{c} - \Delta t\right)\right]\right)^{\frac{N}{2}} \quad (3.160)$$

As $c$ is a function of $\omega$ (Eq. 3.61), $\left|\frac{b}{a}\right|$ will, in general, not be zero. Selecting

$$\Delta t = \frac{e}{c_l} \quad (3.161)$$

and using Eq. 3.61 leads to

$$\left|\frac{b}{a}\right| = 2^{\frac{N}{2}}(1 - \cos[\kappa e(\sqrt{a_o^2 - 1} - a_o)])^{\frac{N}{2}} \quad (3.162)$$

$\left|\frac{b}{a}\right|$ is plotted versus $a_o$ for $\kappa e = 0.1$ in Fig. 3-21. Note that, in contrast to the paraxial boundary (Fig. 3-18), $\left|\frac{b}{a}\right|$ is not equal to 1 at $a_o = 1$. It can be made arbitrarily small by decreasing $e$. Increasing $N$ (and decreasing $e$) improves the accuracy.

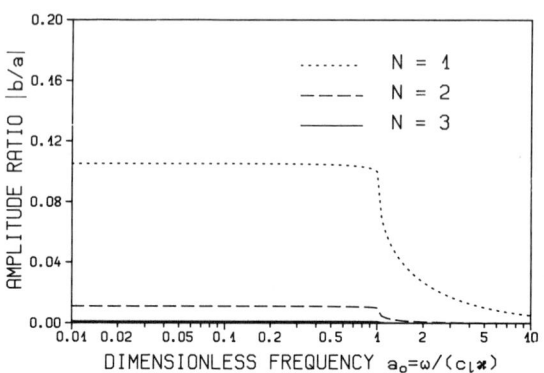

**Figure 3-21** Extrapolation algorithm, case 1.

**Case 2.** Again substituting Eq. 3.157 for $x = 0$ in Eq. 3.155 results in

$$(a + b) \exp(i\omega t) = \sum_{j=1}^{N} (-1)^{j+1} C_j^N (a \exp(ikje) + b \exp(-ikje)) \exp[i\omega(t - j\Delta t)] \quad (3.163)$$

Proceeding analogously to case 1 leads to

$$\frac{b}{a} = \frac{-1 + \sum_{j=1}^{N} (-1)^{j+1} C_j^N \exp\left[i\omega j \left(\frac{e}{c} - \Delta t\right)\right]}{1 - \sum_{j=1}^{N} (-1)^{j+1} C_j^N \exp\left[-i\omega j \left(\frac{e}{c} + \Delta t\right)\right]} \quad (3.164)$$

and to

$$\left|\frac{b}{a}\right| = \left(\frac{1 - \cos\left[\omega\left(\frac{e}{c} - \Delta t\right)\right]}{1 - \cos\left[\omega\left(\frac{e}{c} + \Delta t\right)\right]}\right)^{N/2} \quad (3.165)$$

For an $\omega$ larger than but close to the cutoff frequency, $\omega\left(\frac{e}{c} + \Delta t\right)$ will be considerably smaller than 1. Expanding the denominator into the first two terms of a Taylor's series results in

$$\left|\frac{b}{a}\right| = 2^{N/2}\left(1 - \cos\left[\omega\left(\frac{e}{c} - \Delta t\right)\right]\right)^{N/2} \left[\frac{1}{\omega\left(\frac{e}{c} + \Delta t\right)}\right]^N \quad (3.166)$$

The first two factors of Eq. 3.166 form the amplitude ratio for case 1 (Eq. 3.160). $\left|\frac{b}{a}\right|$ for case 2 will be larger than for case 1 for this frequency range,

Sec. 3.8  Extrapolation Algorithm

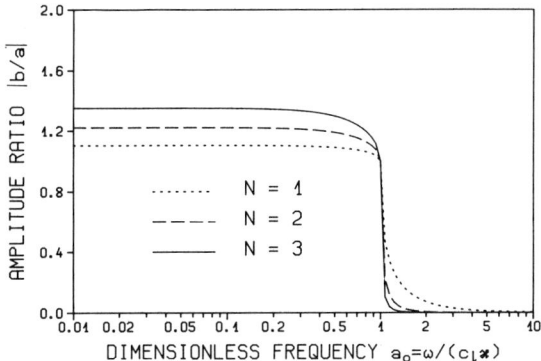

**Figure 3-22** Extrapolation algorithm, case 2.

as expected.

Again, based on Eq. 3.161,

$$\left|\frac{b}{a}\right| = \left(\frac{1 - \cos[\kappa e(\sqrt{a_o^2 - 1} - a_o)]}{1 - \cos[\kappa e(\sqrt{a_o^2 - 1} + a_o)]}\right)^{\frac{N}{2}} \quad (3.167)$$

follows.

$\left|\dfrac{b}{a}\right|$ is plotted versus $a_o$ for $\kappa e = 0.1$ in Fig. 3-22. While increasing $N$ improves the accuracy for $a_o > 1$, this no longer applies for $a_o < 1$.

### 3.8.2 Extension

The extrapolation algorithm concept can be applied to two- and three-dimensional cases with inclined body and surface waves [L1]. An inclined wave with the angle of incidence $\psi$ approaching the artificial boundary, is shown in Fig. 3-23. It travels in the direction of wave propagation $\bar{x}$ with the velocity $c$ ($c_s$ for $S$-waves, $c_p$ for $P$-waves). For any component of the displacement on the artificial boundary $\bar{x} = \bar{x}_l$ for the time station $t$, $u\left(t - \dfrac{\bar{x}_l}{c}\right)$ can be predicted from the values backward along the $\bar{x}$-axis at earlier time stations ($j = 1, 2, \ldots, N$)

$$\begin{aligned}
u\left(t - \frac{\bar{x}_l}{c}\right) &= u\left(t - \Delta t - \frac{\bar{x}_l - c\Delta t}{c}\right) \\
&= u\left(t - 2\Delta t - \frac{\bar{x}_l - c2\Delta t}{c}\right) \\
&= \ldots = u\left(t - j\Delta t - \frac{\bar{x}_l - cj\Delta t}{c}\right)
\end{aligned} \quad (3.168)$$

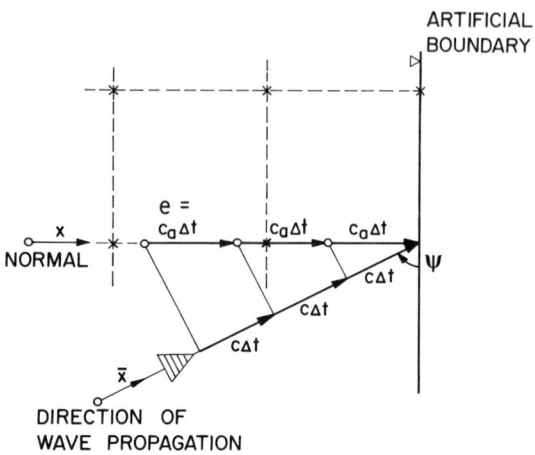

**Figure 3-23** Inclined wave.

As the directions of wave propagation are not available in the numerical procedure, the data points must be introduced on the normal to the artificial boundary, where the waves propagate with the apparent velocity $c_a = c/\sin\psi$. As the amplitudes on the wave front are constant, the extrapolation can be performed on the normal:

$$u\left(t - \frac{x_I}{c_a}\right) = u\left(t - j\Delta t - \frac{x_I - c_a j\Delta t}{c_a}\right) \quad (3.169)$$

In the actual algorithm for a group of waves, the apparent velocity is estimated as a weighted average, taking into account the different angles of incidence, the different wave velocities for P- and S-waves, and the phase velocities for surface waves (which in general depend on frequency). Numerical experiments indicate that the choice $c_a = c_s$ works well [L1]. The data points on the normal are thus separated by a distance $c_a\Delta t$, which is less than the finite-element length resulting from the stability criterion for the explicit algorithm ($> c_p\Delta t$, if P-waves occur). In general, it will thus be necessary to interpolate the sampling points from the nodal values, using, for example, polynomials (Fig. 3-23). The extrapolation equation (Eq. 3.155) is applied in the direction of the normal to the artificial boundary, whereby $e = c_a\Delta t$.

The error measured, for instance, by the reflected wave amplitude can be reduced by either increasing the number $N$ of terms in Eq. 3.155 or by decreasing $\Delta t$ (see Problem 3.6). Stringent tests are discussed in Ref. [L1]. It is shown that the algorithm works well when both P- and S-waves are present and can also successfully process surface waves.

## 3.9 LOCATION OF ARTIFICIAL BOUNDARY

### 3.9.1 Overview

As the transmitting boundaries are generally only approximate representations of the unbounded soil located on the artificial boundaries' exterior, certain reflections will occur. To what degree this will affect the accuracy of the response of the structure depends on many factors, such as the type of results, the material damping in the soil (and thus the level of the excitation), the frequency content of the excitation, the amount of soil–structure interaction which actually occurs, and the type of the transmitting boundary and its distance from the structure. Results for a transient excitation are less sensitive than for a harmonic load arising, for example, from machine vibrations. The larger the internal damping and the smaller the wave length of the excitation are, the more the incident and reflected waves' amplitudes decay. If the fundamental frequency of the (linear) structure–soil system is smaller than that of the soil layer resting on bedrock, the displacements decay exponentially and there is no wave propagation (Section 3.2.1). The soil–structure interaction effects are then small, and it is only necessary to model the static stiffness adequately. Besides, by choosing the type of transmitting boundary, the analyst can influence the accuracy of the results by selecting the artificial boundary's location. The influence of the latter is investigated in this section. The influence of material damping is disregarded.

The transmitting boundaries, which are local in space and time, are only approximations, with the exception of the superposition boundary with a mirror image, which is compatible with the unbounded domain (Section 3.3.1). The performance of a boundary depends strongly on the wave pattern characteristics—in particular on the apparent velocity in the direction perpendicular to the artificial boundary (wave number). The more the apparent velocity deviates from the velocity used to define the transmitting boundary, which in general is the wave velocity (shear or dilatational), the less accurate the results become (Fig. P 3-6). When the two velocities are the same—which corresponds to perpendicular impingement upon the artificial boundary—all transmitting boundaries lead to the exact results.

Intuitively, it is to be expected that the farther away the artificial boundary is placed, the more accurate the results become. The wave pattern close to the vibrating source (the structure) is far more general than that in the far field and is thus more difficult to transmit completely through the artificial boundary. Purely from a geometrical point of view, certain extreme angles of incidence of the waves impinging upon the artificial boundary cannot arise when the latter is far away. This tendency is confirmed in the literature [R3, V1, L1].

In this context, it is instructive to distinguish between body and surface waves. For body waves, addressed in Section 3.9.2, one can show that, in certain situations at a great distance, the waves propagate one-dimensionally in the direction of the normal to the artificial boundary, with the shear-wave velocity for the two tangential displacements and with the dilatational-wave velocity for the normal displacement. In simple cases, the ratio of the reflected and the incident waves' amplitudes depends on the distance of the artificial boundary divided by the wavelength. For surface waves which are dispersive—discussed in Section 3.9.3—the situation is much more complicated and some caution may be advisable. For instance, for another dispersive system—the elastically supported rod (Section 3.2)—the ratio of the amplitudes of the reflected and the incident waves at $x = 0$, $\left|\dfrac{b}{a}\right|$, above the cutoff frequency is independent of the distance $l$ to the artificial boundary with a viscous damper (Fig. 3-24). This is verified, using Eqs. 3.59, 3.60, and 3.113, as

$$\frac{b}{a} = \frac{\sqrt{a_o^2 - 1} - a_o}{\sqrt{a_o^2 - 1} + a_o} \exp[-2i\kappa \sqrt{a_o^2 - 1}\, l] \qquad (3.170)$$

which leads, for $a_o > 1$, to

$$\left|\frac{b}{a}\right| = \frac{\sqrt{a_o^2 - 1} - a_o}{\sqrt{a_o^2 - 1} + a_o} \qquad (3.171)$$

Only for $a_o < 1$, where no waves propagate, does an increase in $l$ improve the accuracy.

The higher the frequency of excitation is, the more the directionality of the waves increases—that is, the stronger the tendency is to transmit energy radiating from a plane source through only part of the medium. In the limit for $\omega \to \infty$, wave propagation occurs only in the direction normal to the vibrating source. The velocities of this one-dimensional wave propagation are equal to the shear-wave and dilatational-wave velocities for the motion,

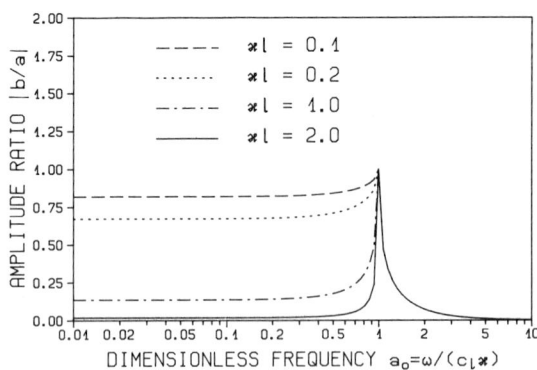

**Figure 3-24** Viscous damper on artificial boundary of rod supported elastically.

with components perpendicular to and coinciding with the normal, respectively. In this case, very simple local transmitting boundaries consisting of dampers, in some cases in addition to springs (Sections 3.1.1 and 3.1.2), can be placed directly on the structure–soil interface and will result in the exact solution. These features are further addressed in Section 6.3.

### 3.9.2 Body Waves

**Spherical wave.** For the sake of illustration, the one-dimensional symmetric spherical wave propagation discussed in Section 2.3 is examined. The artificial boundary is located at the radius $R$, whereby the superposition boundary, the viscous damper, and the extrapolation algorithm are addressed. The ratio of the "amplitudes" of the reflected wave $b$ and of the incident wave $a$ is calculated. Analogous to Eq. 2.22, the radial displacement $u(\omega)$ is formulated as

$$u(\omega) = a\left(-\frac{1}{r^2} - i\frac{k}{r}\right) \exp(-ikr) + b\left(-\frac{1}{r^2} + i\frac{k}{r}\right) \exp(ikr) \quad (3.172)$$

Note that in Eq. 3.172 $a$ is actually the integration constant associated with the wave propagating in the positive $r$-direction. The wave number $k$ equals $\omega/c_p$.

For the superposition boundary (Section 3.3.1), the symmetric case enforces

$$u(R, \omega) = 0 \quad (3.173)$$

leading to

$$\frac{b}{a} = \frac{1 + ikR}{-1 + ikR} \exp(-2ikR) \quad (3.174)$$

and for the antimetric case

$$\sigma_r(R, \omega) = (\lambda + 2G)\, u(R, \omega)_{,r} + \frac{2G}{R} u(R, \omega) = 0 \quad (3.175)$$

applies (with the two Lamé constants $\lambda$ and $G$, Eq. 2.14), resulting in

$$\frac{b}{a} = \frac{4 + 4ikR - \dfrac{c_p^2}{c_s^2} k^2 R^2}{-4 + 4ikR + \dfrac{c_p^2}{c_s^2} k^2 R^2} \exp(-2ikR) \quad (3.176)$$

The average equals

$$\frac{b}{a} = \frac{1}{2}\left(\frac{1 + ikR}{-1 + ikR} + \frac{4 + 4ikR - \dfrac{c_p^2}{c_s^2} k^2 R^2}{-4 + 4ikR + \dfrac{c_p^2}{c_s^2} k^2 R^2}\right) \exp(-2ikR) \quad (3.177)$$

Because the mirror image of the spherical medium with respect to the artificial boundary is not compatible with the unbounded domain on the exterior, the superposition boundary does not lead to the exact result.

For the viscous damper,

$$\sigma_r(R, \omega) + \rho c_p \dot{u}(R, \omega) = 0 \qquad (3.178)$$

is enforced (Eq. 3.108) which leads to

$$\frac{b}{a} = \frac{4 + 4ikR - i\dfrac{c_p^2}{c_s^2} kR}{-4 + 4ikR - \dfrac{c_p^2}{c_s^2}(2k^2R^2 + ikR)} \exp(-2ikR) \qquad (3.179)$$

For the extrapolation algorithm, it is assumed that the displacements in the interior at earlier times consist only of waves propagating in the positive $r$-direction (Case 1). The extrapolation series is then formulated as (Eq. 3.155)

$$u(R, t) = \left[a\left(-\frac{1}{R^2} - \frac{ik}{R}\right)\exp(-ikR) + b\left(-\frac{1}{R^2} + \frac{ik}{R}\right)\exp(ikR)\right]\exp(i\omega t)$$

$$= \sum_{j=1}^{N}(-1)^{j+1}C_j^N a\left(-\frac{1}{(R-je)^2} - \frac{ik}{R-je}\right)\exp[-ik(R-je)]\exp[i\omega(t-j\Delta t)]$$
(3.180)

with $C_j^N$ specified in Eq. 3.156. Selecting the distance between data points $e = c_p \Delta t$, Eq. 3.180 is formulated as

$$\frac{b}{a} = \frac{1 + ikR + \sum_{j=1}^{N}(-1)^{j+1}C_j^N\left(\dfrac{1}{1 - 2j\,e/R + j^2 e^2/R^2} + \dfrac{ikR}{1 - j\,e/R}\right)}{1 - ikR}$$
$$\times \exp(-2ikR) \qquad (3.181)$$

For all three transmitting boundaries, the radius of the artificial boundary $R$ and the wave number $k$ appear only as factors in the multiplication $kR$. This product $kR = R\omega/c_p = 2\pi R/\lambda$ is thus proportional to the ratio of the distance $R$ and the wave length $\lambda$. For $kR \to \infty$, $\left|\dfrac{b}{a}\right|$ converges to zero for the superposition and viscous boundaries and for the extrapolation algorithm only when $e/R$ vanishes.

In Fig. 3-25, $\left|\dfrac{b}{a}\right|$ is plotted versus $kR$ assuming Poisson's ratio $\nu =$

Sec. 3.9  Location of Artificial Boundary                                    129

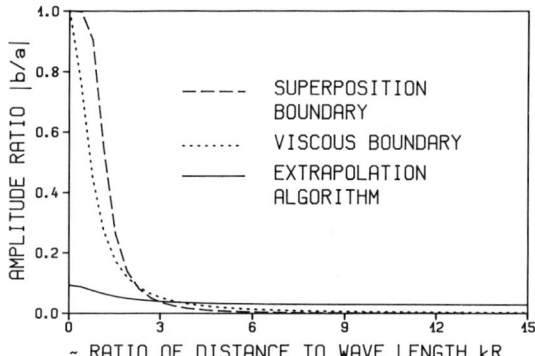

**Figure 3-25** Influence of artificial boundary location on accuracy of results for spherical waves.

1/3 ($c_p = 2c_s$). For the extrapolation algorithm, $N = 2$ and $e/R = 0.1$ are chosen. The extrapolation algorithm performs better than the other two transmitting boundaries for small values of $kR$. One achieves quite a dramatic improvement by selecting the artificial boundary farther away from the structure for the superposition and viscous boundaries.

**Far field.** It is instructive to study the behavior of the equation of motion at a large distance from the vibrating source, in the so-called far field. In the spherical-coordinate system $r$, $\theta$, $\varphi$ shown in Fig. 3-26, the dynamic-equilibrium equations are formulated as

$$\sigma_{r,r} + \frac{1}{r}\tau_{r\theta,\theta} + \frac{1}{r\sin\theta}\tau_{r\varphi,\varphi} + \frac{2\sigma_r - \sigma_\theta - \sigma_\varphi + \cot\theta\,\tau_{r\theta}}{r} - \rho\ddot{u} = 0 \quad (3.182a)$$

$$\frac{1}{r}\sigma_{\theta,\theta} + \tau_{r\theta,r} + \frac{1}{r\sin\theta}\tau_{\theta\varphi,\varphi} + \frac{\cot\theta(\sigma_\theta - \sigma_\varphi) + 3\tau_{r\theta}}{r} - \rho\ddot{v} = 0 \quad (3.182b)$$

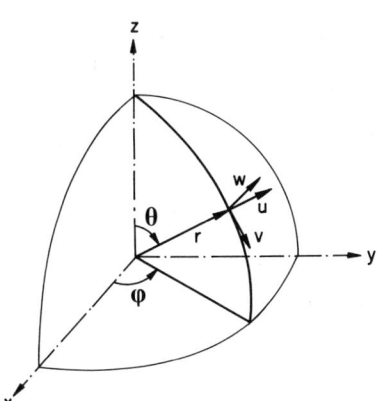

**Figure 3-26** Spherical coordinate system.

$$\frac{1}{r\sin\theta}\sigma_{\varphi,\varphi} + \tau_{r\varphi,r} + \frac{1}{r}\tau_{\theta\varphi,\theta} + \frac{3\tau_{r\varphi} + 2\cot\theta\,\tau_{\theta\varphi}}{r} - \rho\ddot{w} = 0 \qquad (3.182c)$$

All displacements and stresses are functions of $r$, $\theta$, $\varphi$. The strain–displacement relationships are equal to

$$\varepsilon_r = u_{,r} \qquad (3.183a)$$

$$\varepsilon_\theta = \frac{u}{r} + \frac{1}{r}v_{,\theta} \qquad (3.183b)$$

$$\varepsilon_\varphi = \frac{u}{r} + \frac{\cot\theta}{r}v + \frac{1}{r\sin\theta}w_{,\varphi} \qquad (3.183c)$$

$$\gamma_{r\theta} = \frac{1}{r}u_{,\theta} + v_{,r} - \frac{v}{r} \qquad (3.183d)$$

$$\gamma_{r\varphi} = \frac{1}{r\sin\theta}u_{,\phi} + w_{,r} - \frac{w}{r} \qquad (3.183e)$$

$$\gamma_{\theta\varphi} = \frac{1}{r\sin\theta}v_{,\varphi} + \frac{1}{r}w_{,\theta} - \frac{\cot\theta}{r}w \qquad (3.183f)$$

Hooke's law is specified as

$$\sigma_r = (\lambda + 2G)\varepsilon_r + \lambda\varepsilon_\theta + \lambda\varepsilon_\varphi \qquad (3.184a)$$

$$\sigma_\theta = \lambda\varepsilon_r + (\lambda + 2G)\varepsilon_\theta + \lambda\varepsilon_\varphi \qquad (3.184b)$$

$$\sigma_\varphi = \lambda\varepsilon_r + \lambda\varepsilon_\theta + (\lambda + 2G)\varepsilon_\varphi \qquad (3.184c)$$

$$\tau_{r\theta} = G\gamma_{r\theta} \qquad (3.184d)$$

$$\tau_{r\varphi} = G\gamma_{r\varphi} \qquad (3.184e)$$

$$\tau_{\theta\varphi} = G\gamma_{\theta\varphi} \qquad (3.184f)$$

with the two Lamé constants $\lambda$ and $G$ (Eq. 2.14).

For large $r$, all terms with $r$ in the denominator are neglected. Equation 3.182 is thus transformed to

$$\sigma_{r,r} - \rho\ddot{u} = 0 \qquad (3.185a)$$

$$\tau_{r\theta,r} - \rho\ddot{v} = 0 \qquad (3.185b)$$

$$\tau_{r\varphi,r} - \rho\ddot{w} = 0 \qquad (3.185c)$$

and Eq. 3.183 to

$$\varepsilon_r = u_{,r} \qquad (3.186a)$$

$$\varepsilon_\theta = 0 \qquad (3.186b)$$

## Sec. 3.9    Location of Artificial Boundary

$$\varepsilon_\varphi = 0 \tag{3.186c}$$

$$\gamma_{r\theta} = v_{,r} \tag{3.186d}$$

$$\gamma_{r\varphi} = w_{,r} \tag{3.186e}$$

$$\gamma_{\theta\varphi} = 0 \tag{3.186f}$$

Eliminating the strains and the stresses in Eqs. 3.184, 3.185, and 3.186 leads to the equations of motion

$$u_{,rr} - \frac{\ddot{u}}{c_p^2} = 0 \tag{3.187a}$$

$$v_{,rr} - \frac{\ddot{v}}{c_s^2} = 0 \tag{3.187b}$$

$$w_{,rr} - \frac{\ddot{w}}{c_s^2} = 0 \tag{3.187c}$$

with the dilatational- and shear-wave velocities $c_p$ and $c_s$ defined in Eqs. 2.13 and 2.10. Equation 3.187 corresponds to the standard one-dimensional equations in a prismatic rod (Eqs. 3.3 and 3.13). Interpreting $r$ = constant as a boundary surface, Eq. 3.187a describes the radial wave $u$ propagating in the $r$-direction with the velocity $c_p$ and Eqs. 3.187b, c the two tangential waves $v$ and $w$ propagating with the velocity $c_s$.

These conclusions are verified by addressing the analytical solution for the displacements and stresses in an elastic half-space loaded by the constant pressure $p(\omega) \exp(i\omega t)$ acting on a circle of radius $a$ perpendicular to the free surface. The contribution of the body waves leads to the following amplitudes of the far-field displacements $u(\omega)$ and $v(\omega)$ in the $r$- and $\theta$-directions and to the corresponding stresses $\sigma_r(\omega)$ and $\tau_{r\theta}(\omega)$ in the spherical-coordinate system with the origin at the center of the loaded area [M2, M3].

$$u(\omega) = -\frac{a^2 p(\omega)}{2Gr} \theta_1(\theta) \exp\left[i\omega\left(t - \frac{r}{c_p}\right)\right] \tag{3.188a}$$

$$v(\omega) = -\frac{ia^2\mu^3 p(\omega)}{2Gr} \theta_2(\theta) \exp\left[i\omega\left(t - \frac{r}{c_s}\right)\right] \tag{3.188b}$$

$$\sigma_r(\omega) = \frac{ia^2\omega\mu^2 p(\omega)}{2rc_p} \theta_1(\theta) \exp\left[i\omega\left(t - \frac{r}{c_p}\right)\right] \tag{3.188c}$$

$$\tau_{r\theta}(\omega) = -\frac{a^2\omega\mu^4 p(\omega)}{2rc_p} \theta_2(\theta) \exp\left[i\omega\left(t - \frac{r}{c_s}\right)\right] \tag{3.188d}$$

with

$$\mu = \sqrt{\frac{\lambda + 2G}{G}} \tag{3.189a}$$

$$\theta_1(\theta) = \frac{\cos\theta \, (\mu^2 - 2\sin^2\theta)}{F(\sin\theta)} \tag{3.189b}$$

$$\theta_2(\theta) = \frac{\sin 2\theta \, \sqrt{\mu^2 \sin^2\theta - 1}}{F(\mu \sin\theta)} \tag{3.189c}$$

$$F(\xi) = (2\xi^2 - \mu^2)^2 - 4\xi^2 \sqrt{(\xi^2 - \mu^2)(\xi^2 - 1)} \tag{3.189d}$$

These relations, which assume $a \ll r$, satisfy the radiation condition of the one-dimensional wave propagation (Eqs. 3.108 and 3.109):

$$\sigma_r(\omega) + \rho c_p \dot{u}(\omega) = 0 \tag{3.190a}$$

$$\tau_{r\theta}(\omega) + \rho c_s \dot{v}(\omega) = 0 \tag{3.190b}$$

The contribution of the surface wave in the far field, which is specified in Refs. [M2, M3], does not satisfy either Eq. 3.187 or Eq. 3.190.

### 3.9.3 Surface Waves

**Love wave.** To study the influence of the artificial boundary's location on the accuracy of the results in the presence of surface waves, the out-of-plane motion of a homogeneous semi-infinite layer built-in at its base is addressed (Fig. 3-27). As will become apparent, this system has a cutoff frequency at the layer's fundamental natural frequency. At this frequency and at each of the higher natural frequencies, a Love mode starts with a phase velocity (apparent velocity) of infinity which decreases for higher fre-

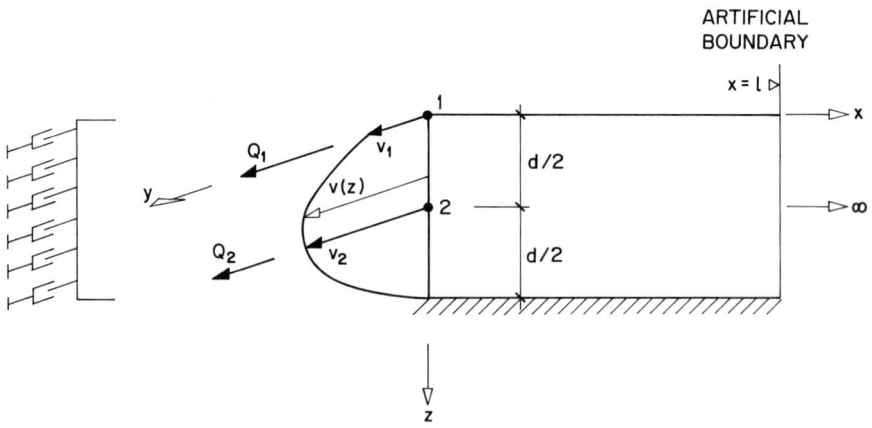

**Figure 3-27** Out-of-plane motion of layer built-in at its base.

## Sec. 3.9  Location of Artificial Boundary

quencies, converging to the shear-wave velocity for an infinite frequency. This case, which often occurs in practice, is thus a very stringent test.

The depth is indicated as $d$, and $G$ and $\rho$ denote the shear modulus and the mass density. The dynamic-stiffness matrix $[S]$ at $x = 0$, corresponding to a parabolic variation of the displacement $v(z)$, is to be calculated. As the displacement is zero at the base, only two nodes with the displacements $v_1$ and $v_2$ collected in the vector $\{v\}$ are introduced.

$$v(z) = [N(z)]\{v\} \qquad (3.191a)$$

where $[N(z)]$ is the shape function

$$[N(z)] = \left[ 2\frac{z^2}{d^2} - 3\frac{z}{d} + 1 \ \bigg| \ -4\frac{z^2}{d^2} + 4\frac{z}{d} \right] \qquad (3.191b)$$

The vector of the point loads $\{Q\}$ made up of $Q_1$ and $Q_2$ is related to $\{v\}$ as

$$\{Q\} = [S]\{v\} \qquad (3.192)$$

The wave equation, which is derived in Section 3.7.1, is specified in Eq. 3.118; its solution is in Eq. 3.119, with the condition indicated in Eq. 3.120. The wave numbers $k_x$ and $k_z$ can have both signs, leading to another three relations similar to the one of Eq. 3.119. Dropping the subscript $x$ in the wave number in the $x$-direction and expressing $k_z$ using Eq. 3.120 as

$$k_z = \sqrt{\frac{\omega^2}{c_s^2} - k^2} = kt \qquad (3.193)$$

with

$$t = \sqrt{\frac{\omega^2}{c_s^2 k^2} - 1} = i\sqrt{1 - \frac{\omega^2}{c_s^2 k^2}} \qquad (3.194)$$

the general solution is written as (Eq. 3.119)

$$v(x, z, t) = ([A \exp(iktz) + B \exp(-iktz)] \exp(-ikx)$$
$$+ [C \exp(iktz) + D \exp(-iktz)] \exp(ikx)) \exp(i\omega t) \qquad (3.195)$$

The first and second halves of the expression on the right-hand side represent waves propagating in the positive and negative $x$-directions, respectively. The expressions with the amplitudes $A$ and $B$ (and $C$ and $D$) correspond to waves propagating in the negative and positive $z$-directions, respectively.

The exact solution is calculated first. As no incoming wave exists, $C = D = 0$. Enforcing the traction-free condition at $z = 0$

$$\tau_{yz}(x) = G v_{,z}(x) = 0 \qquad (3.196)$$

leads to $B = A$. The other boundary condition at $z = d$

$$v(x) = 0 \tag{3.197}$$

results in the characteristic equation

$$\cos ktd = 0 \tag{3.198}$$

which is satisfied for the discrete values

$$k_j t_j d = \frac{(2j-1)\pi}{2} \qquad j = 1, 2, 3, \ldots, \infty \tag{3.199}$$

Deleting the time variation, the complete solution is equal to

$$v(x, z, \omega) = \sum_{j=1}^{\infty} A_j \cos k_j t_j z \exp(-ik_j x) \tag{3.200}$$

$v(z)$ in Eq. 3.191 is expanded into a Fourier series with terms $\cos k_j t_j z$. After using Eq. 3.199, this leads to

$$v(z) = \frac{8}{\pi^2} \sum_{j=1}^{\infty} \frac{1}{(2j-1)^2} \left[ -\frac{(-1)^{j+1} 8}{(2j-1)\pi} + 3 \frac{(-1)^{j+1} 16}{(2j-1)\pi} - 4 \right] \cos k_j t_j z \{v\} \tag{3.201}$$

Formulating Eq. 3.200 for $x = 0$ and comparing it with Eq. 3.201 determines the amplitudes $A_j$ for each $j$.

The corresponding amplitude of the shear stress $\tau_{yx}$ at $x = 0$ (Eq. 3.117a) follows as

$$\tau_{yx}(z) = -iG \sum_{j=1}^{\infty} A_j k_j \cos k_j t_j z \tag{3.202}$$

Integrating the surface traction, which is equal to the negative value of $\tau_{yx}$ at $x = 0$, with the shape function $[N(z)]$, results in the nodal forces $\{Q\}$

$$\{Q\} = -\int_0^d [N(z)]^T \tau_{yx}(z) \, dz \tag{3.203}$$

Substituting Eq. 3.202 into Eq. 3.203 leads to the dynamic-stiffness matrix $[S]$ (Eq. 3.192), whereby the dimensionless frequency $a_o$ is introduced ($c_s = \sqrt{G/\rho}$):

$$a_o = \frac{\omega d}{c_s} \tag{3.204}$$

and

$$t_j = i\sqrt{1 - \frac{a_o^2}{d^2 k_j^2}} = i\sqrt{1 - \frac{a_o^2 t_j^2}{(k_j t_j d)^2}} = \frac{i}{\sqrt{1 - \frac{4a_o^2}{\pi^2(2j-1)^2}}} \tag{3.205}$$

## Sec. 3.9  Location of Artificial Boundary

$$k_j d = \frac{k_j t_j d}{t_j} = -i(2j-1)\frac{\pi}{2}\sqrt{1 - \frac{4}{\pi^2}\frac{a_o^2}{(2j-1)^2}} \qquad (3.206)$$

are used.

$$[S(a_o)] = \frac{16}{\pi^3} G \sum_{j=1}^{\infty} \frac{1}{(2j-1)^3}$$

$$\times \begin{bmatrix} \left[3 - \frac{(-1)^{j+1}8}{(2j-1)\pi}\right]^2 & \left[3 - \frac{(-1)^{j+1}8}{(2j-1)\pi}\right]\left[-4 + \frac{(-1)^{j+1}16}{(2j-1)\pi}\right] \\ \text{symmetric} & \left[-4 + \frac{(-1)^{j+1}16}{(2j-1)\pi}\right]^2 \end{bmatrix}$$

$$\times \sqrt{1 - \frac{4a_o^2}{(2j-1)^2\pi^2}} \qquad (3.207)$$

The static values follow from Eq. 3.207 for $a_o = 0$ as $K_{11} = 0.449\,G$, $K_{12} = -0.241\,G$, $K_{22} = 1.339\,G$.

Nondimensionalizing $S_{ij}$ with $K_{ij}$ and then splitting the coefficient into its real and imaginary parts defines the spring coefficient $k_{ij}(a_o)$ and the damping coefficient $c_{ij}(a_o)$:

$$S_{ij}(a_o) = K_{ij}[k_{ij}(a_o) + ia_o c_{ij}(a_o)] \qquad (3.208)$$

These dimensionless spring and damping coefficients are plotted versus $a_o$ in Fig. 3-28. Below the cutoff frequency $a_o = \pi/2$, which is equal to the fundamental frequency of the layer built-in at its base ($= \pi c_s/(2d)\ d/c_s$), $c_{11} = c_{12} = c_{22}$ are equal to zero. For $a_o \geq (2j-1)\pi/2$, the square root becomes imaginary, leading to dampers. At each natural frequency of the layer an additional square root becomes imaginary. For $a_o = \infty$, all square roots are imaginary and the spring coefficients vanish. The dynamic-stiffness matrix equals

$$[S(a_o = \infty)] = i\omega dc_s\rho \begin{bmatrix} \frac{2}{15} & \frac{1}{15} \\ \text{symmetric} & \frac{8}{15} \end{bmatrix} \qquad (3.209)$$

This corresponds to concentrated dampers, which could also be determined by lumping the evenly distributed dampers having a constant $\rho c_s dz$ (Fig. 3-27).

$$[S(a_o = \infty)] = i\omega c_s \int_o^d [N(z)]^T [N(z)]\,dz \qquad (3.210)$$

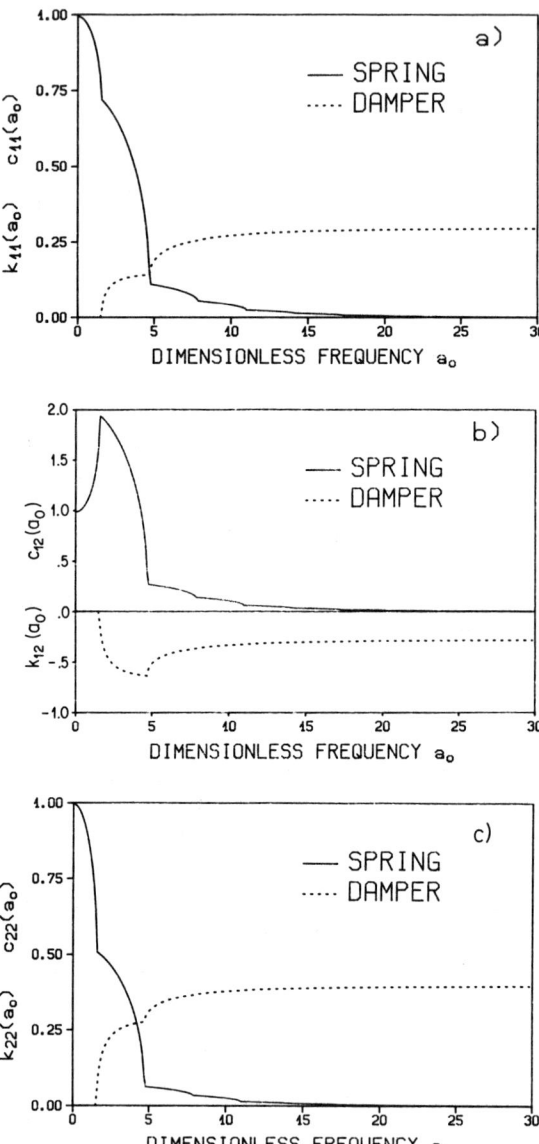

**Figure 3-28** Dynamic-stiffness coefficients. a. Node 1. b. coupling term. c. node 2.

### Sec. 3.9  Location of Artificial Boundary

For $a_o = \infty$, the waves propagate perpendicularly to the surface $x = 0$. This wave propagation in the direction of the positive $x$-axis is thus one-dimensional and the damper with a coefficient $\rho c_s dz$ is the exact representation of the unbounded domain (Eq. 3.12). The dampers can be placed directly at $x = 0$.

The influence of introducing an artificial boundary at $x = l$ is examined next. The condition corresponding to the transmitting boundary is treated analytically. In an actual application, of course, a spatial discretization would be performed.

As the transmitting boundaries are only approximate, some waves propagating in the negative $x$-axis will exist. The general solution specified in Eq. 3.195 applies. Enforcing the traction-free condition at $z = 0$ (Eq. 3.196) leads to $D = C$ in addition to $B = A$. The built-in boundary condition at $z = d$ (Eq. 3.197) results in the same characteristic equation (Eq. 3.198), which is satisfied for the same discrete values (Eq. 3.199). The solution is thus formulated as

$$v(x,z,t) = \left[\sum_{j=1}^{\infty} A_j \cos k_j t_j z \exp(-ik_j x) + \sum_{j=1}^{\infty} C_j \cos k_j t_j z \exp(ik_j x)\right] \exp(i\omega t)$$

(3.211)

The prescribed displacement $v(z)$ at $x = 0$ corresponding to the shape function is expanded into the same Fourier series with terms $\cos k_j t_j z$, as in the case of the exact solution (Eq. 3.201). Formulating Eq. 3.211 for $x = 0$ and comparing it with Eq. 3.201 determines the sum of the amplitudes $A_j + C_j$ for each $j$.

$$A_j + C_j = \frac{8}{(2j-1)^2 \pi^2} \left[ -\frac{(-1)^{j+1} 8}{(2j-1)\pi} + 3 \left| \frac{(-1)^{j+1} 16}{(2j-1)\pi} - 4 \right| \right] \{v\}$$

(3.212)

Deleting the time variation, the corresponding amplitude of the shear stress $\tau_{yx}$ at $x = 0$ (Eq. 3.117a) follows as

$$\tau_{yx}(z) = -iG \sum_{j=1}^{\infty} (A_j - C_j) k_j \cos k_j t_j z$$

(3.213)

Proceeding analogously as for the exact solution (Eq. 3.203) leads to the dynamic-stiffness matrix $[S(a_o)]$

$$[S(a_o)] = \frac{2}{\pi} G \sum_{j=1}^{\infty} \begin{bmatrix} 3 - \dfrac{(-1)^{j+1} 8}{(2j-1)\pi} \\ -4 + \dfrac{(-1)^{j+1} 16}{(2j-1)\pi} \end{bmatrix} \frac{A_j - C_j}{2j-1} \sqrt{1 - \frac{4a_o^2}{\pi^2 (2j-1)^2}}$$

(3.214)

Equation 3.214 expresses $[S(a_o)]$ as a function of $A_j$, $C_j$. The second equation needed to determine $A_j$, $C_j$ (besides Eq. 3.212) follows from the condition enforced on the transmitting boundary. Applying the superposition boundary (Section 3.3.1) results in the exact solution and is thus not addressed any further.

First, the viscous damper is examined. The boundary condition (Eq. 3.109) is formulated at $x = l$ as

$$\tau_{yx}(z) + \rho c_s \dot{v}(z) = 0 \qquad (3.215)$$

or, after using Eq. 3.117a, as

$$v_{,x}(z) + \frac{\dot{v}(z)}{c_s} = 0 \qquad (3.216)$$

Substituting Eq. 3.211 results in

$$C_j = \frac{1 - \dfrac{\omega}{c_s k_j}}{1 + \dfrac{\omega}{c_s k_j}} \exp(-2ik_j l)\, A_j \qquad (3.217)$$

and, after using Eq. 3.206, in

$$\frac{C_j}{A_j} = \frac{\sqrt{(2j-1)^2\pi^2 - 4a_o^2} - 2a_o i}{\sqrt{(2j-1)^2\pi^2 - 4a_o^2} + 2a_o i} \exp\!\left(-i\sqrt{4 - \frac{(2j-1)^2\pi^2}{a_o^2}}\,\frac{l}{d} a_o\right) \qquad (3.218)$$

Second, the paraxial boundary is discussed. The boundary condition (Eq. 3.124) at $x = l$ is repeated for convenience.

$$\frac{1}{c_s} \dot{v}_{,x}(z) - \frac{1}{2} v_{,zz}(z) + \frac{\ddot{v}(z)}{c_s^2} = 0 \qquad (3.219)$$

Substituting Eq. 3.211 leads to

$$\frac{C_j}{A_j} = \frac{\sqrt{(2j-1)^2\pi^2 - 4a_o^2} - 2a_o i + \dfrac{(2j-1)^2\pi^2}{4a_o} i}{\sqrt{(2j-1)^2\pi^2 - 4a_o^2} + 2a_o i - \dfrac{(2j-1)^2\pi^2}{4a_o} i}$$

$$\times \exp\!\left(-i\sqrt{4 - \frac{(2j-1)^2\pi^2}{a_o^2}}\,\frac{l}{d} a_o\right) \qquad (3.220)$$

Finally, the extrapolation algorithm with one term ($N = 1$) is addressed. It is assumed that the displacement in the interior node consists exclusively of

## Sec. 3.9 Location of Artificial Boundary

a wave propagating in the positive x-direction (Case 1 of Section 3.8.1). Substituting Eq. 3.211 in Eq. 3.154a results in

$$[A_j \exp(-ik_j l) + C_j \exp(ik_j l)] \exp(i\omega t)$$
$$= A_j \exp[-ik_j(l - e)] \exp[i\omega(t - \Delta t)] \quad (3.221)$$

where $e$ is the distance of the interior node from the artificial boundary. Assuming $\Delta t = e/c_s$ leads to

$$\frac{C_j}{A_j} = \left(-1 + \exp\left[\left(\frac{\sqrt{(2j-1)^2\pi^2 - 4a_o^2}}{2} - ia_o\right)\frac{e}{d}\right]\right)$$
$$\times \exp\left(-i\sqrt{4 - \frac{(2j-1)^2\pi^2}{a_o^2}}\frac{l}{d}a_o\right) \quad (3.222)$$

The ratio of the reflected and the incident waves' amplitudes shows the same tendency for the viscous damper (Eq. 3.218), the paraxial boundary (Eq. 3.220), and the extrapolation algorithm (Eq. 3.222). For $a_o \geq (2j - 1) \pi/2$, the magnitude $|C_j/A_j|$ is independent of $l/d$—that is, of the artificial boundary's location. Increasing $a_o$ in this frequency range improves the accuracy. For $a_o < (2j - 1) \pi/2$, the larger $l/d$ is, the smaller $|C_j/A_j|$ is; that is, the farther away the boundary is placed, the more accurate the results are. This is illustrated for the paraxial boundary for $j = 1$ in Fig. 3-29. This figure is very similar to that of the rod on an elastic foundation (Fig. 3-24).

The dynamic-stiffness coefficient corresponding to node 1 is plotted for the viscous boundary versus $a_o$ in Fig. 3-30, varying the location of the artificial boundary ($l/d = 0.4, 1, 2,$ and 5). The analytical values for $k_{11}$ and $c_{11}$ are shown as a solid and as a dotted line (Fig. 3-28a). The agreement of the damper $c_{11}$ is much better than that of the spring $k_{11}$. As is apparent from

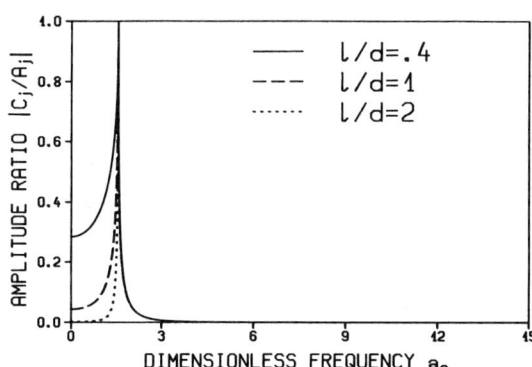

**Figure 3-29** Ratio of amplitudes of reflected and incident waves for paraxial boundary.

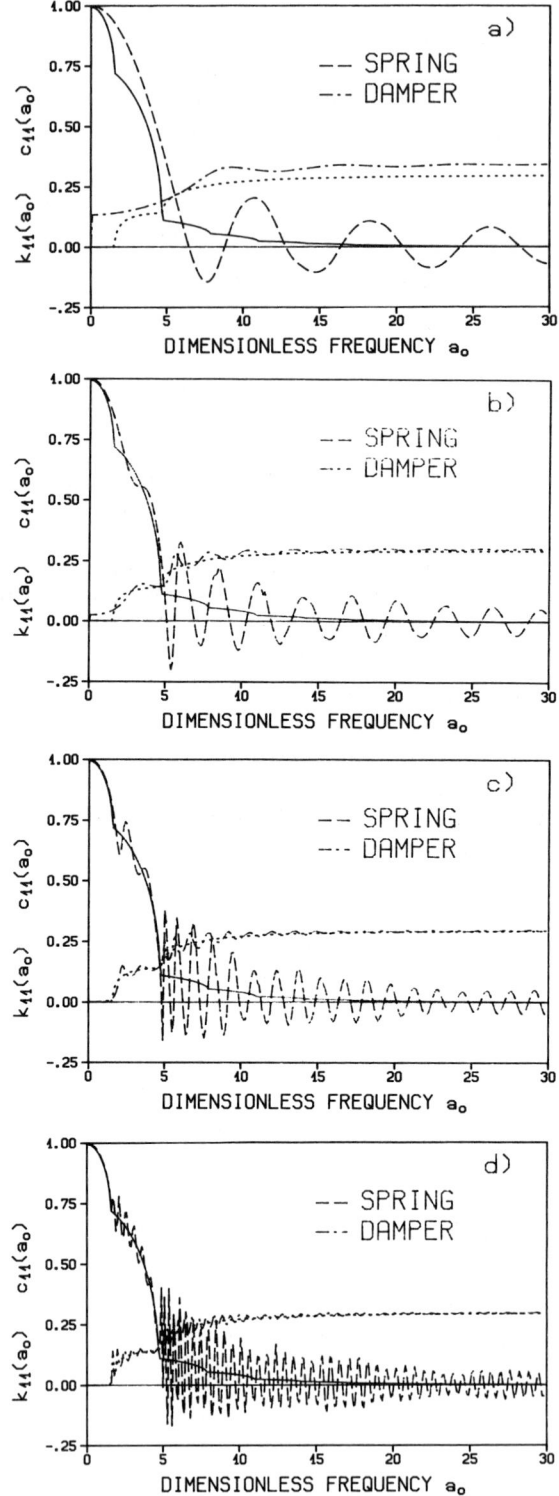

**Figure 3-30** Dynamic-stiffness coefficient for viscous boundary. a. $l/d = 0.4$. b. $l/d = 1$. c. $l/d = 2$. d. $l/d = 5$.

## Sec. 3.9  Location of Artificial Boundary

$C_j/A_j$ in Eq. 3.218, increasing $l/d$ (for $a_o > (2j - 1)\pi/2$) leads to oscillations which vibrate faster with respect to $a_o$, without decreasing in magnitude. This is clearly evident in Fig. 3-30. For harmonic excitation, no improvement can thus be expected by placing the artificial boundary farther away. Below the cutoff frequency $a_o = \pi/2$, where essentially static properties govern, convergence is achieved for increasing $l/d$ (spring and damper). Figure 3-31 shows the same dynamic-stiffness coefficient, but for the paraxial boundary. The result is a dramatic improvement in the accuracy compared to that of the viscous boundary (Fig. 3-30). However, the high-frequency vibrations still exist, although they are less disturbing because the accuracy is better. For the extrapolation algorithm with $N = 1$ (Fig. 3-32), the location of the artificial boundary is held fixed ($l/d = 1$) and the distance of the data point is varied ($e/d = 0.1$, 0.05 and 0.01). While the results are quite poor for $e/d = 0.1$ (Fig. 3-32a)—which is understandable considering that only one data point is selected—decreasing $e/d$ is very effective in increasing the accuracy. This is to be expected from Eq. 3.222. $e/d = 0.01$ is, however, unrealistically small (Fig. 3-32c). (For the second-order extrapolation algorithm see Problem 3.9.)

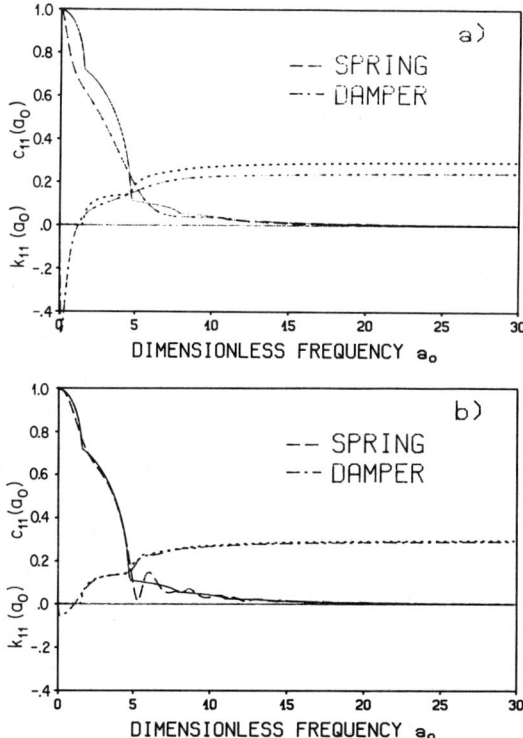

**Figure 3-31** Dynamic-stiffness coefficient for paraxial boundary. a. $l/d = 0.4$. b. $l/d = 1$. c. $l/d = 2$. d. $l/d = 5$.

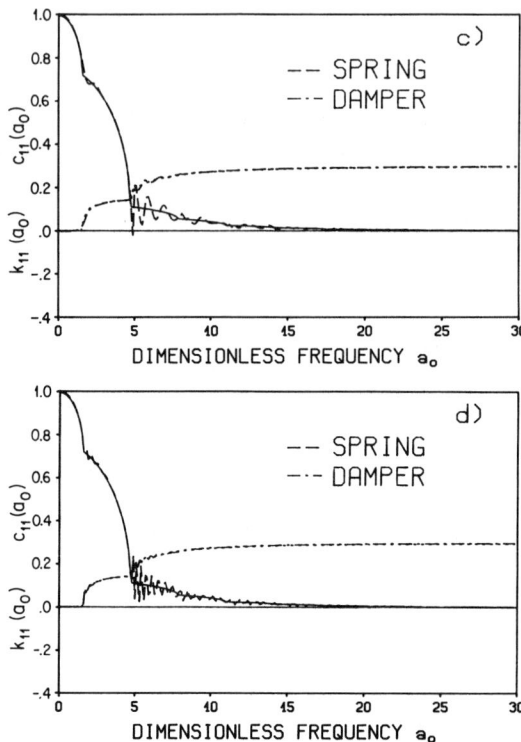

**Figure 3-31** Continued

The question arises: Does the introduction of material damping eliminate the oscillations of the spring coefficient? Hysteretic damping with a ratio $\zeta$ is assumed for this investigation. The correspondence principle applies, which replaces the shear modulus $G$ by the corresponding complex value $G(1 + 2\zeta i)$. The case with the viscous damper on the artificial boundary is addressed, whereby the damping coefficient $\rho c_s$ in Eq. 3.215 is not modified—that is, the correspondence principle does not apply to this term. The results are plotted for $l/d = 1$ in Fig. 3-33. For a damping ratio $\zeta = 0.001$, the solution (Fig. 3-33a) coincides from a practical point of view with that of the undamped case (Fig. 3-30b). Thus, introducing a very small fictitious damping ratio does not eliminate the oscillations. Even for a ratio $\zeta = 0.05$, the oscillations of the spring coefficient still exist (Fig. 3-33b), although the amplitudes are somewhat reduced compared to those of the undamped case.

**Rayleigh wave.** The transmitting boundaries' ability to process a Rayleigh wave propagating in an elastic half-space is addressed next. This is an important test, as the surface wave is dominant in the in-plane motion

Sec. 3.9  Location of Artificial Boundary 143

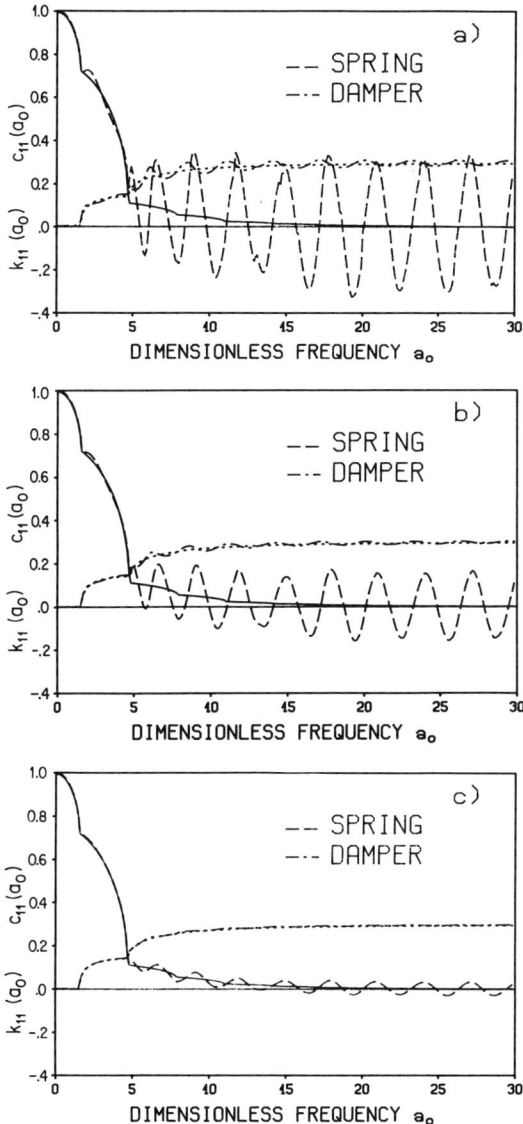

**Figure 3-32** Dynamic-stiffness coefficient for extrapolation algorithm. a. $e/d = 0.1$. b. $e/d = 0.05$. c. $e/d = 0.01$.

of soil–structure interaction. For a harmonic point load acting perpendicularly to the free surface of an elastic half-space, 67% of the total energy is radiated by the Rayleigh wave and 26% and 7% by shear- and dilatational-waves, respectively [M3]. The phase velocity is independent of frequency for a homogeneous half-space. The test is thus not as stringent as for a layered system (see section titled "Love Wave").

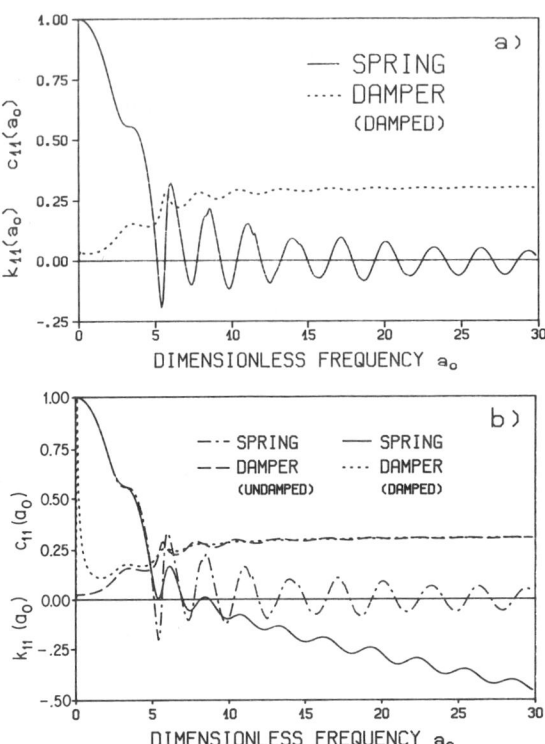

**Figure 3-33** Dynamic-stiffness coefficient for viscous boundary with hysteretic material damping. a. $\zeta = 0.001$. b. $\zeta = 0.05$.

The test is reproduced from Ref. [C4]. The mesh is shown in Fig. 3-34, whereby the free surface is located at $z = 0$, and the depth of the mesh, which is at rest initially, equals two wave lengths $\lambda$. The bottom boundary is fixed. The transmitting boundary is located at $x = 5\lambda/12$. Square finite elements with a side length $= \lambda/12$ are used. An explicit time-integration algorithm is selected. The displacements corresponding to the Rayleigh wave are prescribed on the model's left at $x = 0$. At first, they will generate transient waves, which, at later times will approach the steady-state condition. The horizontal and vertical displacements for a homogeneous half-space with $\nu = 0.25$ are specified as

$$u(z, t) = \left[ \exp\left(-5.324 \frac{z}{\lambda}\right) - 0.577 \exp\left(-2.471 \frac{z}{\lambda}\right) \right] \sin\omega t \quad (3.223a)$$

$$w(z, t) = \left[ -0.8475 \exp\left(-5.324 \frac{z}{\lambda}\right) + 1.4678 \exp\left(-2.471 \frac{z}{\lambda}\right) \right] \cos\omega t$$

$$(3.223b)$$

## Sec. 3.9  Location of Artificial Boundary

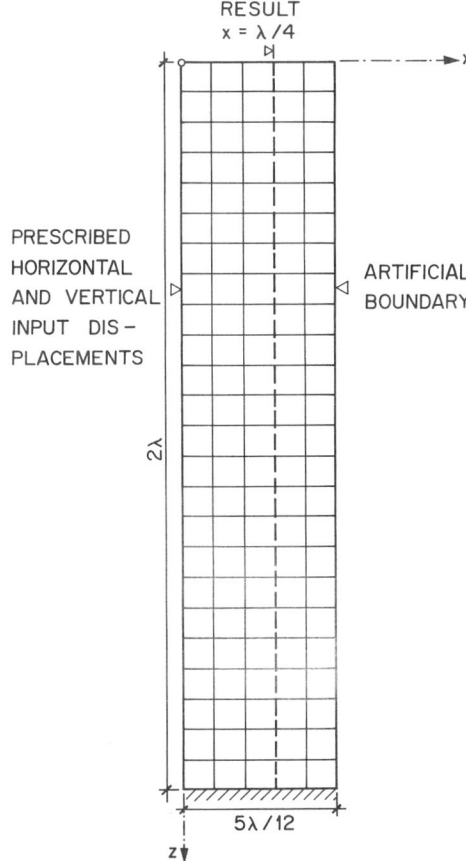

**Figure 3-34** Finite-element mesh with transmitting boundary used for Rayleigh-wave loading.

where the Rayleigh-wave length is denoted as $\lambda$ ($= 2\pi\, c_R/\omega$, with the Rayleigh-wave velocity $c_R = 0.920\, c_s$). $\omega$ is selected as $0.278\, s^{-1}$. The bracketed expressions of Eqs. 3.223a and 3.223b are shown as dashed lines in Figs. 3-35a and 3-35b.

The response consisting of the horizontal and vertical displacements is examined on the line $x = \lambda/4$ for $t = 27$ s. In the ideal case, the profile of these displacements will duplicate the prescribed displacements at $x = 0$. To be able to evaluate the influence of the transient waves and the discretization separately from the approximation of the transmitting boundaries, an analysis with an extended mesh (width = 20 elements) is also performed, for which the reflected waves will arrive at $x = \lambda/4$ only after the time investigated. The response is determined with the mesh shown in Fig. 3-34, using a free boundary, a viscous damper, and a paraxial element located at the artificial boundary. As can be seen from Fig. 3-35, the extended mesh and the two transient boundaries generate displacements which are similar to the pre-

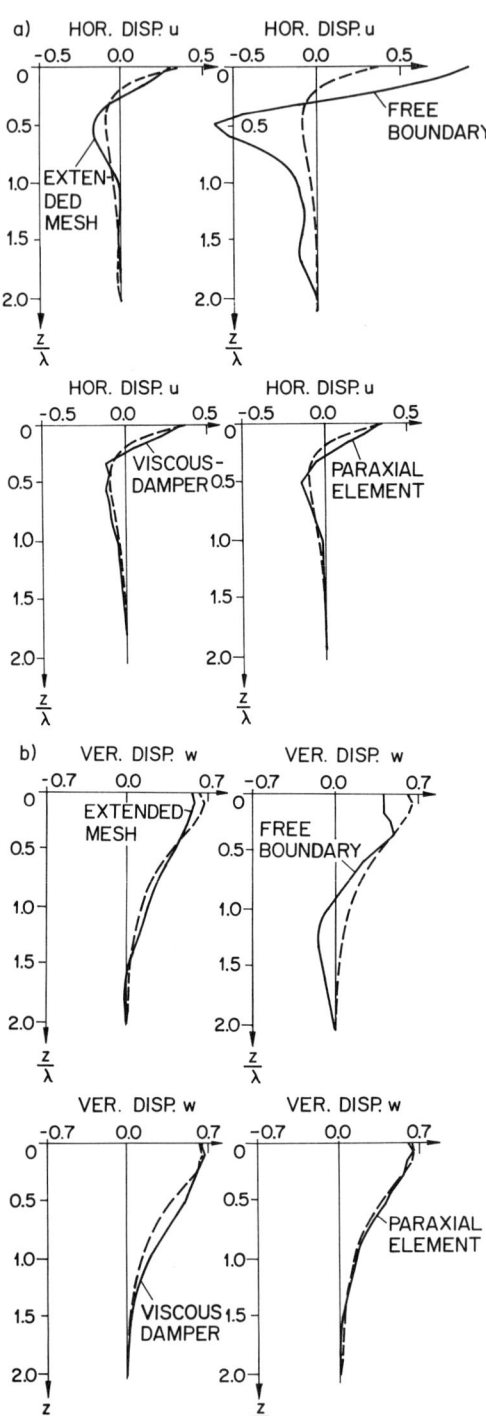

**Figure 3-35** Displacements for Rayleigh-wave loading. a. Horizontal. b. Vertical.

scribed input profile. The approximation introduced by the transmitting boundaries does not introduce large errors. In this case the viscous damper appears to function about as well as the paraxial element. The solution, based on the free boundary—an elementary boundary condition—is unacceptable.

## 3.10 FREE-FIELD LOADING

When the loading is introduced through the soil, the loading environment must be determined first (Section 1.4). It consists of the free-field response on the surface which subsequently coincides with the artificial boundary, or in the region adjacent to this surface which forms the boundary zone. The free-field response consists of the displacements and, for certain formulations of the transmitting boundaries, also of the surface tractions.

The equations developed in the preceding sections for the various transmitting boundaries still apply when a free-field motion is present. All variables now refer to the relative motion—that is, to the total motion minus the free-field one.

It is appropriate to use the viscous damper (Section 3.5) for illustration. For instance, in the boundary's normal direction, the boundary condition (Eq. 3.108) equals

$$\sigma(s) + \rho c_p \dot{u}(s) = 0 \qquad (3.224)$$

where $\sigma(s)$ and $\dot{u}(s)$ are the relative stress (surface traction) and relative velocity, respectively. The superscripts $t$ and $f$ denote the total and free-field quantities. Substituting

$$\sigma(s) = \sigma^t(s) - \sigma^f(s) \qquad (3.225)$$

$$\dot{u}(s) = \dot{u}^t(s) - \dot{u}^f(s) \qquad (3.226)$$

into Eq. 3.224 leads to the boundary condition formulated in total variables

$$\sigma^t(s) + \rho c_p \dot{u}^t(s) = \sigma^f(s) + \rho c_p \dot{u}^f(s) \qquad (3.227a)$$

or

$$\sigma^t(s) = \sigma^f(s) - \rho c_p (\dot{u}^t(s) - \dot{u}^f(s)) \qquad (3.227b)$$

Equation 3.227b expresses that the total surface traction at the artificial boundary equals the surface traction of the free field and that caused by the motion relative to the free field (equals the product of the damper coefficient and the difference between the total and free-field velocities). It follows from Eq. 3.227a that the loading consists of the sum of the free-field surface traction and the damper force determined with the free-field velocity. In discretized form (Fig. 3-14), the right-hand sides of the boundary conditions in total displacements in the normal and tangential directions are equal to $A(\sigma^f + \rho c_p \dot{u}^f)$ and $A(\tau^f + \rho c_s \dot{v}^f)$, respectively.

In the case of the doubly asymptotic approximation (Section 3.6), the right-hand side of the boundary condition in total displacements is equal to that of the viscous damper and the product of the matrix of the static-spring coefficients and the vector of the free-field displacements. For instance, for the benchmark problem, the boundary condition is formulated as (see also Eq. 3.114)

$$N^t + Ku^t + A\rho c_l \dot{u}^t = N^f + Ku^f + A\rho c_l \dot{u}^f \qquad (3.228)$$

where $N^f$ is the normal force of the free field at the artificial boundary.

For the paraxial boundary (Section 3.7), the same general procedure can be followed. For instance, substituting for the out-of-plane motion

$$v = v^t - v^f \qquad (3.229)$$

into Eq. 3.124 results in the inhomogeneous boundary condition in total displacements

$$\frac{1}{c_s}\dot{v}^t_{,x} - \frac{1}{2}v^t_{,zz} + \frac{\ddot{v}^t}{c_s^2} = \frac{1}{c_s}\dot{v}^f_{,x} - \frac{1}{2}v^f_{,zz} + \frac{\ddot{v}^f}{c_s^2} \qquad (3.230)$$

Analogously, for the extrapolation algorithm (Section 3.8), the extrapolation series is formulated as (Eq. 3.155)

$$u^t(l, t) = \sum_{j=1}^{N} (-1)^{j+1} C_j^N u^t(l - je, t - j\Delta t) + u^f(l, t) \\ - \sum_{j=1}^{N} (-1)^{j+1} C_j^N u^f(l - je, t - j\Delta t) \qquad (3.231)$$

Finally, also for the superposition boundary, the equations specified in Section 3.3 apply to the relative motion. For instance, Eq. 3.92 is transformed for total motion to

$$u^t_s(x, t) = f\left(t - \frac{x}{c}\right) - f\left(t + \frac{x}{c}\right) + g_s\left(t + \frac{x}{c}\right) + u^f(x, t) \qquad (3.232a)$$

$$u^t_a(x, t) = f\left(t - \frac{x}{c}\right) + f\left(t + \frac{x}{c}\right) - g_a\left(t + \frac{x}{c}\right) + u^f(x, t) \qquad (3.232b)$$

## 3.11 STRUCTURES WITH ELASTO-PLASTIC BASE ISOLATION PERMITTING UPLIFT AND SLIPPING

### 3.11.1 Computational Procedure

As mentioned in Section 3.6, springs and dampers with frequency-independent coefficients can be obtained by evaluating the dynamic-stiffness

## Sec. 3.11 Structures with Elasto-Plastic Base Isolation

coefficients of the unbounded soil at a specific frequency—for example, at the fundamental frequency of the structure–soil system. Placing the interaction horizon at the structure–soil interface and subdividing it into (boundary) elements, a far-coupled system of springs and dampers represents the unbounded soil. This approximate procedure is thus actually a substructure method. It is well suited to model the basemat's partial uplift, mentioned in Section 2.14, whereby slipping can also be taken into account. A rigid basemat at the surface of the layered half-space is addressed next (Fig. 3-36).

The soil's interaction–force displacement relationship is formulated as

$$\{R\} = [K](\{u^t\} - \{u^f\}) + [C](\{\dot{u}^t\} - \{\dot{u}^f\}) \tag{3.233}$$

$\{R\}$ is the vector of the interaction forces containing two horizontal and one vertical component for each (boundary) element (subdisk) into which the basemat–soil interface is subdivided. $\{u^t\}$ is the vector of the total displacements (two horizontal and one vertical, $w^t$) of the elements; and $\{u^f\}$ is the vector of the free-field motion. $[K]$ and $[C]$ denote the matrices of the spring and damping coefficients. It follows from Eq. 3.233 that the interaction forces are a function of the relative motion (total minus free field), as explained in Section 3.10. For the sake of clarity, in Fig. 3-36, only the vertical components of displacements and forces are presented.

The numerical procedure based on an explicit time-integration scheme with predicted values is examined next [W7]. For the sake of conciseness, it is convenient to describe the algorithm first for the case of no slipping. At time $t = n\Delta t$ (subscript $n$), separate treatment is given to the elements in contact, denoted with a subscript $c$, and those where a gap has formed, denoted with a subscript $u$ (for uplift). Partitioning the vectors and matrices in Eq. 3.233 correspondingly leads to

$$\left\{\begin{matrix}\{R_c\}\\ \{R_u\}\end{matrix}\right\}_n = \begin{bmatrix}[K_{cc}] & [K_{cu}]\\ [K_{uc}] & [K_{uu}]\end{bmatrix}_{n-1} \left(\left\{\begin{matrix}\{\bar{u}_c^t\}\\ \{u_u^t\}\end{matrix}\right\}_n - \left\{\begin{matrix}\{u_c^f\}\\ \{u_u^f\}\end{matrix}\right\}_n\right)$$

$$+ \begin{bmatrix}[C_{cc}] & [C_{cu}]\\ [C_{uc}] & [C_{uu}]\end{bmatrix}_{n-1} \left(\left\{\begin{matrix}\{\dot{\bar{u}}_c^t\}\\ \{\dot{u}_u^t\}\end{matrix}\right\}_n - \left\{\begin{matrix}\{\dot{u}_c^f\}\\ \{\dot{u}_u^f\}\end{matrix}\right\}_n\right) \tag{3.234}$$

where, for the elements in contact, the rigid-body constraints apply:

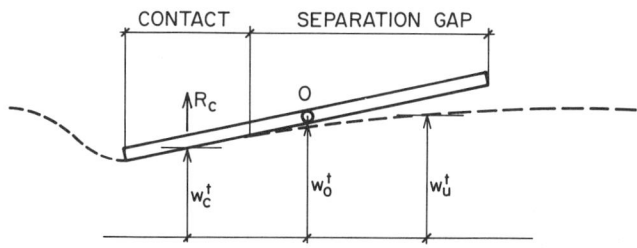

**Figure 3-36** Discretized structure–soil interface.

$$\{\bar{u}_c^t\}_n = [A]_{n-1} \{\bar{u}_o^t\}_n \tag{3.235a}$$

$$\{\tilde{u}_c^t\}_n = [A]_{n-1} \{\tilde{u}_o^t\}_n \tag{3.235b}$$

$\{\bar{u}_o^t\}_n$ denotes the rigid basemat's total (predicted) displacements at point 0 at time $n\Delta t$. $[A]_{n-1}$ represents the compatibility condition of the elements in contact along the basemat at time $(n - 1)\Delta t$; it depends on the coordinates of the elements only.

As the degrees of freedom of the elements without contact do not have any mass associated with them—thus not allowing the standard explicit algorithm to be used—a special treatment is necessary to calculate $\{u_u^t\}_n$ and $\{\ddot{u}_u^t\}_n$. As in the central-difference formula,

$$\{u_u^t\}_n = \{u_u^t\}_{n-1} + \frac{\Delta t}{2}(\{\dot{u}_u^t\}_{n-1} + \{\dot{u}_u^t\}_n) \tag{3.236}$$

is assumed. Substitution of Eq. 3.236 in Eq. 3.234 leads to

$$\begin{Bmatrix}\{R_c\}\\\{R_u\}\end{Bmatrix}_n = \begin{bmatrix}[K_{cc}] & [K_{cu}]\\[K_{uc}] & [K_{uu}]\end{bmatrix}_{n-1} \left(\begin{Bmatrix}\{\bar{u}_c^t\}_n\\\{u_u^t\}_{n-1} + \frac{\Delta t}{2}\{\dot{u}_u^t\}_{n-1}\end{Bmatrix} - \begin{Bmatrix}\{u_c^f\}\\\{u_u^f\}\end{Bmatrix}_n\right)$$

$$+ \begin{bmatrix}[C_{cc}] & [C_{cu}]\\[C_{uc}] & [C_{uu}]\end{bmatrix}_{n-1}\left(\begin{Bmatrix}\{\tilde{u}_c^t\}\\\{0\}\end{Bmatrix}_n - \begin{Bmatrix}\{\dot{u}_c^f\}\\\{\dot{u}_u^f\}\end{Bmatrix}_n\right)$$

$$+ \left(\frac{\Delta t}{2}\begin{bmatrix}[K_{cu}]\\[K_{uu}]\end{bmatrix}_{n-1} + \begin{bmatrix}[C_{cu}]\\[C_{uu}]\end{bmatrix}_{n-1}\right)\{\dot{u}_u^t\}_n \tag{3.237}$$

which can also be written as

$$\begin{Bmatrix}\{R_c\}\\\{R_u\}\end{Bmatrix}_n = \begin{Bmatrix}\{\bar{R}_c\}\\\{\bar{R}_u\}\end{Bmatrix}_n + \begin{bmatrix}[\bar{C}_{cu}]\\[\bar{C}_{uu}]\end{bmatrix}_{n-1}\{\dot{u}_u^t\}_n \tag{3.238}$$

The first vector on the right-hand side is known. As $\{R_u\}_n = 0$, a partial inversion results in

$$\begin{Bmatrix}\{R_c\}\\\{\dot{u}_u^t\}\end{Bmatrix}_n = \begin{Bmatrix}\{\bar{R}_c\}\\\{0\}\end{Bmatrix}_n - \begin{bmatrix}[\bar{C}_{cu}][\bar{C}_{uu}]^{-1}\\[\bar{C}_{uu}]^{-1}\end{bmatrix}_{n-1}\{\bar{R}_u\}_n \tag{3.239}$$

Problems are inherent in this partial inversion. These result from selecting frequency-independent springs and dampers—especially the choice of ω. In all calculations performed in this section, however, no problems are encountered.

In detail, the algorithm proceeds in the $n$th time step from $(n - 1)\Delta t$ to $n\Delta t$ as follows:

a. $\{\bar{u}_o^t\}_n$ and $\{\tilde{u}_o^t\}_n$ are predicted from the corresponding motion at time $(n - 1)\Delta t$ (Eq. 2.60).

Sec. 3.11   Structures with Elasto-Plastic Base Isolation   151

b. $\{\bar{u}_c^t\}_n$ and $\{\tilde{\ddot{u}}_c^t\}_n$ are calculated from Eq. 3.235.
c. $\{R_c\}_n$ and $\{\ddot{u}_u^t\}_n$ are determined from Eq. 3.239, whereby $\{\overline{R}_c\}_n$, $\{\overline{R}_u\}_n$, $[\overline{C}_{cu}]_{n-1}$, and $[\overline{C}_{uu}]_{n-1}$ are known.
d. $\{u_u^t\}_n$ is calculated from Eq. 3.236.
e. The elements are classified as follows:
   —In contact (subscript $c$): compression in $\{R_c\}_n$ and penetration from $\{u_u^t\}_n$.
   —With gap (subscript $u$): tension in $\{R_c\}_n$ and gap from $\{u_u^t\}_n$.
f. As a preparation for the next time step, $[A]_n$ and $[\overline{C}_{uu}]_n^{-1}$, $[\overline{C}_{cu}]_n$ are calculated with a partial inversion of the order of the degrees of freedom whose subscripts have changed.
g. $\{\ddot{u}_o^t\}_n$ is determined from the total dynamic system's equations of motion, with $\{R_c\}_n$ and the corresponding contribution from the structure.
h. The corrected values $\{u_o^t\}_n$ and $\{\dot{u}_o^t\}_n$ are calculated using $\{\ddot{u}_o^t\}_n$ (Eq. 2.62).

For slipping, the algorithm proceeds analogously. The degrees of freedom for which slipping occurs are included in $\{u_u^t\}_n$, and the corresponding elements in $\{R_u\}_n$ are equal to the vertical components multiplied by the friction coefficient.

The incorporation of dead load is straightforward (see also Section 5.4.3).

### 3.11.2 Reactor Building Permitting Uplift and Slipping

As an example, a reactor building with potential uplift and slipping is analyzed for horizontally propagating waves [W7]. The principal dimensions are specified in Fig. 3-37, whereby the base-isolation mechanism is omitted in this section. The structure with a rigid basemat is bonded to the surface of an elastic layered medium having a shear-wave velocity—compatible with the strain developed by the earthquake—varying between 230 m/s and 540 m/s. The friction coefficient between the concrete of the basemat and the soil equals 0.578.

The reactor building is modeled with four vertical beams, introducing 78 dynamic degrees of freedom (Fig. 3-37). The basemat–soil interface is discretized with 113 (boundary) elements. The order of the $\{u\}-$ and $\{u_o\}$ vectors, as described in Section 3.11.1, equals 339 and 6, respectively. The $[K]$- and $[C]$-matrices defined in Eq. 3.233 are calculated for $\omega = 2\pi \cdot 1.95$. 1.95 Hz represents the fundamental frequency of the structure–soil system in horizontal motion. The time step $\Delta t$ used in the explicit integration equals 1/600s.

The record of the 1971 San Fernando earthquake known as the J-145 record is used as the free-field motion. The two horizontal-acceleration

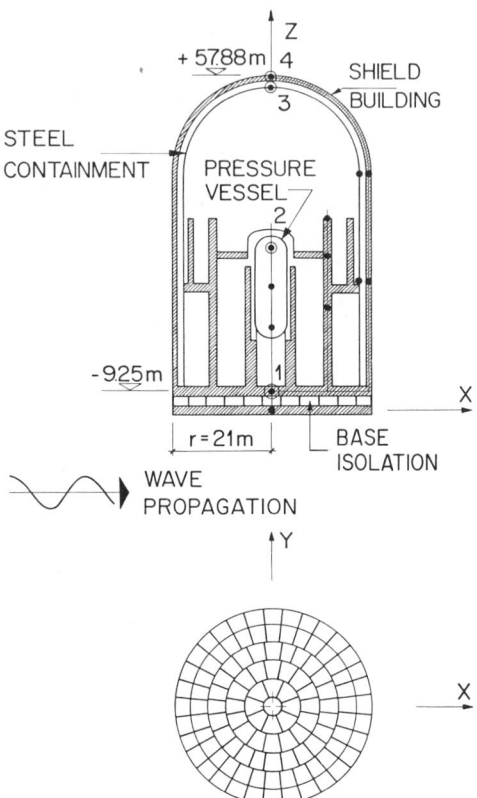

**Figure 3-37** Dynamic model of reactor building and discretization of basemat–soil interface.

records are projected in the radial direction ($\ddot{u}^f_{x\,\max} = 0.114$ g), running from the epicenter to the station, and in the tangential direction ($\ddot{u}^f_{y\,\max} = 0.094$ g). The radial direction is assumed to coincide with the $x$-axis. In the vertical direction, $\ddot{u}^f_{z\,\max} = 0.106$ g. The calculations involving traveling waves are performed for the $2 \times$ J-145 record, resulting in a realistic maximum acceleration for this safe-shutdown earthquake, which also leads to considerable uplift and slipping. When the vertical displacement is plotted versus the horizontal, the strong retrograde motion, which can be associated with Rayleigh waves, is apparent in Fig. 3-38. No experimental information on the apparent velocity of the horizontally propagating wave is available. For the alluvial soil conditions present at the site, an average apparent velocity $c_a$ for the Rayleigh and Love waves of between 200 m/s and 500 m/s might be appropriate. $c_a = 250$ m/s is selected. An apparent velocity $c_a = \infty$ corresponds to vertically incident waves.

The results of all analyses are summarized in the first four columns of Table 3-2. Besides the nonlinear case with uplift and slipping occurring for horizontally propagating waves (column 4), the linear cases for vertically incident waves (column 1) and for horizontally propagating waves (column

## Sec. 3.11 Structures with Elasto-Plastic Base Isolation

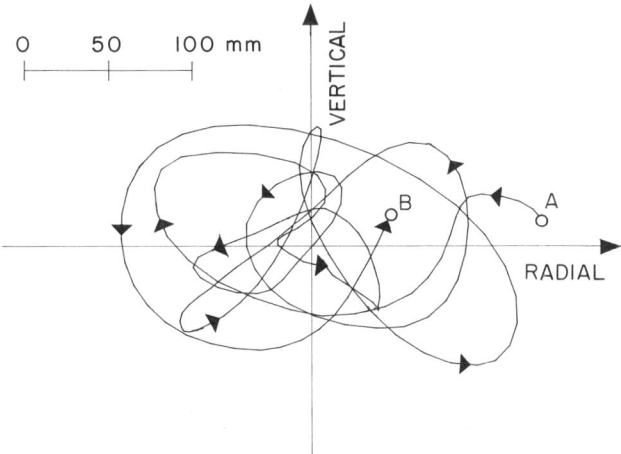

**Figure 3-38** Vertical versus radial displacement, San Fernando earthquake, corrected record J-145, 10-28 s.

2), as well as the nonlinear case for vertically incident waves (column 3), are presented. Maximum total accelerations in the indicated locations and the extreme total soil reactions exerted on the basemat are listed. The results of the analyses are systematically compared in the following.

As is apparent from the first and second columns of Table 3-2, the traveling wave's self-canceling effect results in a smaller response in all directions at the basemat and throughout the structure in the $y$-direction for points located on the axis. The propagating wave's additional rotational input leads to a considerable torsional response and to an increase in the $x$-direction for locations at a higher level. This also applies to the overturning moment and the base shear. The influence of uplift and slipping for vertically incident waves is discussed next. As one can discern from columns 1 and 3, the response in the $x$-direction is reduced when uplift occurs. Turning to the influence of uplift and slipping on the results for horizontally traveling waves, columns 2 and 4 are compared. Significant nonlinear effects are observed. The horizontal response in the $x$-direction is significantly reduced. The extreme vertical soil pressure is shown for horizontally propagating waves at $t = 5.1$ s in Fig. 3-39. Also indicated are the parts of the basemat where uplift and slipping occur.

### 3.11.3 Reactor Building with Elasto-Plastic Base Isolation

As another example, the same reactor building resting on soil as analyzed in Section 3.11.2, but with a base-isolation system (Fig 3-37), is examined

TABLE 3-2  Maximum Response of Reactor Building, 2 × J-145 Earthquake

| Base Isolation | | No | No | No | No | Yes |
|---|---|---|---|---|---|---|
| Uplift/Slipping | | No | No | Yes | Yes | No |
| Apparent Velocity $c_a$ [m/s] | | ∞ | 250 | ∞ | 250 | ∞ |
| Total Acceleration [g] | | | | | | |
| Top Shield Building   Point 4 | $\ddot{u}_z^t$ | 0.900 | 1.354 | 0.642 | 1.060 | 0.376 |
|  | $\ddot{u}_x^t$ | 0.776 | 0.586 | 0.768 | 0.659 | 0.320 |
| Top Steel Containment Point 3 | $\ddot{u}_z^t$ | 0.714 | 1.278 | 0.622 | 0.921 | 0.315 |
|  | $\ddot{u}_x^t$ | 0.770 | 0.492 | 0.769 | 0.725 | 0.284 |
| Top Pressure Vessel   Point 2 | $\ddot{u}_z^t$ | 1.038 | 1.082 | 0.910 | 1.020 | 0.390 |
|  | $\ddot{u}_x^t$ | 0.796 | 0.330 | 0.979 | 0.824 | 0.242 |
| Center Basemat       Point 1 | $\ddot{u}_z^t$ | 0.304 | 0.290 | 0.295 | 0.332 | 0.226 |
|  | $\ddot{u}_x^t$ | 0.284 | 0.162 | 0.272 | 0.264 | 0.202 |
| Overturning Moment [GNm] | y-rot | 9.26 | 14.38 | 6.56 | 8.08 | 3.81 |
| Shear Force at Base [GN] | x-dir | 0.300 | 0.416 | 0.211 | 0.289 | 0.121 |

### Sec. 3.11 Structures with Elasto-Plastic Base Isolation

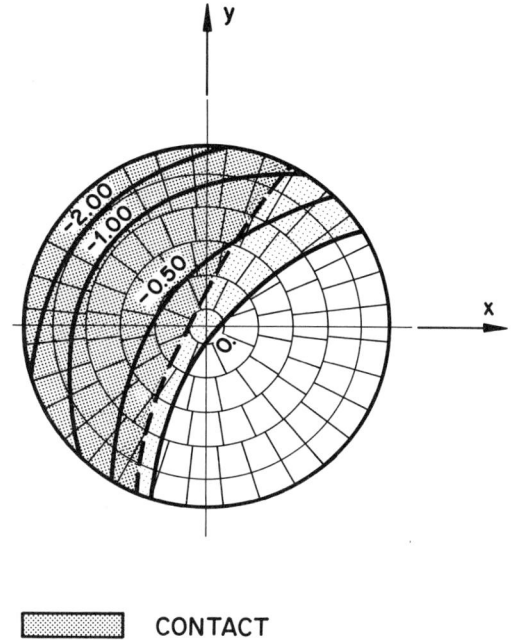

**Figure 3-39** Vertical soil pressure at 5.1 s, horizontally propagating waves [MN/m²].

CONTACT
SLIPPING
SEPARATION

[W8]. These aseismic bearings are located between the lower and upper rafts. Each individual bearing consists of a Neoprene pad, which is flexible in the horizontal direction only, and of a friction plate, which limits the horizontal force transmitted to the structure. This results in a nonlinear elasto-ideal plastic-isolation mechanism acting in the horizontal direction only. The corresponding fixed-base frequency of the base-isolation mechanism with a rigid structure equals (in the horizontal direction) 1 Hz. The plates' friction coefficient is selected as 0.17.

The modeling of the structure and of the basemat–soil interface is discussed in Section 3.11.2. The aseismic bearings with elasto-plastic behavior must be discretized. One uses the same discretization as for the basemat–soil interface (Fig. 3-37), adding (a maximum) 226 degrees of freedom (in the two horizontal directions) of the isolation mechanism's subregions to the 81 dynamic degrees of freedom (78 for the structure and 3 for the lower raft) present in the model of the reactor building with an elastic horizontal base isolation. The lowest natural frequency of the dynamic system equals 0.91 Hz.

Again, the J-145 record (Fig. 3-38), scaled by a factor of 2, is used as the earthquake excitation.

The dynamic response of the reactor building with the elasto-plastic base-isolation mechanism for vertically incident waves is summarized in the last column of Table 3-2. Compared to the corresponding values of the same structure without base isolation (column 1), a significant reduction occurs; in addition, the increase of the accelerations from the basemat to the top of the structure is reduced.

Finally, the time-history of the total displacement at the center of the basemat (point 1 in Fig. 3-37) is examined. In Figs. 3-40a and b, the two horizontal displacements $u_x$ (radial) and $u_y$ (tangential) are plotted. The total motion (solid line) consists of the plastic (dashed line) and elastic portions, the latter being limited by the friction coefficient of the plates. In general, each subregion of the discretized base-isolation mechanism exhibits

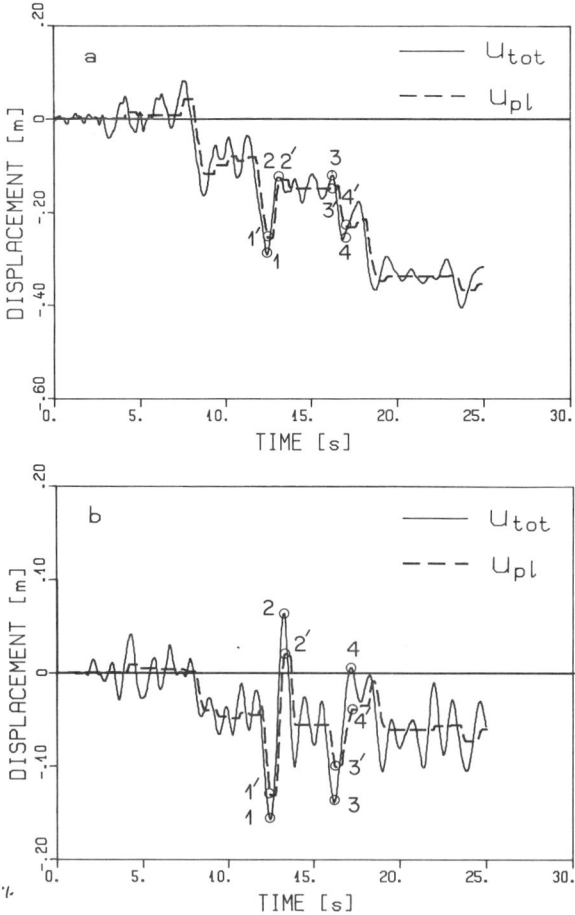

**Figure 3-40** Time history of displacement of basemat center of reactor building with elasto-plastic base isolation. a. $x$-direction. b. $y$-direction. c. $x$-$y$-plane

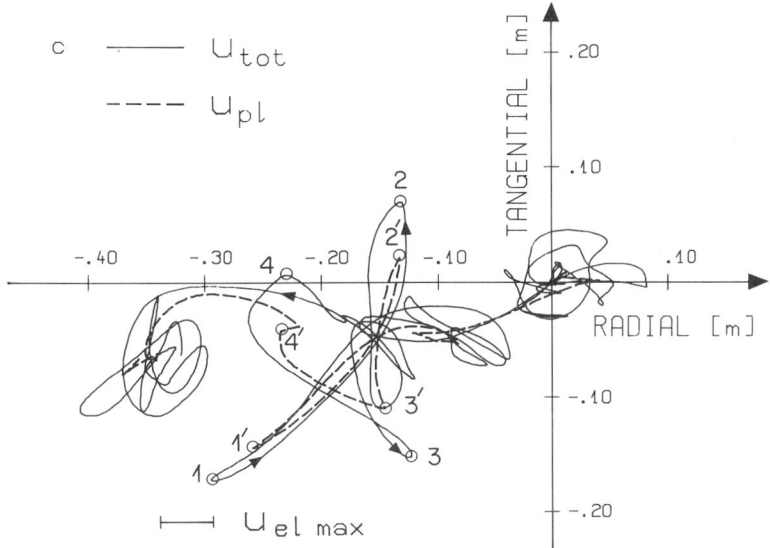

**Figure 3-40** Continued

a different horizontal friction force, because the vertical force is modified by the rocking and vertical motions. To more easily follow the motion in Fig. 3-40c, individual points in time have been assigned corresponding numbers. $u_{el\,max}$ represents the average maximum elastic displacement of all subregions.

## SUMMARY

1. Exact models, to be applied at the artificial boundary located at a finite distance from the vibrating source, exist for one-dimensional wave propagation towards infinity: for the prismatic rod, a damper; for the truncated conical rod in shear, a damper and a spring with the static coefficient; for the truncated conical rod in torsion and for the spherical symmetric wave, a spring with the static-coefficient and a mass with its own degree of freedom, which is connected to the artificial boundary with a damper (discrete model). All these coefficients are frequency-independent.
2. In the superposition-boundary concept, two overlapping narrow boundary zones with prescribed velocities or enforced forces on the artificial boundaries are attached to the main mesh. Through frequent averaging of the results before the reflected waves can reach the main mesh, the latter are canceled as they occur. This superposition also determines the prescribed velocities and forces applied for the following few time steps up to the next averaging. Because the procedure is based on

considerations of symmetry and antimetry with the corresponding amplitudes of the reflected waves which differ in sign, it is "rigorous" for all types of waves, including inclined body waves and surface waves. However, the mirror image (with respect to the artificial boundary) must be compatible with the unbounded domain on the exterior. A significant programming effort is necessary to incorporate the superposition boundary—which is well suited for an explicit time-integration scheme—into a general-purpose program.

3. Lumped viscous dampers, with the coefficients equal to the products of the applicable area, the mass density, and the dilational- and shear-wave velocities for the directions normal and tangential, respectively, to the artificial boundary, are exact transmitting boundaries for body waves impinging perpendicularly. The viscous dampers can be straightforwardly considered in a general-purpose program without any programming effort.

4. The doubly asymptotic approximation uses the (lumped) viscous dampers and the static springs with generally fully coupled coefficients. It is asymptotically exact at both high and low frequencies.

5. In the paraxial-boundary concept, a differential equation similar to that of the wave equation is constructed and enforced as a boundary condition, which only allows approximately outgoing waves to propagate. The incoming wave pattern can be separated from the outgoing one, and the corresponding dispersion relation involving the wave numbers can be expanded in a Taylor series, for which a differential equation can be identified. The paraxial boundary can be regarded as a generalization of the viscous damper, leading, for a transient excitation, to somewhat more accurate results. For body waves not impinging at a right angle upon the artificial boundary, and for surface waves, the paraxial boundary is an approximation. The implementation in a finite-element scheme is not straightforward, making it necessary to use special interface elements between the region governed by the wave equation and the paraxial one and possibly to use other concepts as well.

6. In the extrapolation algorithm used with an explicit time-integration scheme, the displacements of a node on the artificial boundary for the next time step are extrapolated from data at earlier times along a line normal to the boundary in the interior. The wave propagation in one direction only can thus be enforced, whereby some weighted average of the apparent velocities in the direction of the normal is selected for a transient excitation. As for the viscous damper and the paraxial boundary, the apparent velocity in the direction of the normal is the decisive parameter. The accuracy can be increased by selecting either

a higher order extrapolation or a smaller distance between the data points (finer mesh).
7. For body waves, the farther away from the vibrating source the artificial boundary is placed, the more accurate the transmitting boundary is. In the far field the equations of motion asymptotically describe the one-dimensional wave propagations in the direction normal to the artificial boundary, with the dilatational-wave velocity, and in the tangential directions, with the shear-wave velocity. For surface waves, some caution is advisable. When the apparent velocity in the direction perpendicular to the artificial boundary does not vary strongly and is comparatively close to the velocity on which the transmitting boundary is based, acceptable results are calculated using the transmitting boundary. For example, this is the case for Rayleigh waves in a homogeneous half-space. When the apparent velocity varies significantly—for example, in the case of the modes of the surface waves of a layer built-in at its base—the viscous damper can lead to unacceptable errors. The paraxial boundary and the extrapolation algorithm with at least a linear extrapolation and a fine mesh perform quite well. For surface waves, placing the artificial boundary farther away from the vibrating source does not improve the accuracy for the viscous damper, the paraxial boundary, and the extrapolation algorithm.
8. The transmitting boundaries can also be used when the loading is introduced through the soil. The same equations still apply, whereby the variables refer to the relative motion—that is, to the total motion minus the free-field motion. The equations of motion can be formulated in total displacements by straightforward substitution. For instance, for the viscous damper, the loading consists of the free field's surface traction and the damper force determined with the free-field velocity.
9. Discretizing the structure–soil interface into (boundary) elements and evaluating the dynamic-stiffness coefficients of the soil at a specific frequency leads to a far-coupled system of springs and dampers. This approximate procedure can be used to model the basemat's partial uplift and slipping and to analyze a structure with an elasto-plastic base-isolation mechanism, taking into account the interaction effects with the soil. The computational procedure divides the elements into two groups: those which are in contact and those where a gap has formed or slipping has occurred. After checking the conditions for these two states, a partial inversion of the order of the degrees of freedom which have changed groups is performed for each time step. The change in the response arising from uplift and slipping is opposite to that caused by horizontally propagating waves, when compared with the result of a linear analysis for vertical incidence.

## PROBLEMS

**3.1** A rigid massless disk of radius $a$ resting on the free surface of a half-space with shear modulus $G$ and shear velocity $c_s$ is subjected to a prescribed torsional motion (Fig. P 3-1a)

$$\varphi_o(\bar{t}) = \frac{\varphi_o}{2}\left[1 - \cos\left(2\pi \frac{\bar{t}}{\bar{t}_o}\right)\right] \quad 0 < \bar{t} < \bar{t}_o$$

$$\varphi_o(\bar{t}) = 0 \quad \bar{t} > \bar{t}_o$$

with the dimensionless time $\bar{t} = tc_s/a$ and $\bar{t}_o = 2$. Using the discrete model, with an additional degree of freedom $\varphi_1$ corresponding to the conical rod (without curve fitting), calculate the reaction force $P_o(\bar{t})$ and the rotation $\varphi_1(\bar{t})$ in the time domain.

*Result*

With $K = 16\, Ga^3/3$ (Table 2-2), $\gamma_1 = 3\pi/32$, and $\mu_1 = (9\pi/32)^2/3$ (Section 3.1.4), $M_1$ and $C_1$ follow from Eq. 2.33. Using the explicit time-integration algorithm with $\Delta t$

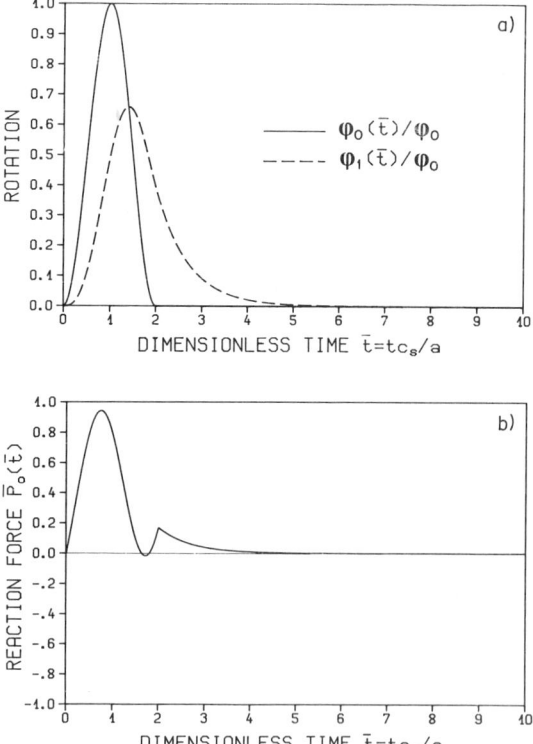

**Figure P 3-1** Response of discrete model representing disk in torsion on half-space. a. Rotation. b. Reaction force.

= 0.05, $\overline{P}_o(\bar{t}) = P_o(\bar{t})/(K\varphi_o)$ and $\varphi_1(\bar{t})/\varphi_o$ are calculated (Fig. P 3-1). Note that $\varphi_1(\bar{t})$ (and also $P_o(\bar{t})$) are not zero for $\bar{t} > 2$.

**3.2** Analytically verify, in the frequency domain, that the superposition-boundary concept eliminates all reflections when a P-wave impinges on an artificial boundary.

*Solution*

In Fig. P 3-2, the incident P-wave with an amplitude $A_P$ propagates with an angle of incidence $\psi_P$ towards the boundary $z = 0$. $B_P$ denotes the amplitude of the reflected P-wave, and $B_{SV}$ denotes that of the reflected SV-wave propagating with an angle of incidence $\psi_{SV}$. The horizontal and vertical in-plane amplitudes of the displacements for $z = 0$ are specified as ([W11], Eq. 4.26)

$$u(x) = (\cos\psi_P A_P + \cos\psi_P B_P - \sin\psi_{SV} B_{SV}) \exp(-i\omega x \cos\psi_P/c_p)$$

$$w(x) = (\sin\psi_P A_P - \sin\psi_P B_P - \cos\psi_{SV} B_{SV}) \exp(-i\omega x \cos\psi_P/c_p)$$

and those of the stresses for $z = 0$

$$\sigma(x) = \left[\frac{i\omega}{c_p} \cos^2\psi_P (1 - \tan^2\psi_{SV}) G(A_P + B_P)\right.$$

$$\left. - \frac{i2\omega}{c_s} \sin\psi_{SV} \cos\psi_{SV} G B_{SV}\right] \exp(-i\omega x \cos\psi_P/c_p)$$

$$\tau(x) = \left[\frac{i2\omega}{c_p} \sin\psi_P \cos\psi_P G(-A_P + B_P)\right.$$

$$\left. + \frac{i\omega}{c_s} \cos^2\psi_{SV} (1 - \tan^2\psi_{SV}) G B_{SV}\right] \exp(-i\omega x \cos\psi_P/c_p)$$

whereby $c_p$ and $c_s$ are the dilatational- and shear-wave velocities, and $G$ denotes the shear modulus. The frequency is omitted as argument. Snell's law applies:

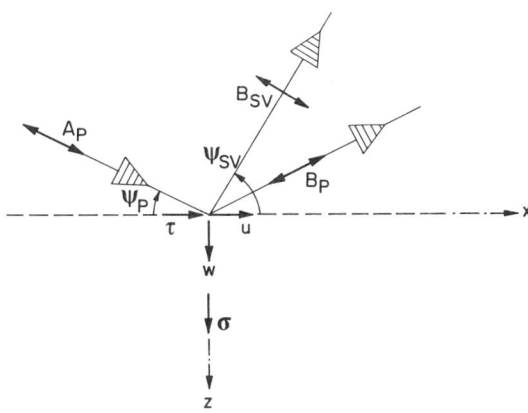

**Figure P 3-2** Nomenclature for in-plane motion.

$$\frac{c_p}{\cos\psi_p} = \frac{c_s}{\cos\psi_{SV}}$$

The boundary condition for the symmetric zone equals

$$w(x) = 0$$
$$\tau(x) = 0$$

which leads to

$$B_P = A_P$$
$$B_{SV} = 0$$

that is, to a reflected P-wave in phase with the incident one and no reflected SV-wave.

The antimetric case

$$u(x) = 0$$
$$\sigma(x) = 0$$

results in

$$B_P = -A_P$$
$$B_{SV} = 0$$

that is, in a reflected P-wave out of phase with the incident one and no reflected SV-wave.

The averaging of the two solutions leads to

$$B_P = B_{SV} = 0$$

that is, to no reflections occurring.

Analogously, it can be shown that an incident SV-wave leads to two reflected SV-waves which cancel each other when superimposed and to no P-wave.

**3.3** Sketch the set of boundary conditions of the superposition boundary to be enforced at the two sides which meet at a corner to eliminate all reflections arising from an incident P-wave [S3] (see also Problem 3.2).

*Result*

As shown in Fig. P 3-3, four solutions are required. After enforcing the symmetric and antimetric boundary conditions (first two parts), the secondary reflections' amplitudes are of the same sign, so that another two must be added. The four solutions are all possible combinations of symmetric and antimetric boundary conditions on the two sides. Note that the same set also eliminates all reflections arising from an incident SV-wave.

**3.4** Averaging the displacements in the two boundary zones before the time the error function propagating in the direction of the reflected waves reaches the main mesh, leads to the incident waves only and thus to the exact solution. It is worthwhile to examine the wave pattern which develops when this superposition is not performed

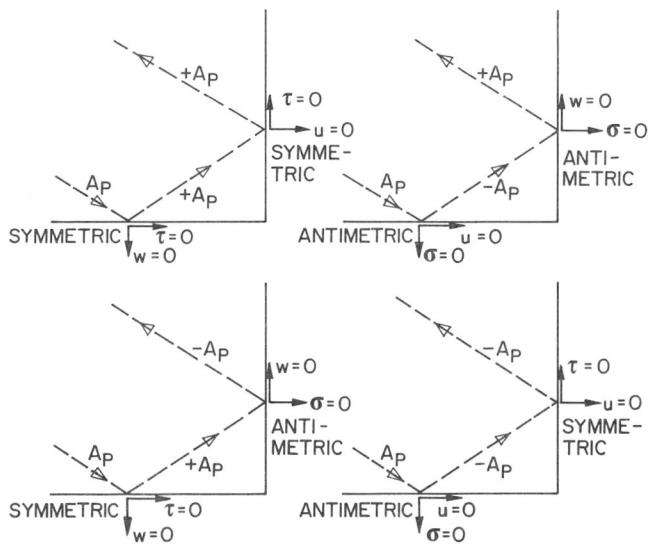

**Figure P 3-3** Four solutions necessary to cancel all reflections at corner.

[D2, K2]. The error functions $+e(t + x/c)$ and $-e(t + x/c)$ propagate in the symmetric and antimetric boundary zones, respectively, towards Point A (Eq. 3.106, Fig. 3-12). Because of the opposite signs, this point acts as a fixed boundary, leading to reflections, with the same waves propagating towards the artificial boundaries with opposite signs. This new pattern, consisting of $-e(t - x/c)$ and $+e(t - x/c)$ in the symmetric and antimetric boundary zones, respectively, represents the new incident waves (disregarding a time shift $l/c$, where $l$ is the distance of $A$ from the artificial boundary, Fig. 3-12) propagating towards the artificial boundaries. Study the wave pattern after the subsequent reflections at the artificial boundaries occur.

*Solution*

Analogously to Eq. 3.92, the wave patterns after the reflections at the artificial boundaries are specified for the symmetric and antimetric boundary zones as (time shift $2l/c$)

$$u_s(x, t) = -e\left(t - \frac{x}{c}\right) + e\left(t + \frac{x}{c}\right) - h_s\left(t + \frac{x}{c}\right)$$

$$u_a(x, t) = e\left(t - \frac{x}{c}\right) + e\left(t + \frac{x}{c}\right) - h_a\left(t + \frac{x}{c}\right)$$

The first terms on the right-hand side of these two equations represent the incident waves; the second represent the reflected waves for homogeneous boundary conditions; and the third represent the generated waves for non-zero boundary conditions (taking the signs of the incident waves into account). The velocities and the forces at the fixed and free boundaries are formulated as

$$\bar{u}_s(t) = -h'_s(t)$$

$$P_s(t) = \frac{EA}{c}[2e'(t) - h'_s(t)]$$

$$\dot{u}_a(t) = 2e'(t) - h'_a(t)$$

$$\overline{P}_a(t) = -\frac{EA}{c}h'_a(t)$$

Substituting into the forward-difference approximation (Eq. 3.98) leads to

$$h'_s(t) = \frac{1}{2}[h'_s(t - \Delta t) - 2e'(t - \Delta t) + h'_a(t - \Delta t)]$$

$$h'_a(t) = \frac{1}{2}[h'_a(t - \Delta t) - 2e'(t - \Delta t) + h'_s(t - \Delta t)]$$

and after observing $h'_s(t) = h'_a(t)$ and substituting Eq. 3.105

$$h'_s(t) - h'_s(t - \Delta t) = -e'(t - \Delta t) = f'(t - \Delta t) - f'(t - 2\Delta t)$$

results. Thus with homogeneous boundary conditions

$$h_s(t) = h_a(t) = f(t - \Delta t)$$

holds. The wave pattern is thus formulated as

$$u_s(x, t) = -f\left(t - \Delta t - \frac{x}{c}\right) + f\left(t - \frac{x}{c}\right) - f\left(t + \frac{x}{c}\right)$$

$$u_a(x, t) = f\left(t - \Delta t - \frac{x}{c}\right) - f\left(t - \frac{x}{c}\right) - f\left(t + \frac{x}{c}\right)$$

The average equals

$$u(x, t) = -f\left(t + \frac{x}{c}\right)$$

By not superimposing the solutions after the first reflections have occurred, the original incident wave $f$ is thus regenerated (with a change in sign) at the artificial boundaries when the second reflections occur. The error function, however small in magnitude, carries all the information necessary. These multiple reflections at the same boundary also cause the procedure of Section 3.3.1 to fail, where the final solutions are averaged.

**3.5** The semi-infinite rod with exponentially increasing area (Fig. P 3-5a)

$$A(x) = A_o \exp\left(\frac{x}{f}\right)$$

is a dispersive system with a cutoff frequency [W11]. The length $f$ represents the length at which the area equals $A_o \exp(1)$. $E$ is the modulus of elasticity and $\rho$ the mass density.

**a.** Formulate the equation of motion for longitudinal waves and solve it in the frequency domain, introducing the dimensionless frequency $a_o = \omega f/c_l$ (rod velocity of prismatic bar $c_l = \sqrt{E/\rho}$).

**b.** Determine the ratio of the amplitudes of the reflected and incident waves $\left|\frac{b}{a}\right|$ for

an artificial boundary at $x = l$, consisting of a superposition boundary. Plot $\left|\dfrac{b}{a}\right|$ as a function of $a_o$ for $l = 0$, $l = f$, and $l = 2f$.
c. Repeat (b) for the viscous damper.
d. Repeat (b) for the doubly asymptotic approximation.
e. Determine the first and second approximations, in the form of differential equations of the paraxial boundary, for the frequency range below and above the cutoff frequency. Plot $\left|\dfrac{b}{a}\right|$ versus $a_o$ for $l = f$.

*Solution*

a. Formulating the equilibrium equation of the infinitesimal element (Fig. P 3-5b)

$$N_x \, dx - \rho A \, dx \, \ddot{u} = 0$$

and substituting the normal force–displacement relationship

$$N = EAu_{,x}$$

leads to the equation of motion (Eq. 3.23a)

$$u_{,xx} + \frac{A_{,x}}{A} u_{,x} - \frac{\ddot{u}}{c_l^2} = 0$$

or

$$u_{,xx} + \frac{1}{f} u_{,x} - \frac{\ddot{u}}{c_l^2} = 0$$

A solution of the form $u = u(\omega) \exp(i\omega t)$ with $u(\omega) = \exp(i\gamma x)$ is considered, resulting in

$$\gamma^2 - \frac{i\gamma}{f} - \frac{a_o^2}{f^2} = 0$$

$$\gamma_1 = \frac{i}{2f} - \frac{1}{2f} \sqrt{4a_o^2 - 1}$$

$$\gamma_2 = \frac{i}{2f} + \frac{1}{2f} \sqrt{4a_o^2 - 1}$$

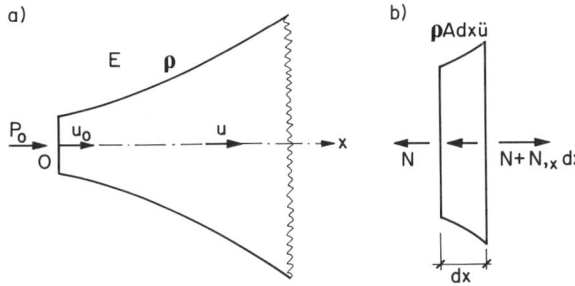

**Figure P 3-5** Semi-infinite rod with exponentially increasing area. a. System. b. Infinitesimal element. c. Superposition boundary. d. Viscous damper. e. Doubly asymptotic approximation. f. Paraxial boundary.

The solution can be formulated as

$$u(\omega) = a \exp\left(-\frac{x}{2f}\right) \exp\left(-\frac{i\omega x}{c}\right) + b \exp\left(-\frac{x}{2f}\right) \exp\left(\frac{i\omega x}{c}\right)$$

with the phase velocity

$$c = \frac{2a_o c_l}{\sqrt{4a_o^2 - 1}}$$

$c$ is imaginary for $a_o < 0.5$, the cutoff frequency. The displacement decays exponentially

$$u(\omega) = a \exp\left(-\frac{x}{2f}\right) \exp\left(-\frac{\sqrt{1 - 4a_o^2}\, x}{2f}\right) + b \exp\left(-\frac{x}{2f}\right) \exp\left(+\frac{\sqrt{1 - 4a_o^2}\, x}{2f}\right)$$

For $a_o > 0.5$, waves propagate in the dispersive system, with $a$ denoting the amplitude of the outgoing wave and $b$ that of the incoming wave.

**b.** Superposition boundary at $x = l$. Symmetric boundary condition $u = 0$:

$$\frac{b}{a} = -\exp\left(-\sqrt{1 - 4a_o^2}\,\frac{l}{f}\right)$$

Antimetric boundary condition $N = 0$:

$$\frac{b}{a} = -\frac{1 + \sqrt{1 - 4a_o^2}}{1 - \sqrt{1 - 4a_o^2}} \exp\left(-\sqrt{1 - 4a_o^2}\,\frac{l}{f}\right)$$

Averaging:

$$\frac{b}{a} = -\frac{1}{2}\left(1 + \frac{1 + \sqrt{1 - 4a_o^2}}{1 - \sqrt{1 - 4a_o^2}}\right) \exp\left(-\sqrt{1 - 4a_o^2}\,\frac{l}{f}\right)$$

$\left|\dfrac{b}{a}\right|$ is plotted in Fig. P 3-5c. For $a_o > 0.5$, $\left|\dfrac{b}{a}\right|$ is independent of $l/f$.

**c.** Viscous damper at $x = l$:

$$u_{,x} + \frac{\dot{u}}{c_l} = 0$$

$$\frac{b}{a} = -\frac{1 - 2a_o i + \sqrt{1 - 4a_o^2}}{1 - 2a_o i - \sqrt{1 - 4a_o^2}} \exp\left(-\sqrt{1 - 4a_o^2}\,\frac{l}{f}\right)$$

$\left|\dfrac{b}{a}\right|$ is plotted in Fig. P 3-5d.

**d.** Doubly asymptotic approximation:

$$u_{,x} + \frac{u}{f} + \frac{\dot{u}}{c_l} = 0$$

$$\frac{b}{a} = -\frac{1 + 2a_o i - \sqrt{1 - 4a_o^2}}{1 + 2a_o i + \sqrt{1 - 4a_o^2}} \exp\left(-\sqrt{1 - 4a_o^2}\,\frac{l}{f}\right)$$

$\left|\dfrac{b}{a}\right|$ is plotted in Fig. P 3-5e.

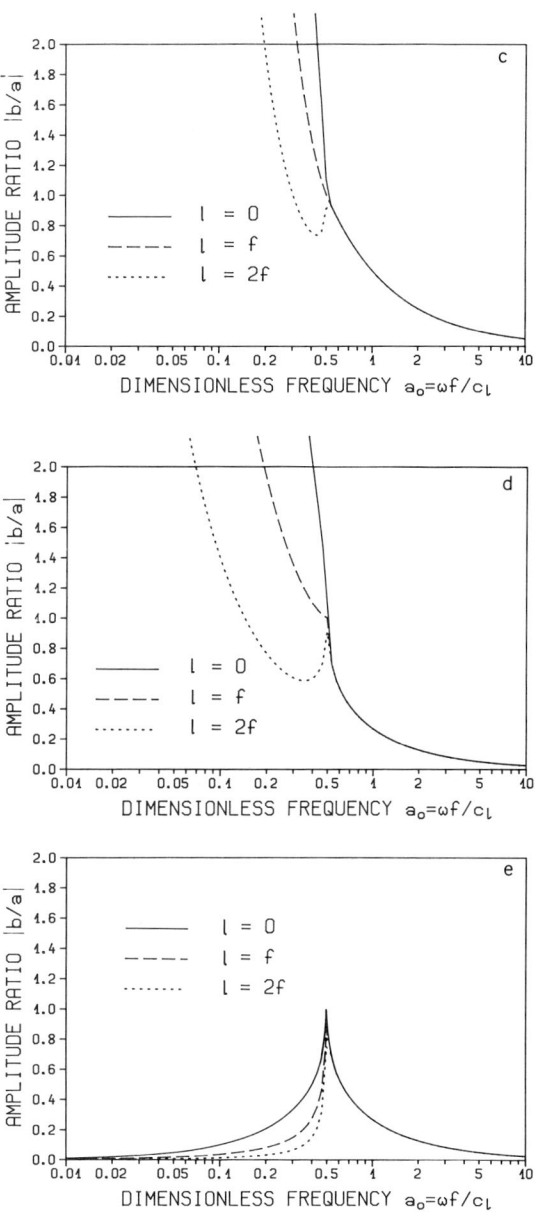

**Figure P 3-5** Continued

**e.** Paraxial boundary:

$$(\gamma - \gamma_1)(\gamma - \gamma_2) = 0$$

Outgoing waves $\gamma - \gamma_1 = 0$:

$\underline{a_o < 0.5}$

$$\gamma - \frac{i}{2f} - \frac{i}{2f}\sqrt{1 - 4a_o^2} = 0$$

Taylor series expansion:

$$\gamma - \frac{i}{2f} - \frac{i}{2f}(1 - 2a_o^2) = 0$$

First approximation:

$$\gamma - \frac{i}{f} = 0$$

Corresponding differential equation:

$$u_x + \frac{u}{f} = 0$$

$$\frac{b}{a} = -\frac{1 - \sqrt{1 - 4a_o^2}}{1 + \sqrt{1 - 4a_o^2}} \exp\left(-\sqrt{1 - 4a_o^2}\,\frac{l}{f}\right)$$

Second approximation:

$$\gamma - \frac{i}{f} + \frac{a_o^2 i}{f} = 0$$

Corresponding differential equation:

$$u_x + \frac{u}{f} + \frac{f}{c_l^2}\ddot{u} = 0$$

$$\frac{b}{a} = -\frac{1 - 2a_o^2 - \sqrt{1 - 4a_o^2}}{1 - 2a_o^2 + \sqrt{1 - 4a_o^2}} \exp\left(-\sqrt{1 - 4a_o^2}\,\frac{l}{f}\right)$$

$\underline{a_o > 0.5}$

$$\gamma - \frac{i}{2f} + \frac{a_o}{f}\sqrt{1 - \frac{1}{4a_o^2}} = 0$$

Taylor series expansion:

$$\gamma - \frac{i}{2f} + \frac{a_o}{f}\left(1 - \frac{1}{8a_o^2}\right) = 0$$

First approximation:

$$\gamma - \frac{i}{2f} + \frac{\omega}{c_l} = 0$$

## Chap. 3 Problems

Corresponding differential equation:

$$u_x + \frac{u}{2f} + \frac{\dot{u}}{c_l} = 0$$

$$\frac{b}{a} = -\frac{2a_o - \sqrt{4a_o^2 - 1}}{2a_o + \sqrt{4a_o^2 - 1}} \exp\left(-i\sqrt{4a_o^2 - 1}\,\frac{l}{f}\right)$$

Second approximation:

$$-\gamma\omega + i\frac{\omega}{2f} - \frac{\omega^2}{c_l} + \frac{c_l}{8f^2} = 0$$

Corresponding differential equation:

$$\dot{u}_x + \frac{\dot{u}}{2f} + \frac{\ddot{u}}{c_l} + \frac{c_l}{8f^2} u = 0$$

$$\frac{b}{a} = -\frac{2a_o - \sqrt{4a_o^2 - 1} - \dfrac{1}{4a_o}}{2a_o + \sqrt{4a_o^2 - 1} - \dfrac{1}{4a_o}} \exp\left(-i\sqrt{4a_o^2 - 1}\,\frac{l}{f}\right)$$

$\left|\dfrac{b}{a}\right|$ is plotted in Fig. P 3-5f (first approximations: dashed and solid lines; second approximations: dotted and dash-dotted lines).

It is worth mentioning that the paraxial boundary's differential equation can also be derived starting with the force-displacement relationship which involves the dynamic-stiffness coefficient $S(a_o)$.

$$P_o(a_o) = S(a_o)u_o(a_o)$$

with (Problem 6.5)

$$S(a_o) = \frac{EA_o}{f}\frac{1}{2}(1 + \sqrt{1 - 4a_o^2})$$

Substituting

$$P_o(a_o) = -EA_o u_{,x}(a_o)$$

and expanding $\sqrt{1 - 4a_o^2}$ into a Taylor series leads to the differential equation.

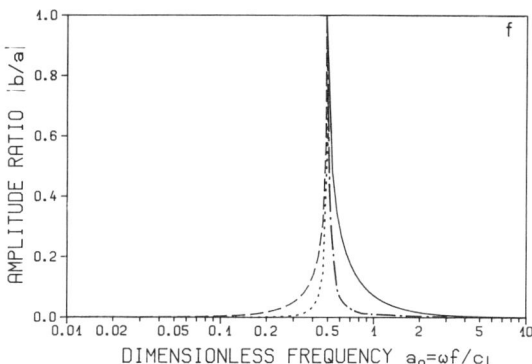

Figure P 3-5 Continued

**3.6** An incident wave of the out-of-plane motion $v(x, z)$ with the amplitude $A$ impinges upon the artificial boundary $x = 0$ with an angle $\psi$ (Fig. 3-17). With $B$ denoting the reflected wave's amplitude, determine $\left|\dfrac{B}{A}\right|$ for the extrapolation algorithm with $N$ terms for a frequency $\omega$. $e$ denotes the distance of the sampling points on the normal to the boundary, and $c_s$ denotes the shear-wave velocity. Assuming $e = c_s \Delta t$, and for 10 sampling points per wave length ($10e = 2\pi c_s/\omega$), plot $\left|\dfrac{B}{A}\right|$ versus $\psi$ for $N = 2$ and compare this with the corresponding values of the viscous damper (Eq. 3.131) and of the paraxial boundary (Eq. 3.130) [L1]. Calculate $\left|\dfrac{B}{A}\right|$ for $\psi = 0°$ selecting $N = 1, 2,$ and 3.

*Solution*

The out-of-plane motion is specified as (Eq. 3.129)

$$v(x,z,t) = [A \exp(-i\omega \sin\psi\, x/c_s) + B \exp(i\omega \sin\psi\, x/c_s)] \exp(-i\omega \cos\psi\, z/c_s) \exp(i\omega t)$$

Equation 3.155 is formulated as

$$(A + B) \exp(i\omega t) = \sum_{j=1}^{N} (-1)^{j+1} C_j^N A \exp(i\omega \sin\psi\, j\, e/c_s) \exp[i\omega(t - j\Delta t)]$$

whereby it is assumed that only the incident wave appears on the right-hand side (Case 1 of Section 3.8.1). Simplifying leads to

$$\left|\frac{B}{A}\right| = 2^{N/2} [1 - \cos(\omega(\sin\psi \frac{e}{c_s} - \Delta t))]^{N/2}$$

or

$$\left|\frac{B}{A}\right| = 2^{N/2} [1 - \cos(\omega \Delta t(\sin\psi \frac{e}{c_s \Delta t} - 1))]^{N/2}$$

Decreasing $\Delta t$ leads to smaller reflections.

Setting $N = 2$, $e = c_s \Delta t$, and $\omega = 2\pi c_s/(10e)$ results in the curve shown in Fig. P 3-6. Note that, in contrast to the viscous and paraxial boundaries, $B$ is considerably smaller than $A$ for glazing waves ($\psi \to 0°$). $\left|\dfrac{B}{A}\right|$ equals 0.62, 0.38, and 0.24 for $N = 1, 2,$ and 3, respectively for $\psi = 0°$.

**3.7** One-dimensional cylindrical wave propagation with radial symmetry occurs, for instance, when the wall of a circular cavity embedded in a full plane is loaded uniformly. For a circular artificial boundary at the radius $R$ (Fig. P 3-7a), determine the ratio of the amplitudes of the reflected and the incident waves $b/a$ as a function of $kR$ ($k = \omega/c_p$). Examine the superposition boundary, the viscous damper, and the extrapolation algorithm (Case 1). Plot $\left|\dfrac{b}{a}\right|$ versus $kR$ for $\nu = 1/3$, assuming for the extrapolation algorithm $e = c_p \Delta t$, $N = 2$, and $e/R = 0.2$ ($e$ = distance between data points).

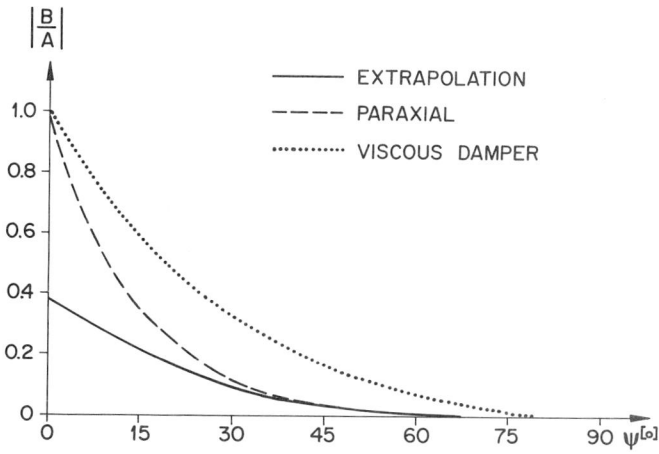

**Figure P 3-6** Reflected wave for inclined SH-wave.

*Solution*

Only a radial displacement with an amplitude $u(\omega)$ arises. The radial coordinate $r$ is the independent variable. The stress–displacement relationships in cylindrical coordinates

$$\sigma_r(\omega) = (\lambda + 2G)u(\omega)_{,r} + \lambda \frac{u(\omega)}{r}$$

$$\sigma_\theta(\omega) = \lambda u(\omega)_{,r} + (\lambda + 2G)\frac{u(\omega)}{r}$$

substituted into the equilibrium equation in the radial direction

$$\sigma_r(\omega)_{,r} + \frac{\sigma_r(\omega) - \sigma_\theta(\omega)}{r} + \rho\omega^2 u(r) = 0$$

leads to the equation of motion:

$$r^2 u(\omega)_{,rr} + r u(\omega)_{,r} + (k^2 r^2 - 1)u(\omega) = 0$$

This differential equation is the Bessel equation of order 1 with the parameter $k$. The solution is specified as

$$u(\omega) = a H_1^{(2)}(kr) + b H_1^{(1)}(kr)$$

where $H_1^{(2)}$ and $H_1^{(1)}$ are the first-order Hankel functions of the second and first kind, respectively. Examining this equation for large arguments

$$H_1^{(2)}(kr) \sim \sqrt{\frac{2}{\pi k r}} \exp\left[-i\left(kr - \frac{3\pi}{4}\right)\right]$$

$$H_1^{(1)}(kr) \sim \sqrt{\frac{2}{\pi k r}} \exp\left[i\left(kr - \frac{3\pi}{4}\right)\right]$$

shows that the first and second terms on the right-hand side represent waves propagating in the positive and negative $r$-directions, respectively. $a$ and $b$ can thus be interpreted as the amplitudes of the incident and reflected waves, respectively.

The superposition boundary averages the results for $u(R,\omega) = 0$ and for $\sigma_r(R,\omega) = 0$, leading to

$$\frac{b}{a} = -\frac{1}{2}\left[\frac{H_o^{(2)}(kR) + H_2^{(2)}(kR)}{H_o^{(1)}(kR) + H_2^{(1)}(kR)} + \frac{\left(1 - \frac{c_s^2}{c_p^2}\right)H_o^{(2)}(kR) - \frac{c_s^2}{c_p^2}H_2^{(2)}(kR)}{\left(1 - \frac{c_s^2}{c_p^2}\right)H_o^{(1)}(kR) - \frac{c_s^2}{c_p^2}H_2^{(1)}(kR)}\right]$$

The viscous damper enforces

$$\sigma(R, \omega) + \rho c_p \dot{u}(R, \omega) = 0$$

which results in

$$\frac{b}{a} = -\frac{\left(ikR - 2\frac{c_s^2}{c_p^2}\right)H_1^{(2)}(kR) + kR\,H_o^{(2)}(kR)}{\left(ikR - 2\frac{c_s^2}{c_p^2}\right)H_1^{(1)}(kR) + kR\,H_o^{(1)}(kR)}$$

The extrapolation algorithm with $\Delta t = e/c_p$ leads to

$$\frac{b}{a} = \frac{1}{H_1^{(1)}(kR)}\left[\sum_{j=1}^{N}(-1)^{j+1}C_j^N H_1^{(2)}\left\{kR\left(1 - j\frac{e}{R}\right)\right\}\exp\left[-\frac{ijkRe}{R}\right] - H_1^{(2)}(kR)\right]$$

The amplitude ratio $\left|\frac{b}{a}\right|$ is plotted as a function of $kR$ (which is proportional to the ratio of the artificial boundary's distance to the wave length $c_p\, 2\pi/\omega$) in Fig. P 3-7b.

a)

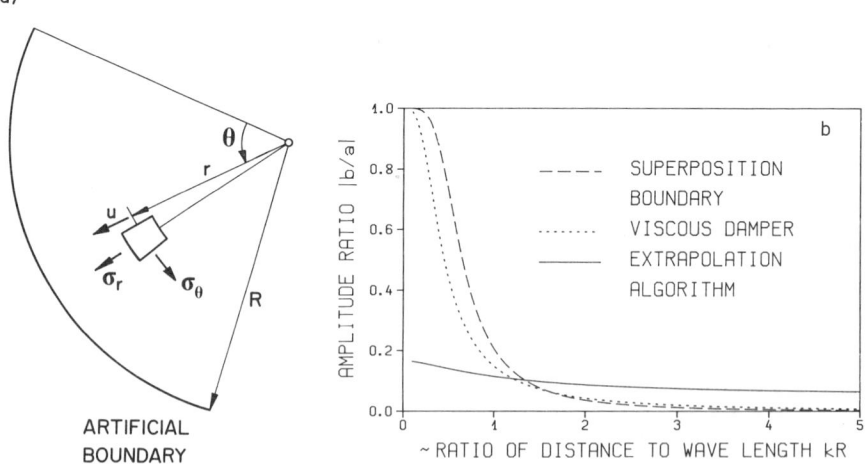

**Figure P 3-7** Axisymmetric one-dimensional waves in cylindrical coordinates. a. Nomenclature. b. Influence of artificial boundary's location on accuracy of results.

**3.8** Verify that the two-dimensional equations of motion in cylindrical coordinates correspond in the far field to the one-dimensional wave equations for the radial component $u$ propagating in the $r$-direction with the dilatational-wave velocity $c_p$ and the tangential component $v$ propagating with the shear-wave velocity $c_s$.

*Solution*

Dynamic-equilibrium equations in the radial $r$ and circumferential directions $\theta$:

$$\sigma_{r,r} + \frac{1}{r}\tau_{r\theta,\theta} + \frac{\sigma_r - \sigma_\theta}{r} - \rho\ddot{u} = 0$$

$$\tau_{r\theta,r} + \frac{1}{r}\sigma_{\theta,\theta} + \frac{2}{r}\tau_{r\theta} - \rho\ddot{v} = 0$$

Strain–displacement relations:

$$\varepsilon_r = u_{,r}$$

$$\varepsilon_\theta = \frac{u}{r} + \frac{1}{r}v_{,\theta}$$

$$\gamma_{r\theta} = \frac{1}{r}u_{,\theta} + v_{,r} - \frac{v}{r}$$

Hooke's law:

$$\sigma_r = (\lambda + 2G)\varepsilon_r + \lambda\varepsilon_\theta$$

$$\sigma_\theta = \lambda\varepsilon_r + (\lambda + 2G)\varepsilon_\theta$$

$$\tau_{r\theta} = G\gamma_{r\theta}$$

Far field ($r$ large):

$$\sigma_{r,r} - \rho\ddot{u} = 0$$

$$\tau_{r\theta,r} - \rho\ddot{v} = 0$$

$$\varepsilon_r = u_{,r}$$

$$\varepsilon_\theta = 0$$

$$\gamma_{r\theta} = v_{,r}$$

Equations of motion:

$$u_{,rr} - \frac{\ddot{u}}{c_p^2} = 0$$

$$v_{,rr} - \frac{\ddot{v}}{c_s^2} = 0$$

**3.9** In Section 3.9.3, the dynamic-stiffness matrix $[S]$ for the out-of-plane motion, corresponding to the nodes with parabolic shape functions shown in Fig. 3-27, is calculated. The extrapolation algorithm (Case 1) with one term ($N = 1$) is applied to the artificial boundary at a distance $l$. Calculate the ratio of the amplitudes of the reflected and the incident waves $|C_j/A_j|$ for the extrapolation algorithm, with two terms used in the extrapolation series ($N = 2$) and assuming $\Delta t = e/c_s$ ($e$ = distance of data points). Plot the dimensionless spring coefficient $k_{11}(a_o)$ and the damping coefficient $c_{11}(a_o)$ versus $a_o$ for $l/d = 1$ and $e/d = 0.1$. Use the exact static value $K_{11} = 0.449$ G to nondimensionalize $S_{11}$.

*Solution*

Substituting Eq. 3.211 in Eq. 3.154b results in

$[A_j \exp(-ik_j l) + C_j \exp(ik_j l)] \exp(i\omega t)$

$= 2A_j \exp[-ik_j(l - e)] \exp[i\omega(t - \Delta t)] - A_j \exp[-ik_j(l - 2e)] \exp[i\omega(t - 2\Delta t)]$

and after introducing $\Delta t = e/c_s$

$$\frac{C_j}{A_j} = -\left[-1 + \exp\left[\left(\frac{\sqrt{(2j-1)^2\pi^2 - 4a_o^2}}{2} - ia_o\right)\frac{e}{d}\right]\right]^2$$

$$\times \exp\left[-i\sqrt{4 - \frac{(2j-1)^2\pi^2}{a_o^2}}\frac{l}{d}a_o\right]$$

Solving this equation and Eq. 3.212 for $A_j$, $C_j$ and substituting in Eq. 3.214 leads to the dynamic-stiffness matrix $[S]$. $k_{11}(a_o)$ and $c_{11}(a_o)$ are plotted for $l/d = 1$ and $e/d = 0.1$ in Fig. P 3-9. Compared to the corresponding values for $N = 1$ (Fig. 3-32a), a significant improvement is achieved.

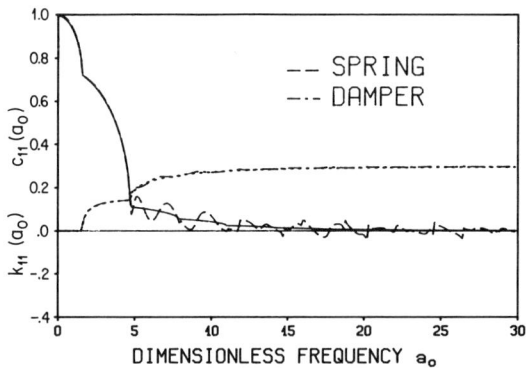

**Figure P 3-9** Dynamic-stiffness coefficient for extrapolation algorithm $N = 2$, $l/d = 1$, $e/d = 0.1$.

# 4
# *Summary of Substructure Method in Frequency Domain*

The fundamental equations of the substructure method working in the frequency domain are summarized in this chapter. Only those relations which are actually used in Chapters 5, 6, and 7, where the substructure method in the time domain is addressed, are specified. For a thorough treatment of this subject—including all derivations and a discussion of the assumptions and limitations, as well as illustrative examples from actual practice—the textbook [W11] and review articles [L7, L4] in linear soil–structure-interaction analysis should be studied. One objective is to define the nomenclature used in the frequency-domain analysis. This will make it easier to demonstrate the analogies between the time-domain procedures of the following chapters and the familiar frequency-domain methods.

Using a discrete Fourier transformation, the loading is represented by a Fourier series, where each term is associated with a specific frequency. In this method of complex response, the response for each term in the series is determined independently and the result in the time domain follows from an inverse discrete Fourier transformation.

The basic equation of motion in total displacements is specified in Section 4.1. The dynamic-stiffness matrix of the soil and the free-field motion appear in this equation. To be able to calculate these two quantities, the site (free field) must be analyzed for dynamic excitation. The corresponding dynamic-stiffness matrix is discussed in Section 4.2, where certain wave propagation fundamentals are also addressed. The determination of the site's free-field response is outlined in Section 4.3. The calculation of the soil's dynamic-stiffness matrix is examined in Section 4.4. It is based on the boundary-element method. Using the same concept, the scattered motion (see Section 4.5) can be determined.

The variables used in this chapter are a function of the (discrete) frequency and are thus actually (complex) amplitudes. For the sake of conciseness and clarity, the argument ω is omitted when not needed.

## 4.1 BASIC EQUATION OF MOTION

The basic equation of motion to analyze soil–structure interaction is specified for the discretized dynamic system shown in Fig. 4-1. The latter consists of the (generalized) structure, which can include an irregular bounded soil region adjacent to the structure and of a regular unbounded soil. The word *regular* refers to the horizontally layered half-space with excavation. (The word *generalized* is dropped in the following.) Both substructures are linear. The subscript $b$ (for base) denotes the nodes lying on the structure–soil interface; $s$ (for structure) denotes the remaining ones of the structure. The unbounded soil with excavation is denoted by the superscript $g$ (for ground), the unbounded soil without excavation (virgin soil) by $f$ (for free field), and the excavated soil by $e$ (Fig. 4-2).

In the frequency domain, the discretized equation of motion of the structure–soil system for an excitation introduced via the soil equals [W11]

$$\begin{bmatrix} [S_{ss}(\omega)] & [S_{sb}(\omega)] \\ [S_{bs}(\omega)] & [S_{bb}(\omega)] + [S^g_{bb}(\omega)] \end{bmatrix} \begin{Bmatrix} \{u^t_s(\omega)\} \\ \{u^t_b(\omega)\} \end{Bmatrix} = \begin{Bmatrix} \{0\} \\ [S^g_{bb}(\omega)]\{u^g_b(\omega)\} \end{Bmatrix}$$
$$= \begin{Bmatrix} \{0\} \\ [S^f_{bb}(\omega)]\{u^f_b(\omega)\} \end{Bmatrix}$$

(4.1)

$\{u^t\}$ is the vector of the total-displacement amplitudes. The dynamic-stiffness matrix $[S]$ of the structure equals

$$[S(\omega)] = [K](1 + 2\zeta i) - \omega^2[M]$$ (4.2)

where $[K]$ and $[M]$ are the static-stiffness and mass matrices, and $\zeta$ is the

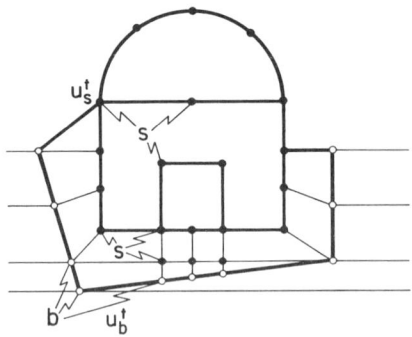

**Figure 4-1** Structure–soil system.

Sec. 4.1    Basic Equation of Motion                                                    177

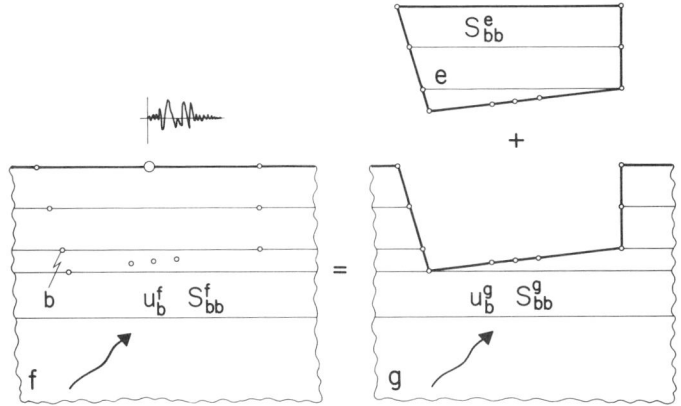

**Figure 4-2** Reference soil systems.

hysteretic-damping ratio. $[S_{bb}^f]$ represents the dynamic-stiffness matrix of the continuous soil discretized in those nodes which subsequently will lie on the structure–soil interface; $\{u_b^f\}$ represents the amplitudes of the corresponding motion in the same nodes. $[S_{bb}^g]$ and $\{u_b^g\}$ are the corresponding quantities for the soil system ground (Fig. 4-2). The free-field motion, determined in the nodes on the structure–soil interface, thus characterizes the load vector.

The amplitudes of the interaction forces $\{R_b\}$ of the unbounded soil (system $g$) acting at the nodes $b$ (which determines the unbounded soil's contribution to the basic equation of motion) are expressed as

$$\{R_b(\omega)\} = [S_{bb}^g(\omega)] \left(\{u_b^t(\omega)\} - \{u_b^g(\omega)\}\right) \tag{4.3}$$

$\{R_b\}$ is a function of the motion relative to the ground. By using

$$[S_{bb}^g(\omega)] + [S_{bb}^e(\omega)] = [S_{bb}^f(\omega)] \tag{4.4}$$

the identity

$$[S_{bb}^g(\omega)] \{u_b^g(\omega)\} = [S_{bb}^f(\omega)] \{u_b^f(\omega)\} \tag{4.5}$$

is derived.

If loads are applied directly to the structure, the corresponding load amplitudes $\{R_s(\omega)\}$ and $\{R_b^s(\omega)\}$ (superscript $s$ for structure) are added to the right-hand side of Eq. 4.1.

After generalizing the meaning of the various terms, the basic equation of motion (Eq. 4.1) also applies to a structure embedded in a more general site, as shown in Fig. 1-2b.

If the base consisting of the basemat and the adjacent walls can be assumed to be rigid, the following compatibility constraint

$$\{u_b^t(\omega)\} = [A] \{u_o^t(\omega)\} \tag{4.6}$$

and the equilibrium condition

$$\{R_o(\omega)\} = [A]^T \{R_b(\omega)\} \tag{4.7}$$

are introduced into the basic equation of motion (Eq. 4.1). The kinematic transformation matrix $[A]$ contains geometric quantities. $\{u'_o(\omega)\}$ are the amplitudes of the rigid body motion in point 0 of the base, and $\{R_o(\omega)\}$ are the amplitudes of the resultant interaction forces.

In the basic equation of motion formulated in total displacements, the loads for an excitation introduced through the soil (such as a seismic excitation) are applied only at the nodes located on the structure–soil interface (Eq. 4.1). Many other equivalent formulations exist [W11]. They are derived by defining the unknown motion relative to an appropriately chosen known motion—that is, by a transformation of variables. In this case, all nodes of the discretized system will be loaded.

The generalized scattered motion $\{u_b^g(\omega)\}$ can be calculated, based on Eq. 4.5, as

$$\{u_b^g(\omega)\} = [S_{bb}^g(\omega)]^{-1} [S_{bb}^f(\omega)] \{u_b^f(\omega)\} \tag{4.8}$$

The expression *generalized scattered motion* thus describes the sum of the components of the free-field and of the additional waves which are produced when the incident free-field waves encounter the excavation (Fig. 4-2).

## 4.2 DYNAMIC STIFFNESS OF SITE

### 4.2.1 Three-Dimensional Wave Equation in Cartesian Coordinates

It is appropriate first to outline the fundamentals of wave propagation in a full infinite homogeneous and isotropic space.

The equations of motion can be decomposed into that governed by the volumetric strain leading to the P-wave and those with rotation strains resulting in the S-waves.

For a P-wave, the amplitudes of the displacements $u_p$, $v_p$, and $w_p$ in the directions of the coordinate system $x$, $y$, $z$ are equal to

$$u_p = l_x A_P \exp\left[\frac{i\omega}{c_p}(-l_x x - l_y y - l_z z)\right] \tag{4.9a}$$

$$v_p = l_y A_P \exp\left[\frac{i\omega}{c_p}(-l_x x - l_y y - l_z z)\right] \tag{4.9b}$$

$$w_p = l_z A_P \exp\left[\frac{i\omega}{c_p}(-l_x x - l_y y - l_z z)\right] \tag{4.9c}$$

## Sec. 4.2 Dynamic Stiffness of Site

where

$$l_x^2 + l_y^2 + l_z^2 = 1 \tag{4.10}$$

applies (Fig. 4-3a). The three scalars $l_x$, $l_y$, and $l_z$ are the direction cosines of a straight line with the coordinate $s$ ($= l_x x + l_y y + l_z z$) along which the wave propagates with the dilatational-wave velocity $c_p$:

$$c_p^2 = \frac{\lambda + 2G}{\rho} \tag{4.11}$$

The Lamé constants are $\lambda$, $G$, and the mass density is $\rho$. $A_P$ denotes the amplitude. P-wave's particle motion takes place along the direction of propagation and is constant over a plane perpendicular to it.

For an S-wave, the displacement amplitudes equal

$$u_s = \frac{m_x m_z A_{SV} - m_y A_{SH}}{\sqrt{m_x^2 + m_y^2}} \exp\left[\frac{i\omega}{c_s}(-m_x x - m_y y - m_z z)\right] \tag{4.12a}$$

$$v_s = \frac{m_y m_z A_{SV} + m_x A_{SH}}{\sqrt{m_x^2 + m_y^2}} \exp\left[\frac{i\omega}{c_s}(-m_x x - m_y y - m_z z)\right] \tag{4.12b}$$

$$w_s = -\sqrt{m_x^2 + m_y^2}\, A_{SV} \exp\left[\frac{i\omega}{c_s}(-m_x x - m_y y - m_z z)\right] \tag{4.12c}$$

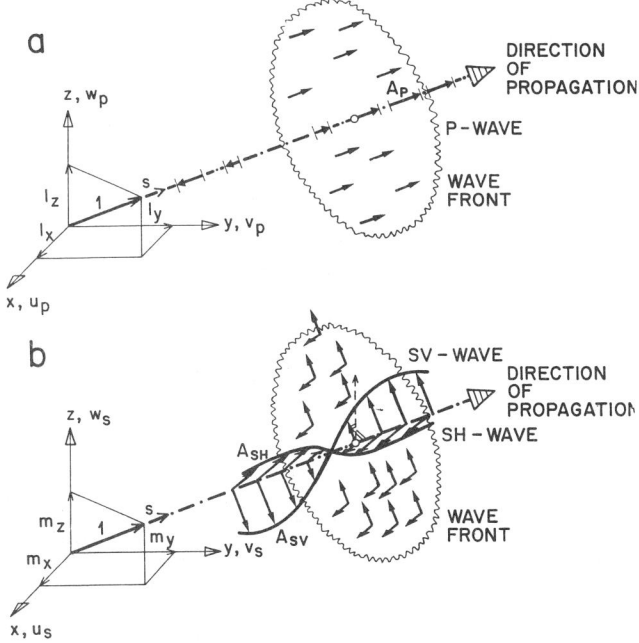

**Figure 4-3** Displacements associated with body waves. a. P-wave. b. S-wave.

with
$$m_x^2 + m_y^2 + m_z^2 = 1 \tag{4.13}$$
applying (Fig. 4-3b). The direction of propagation, which occurs with the shear-wave velocity $c_s$ equal to
$$c_s^2 = \frac{G}{\rho} \tag{4.14}$$
is specified by the direction cosines $m_x$, $m_y$, and $m_z$. The particle motion, with the horizontal and vertical amplitudes $A_{SH}$ and $A_{SV}$ of the S-wave, lies in a plane perpendicular to the direction of propagation and is constant over this plane.

The effect of material damping is introduced by applying the correspondence principle. The material properties appearing in Eqs. 4.11 and 4.14 are replaced by the corresponding complex values.
$$\lambda^* + 2G^* = (\lambda + 2G)(1 + 2\zeta_p i) \tag{4.15a}$$
$$G^* = G(1 + 2\zeta_s i) \tag{4.15b}$$
with the linear hysteretic damping ratios of the P- and S-waves $\zeta_p$ and $\zeta_s$. The complex wave velocities are denoted as $c_p^*$ and $c_s^*$.

It is reasonable to assume that the P- and S-waves' directions of propagation lie in the same vertical plane $x$-$z$. With $l_y = m_y = 0$ and taking material damping into consideration leads to the total motion

$$u = l_x A_P \exp\left[i\omega\left(-\frac{l_x x}{c_p^*} - \frac{l_z z}{c_p^*}\right)\right] + m_z A_{SV} \exp\left[i\omega\left(-\frac{m_x x}{c_s^*} - \frac{m_z z}{c_s^*}\right)\right]$$
$$\tag{4.16a}$$

$$v = A_{SH} \exp\left[i\omega\left(-\frac{m_x x}{c_s^*} - \frac{m_z z}{c_s^*}\right)\right] \tag{4.16b}$$

$$w = l_z A_P \exp\left[i\omega\left(-\frac{l_x x}{c_p^*} - \frac{l_z z}{c_p^*}\right)\right] - m_x A_{SV} \exp\left[i\omega\left(-\frac{m_x x}{c_s^*} - \frac{m_z z}{c_s^*}\right)\right]$$
$$\tag{4.16c}$$

The in-plane displacements with the amplitudes $u$ and $w$ depend only on the P- and SV-waves. The out-of-plane displacement with the amplitude $v$ is caused by the SH-wave and is thus independent of $u$ and $w$.

To analyze the layered half-space used as the site's model, boundary conditions (at the free surface) and continuity requirements between adjacent layers and between the bottom layer and the half-space must be formulated. Excitations, either as prescribed motions or as specified external loads, must be processed. This can be achieved, as in conventional structural analysis, forming the total system by assembling the contributions of the individual

## Sec. 4.2 Dynamic Stiffness of Site

structural elements. It is thus necessary to calculate the dynamic-stiffness matrices and the consistent-load vectors of a layer and of a half-space, which form the structural elements in this case. This is performed in the remainder of this section.

### 4.2.2 Out-of-Plane Motion

For a homogeneous isotropic layer of depth $d$, shown in Fig. 4-4, the amplitude of the out-of-plane displacement $v(x, z)$ is specified as

$$v(x, z) = v(z) \exp(-ikx) \tag{4.17}$$

with $v(z)$ propagating in the positive $x$-direction

$$v(z) = A_{SH} \exp(iktz) + B_{SH} \exp(-iktz) \tag{4.18}$$

$k$ denotes the wave number

$$k = \frac{\omega}{c} \tag{4.19}$$

where the phase velocity $c$ is defined as

$$c = \frac{c_s^*}{m_x} \tag{4.20}$$

with $m_x$ equal to $\cos\psi_{SH}$. $\psi_{SH}$ is the angle of incidence of the waves with amplitudes $A_{SH}$ and $B_{SH}$ traveling in the negative and positive $z$-directions, respectively (Fig. 4-4). $t$ is equal to $\tan\psi_{SH}$

$$t = \sqrt{\frac{1}{m_x^2} - 1} \tag{4.21}$$

The dynamic-stiffness matrix reflects the variation of the variables with depth.

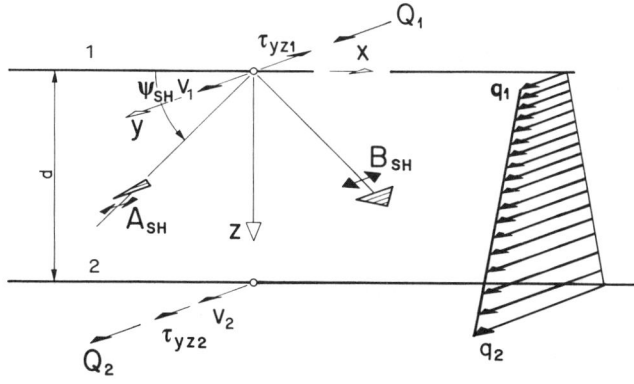

**Figure 4-4** Nomenclature of layer for out-of-plane motion.

The two faces of the layer are denoted with the subscripts 1 and 2, with $\tau_{yz}$ being the shear-stress amplitude. The force amplitudes $Q_1 = -\tau_{yz1}$ and $Q_2 = \tau_{yz2}$, defined in the global coordinate system, are related to the displacement amplitudes $v_1$ and $v_2$ as [K1]

$$\begin{Bmatrix} Q_1 \\ Q_2 \end{Bmatrix} = \frac{ktG^*}{\sin ktd} \begin{bmatrix} \cos ktd & -1 \\ -1 & \cos ktd \end{bmatrix} \begin{Bmatrix} v_1 \\ v_2 \end{Bmatrix} \tag{4.22}$$

with the coefficient matrix on the right-hand side representing the (symmetric) dynamic-stiffness matrix of the layer $[S_{SH}^L]$ (superscript L for layer). The corresponding force–displacement relationship at the free surface (subscript $o$) of the half-space equals

$$Q_o = iktG^* v_o \tag{4.23}$$

with the dynamic-stiffness coefficient of the half-space $S_{SH}^R$ equal to $iktG^*$ (superscript R for rock). The dynamic-stiffness matrices of an undamped layer and of an undamped half-space are real and imaginary, respectively—the latter indicating that energy is radiated towards infinity.

It should be stressed that the amplitudes of the displacements $v$ (Eq. 4.18) and of the forces $Q$ (or stresses $\tau_{yz}$) vary as $\exp(-ikx)$ in the horizontal direction. An expansion in the wave-number domain is thus performed, which is actually a Fourier transformation. The following transforms, for, for example, $v(x)$, apply.

$$v(k) = \frac{1}{2\pi} \int_{-\infty}^{+\infty} v(x) \exp(ikx)\, dx \tag{4.24a}$$

$$v(x) = \int_{-\infty}^{+\infty} v(k) \exp(-ikx)\, dk \tag{4.24b}$$

The response for a linearly varying distributed load acting on an inclined line in the out-of-plane direction (Fig. 4-4) is specified in Ref. [W10].

### 4.2.3 In-Plane Motion

For a homogeneous isotropic layer of depth $d$, shown in Fig. 4-5, the amplitudes of the in-plane displacements $u(x, z)$, $w(x, z)$ are specified as

$$u(x, z) = u(z) \exp(-ikx) \tag{4.25}$$
$$w(x, z) = w(z) \exp(-ikx)$$

where the amplitudes of the wave traveling in the positive $x$-direction are equal to

$$u(z) = l_x[A_P \exp(iksz) + B_P \exp(-iksz)] - m_x t [A_{SV} \exp(iktz)$$
$$- B_{SV} \exp(-iktz)] \tag{4.26a}$$

Sec. 4.2    Dynamic Stiffness of Site                                                                 183

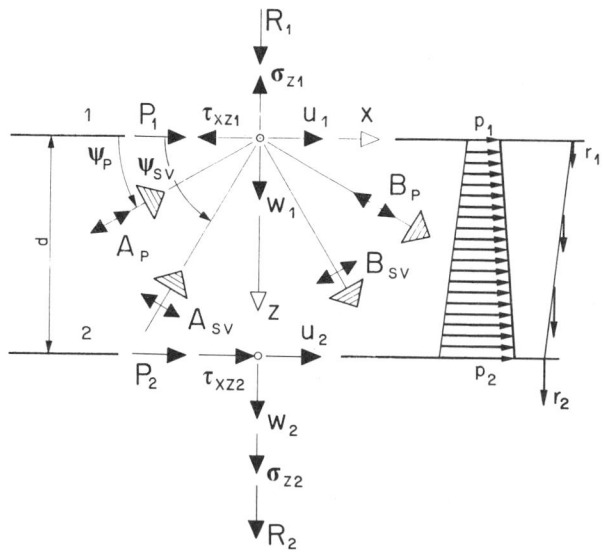

**Figure 4-5**  Nomenclature of layer for in-plane motion.

$$w(z) = -l_x s [A_P \exp(iksz) - B_P \exp(-iksz)] - m_x [A_{SV} \exp(iktz)$$
$$+ B_{SV} \exp(-iktz)] \quad (4.26b)$$

$k$ denotes the wave number (Eq. 4.19). The phase velocity $c$ is the same for P- and SV-waves

$$c = \frac{c_P^*}{l_x} = \frac{c_s^*}{m_x} \quad (4.27)$$

with $l_x = \cos\psi_P$ and $m_x = \cos\psi_{SV}$ ($\psi$ equals the corresponding wave's angle of incidence). The amplitudes of the P-waves propagating in the negative and positive $z$-directions are denoted as $A_P$ and $B_P$, respectively. An analogous nomenclature applies to the SV-waves. $s$ and $t$ are defined as

$$s = \sqrt{\frac{1}{l_x^2} - 1} \quad (4.28a)$$

$$t = \sqrt{\frac{1}{m_x^2} - 1} \quad (4.28b)$$

The shear and normal stresses' amplitudes are denoted as $\tau_{xz}$ and $\sigma_z$, with the subscripts 1 and 2 to identify the layer's face. The force amplitudes $P_1 = -\tau_{xz1}$, $R_1 = -\sigma_{z1}$, $P_2 = \tau_{xz2}$, and $R_2 = \sigma_{z2}$, defined in the global coordinate system, are related to the displacement amplitudes $u_1$, $w_1$, $u_2$, and $w_2$, as shown in Eq. 4.29 (Table 4-1). To achieve symmetry of the dynamic-stiffness matrix of the layer $[S^L_{P-SV}]$, $R_1$, $R_2$, $w_1$, and $w_2$ are multiplied by $i$ [K1]. The corresponding force–displacement relationship at the free surface of a half-space equals

$$\begin{Bmatrix} P_o \\ iR_o \end{Bmatrix} = kG^* \begin{bmatrix} \dfrac{is(1 + t^2)}{1 + st} & 2 - \dfrac{1 + t^2}{1 + st} \\ 2 - \dfrac{1 + t^2}{1 + st} & \dfrac{it(1 + t^2)}{1 + st} \end{bmatrix} \begin{Bmatrix} u_o \\ iw_o \end{Bmatrix} \qquad (4.30)$$

The coefficient matrix represents the dynamic-stiffness matrix of the half-space $[S^R_{P-SV}]$.

As for the out-of-plane case, the amplitudes of the displacements and of the forces (or stresses) can be expanded in the horizontal direction into a Fourier integral with terms $\exp(-ikx)$ (Eq. 4.24).

The response for a linearly varying distributed load acting on an inclined line in the in-plane directions (Fig. 4-5) is specified in Ref. [W10].

### 4.2.4 Three-Dimensional Wave Equation in Cylindrical Coordinates

In a cylindrical coordinate system $r$, $\theta$, $z$ (Fig. 4-6), the amplitudes of the displacements and of the forces (or stresses) are expanded in a Fourier series in the circumferential direction (integer $n = 0, 1, 2, \ldots$) and in Bessel functions involving the wave number $k$ in the radial direction. For the am-

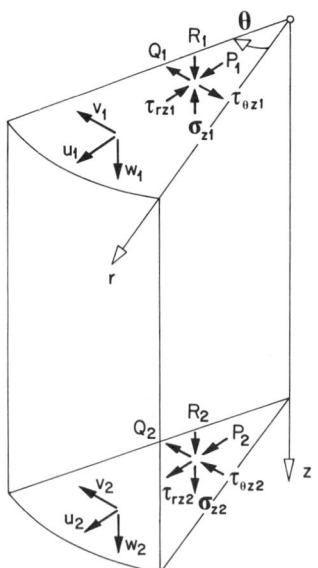

**Figure 4-6** Nomenclature of layer in cylindrical coordinate system.

**TABLE 4-1  Dynamic-Stiffness Matrix of Layer, Eq. 4.29**

$$\begin{Bmatrix} P_1 \\ iR_1 \\ P_2 \\ iR_2 \end{Bmatrix} = \frac{(1+t^2)kG^*}{D} \begin{bmatrix} \dfrac{1}{t}\cos ksd \sin ktd \\ +s\sin ksd \cos ktd & \dfrac{3-t^2}{1+t^2}(1-\cos ksd \cos ktd) \\ & +\dfrac{1+2s^2t^2-t^2}{st(1+t^2)}\sin ksd \sin ktd & -s\sin ksd & \cos ksd \\ & \dfrac{1}{s}\sin ksd \cos ktd \\ & +t\cos ksd \sin ktd & -\dfrac{1}{t}\sin ktd & -\cos ktd \\ -s\sin ksd & -\cos ksd & \dfrac{1}{t}\cos ksd \sin ktd & \dfrac{t^2-3}{1+t^2}(1-\cos ksd \cos ktd) \\ & & +s\sin ksd \cos ktd & +\dfrac{t^2-2s^2t^2-1}{st(1+t^2)}\sin ksd \sin ktd \\ -\dfrac{1}{t}\sin ktd & +\cos ktd & \dfrac{t^2-3}{1+t^2}(1-\cos ksd \cos ktd) & \dfrac{1}{s}\sin ksd \\ \cos ksd & -\dfrac{1}{s}\sin ksd & +\dfrac{t^2-2s^2t^2-1}{st(1+t^2)}\sin ksd \sin ktd & \\ -\cos ktd & -t\sin ktd & \dfrac{1}{s}\sin ksd \cos ktd \\ & & +t\cos ksd \sin ktd \end{bmatrix} \begin{Bmatrix} u_1 \\ iw_1 \\ u_2 \\ iw_2 \end{Bmatrix}$$

where

$$D = 2(1 - \cos ksd \cos ktd) + \left(st + \dfrac{1}{st}\right)\sin ksd \sin ktd$$

plitudes of the displacements $u$, $v$, $w$ for the symmetric case with respect to $\theta = 0$, the following Bessel-transform pair applies:

$$\begin{Bmatrix} u(k,n) \\ v(k,n) \\ w(k,n) \end{Bmatrix} = a_n \int_{r=0}^{\infty} r \begin{bmatrix} \frac{1}{k}J_n(kr)_{,r} & \frac{n}{kr}J_n(kr) & \\ \frac{n}{kr}J_n(kr) & \frac{1}{k}J_n(kr)_{,r} & \\ & & -J_n(kr) \end{bmatrix} \int_{\theta=0}^{2\pi} \begin{bmatrix} \cos n\theta & & \\ & -\sin n\theta & \\ & & \cos n\theta \end{bmatrix} \cdot \begin{Bmatrix} u(r,\theta) \\ v(r,\theta) \\ w(r,\theta) \end{Bmatrix} d\theta\, dr \quad (4.31\text{a})$$

$$\begin{Bmatrix} u(r,\theta) \\ v(r,\theta) \\ w(r,\theta) \end{Bmatrix} = \sum_{n=0}^{\infty} \begin{bmatrix} \cos n\theta & & \\ & -\sin n\theta & \\ & & \cos n\theta \end{bmatrix} \int_{k=0}^{\infty} k \begin{bmatrix} \frac{1}{k}J_n(kr)_{,r} & \frac{n}{kr}J_n(kr) & \\ \frac{n}{kr}J_n(kr) & \frac{1}{k}J_n(kr)_{,r} & \\ & & -J_n(kr) \end{bmatrix} \cdot \begin{Bmatrix} u(k,n) \\ v(k,n) \\ w(k,n) \end{Bmatrix} dk \quad (4.31\text{b})$$

$J_n(kr)$ denotes the Bessel function of order $n$ of the first kind. $a_n$ is the normalization factor, which is equal to $1/(2\pi)$ for $n = 0$, and $n = 1/\pi$ for $n \neq 0$. For the antimetric case, the diagonal elements describing the variation in the circumferential direction are equal to $\sin n\theta$, $\cos n\theta$ and $\sin n\theta$.

The variations with depth $v(z)$ and $u(z)$, $w(z)$ in cylindrical coordinates are identical to those arising for plane waves in out-of-plane (Eq. 4.18) and in-plane motions (Eq. 4.26), respectively. This then leads to the same force–displacement relationships in the wave-number domain. The dynamic-stiffness matrices for a layer specified for the out-of-plane motion $[S_{\text{SH}}^L]$ in Eq. 4.22 and for the in-plane motion $[S_{\text{P-SV}}^L]$ in Eq. 4.29 thus still apply. They are independent of $n$ and are the same for the symmetric and antimetric cases. With the nomenclature illustrated in Fig. 4-6,

$$\begin{Bmatrix} Q_1 \\ Q_2 \end{Bmatrix} = \begin{Bmatrix} -\tau_{\theta z 1} \\ \tau_{\theta z 2} \end{Bmatrix} = [S_{\text{SH}}^L] \begin{Bmatrix} v_1 \\ v_2 \end{Bmatrix} \quad (4.32\text{a})$$

Sec. 4.2  Dynamic Stiffness of Site                                         187

$$\begin{Bmatrix} P_1 \\ R_1 \\ P_2 \\ R_2 \end{Bmatrix} = \begin{Bmatrix} -\tau_{rz1} \\ -\sigma_{z1} \\ \tau_{rz2} \\ \sigma_{z2} \end{Bmatrix} = \begin{bmatrix} S_{P-SV}^L \end{bmatrix} \begin{Bmatrix} u_1 \\ w_1 \\ u_2 \\ w_2 \end{Bmatrix} \qquad (4.32b)$$

applies, whereby the argument $(k, n)$ is omitted. In cylindrical coordinates, the displacement and stress amplitudes are not multiplied by $i$.

For the half-space, analogous expressions apply.

### 4.2.5 Assemblage of Dynamic-Stiffness Matrix

In the case of a site which consists of several layers resting on a half-space, the global dynamic-stiffness matrix in the wave-number domain is constructed by overlapping the contributions of the layer matrices and of the half-space matrix at each "node" (interface) of the layered half-space. Thus, at each interface (including the free surface), three equations are formulated (one for the out-of-plane and two for the in-plane motions), and three unknown displacement amplitudes are present. The global load vector consists of the prescribed exterior loading amplitudes. This procedure is rigorous, as the exact dynamic-stiffness matrices of the layers with finite depth and of the half-space and the consistent-load vector are used. The assemblage and the solution procedures to analyze the layered site (free field) are thus analogous to the familiar direct stiffness method of structural analysis.

## 4.3 FREE-FIELD RESPONSE OF SITE

### 4.3.1 Definition of Task

As can be seen from the load-amplitude vector in the basic equation of motion (Eq. 4.1), the free-field motion $\{u_b^f\}$ must be determined in the nodes $b$ which subsequently lie on the structure–soil interface. The latter are shown in Fig. 4-7 for an embedded structure founded on piles.

For seismic excitation, the following approach is followed. When determining the free field's seismic environment, three interrelated aspects are considered. The first covers the selection of the control motion, which can, for example, consist of a historic earthquake record. The second governs the selection of the control point where the motion is applied. The selected control point should be on the free field's ground surface (point $A$ in Fig. 4-7) or at an assumed rock outcrop (point $B$) that is on the level of the rock, assuming there is no soil on top. The third addresses the wave pattern—

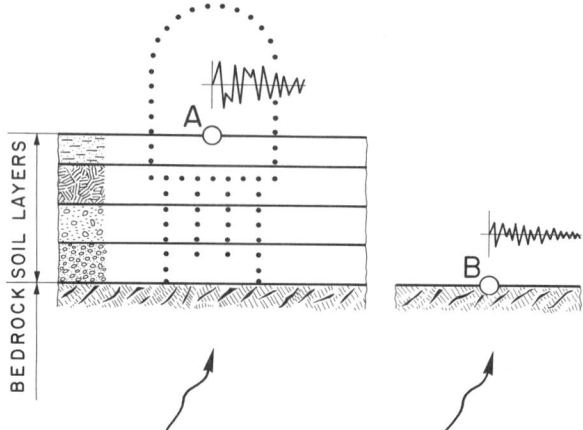

**Figure 4-7** Selection of control point for seismic input.

that is, prescribing the types of waves, such as vertically incident or inclined body waves or surface waves, of which the motion consists.

### 4.3.2 Body Wave

The equations governing the out-of-plane and in-plane motions are uncoupled (Section 4.2.1). This allows the computational procedure that determines the spatial variation of the free-field motion to be summarized using the out-of-plane motion only (Fig. 4-8).

The wave pattern is governed by the angle of incidence $\psi_{SH}$ of the body wave in the rock on which the layers rest. It determines the phase velocity $c$ (Eq. 4.20) and thus the wave number $k$ (Eq. 4.19). Together with the depths of the layers and the material properties, the stiffness matrices of the layers $[S_{SH}^L]$ (Eq. 4.22) and of the rock $S_{SH}^R$ (Eq. 4.23) are thus known. If

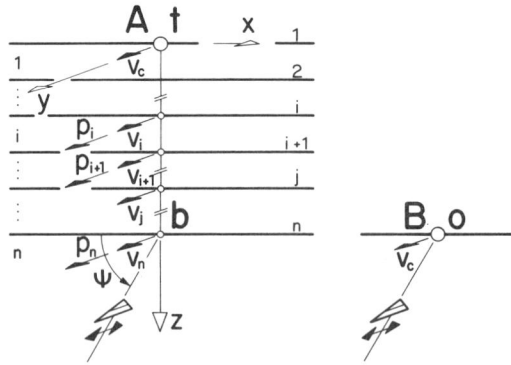

**Figure 4-8** Layered site and reference soil system with nomenclature for out-of-plane motion.

## Sec. 4.4   Dynamic Stiffness of Embedded Foundation

assembled form, the discretized dynamic-equilibrium equations of the site equal

$$[S_{SH}]\{v\} = \{Q\} \tag{4.33}$$

The total system's dynamic-stiffness matrix is denoted by $[S_{SH}]$; $\{v\}$ is the vector of the displacement amplitudes $v_1$ to $v_n$; and $\{Q\}$ is the external load amplitudes' vector.

For body waves, the control point can be selected either at the free surface (point $A$) or at rock outcrop (point $B$), which influences only $\{Q\}$ in Eq. 4.33. For a prescribed control motion $v_c$ in point $A$, all exterior load amplitudes $\{Q\}$ are equal to zero (with the exception of the $n$th element which is not needed in the calculation). Setting $v_1 = v_c$ and using, for example, Gaussian elimination, the free-field displacement amplitude $v_j$ in node $j$ results (Fig. 4-8). The amplitude $v_j$ is in this case independent of the properties of the system below node $j$.

For a prescribed control motion $v_c$ in point $B$, $\{Q\}$ can be calculated analogously as the load vector in the basic equation of soil–structure interaction, Eq. 4.1. For the site, the layers and bedrock represent the two substructures. The $n$th element of $\{Q\}$ can be formulated as

$$Q_n = S_{SH}^R v_c \tag{4.34}$$

with all other elements being zero. The displacement amplitude $v_j$, which again follows from Eq. 4.33, is a function of the total system's properties.

### 4.3.3 Surface Wave

For surface waves, the control point can be selected only at the free surface (point $A$). Setting $\{Q\} = 0$ in Eq. 4.33, an eigenvalue problem arises. The surface-wave motion is thus equal to the site's natural modes of wave propagation. For a given frequency $\omega$, the "free" parameter is the phase velocity $c$ (or the wave number $k$); only distinct values of $c$ associated with the various modes exist. For these values of $c(\omega)$, the determinant of $[S_{SH}]$ vanishes, leading to nontrivial solutions for the displacement amplitudes $\{v\}$. These are then scaled to form the control motion. Surface waves are, in general, dispersive.

## 4.4 DYNAMIC STIFFNESS OF EMBEDDED FOUNDATION

### 4.4.1 Green's Function

In all boundary-element methods used to calculate the unbounded soil's dynamic-stiffness matrix, Green's functions (constructed from fundamental solutions)—that is, amplitudes of displacements and surface tractions for amplitudes of prescribed fictitious loads—must be determined in the layered half-space (free field). Before addressing this system, the homogeneous isotropic full infinite space is examined.

**Full infinite space.** A concentrated load of unit amplitude is applied in the $x_j$-direction at the source specified by the position vector $\vec{x}'$ (coordinate $x'$). The amplitude of the displacement component in the $x_i$-direction at the receiver $\vec{x}$ (coordinate $x$) $u_{ij}(x, x', \omega)$ in this three-dimensional case is specified as $(i, j = 1, 2, 3)$

$$u_{ij}(x, x', \omega) = \frac{1}{\alpha \pi \rho c_s^2} (\psi \delta_{ij} - \chi l_i l_j) \tag{4.35}$$

where

$$\psi = \left(1 - i\frac{1}{a_o} - \frac{1}{a_o^2}\right) \frac{\exp(-ia_o)}{r} + \left(i\frac{c_s}{c_p a_o} + \frac{1}{a_o^2}\right) \frac{\exp\left(-i\frac{c_s}{c_p} a_o\right)}{r} \tag{4.36}$$

$$\chi = \left(1 - i\frac{3}{a_o} - \frac{3}{a_o^2}\right) \frac{\exp(-ia_o)}{r} - \left(\frac{c_s^2}{c_p^2} - i\frac{3c_s}{c_p a_o} - \frac{3}{a_o^2}\right) \frac{\exp\left(-i\frac{c_s}{c_p} a_o\right)}{r} \tag{4.37}$$

with

$$a_o = \frac{\omega r}{c_s} \tag{4.38}$$

$l_i$ denotes the components of the unit vector (direction cosines) from the source point $\vec{x}'$ to the receiver point $\vec{x}$, and $r = |\vec{x} - \vec{x}'|$ is the distance between these two points. $\alpha = 4$ and $\delta_{ij}$ is the Kronecker delta function.

The amplitude of the surface traction component in the $x_i$-direction at the receiver point on a surface with unit external normal $\vec{n}(n_k)$ is equal to

$$t_{ij}(x, n, x', \omega) = \frac{1}{\alpha \pi} \left[ \left(\frac{d\psi}{dr} - \frac{\chi}{r}\right)\left(\delta_{ij}\frac{\partial r}{\partial n} + l_j n_i\right) - \frac{2\chi}{r}\left(l_i n_j - 2l_i l_j \frac{\partial r}{\partial n}\right) \right.$$
$$\left. - 2l_i l_j \frac{d\chi}{dr}\frac{\partial r}{\partial n} + \left(\frac{c_p^2}{c_s^2} - 2\right) l_i n_j \left(\frac{d\psi}{dr} - \frac{d\chi}{dr} - \frac{\alpha \chi}{2r}\right) \right] \tag{4.39}$$

**Full infinite plane.** For a plane-strain problem (in-plane motion), the same expressions for the Green's functions apply with $i, j = 1, 2$ ( Eqs. 4.35 and 4.39), but $\psi$ and $\chi$ are specified for the two-dimensional case as

## Sec. 4.4  Dynamic Stiffness of Embedded Foundation

$$\psi = K_o(ia_o) - \frac{i}{a_o}\left[K_1(ia_o) - \frac{c_s}{c_p}K_1\left(i\frac{c_s}{c_p}a_o\right)\right] \quad (4.40)$$

$$\chi = K_2(ia_o) - \frac{c_s^2}{c_p^2}K_2\left(i\frac{c_s}{c_p}a_o\right) \quad (4.41)$$

and $\alpha = 2$. $K_\nu$ is the modified Bessel function of order $\nu$ and of the second kind.

**Axisymmetric layered half-space.** As an example of a Green's function calculated in a layered half-space, a constant distributed loading acting on a disk of radius $\Delta a$ at the node $j$ (interface) is addressed (Fig. 4-9). The displacements on the level of node $i$ represent the Green's functions.

The vertical distributed load with amplitude $r_o$ is examined first (Fig. 4-9a). Only the zeroth symmetric Fourier term ($n = 0$) arises. Formulating the third line of Eq. 4.31a for the load's amplitude leads to

$$r(k) = -\frac{1}{2\pi}\int_{r=0}^{\Delta a} rJ_o(kr)\int_{\theta=0}^{2\pi} r_o\, d\theta\, dr$$

$$= -r_o \int_{r=0}^{\Delta a} rJ_o(kr)\, dr = -\frac{r_o\Delta a}{k}J_1(k\Delta a) \quad (4.42)$$

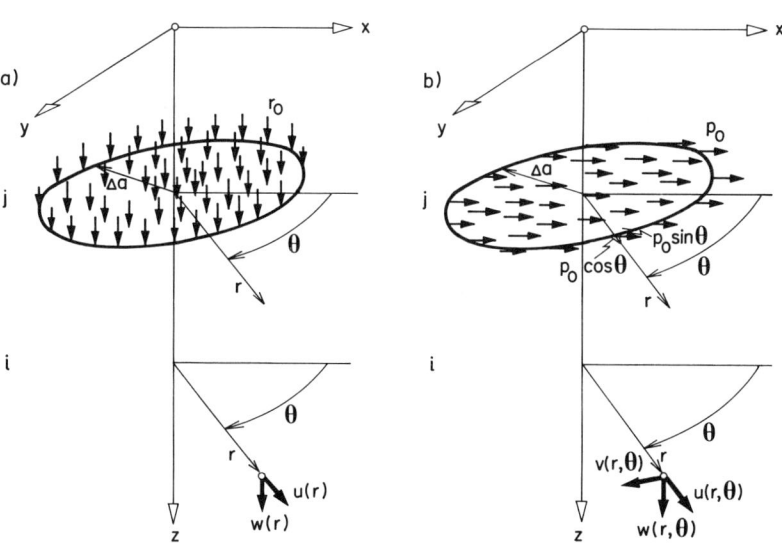

**Figure 4-9** Nomenclature for calculation of Green's functions in cylindrical coordinates. a. Vertical. b. Horizontal.

Assembling the layered half-space's dynamic-stiffness matrix $[S_{P-SV}]$ using the matrices of the individual layers (Eq. 4.32b) and of the rock, and after making eliminations, the displacement amplitudes at node $i$ in the wavenumber domain can be expressed as

$$\begin{Bmatrix} u(k) \\ w(k) \end{Bmatrix} = \begin{Bmatrix} F_{uw}(k) \\ F_{ww}(k) \end{Bmatrix} r(k) \qquad (4.43)$$

$F$ represents the (condensed) dynamic-flexibility coefficient.

The inverse transformation leading to $u(r)$ and $w(r)$ follows from the first and third lines of Eq. 4.31b, formulated for the displacement amplitudes:

$$\begin{Bmatrix} u(r) \\ w(r) \end{Bmatrix} = \int_{k=0}^{\infty} \begin{bmatrix} J_o(kr)_{,r} \\ -kJ_o(kr) \end{bmatrix} \begin{Bmatrix} u(k) \\ w(k) \end{Bmatrix} dk \qquad (4.44)$$

Substituting Eqs. 4.42 and 4.43 leads to

$$\begin{Bmatrix} u(r) \\ w(r) \end{Bmatrix} = \Delta a \left[ \int_{k=0}^{\infty} J_1(k\Delta a) \begin{Bmatrix} F_{uw}(k)J_1(kr) \\ F_{ww}(k)J_o(kr) \end{Bmatrix} dk \right] r_o \qquad (4.45)$$

Because integrals over all possible wave numbers are involved, all types of waves are taken into account.

The horizontal distributed load acting in the $x$-direction, which is assumed to have a constant amplitude $p_o$, is discussed next (Fig. 4-9b). The radial and circumferential distributions vary as $p_o \cos\theta$ and $-p_o \sin\theta$, respectively. The first symmetric Fourier term is thus involved. The corresponding load amplitudes in the $k$-domain are calculated from the first two lines of Eq. 4.31a, formulated for the load amplitudes:

$$\begin{Bmatrix} p(k) \\ q(k) \end{Bmatrix} = \frac{p_o}{k} \int_{r=0}^{\Delta a} \begin{Bmatrix} rJ_1(kr)_{,r} + J_1(kr) \\ J_1(kr) + rJ_1(kr)_{,r} \end{Bmatrix} dr = \begin{Bmatrix} \dfrac{p_o \Delta a}{k} J_1(k\Delta a) \\ \dfrac{p_o \Delta a}{k} J_1(k\Delta a) \end{Bmatrix} \qquad (4.46)$$

The corresponding displacement amplitudes in the $k$-domain are determined from the flexibility matrix (condensed at node $i$) which follows from the dynamic-stiffness matrix of the layered half-space:

$$\begin{Bmatrix} u(k) \\ v(k) \\ w(k) \end{Bmatrix} = \begin{bmatrix} F_{uu}(k) \\ & F_{vv}(k) \\ F_{wu}(k) \end{bmatrix} \begin{Bmatrix} p(k) \\ q(k) \end{Bmatrix} \qquad (4.47)$$

Applying the inverse transformation of Eq. 4.31b to the displacements leads to the displacement amplitudes $u(r, \theta)$, $v(r, \theta)$, and $w(r, \theta)$. Substituting

## Sec. 4.4  Dynamic Stiffness of Embedded Foundation

Eqs. 4.46 and 4.47,

$$\begin{Bmatrix} u(r, \theta) \\ v(r, \theta) \\ w(r, \theta) \end{Bmatrix} = \frac{\Delta a}{2} \begin{bmatrix} \cos\theta & & \\ & -\sin\theta & \\ & & \cos\theta \end{bmatrix}$$

$$\times \left( \int_{k=0}^{\infty} J_1(k\Delta a) \begin{bmatrix} J_o(kr) - J_2(kr) & J_o(kr) + J_2(kr) & \\ J_o(kr) + J_2(kr) & J_o(kr) - J_2(Kr) & \\ & & -2 J_1(kr) \end{bmatrix} \begin{Bmatrix} F_{uu}(k) \\ F_{vv}(k) \\ F_{wu}(k) \end{Bmatrix} dk \right) p_o \quad (4.48)$$

results.

The coefficients in Eqs. 4.45 and 4.48 are the Green's functions.

Green's functions can also be determined for loads which do not act on a horizontal plane. Solutions for loads applied to a cylinder's surface are presented in Ref. [K6].

**Layered half-plane.** The Green's functions consisting of the displacements on the level $i$ are calculated in a layered half-plane for a constant distributed loading acting on a strip of width $2\Delta b$ at the node $j$ (Fig. 4-10).

The out-of-plane motion arising from the amplitude $q_o$ is addressed first. Formulating Eq. 4.24a for the load leads to its amplitude in the $k$-domain:

$$q(k) = \frac{q_o}{2\pi} \int_{-\Delta b}^{+\Delta b} \exp(ikx) \, dx = \frac{q_o}{\pi k} \sin k\Delta b \quad (4.49)$$

Solving the site's equation of motion for the displacement amplitude in the

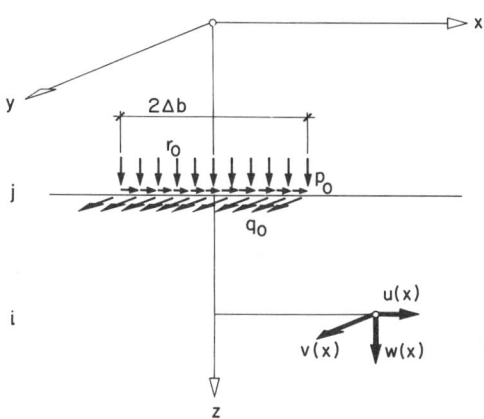

**Figure 4-10** Nomenclature for calculation of Green's functions in Cartesian coordinates.

$k$-domain leads to

$$v(k) = F_{vv}(k)q(k) \tag{4.50}$$

The element $F_{vv}(k)$ is the flexibility coefficient (condensed at node $i$) for the site's out-of-plane motion. The inverse transformation of Eq. 4.24b is equal to

$$v(x) = \int_{-\infty}^{+\infty} v(k) \exp(-ikx) \, dk \tag{4.51}$$

The matrices $[S_{SH}^L]$ (Eq. 4.22) and $S_{SH}^R$ (Eq. 4.23) depend on $kt$, which is an even function of $k$. The same thus also applies to $F_{vv}(k)$. Substituting Eqs. 4.49 and 4.50 in Eq. 4.51 leads to

$$v(x) = \frac{2}{\pi} \left[ \int_0^\infty \frac{\sin k\Delta b}{k} F_{vv}(k) \cos kx \, dk \right] q_o \tag{4.52}$$

For the in-plane motion, the load amplitudes $p(k)$ and $r(k)$ follow from Eq. 4.49, substituting the corresponding load amplitudes and intensities $p_o$ and $r_o$. Analogously, the flexibility equation in the $k$-domain formulated at the node $i$ is equal to

$$\begin{Bmatrix} u(k) \\ iw(k) \end{Bmatrix} = \begin{bmatrix} F_{uu}(k) & F_{uw}(k) \\ F_{wu}(k) & F_{ww}(k) \end{bmatrix} \begin{Bmatrix} p(k) \\ ir(k) \end{Bmatrix} \tag{4.53}$$

The elements $F_{uu}(k)$, $F_{ww}(k)$, and $F_{uw}(k)$ are even and odd functions of $k$, respectively. The inverse transformations leading to $u(x)$ and $w(x)$ follow from Eq. 4.51, replacing the variables. Performing the appropriate substitutions leads to

$$\begin{Bmatrix} u(x) \\ w(x) \end{Bmatrix} = \frac{2}{\pi} \left( \int_0^\infty \frac{\sin k\Delta b}{k} \begin{bmatrix} F_{uu}(k) \cos kx & F_{uw}(k) \sin kx \\ -F_{wu}(k) \sin kx & F_{ww}(k) \cos kx \end{bmatrix} dk \right) \begin{Bmatrix} p_o \\ r_o \end{Bmatrix} \tag{4.54}$$

The coefficients in Eqs. 4.52 and 4.54 are the Green's functions.

The Green's functions for a load which acts on an inclined line to the horizontal are specified in Ref. [W10].

### 4.4.2 Boundary-Integral Equation

**Reciprocity theorem of Maxwell-Betti.** The dynamic-reciprocity theorem specifies a relationship between a pair of solutions corresponding to two different loadings and boundary conditions. The displacements of the first state with amplitudes $\{u(x, \omega)\}$ are caused by the loading with amplitudes $\{f(x, \omega)\}$ acting in the volume $V$, denoted as $x$, and by the boundary conditions on $S$ (coordinate $s$) with the amplitudes of the displacements $\{u(s, \omega)\}$ and surface tractions $\{t(s, \omega)\}$. The second state is denoted with an asterisk. The

Sec. 4.4   Dynamic Stiffness of Embedded Foundation

dynamic-reciprocity theorem is stated as

$$\int_V \{f(x, \omega)\}^T \{u^*(x, \omega)\} \, dv + \int_S \{t(s, \omega)\}^T \{u^*(s, \omega)\} \, ds$$
$$= \int_V \{f^*(x, \omega)\}^T \{u(x, \omega)\} \, dv + \int_S \{t^*(s, \omega)\}^T \{u(s, \omega)\} \, ds \quad (4.55)$$

**Representation theorem.** To express the displacement at a location within $V$ as a function of the applied loading and of the boundary conditions, Green's functions are used for the state denoted by an asterisk. Applying a unit load amplitude at the location $x'$ in the direction $j$ (where the displacement amplitude is $u_j(x', \omega)$) and denoting the Green's functions (fundamental solutions) for the displacement amplitudes as $\{g_u(x, \omega)\}$ in $V$ and $\{g_u(s, \omega)\}$ on $S$ and for the traction amplitudes as $\{g_t(s, \omega)\}$ on $S$ transforms Eq. 4.55 to

$$u_j(x', \omega) = \int_V \{f(x, \omega)\}^T \{g_u(x, \omega)\} \, dv$$
$$+ \int_S \{t(s, \omega)\}^T \{g_u(s, \omega)\} \, ds - \int_S \{u(s, \omega)\}^T \{g_t(s, \omega)\} \, ds \quad (4.56)$$

For when the point $x'$ lies outside the volume $V$ (and also not on the boundary $S$) and for vanishing body loads,

$$\int_S \{g_u(s, \omega)\}^T \{t(s, \omega)\} \, ds = \int_S \{g_t(s, \omega)\}^T \{u(s, \omega)\} \, ds \quad (4.57)$$

applies.

Through a limiting process, selecting the point $x'$ on the boundary $S$, the following representation theorem for a smooth boundary is derived:

$$\frac{1}{2} u_j(s', \omega) = \int_V \{f(x, \omega)\}^T \{g_u(x, \omega)\} \, dv$$
$$+ \int_S \{t(s, \omega)\}^T \{g_u(s, \omega)\} \, ds - \int_S \{u(s, \omega)\}^T \{g_t(s, \omega)\} \, ds \quad (4.58)$$

For vanishing body loads, this singular integral equation relates the amplitudes of the displacements and of the surface tractions on the boundary.

### 4.4.3 Different Formulations of Boundary-Integral Equations and Their Spatial Discretization

The calculation of the unbounded soil's (system ground's) dynamic-stiffness matrix can be based on the reciprocity theorem or on the limiting case of the representation theorem. Before this concept is discussed, the so-called weighted-residual method is outlined, which represents an application of the superposition principle. Other approaches based on modified boundary-integral equations also exist [L4].

In all these formulations Green's functions will appear. As these analytical solutions satisfy the radiation condition, as well as the boundary conditions at the free surface and the interfaces (nodes) of the layered halfspace (system free field), the discretization is restricted to the structure–soil interface [W9, W10].

**Weighted-residual method.** In Fig. 4-11, the nomenclature is illustrated for the calculation of $[S_{bb}^g]$ for the in-plane motion. The embedded foundation's structure–soil interface $S$ is discretized using boundary elements compatible with the adjacent finite-element model of the structure. The prescribed displacement amplitudes on $S$ are specified as

$$\{u(s)\} = [N(s)] \{u_b\} \tag{4.59}$$

where $\{u(s)\}$ contains the three elements $u(s)$, $v(s)$, and $w(s)$ in the directions of the three coordinate axes $x$, $y$, $z$. $[N(s)]$ are the shape functions and $\{u_b\}$ denotes the nodal values. The distribution of load amplitudes $\{p(s')\}$, with the three components $p(s')$, $q(s')$, and $r(s')$ that have initially unknown intensities $\{p\}$, is assumed to act on a source surface $S'$. $S'$ is always offset towards the soil region to be excavated, in the limit by an infinitesimal amount. The loads act on the dynamic system consisting of the continuous soil—that is, on the layered half-space without excavation (free field). $\{p(s')\}$ is expressed, as a function of the source parameters,

$$\{p(s')\} = [L(s')] \{p\} \tag{4.60}$$

where $[L(s')]$ are the interpolation functions. Discontinuities can be introduced as shown in Fig. 4-11c. The number of source parameters must be larger than or equal to the order of $\{u_b\}$. The amplitudes of the displacement $\{u_p(s)\}$, with the components $u_p(s)$, $v_p(s)$, and $w_p(s)$, and the amplitudes of the surface traction $\{t_p(s)\}$, with the components $t_{px}(s)$, $t_{py}(s)$, and $t_{pz}(s)$ on the surface $S$, which subsequently will form the structure–soil interface, are formulated as

$$\{u_p(s)\} = [g_u(s)] \{p\} \tag{4.61}$$

$$\{t_p(s)\} = [g_t(s)] \{p\} \tag{4.62}$$

$[g_u(s)]$ and $[g_t(s)]$ are the Green's functions of the continuous layered half-space (free field).

The displacement-boundary condition on $S$ can be satisfied only in an average sense as

$$\int_S [W(s)]^T (\{u_p(s)\} - \{u(s)\}) \, ds = \{0\} \tag{4.63}$$

Sec. 4.4    Dynamic Stiffness of Embedded Foundation    197

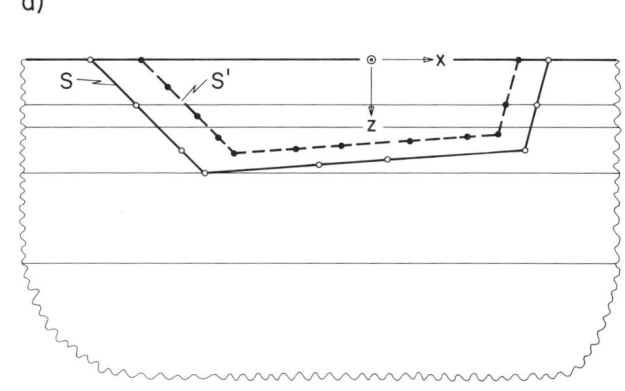

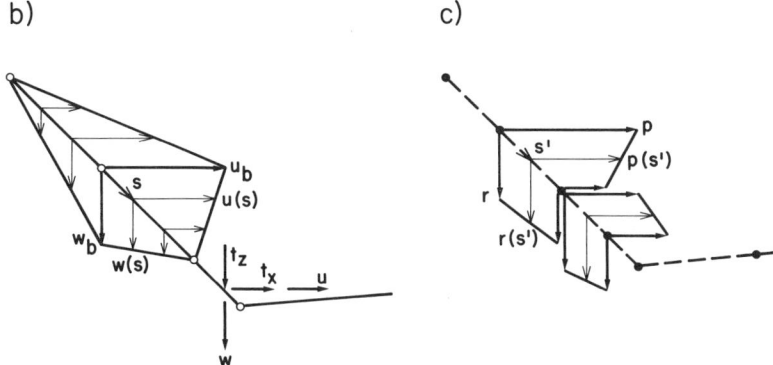

**Figure 4-11** Elements of discretization.  a. Source surface and structure–soil interface.  b. Prescribed displacement.  c. Selected load distribution.

with the weighting-function matrix $[W(s)]$. Substituting Eq. 4.61 in Eq. 4.63 leads to the generalized stress–displacement relationship

$$[G]\{p\} = [U]\{u_b\} \tag{4.64}$$

with the strain–displacement matrix specified as

$$[U] = \int_S [W(s)]^T [N(s)]\, ds \tag{4.65}$$

and the (nonsymmetric) flexibility matrix as

$$[G] = \int_S [W(s)]^T [g_u(s)]\, ds \tag{4.66}$$

The concentrated interaction forces with amplitudes $\{R_b\}$ are obtained as

$$\{R_b\} = \int_S [N(s)]^T \{t_p(s)\} \, ds \qquad (4.67)$$

Solving for $\{p\}$ in Eq. 4.64 and substituting in Eq. 4.62 and in Eq. 4.67 results in

$$\{R_b\} = [T]^T [G]^{-1} [U] \{u_b\} \qquad (4.68)$$

with

$$[T] = \int_S [g_t(s)]^T [N(s)] \, ds \qquad (4.69)$$

The dynamic-stiffness matrix $[S_{bb}^g]$ is thus equal to

$$[S_{bb}^g] = [T]^T [G]^{-1} [U] \qquad (4.70)$$

which will, in general, not be symmetric.

**Indirect boundary-element method.** As a special case of the weighted-residual technique, the weighting function's matrix $[W(s)]$ is chosen to be equal to the matrix of the Green's functions for the surface tractions $[g_t(s)]$. The matrix $[U]$ (Eq. 4.65) is equal to $[T]$ (Eq. 4.69), and $[G]$ is specialized as

$$[G] = \int_S [g_t(s)]^T [g_u(s)] \, ds \qquad (4.71)$$

The flexibility matrix is symmetric, as is easily verified using the reciprocity theorem.

The dynamic-stiffness matrix is equal to

$$[S_{bb}^g] = [T]^T [G]^{-1} [T] \qquad (4.72)$$

This choice of the weighting function thus guarantees the symmetry of $[S_{bb}^g]$

The same dynamic-stiffness matrix can also be derived starting with the reciprocity theorem (Eq. 4.57, modified slightly)

$$\int_S \{t(s)\}^T \{u_p(s)\} \, ds = \int_S \{t_p(s)\}^T \{u(s)\} \, ds \qquad (4.73)$$

$\{t(s)\}$ denotes the surface tractions' amplitudes corresponding to the prescribed displacement amplitudes $\{u(s)\}$. Assuming that this $\{t(s)\}$ can be expressed as a function of the same source load amplitudes

$$\{t(s)\} = [g_t(s)] \{p\} \qquad (4.74)$$

and making the appropriate substitutions in Eq. 4.73 leads to Eq. 4.64.

## Sec. 4.4  Dynamic Stiffness of Embedded Foundation

**Direct boundary-element method.** The starting point of this formulation is again the reciprocity theorem (Eq. 4.57). The surface tractions, with amplitudes $t(s)$ corresponding to the prescribed displacements, are interpolated as

$$\{t(s)\} = [M(s)]\,\{t\} \qquad (4.75)$$

where $\{t\}$ contains $t_x$, $t_y$, and $t_z$ in all nodal points. The order of $\{t\}$ must be equal to that of $\{p\}$. In principle, the nodes for the surface tractions can be selected differently from those associated with the dynamic-stiffness matrix.

Substituting Eqs. 4.75, 4.59, 4.61, and 4.62 in Eq. 4.57 leads to the stress–displacement relationship

$$[H]\,\{t\} = [T]\,\{u_b\} \qquad (4.76)$$

with the (nonsymmetric) flexibility matrix

$$[H] = \int_S [g_u(s)]^T\,[M(s)]\,ds \qquad (4.77)$$

Solving Eq. 4.76 for $\{t\}$ and integrating the surface tractions as

$$\{R_b\} = \int_S [N(s)]^T\,\{t(s)\}\,ds \qquad (4.78)$$

results in the (nonsymmetric) dynamic-stiffness matrix

$$[S^g_{bb}] = [V]^T\,[H]^{-1}\,[T] \qquad (4.79)$$

with

$$[V] = \int_S [M(s)]^T\,[N(s)]\,ds \qquad (4.80)$$

The standard direct boundary-element method based on the singular integral equation (Eq. 4.58 with vanishing interior loading) is contained in this formulation (see Ref. [W9] for details). Working directly on the boundary, with the Green's functions (fundamental solutions) for concentrated loads, leads to singularities, which can, however, be removed [R2]. The generalized strain–displacement matrix is evaluated in the Cauchy principle sense.

The weighted-residual method, and the indirect and direct boundary-element methods are summarized in Table 4-2.

A comparison between the accuracy of the three methods in Ref. [W9] indicates that the indirect boundary-element method leads in many cases to the most accurate results. The guaranteed symmetry in more complicated cases and the fact that the displacements arising from the applied loads can be calculated (Eq. 4.61) and compared to the prescribed displacements (Eq. 4.59) makes the indirect boundary-element method especially attractive for calculating the soil's dynamic-stiffness matrix.

**TABLE 4-2 Various Formulations of the Boundary-Element Method to Calculate the Soil's Dynamic-Stiffness Matrix**

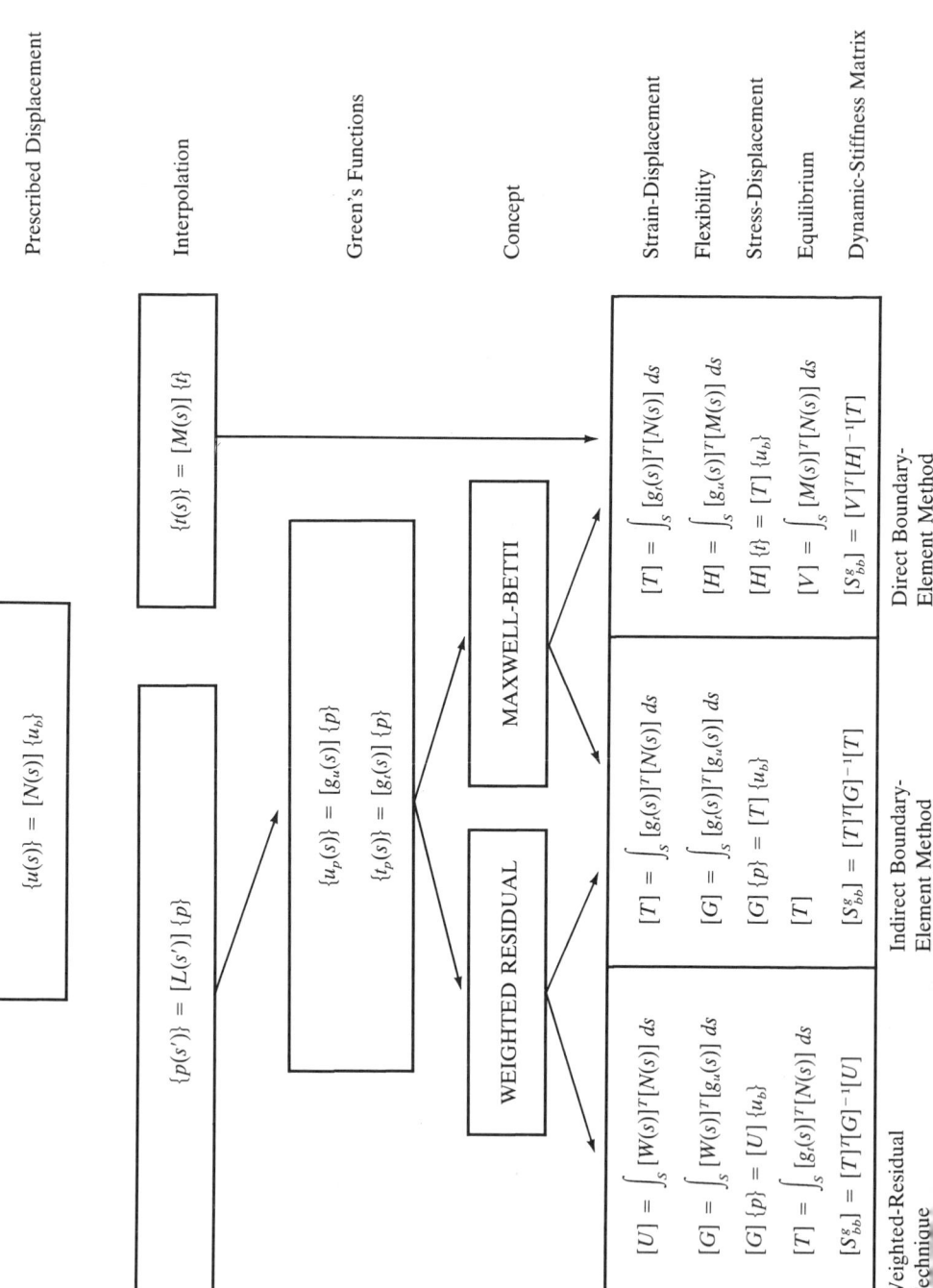

Sec. 4.5    Scattered Motion    201

Instead of calculating the embedded foundation's (system ground's) dynamic-stiffness matrix with the boundary-element method, it can be determined as the difference between those of the free field and of the excavated part (Eq. 4.4). The excavated part's dynamic stiffness can be calculated by the finite-element method; the free field's can be determined by the boundary-element method, as discussed next.

**Dynamic stiffness of free field.** Although for the continuous system $f$ the line $S$ is not a boundary, the concept of the boundary-element method can still be applied. The loaded source surface $S'$ must coincide with the line $S$ to determine $[S_{bb}^f]$. The applied loads $\{p(s)\}$ specified in Eq. 4.60 are integrated with the shape function $[N(s)]$ to determine the concentrated forces $\{R_b\}$ for the weighted-residual technique and the indirect boundary-element method. $[g_t(s)]$ in Eq. 4.62 is thus replaced by $[L(s)]$, which affects the $[T]$ matrix (Eq. 4.69), for the weighted-residual technique, and the $[T]$ matrix and the $[G]$ matrix (Eq. 4.71), for the indirect boundary-element method. $\{t_p(s)\}$ appearing in Eq. 4.62 is thus replaced by the load $\{p(s)\}$. For the direct boundary-element method, $[g_t(s)]$ in Eq. 4.69 is again replaced by $[L(s)]$. Also, $\{t(s)\}$ in Eq. 4.75 is actually a load. For easy reference, $[S_{bb}^f]$ is specified

$$[S_{bb}^f] = \left( \int_S [L(s)]^T [N(s)] \, ds \right)^T \left( \int_S [L(s)]^T [g_u(s)] \, ds \right)^{-1} \int_S [L(s)]^T [N(s)] \, ds \tag{4.81}$$

It is important to realize that, for the calculation of $[S_{bb}^f]$, $[g_t(s)]$ is not needed and thus does not need to be calculated. Only one Green's function $([g_u(s)])$ is determined in this case.

## 4.5 SCATTERED MOTION

### 4.5.1 Introductory Remarks

When an incident wave propagating in a medium encounters an obstacle with either zero tractions or zero displacements prescribed on its surface, additional waves are generated. This phenomenon is called the *scattering* (or *diffraction*) *of waves*. In the context of the basic equation of motion, the free-field motion in the layered half-space corresponds to the incident wave, and the excavation corresponds to a free surface to the obstacle. The total motion which occurs—that is, the sum of the free field one and the additional one—is called the generalized scattered motion. It is denoted as $\{u_b^g\}$ in Eq. 4.8 and can easily be determined once $[S_{bb}^g]$ and $[S_{bb}^f]$ are known.

The generalized scattered motion need not be calculated to analyze soil–structure interaction, because the loading-amplitude vector can be expressed as $[S_{bb}^f]\{u_b^f\}$ (Eq. 4.1)—that is, the free-field motion $\{u_b^f\}$ appears. However, valuable physical insight can be gained. For a more general site than the horizontally layered half-space, such as is shown in Fig. 1-2b, the motion before excavation (which corresponds to the free-field motion) must be determined.

The procedures to calculate such an irregular site's motion are also addressed in this section.

The dynamic-reciprocity theorem in the form of the representation theorem with an incident wave field (free-field motion) is stated in Section 4.5.2. The different formulations, such as the weighted-residual method and the indirect and direct boundary-element methods, to calculate the generalized scattered motion directly (and not by using Eq. 4.8) are summarized in Section 4.5.3. To perform a soil–structure interaction analysis, the product $[S_{bb}^f]\{u_b^f\}$ is needed. Based on the weighted-residual and indirect boundary-element methods, this load vector can be expressed in a compact form using the same matrices as already calculated for the dynamic-stiffness matrix. This concept is examined in Section 4.5.4 where, in addition, the relations between the scattering and the corresponding radiation problems are developed. In a radiation problem, non-zero displacements or tractions are prescribed on the obstacle's surface in the absence of incident waves. The calculation of the dynamic-stiffness matrix thus represents such a radiation problem.

### 4.5.2 Boundary-Integral Equation

In the presence of a free-field motion (or incident waves), the representation theorem for a source on the surface $S$ corresponding to Eq. 4.58 for vanishing body loads is formulated as [R2]

$$\frac{1}{2} u_j^t(s') = \int_S \{g_u(s)\}^T \{t^t(s)\} \, ds - \int_S \{g_t(s)\}^T \{u^t(s)\} \, ds + u_j^f(s') \qquad (4.82)$$

$u_j^f(s')$ represents the amplitude of the $j$th component of the free-field displacement at the source with coordinate $s'$ located on the smooth boundary $S$.

$\{u^t(s)\}$ and $\{t^t(s)\}$ denote the amplitudes of the total displacements and surface tractions on $S$. For the scattering problem, $\{u^t(s)\}$ will be equal to the generalized scattered motion $\{u^g(s)\}$ (free field and additional), and $\{t^t(s)\}$ will vanish on the free surface.

If the source point lies in the region's interior and not on its boundary, Eq. 4.82 still applies, with the factor 0.5 on the left-hand side replaced by 1 (analogous to Eq. 4.56).

The contribution of the unbounded soil's interaction forces to the basic equation of motion can be derived based on Eq. 4.82. Formulating this equation repeatedly (order of $\{u_b^t\}$ times), the corresponding surface tractions can be expressed in discretized form as a function of $\{u_b^t\}$ and of $\{u_b^f\}$ (direct boundary-element method). After integrating the surface tractions to form the concentrated interaction forces $\{R_b\}$, the coefficient matrix of $\{u_b^t\}$ will form the dynamic-stiffness matrix $[S_{bb}^g]$, and that of $\{u_b^f\}$ will be equal to $-[S_{bb}^f]$. The interaction force–displacement relationship presented in Eq. 4.3 (using Eq. 4.5) is thus derived.

### 4.5.3 Different Formulations of Boundary-Integral Equations and Their Spatial Discretization

The procedures to determine the generalized scattered motion are analogous to those described to calculate the dynamic-stiffness matrix of an embedded foundation (Section 4.4.3). The equations are developed for a layered half-space with excavated soil, as shown in Fig. 4-11. As the Green's functions satisfy the conditions on the free surface and on the layers' interfaces, the discretization is restricted to the structure–soil interface $S$. The amplitudes of the free-field motion's displacements and surface tractions are denoted as $\{u^f(s)\}$ and $\{t^f(s)\}$. On this surface, the resulting tractions (free-field and additional) must vanish. An application to a more general site is also addressed towards the end of this section.

**Weighted-residual method.** Fictitious loads with amplitudes $\{p(s')\}$ acting on the source surface $S'$ are introduced in the weighted-residual method. The interpolation relation (Eq. 4.60) and the definitions of the Green's functions on $S$ ( Eqs. 4.61, 4.62) still apply. The surface-traction boundary condition on $S$ is formulated as

$$\int_S [W(s)]^T (\{t_p(s)\} + \{t^f(s)\}) \, ds = 0 \tag{4.83}$$

with the weighting-function matrix $[W(s)]$. Substitution of Eq. 4.62 in Eq. 4.83 results in

$$[E]\{p\} = \{B\} \tag{4.84}$$

with

$$[E] = \int_S [W(s)]^T [g_t(s)] \, ds \tag{4.85}$$

$$\{B\} = -\int_S [W(s)]^T \{t^f(s)\} \, ds \tag{4.86}$$

Solving for $\{p\}$ in Eq. 4.84, and substituting in

$$\{u^g(s)\} = \{u^f(s)\} + \{u_p(s)\} \tag{4.87}$$

and using Eq. 4.61 leads to the generalized scattered motion

$$\{u^g(s)\} = \{u^f(s)\} + [g_u(s)] [E]^{-1} \{B\} \tag{4.88}$$

**Indirect boundary-element method.** As a special case of the weighted-residual technique, $[W(s)]$ is chosen as $[g_t(s)]$. Equation 4.88 still applies, with

$$[E] = \int_S [g_t(s)]^T [g_t(s)] \, ds \tag{4.89}$$

$$\{B\} = -\int_S [g_t(s)]^T \{t^f(s)\} \, ds \tag{4.90}$$

As an alternative, $[W(s)]$ can be selected as $[g_u(s)]$. In both cases, the matrix $[E]$ is symmetric.

**Direct boundary-element method.** The direct boundary-element method can be derived starting with the representation theorem. Equation 4.57 is written in slightly modified form as

$$\int_S \{t_p(s)\}^T \{u(s)\} \, ds = \int_S \{u_p(s)\}^T \{t(s)\} \, ds \tag{4.91}$$

Substituting the surface-traction amplitudes

$$\{t(s)\} = -\{t^f(s)\} \tag{4.92}$$

into Eq. 4.91 and denoting the corresponding displacement amplitudes as $\{u(s)\}$, which are interpolated as

$$\{u(s)\} = [N(s)] \{u_b\} \tag{4.93}$$

leads to

$$[E] \{u_b\} = \{C\} \tag{4.94}$$

with

$$[E] = \int_S [g_t(s)]^T [N(s)] \, ds \tag{4.95}$$

$$\{C\} = -\int_S [g_u(s)]^T \{t^f(s)\} \, ds \tag{4.96}$$

Solving for $\{u_b\}$ in Eq. 4.94, using Eq. 4.93, and substituting in Eq. 4.87 with the subscript $p$ deleted, results in

$$\{u^g(s)\} = \{u^f(s)\} + [N(s)] [E]^{-1} \{C\} \tag{4.97}$$

Alternatively, the scattered motion can be determined using Eq. 4.82 (see next section).

## Sec. 4.5 Scattered Motion

**Irregular site.** The application of the indirect and direct boundary-element methods to the analysis of the motion in an irregular site as shown in Fig. 1-2b is sketched in the following.

As a first example, the calculation of the scattered motion in a site consisting of two layered half-spaces displaced with respect to one another along an inclined interface (Fig. 4-12a) is addressed [W1]. The indirect boundary-element method is applied.

The truncated half-space 1 is augmented to form a (horizontally layered) complete half-space for which Green's functions can be calculated (Fig. 4-12b). On the source surface $S_1'$, offset by an infinitesimal amount from the surface $S$ on which the boundary conditions between the two half-spaces are formulated, fictitious source loads with amplitudes $\{p_1(s')\}$ are applied. These are expressed as a function of the source parameters $\{p_1\}$ associated with the

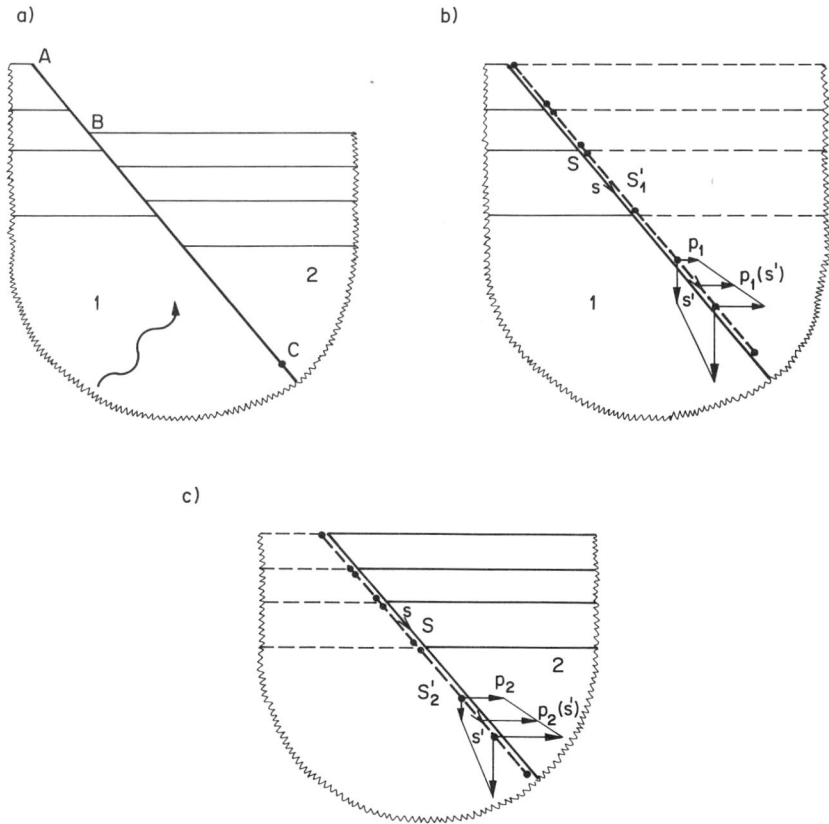

**Figure 4-12** Response of irregular site.  a. Site consisting of two truncated half-spaces with ground surfaces at different levels.  b. First half-space with source loads.  c. Second half-space with source loads.

nodes on $S_1'$. Since the boundary conditions between the two half-spaces cannot be enforced up to infinity, but only up to Point C selected at a sufficient distance, the source surface $S_1'$ is also finite. The amplitudes of the displacements $\{u_{p1}(s)\}$ and of the surface tractions $\{t_{p1}(s)\}$ on $S$ are formulated as

$$\{u_{p1}(s)\} = [g_u^1(s)] \{p_1\} \tag{4.98}$$

$$\{t_{p1}(s)\} = [g_t^1(s)] \{p_1\} \tag{4.99}$$

where $[g_u^1(s)]$ and $[g_t^1(s)]$ are the corresponding Green's functions calculated for the complete half-space 1, using the procedure of Section 4.4.1.

The amplitudes of the free-field displacements and surface tractions on $S$, again determined for the complete half-space 1 (Section 4.3), are denoted as $\{u_1^f(s)\}$ and $\{t_1^f(s)\}$. The half-space 2 is treated analogously (Fig. 4-12c).

The boundary conditions formulated on the surface segments $A$-$B$ and $B$-$C$ are specified as

$$A\text{-}B: \quad \{t_1^f(s)\} + [g_t^1(s)] \{p_1\} = 0 \tag{4.100}$$

$$B\text{-}C: \quad \{u_1^f(s)\} + [g_u^1(s)] \{p_1\} = \{u_2^f(s)\} + [g_u^2(s)] \{p_2\} \tag{4.101}$$

$$B\text{-}C: \quad \{t_1^f(s)\} + [g_t^1(s)] \{p_1\} + \{t_2^f(s)\} + [g_t^2(s)] \{p_2\} = 0 \tag{4.102}$$

This system of equations can be solved by transforming it to a weighted-residual statement. After determining the unknowns $\{p_1\}$ and $\{p_2\}$, the resulting displacements and surface tractions in any point of the system can be calculated by superimposing on the free-field response the response governed by the fictitious loads, which is determined by using the Green's functions of the corresponding complete half-space.

As a second example, an irregular site consisting of a horizontally layered half-space, with an infilled cut-out and an inclusion (Fig. 4-13) [K5], is addressed. The direct boundary-element method based on Eq. 4.82 is applied.

Formulating Eq. 4.82 for the layered half-space 1 repeatedly in all directions in the nodes corresponding to the elements located on the boundaries $S$ between regions 1 and 2 and between 1 and 3 leads to

$$\frac{1}{2}\{u_1\} = \int_S [g_u^1(s)]^T \{t_1(s)\} \, ds - \int_S [g_t^1(s)]^T \{u_1(s)\} \, ds + \{u_1^f\} \tag{4.103}$$

$[g_u^1(s)]$ and $[g_t^1(s)]$ represent the Green's functions for the amplitudes of the displacements and surface tractions on the boundaries determined for the continuous (complete) layered half-space 1. $\{u_1^f\}$ denotes the amplitudes of the free-field displacements in the various points on the boundaries, determined for the half-space 1.

Proceeding analogously for the infilled cut-out in the same points on

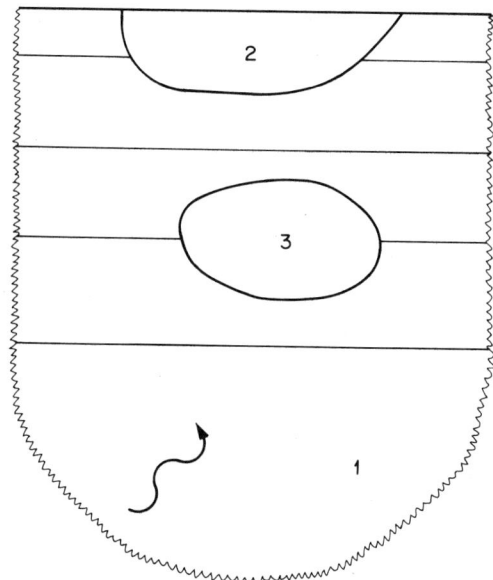

**Figure 4-13** Irregular site with infilled cut-out and inclusion.

the boundary between regions 2 and 1 results in

$$\frac{1}{2}\{u_2\} = \int_S [g_u^2(s)]^T \{t_2(s)\}\, ds - \int_S [g_t^2(s)]^T \{u_2(s)\}\, ds \qquad (4.104)$$

$[g_u^2(s)]$ and $[g_t^2(s)]$ are the Green's functions for a half-space with the material properties of the infilled cut-out, which avoids the discretization of the infilled cut-out's free surface. No term describing the free-field motion appears in Eq. 4.104.

Formulating Eq. 4.82 analogously for the inclusion 3 in the same points on the boundary between regions 3 and 1,

$$\frac{1}{2}\{u_3\} = \int_S [g_u^3(s)]^T \{t_3(s)\}\, ds - \int_S [g_t^3(s)]^T \{u_3(s)\}\, ds \qquad (4.105)$$

is obtained. $[g_u^3(s)]$ and $[g_t^3(s)]$ are the Green's functions for the full infinite space with the material properties of the inclusion.

The conditions are formulated for the boundary between regions 1 and 2 as

$$\{u_1(s)\} = \{u_2(s)\} \qquad (4.106)$$

$$\{t_1(s)\} + \{t_2(s)\} = 0 \qquad (4.107)$$

and for the boundary between regions 1 and 3 as

$$\{u_1(s)\} = \{u_3(s)\} \qquad (4.108)$$

$$\{t_1(s)\} + \{t_3(s)\} = 0 \qquad (4.109)$$

After interpolating the amplitudes of the displacements and of the surface tractions analogously to the procedure used in Eqs. 4.93 and 4.75, and after performing the corresponding integrations, Eqs. 4.103 to 4.109 can be solved for the unknowns, which are the nodal values of the amplitudes of the generalized scattered motion's displacements and surface tractions. The amplitudes of the displacements and surface tractions on the boundaries are thus known.

To determine the motion in any point in the region's interior—for example, in the layered half-space (amplitude $u_j^l$), Eq. 4.82 is again formulated, but with the factor 0.5 on the left-hand side replaced by 1

$$u_j^l = \int_S \{g_u^l(s)\}^T \{t_1(s)\} \, ds - \int_S \{g_t^l(s)\}^T \{u_1(s)\} \, ds + u_j^f \qquad (4.110)$$

The pair $\{t_1(s)\}$ and $\{u_1(s)\}$ are now known. The amplitude of the free-field displacement in the $j$th direction in the interior point is denoted as $u_j^f$.

### 4.5.4 Load Vector

As already pointed out, the generalized scattered motion $\{u_b^g\}$ need not actually be calculated to determine the loading-amplitude vector of the basic equation of motion (Eq. 4.1). Only the product $[S_{bb}^f] \{u_b^f\}$ appears on the right-hand side. If the dynamic-stiffness matrix of the soil $[S_{bb}^g]$ is calculated based on the indirect boundary-element method using Eq. 4.72 with the matrices $[G]$ (Eq. 4.71) and $[T]$ (Eq. 4.69), the loading term can be expressed directly using the same matrices $[G]$ and $[T]$ ([L5]).

The vector $[S_{bb}^f] \{u_b^f\}$ can be interpreted as follows. Each row of $[S_{bb}^f]$ represents the concentrated loads' amplitudes in the nodes $b$ for a unit amplitude of the displacement corresponding to the row. An element of the product $[S_{bb}^f] \{u_b^f\}$ is equal to the work of these concentrated load amplitudes and the free-field displacement amplitudes in the same nodes. From virtual work considerations (or from applying the dynamic-reciprocity theorem), it follows that this work is equal to that calculated using the amplitudes of the source loads $\{p(s')\}$ (which lead to the unit amplitude of the displacement) and the corresponding free-field displacement amplitudes on the source surface $S'$ (Fig. 4-11).

$$[S_{bb}^f] \{u_b^f\} = \int_{S'} \{p(s')\}^T \{u^f(s')\} \, ds' \qquad (4.111)$$

Solving for $\{p\}$ in Eq. 4.64, with $[T]$ replacing $[U]$ for a unit displacement amplitude and substituting in Eq. 4.60, transforms Eq. 4.111 to

$$[S_{bb}^f] \{u_b^f\} = \int_{S'} [T]^T [G]^{-1} [L(s')]^T \{u^f(s')\} \, ds' \qquad (4.112)$$

When the source surface $S'$ is offset from the soil–structure interaction $S$ by an infinitesimal amount, the integration in Eq. 4.112 is performed over $S$ using the free-field motion $\{u^f(s)\}$.

In this formulation, the amplitudes of the free-field tractions $\{t^f\}$ need not be calculated.

If the dynamic-stiffness matrix $[S^f_{bb}]$ is calculated using the indirect boundary-element method as specified in Eq. 4.81, then $[g_t(s)]$ is replaced by $[L(s)]$ in the definitions of $[G]$ (Eq. 4.71) and $[T]$ (Eq. 4.69). Equation 4.112 still applies.

## SUMMARY

1. By merging the two substructures—the actual structure and the soil with excavation—the basic equation of motion in the frequency domain is derived. The total displacements' amplitudes are associated with nodes within the structure and on the structure–soil interface. The corresponding coefficient matrix is formed by assembling the dynamic-stiffness matrices of the discretized structure and of the unbounded soil with excavation with degrees of freedom along the structure–soil interface (base). The corresponding load vector equals the product of the dynamic-stiffness matrix of the soil (free field) and the free-field-motion vector, both determined only in those nodes which subsequently will lie on the structure–soil interface. Alternatively, the load vector can also be determined as the product of the dynamic-stiffness matrix of the soil with excavation and the vector of the corresponding generalized scattered-input motion.

2. In a layered half-space, two types of (inclined) body waves exist: $P$- and $S$-waves. For a $P$-wave, involving a volumetric strain only, the particle motion coincides with the direction along which the wave propagates with the (material-dependent) dilatational-wave velocity and is constant over a plane perpendicular to it. For an $S$-wave, with a distortional strain only, the particle motion takes place in a plane perpendicular to the direction of propagation and is constant over this plane. This motion, which propagates with the (material-dependent) shear-wave velocity, can be decomposed into a horizontal $SH$-wave and a vertical $SV$-wave. Assuming the directions of propagation of the $P$- and $S$-waves to lie in a vertical plane, the $SH$-wave will result in a (horizontal) out-of-plane displacement, which is independent of the in-plane displacements caused by the $P$- and $SV$-waves. The phase velocity (and wave number) is common to the $P$- and $SV$-waves. Introducing material damping results in complex wave velocities.

   In the horizontal direction, all displacements and stresses for plane waves expressed in Cartesian coordinates vary exponentially, whereby an expansion in the wave-number domain (Fourier transform) is performed. In cylindrical coordinates, the displacements and stresses are decomposed circumferentially in a Fourier series and radially in a Bessel function involving the wave number. In the vertical direction, the variation is the same for Cartesian coordinates (plane waves) and for cylindrical coordinates.

3. Separately, for the out-of-plane and in-plane motions, the (symmetric) dynamic-stiffness matrices of a layer and of the half-space can be established. For a given frequency, they depend on the wave number (or the phase velocity), the material properties, and the layer's depth. The same dynamic-stiffness matrices apply to Cartesian and to cylindrical coordinates. Assembling the dynamic-stiffness matrices of the layers and of the half-space leads to the site's rigorous dynamic-stiffness matrix.

4. Starting from the prescribed control motion acting in the selected control point and assuming the nature of the wave pattern, the free-field motion is determined on the line which subsequently will form the structure–soil interface. The left-hand side of the discretized equations of motion is equal to the site's dynamic-stiffness matrix. The right-hand side is zero for the control point at the free surface. For the control point at the outcrop of the rock, the right-hand side is equal to the product of the half-space's dynamic-stiffness matrix and the control motion's vector. Selecting the angle of incidence of the inclined body waves in the half-space determines the apparent velocity (and the wave number). For surface waves, the site's dynamic-stiffness matrix is singular. For a specific frequency, this condition is satisfied only for distinct phase velocities associated with the different modes. Surface waves are thus, in general, dispersive.

5. In all boundary-element methods, Green's functions—that is, displacements and surface tractions—for prescribed fictitious loads are calculated. For the layered half-space, the specified distribution of the load is transformed in the two-dimensional case into a Fourier integral in the wave-number domain; and in the three-dimensional case formulated in cylindrical coordinates, it is expanded in a Fourier series in the circumferential direction, and into Bessel functions involving the wave-number domain in the radial direction. This determines the right-hand side of the site's discretized equations of motion, from which the displacement amplitudes in the wave-number domain follow. The inverse transformation leads to the Green's functions. As these Green's functions satisfy the conditions at the free surface and at the interfaces of the layered half-space, the discretization in the boundary-element method is confined to the (finite) structure–soil interface.

## Chap. 4  Summary

6. The various boundary-element methods to calculate an embedded foundation's dynamic-stiffness matrix are systematically developed. They are the weighted-residual technique using the principle of superposition, the indirect boundary-element method based on a weighted-residual statement, and the direct boundary-element method derived using the dynamic-reciprocity theorem. In the first two methods, fictitious loads with initially unknown intensities are applied on a source surface located outside the soil region to be investigated (excavated part). A weighting function is used; in the case of the indirect method, it is selected as the Green's function for the surface traction, which guarantees the symmetry of the dynamic-stiffness matrix. In the third method, the surface traction along the structure–soil interface is interpolated. The same type of boundary matrices which have a clear physical interpretation are identified in the three formulations.
7. Instead of calculating the embedded foundation's (system ground's) dynamic-stiffness matrix with the boundary-element method, it can be determined as the difference between the matrices of the regular free field and the excavated part. The calculation of the former using the boundary-element method does not require the Green's function for the surface traction. The excavated part's dynamic stiffness can be calculated by the finite-element method.
8. The generalized scattered motion can be calculated based on the same three boundary-element formulations. The basic equation of motion's load vector can be determined directly using the source loads calculated in the analysis of the soil's dynamic-stiffness matrix.

# 5

# Hybrid Frequency–Time-Domain Analysis

Rigorous procedures, based on the substructure method, to analyze nonlinear soil–structure interaction systems in the time domain are described in Chapters 6 and 7. In these methods, the unbounded soil's contribution to the equations of motion involves convolution integrals of, for example, the dynamic-stiffness coefficients in the time domain and the corresponding motions. The formulation is global in space and time—that is, all degrees of freedom on the (generalized) structure–soil interface contribute from the start of the excitation to the equations of motion at a specific time. Its computational effort is thus significant.

An attractive alternative to the procedure based on convolution integrals is a method where the nonlinearities only affect the right-hand side of the equations of motion through the loads calculated from the so-called pseudo-forces, but do not affect the coefficient matrix representing the dynamic stiffnesses. In this so-called hybrid frequency–time-domain analysis [K4], a series of linear analyses is performed with pseudo-loads which are recalculated after each analysis. The linear analyses are performed in the frequency domain, making use of the methods and the corresponding computer programs developed for linear soil–structure interaction calculations. The unbounded soil's (frequency-dependent) dynamic-stiffness, which takes the radiation condition of the semi-infinite domain and the hysteretic material damping into consideration, is thus rigorously taken into account. The nonlinearities are evaluated in the time domain as in the standard time-integration schemes. By running through the response in time, the pseudo-forces, needed to bring the linear forces in the system into coincidence with the true nonlinear forces and following from the material law for the same strains, are calculated. The pseudo-forces lead to pseudo-loads, which are then added to the prescribed exterior loads in the next iteration's linear analysis. The hybrid method is thus similar to the initial-stress approach for nonlinear dynamic analysis, as used in the finite-element method (constant-stiffness approach). However, in the former, the pseudo-forces are recalculated after each linear analysis for the total time history, and convergence is achieved, in principle, for the total time history; while in the latter, the pseudo-forces are only determined for the current time step, for which convergence is achieved before processing the next time step. The rigorous hybrid frequency–time-domain analysis procedure can be applied when otherwise only approximate methods would be used, such as working with frequency-independent springs and dampers (Sections 2.14, 3.6, 3.11).

The formulation of the hybrid frequency–time domain analysis is described in Section 5.1. As an illustrative example, a mass connected by a nonlinear spring to a semi-infinite rod on an elastic foundation is examined in Section 5.2. This simple system can be interpreted as a nonlinear structure interacting with a linear unbounded soil. In an iterative scheme such as the hybrid frequency–time-domain one, the conditions for which stability is achieved are of paramount importance. For the sake of illustration, the stability criterion for systems excited by harmonic loads is derived first for a one-degree-of-freedom system and then in Section 5.3, for a multiple-degree-of-freedom system. The rate of convergence is also vital. To prevent divergence from occurring, methods must be developed which limit the extent by which the intermediate results depart from the exact solution. It is possible to perform the iterations on only part of the total time interval at a time. This segmenting approach is examined in Section 5.4. In particular, the computational effort is discussed. Other issues of implementation are addressed. The stability criterion for systems excited by a transient loading is derived in Section 5.5.

## 5.1 FORMULATION

### 5.1.1 Basic Procedure

The hybrid frequency–time-domain solution scheme is explained for a system with material-type nonlinearity, whereby the constitutive law is assumed to be formulated in total stresses and strains. The dynamic system is the same as that discussed for the linear case, with loads applied directly to the structure, and the loading environment also specified by the free-field motion in nodes which will subsequently lie on the (generalized) structure–soil interface. The nomenclature is introduced in Section 4.1. (Fig. 4-1).

The procedure is summarized in Table 5-1. Steps 1 to 3 are performed initially, then steps 4 to 9 are performed for each iteration $j$ until the convergence criterion is satisfied.

**TABLE 5-1  Hybrid Frequency–Time-Domain Analysis**

*Initial Calculations*

1. Transform loads applied to structure and free-field motion from time to frequency domain:

$$\{R_s(t)\} \rightarrow \{R_s(\omega)\}$$
$$\{R_b^s(t)\} \rightarrow \{R_b^s(\omega)\}$$
$$\{u_b^f(t)\} \rightarrow \{u_b^f(\omega)\}$$

2. Form dynamic-stiffness and loads of total system in frequency domain:

$$[S_o(\omega)] = \begin{bmatrix} [S_{ss}(\omega)] & | & [S_{sb}(\omega)] \\ [S_{bs}(\omega)] & | & [S_{bb}(\omega)] + [S_{bb}^g(\omega)] \end{bmatrix}$$

$$\{R(\omega)\} = \begin{Bmatrix} \{R_s(\omega)\} \\ \{R_b^s(\omega)\} + [S_{bb}^f(\omega)]\{u_b^f(\omega)\} \end{Bmatrix}$$

3. Triangularize dynamic-stiffness in frequency domain (transfer function):

$$[S_o(\omega)] \rightarrow [S_o(\omega)]^{-1}$$

*For Each Iteration*

4. Calculate pseudo-forces in time domain:

$$\{P_o^j(t)\} = \{P_{nl}^{j-1}(t)\} - \{P_l^{j-1}(t)\} \quad (5.1)$$
$$(j = 0: \quad \{P_o^j(t)\} = \{0\})$$

5. Calculate pseudo-loads in time domain:

$$\{P_o^j(t)\} \rightarrow \{Q^j(t)\}$$

6. Transform pseudo-loads from time to frequency domain:

$$\{Q^j(t)\} \rightarrow \{Q^j(\omega)\}$$

7. Solve for total displacements in frequency domain:

$$\{u^j(\omega)\} = [S_o(\omega)]^{-1}(\{R(\omega)\} + \{Q^j(\omega)\}) \quad (5.2)$$

8. Transform displacements from frequency to time domain:

$$\{u^j(\omega)\} \rightarrow \{u^j(t)\}$$

9. Calculate internal forces in time domain:

$$\{P^j(t)\} = \{P_l^j(t)\} + \{P_o^j(t)\} \quad (5.3)$$

## Steps 1–3:

The transformation from the time to the frequency domain (using a Fast-Fourier-Transform algorithm) and the calculation in the frequency domain are governed by the same criteria as in the linear case. These address the need to include an interval of zero load to achieve a periodic response with vanishing initial conditions, the interpolation in the frequency domain of the transfer functions, and the choice of the time step and thus of the highest frequency present which will be determined by the load's Fourier transform and by the highest mode's natural frequency. In addition, it must be taken into consideration that, in a nonlinear system, hyperharmonic components corresponding to multiples of the excitation frequency can be generated. The latter will originate from the pseudo-loads in the hybrid frequency–time-domain analysis. In such a case, a smaller time step than in the linear case must be chosen. Nonlinear systems can often display unexpected behavior!

In Step 3, the dynamic-stiffness matrix of the total pseudo-linear system is triangularized—that is, decomposed into triangular matrices and a diagonal one. This is indicated symbolically in Table 5-1 by the inverse, which determines the transfer function. This operation is performed for those frequencies which serve as anchor points for the interpolation functions.

Steps 4 and 5:

Processing the total time-history, the internal nonlinear forces $\{P_{nl}^{j-1}(t)\}$ (subscript $nl$ for nonlinear) are determined for each time step, using the constitutive law corresponding to the total strains of the $(j-1)$th iteration. Subtracting the linear internal forces $\{P_{l}^{j-1}(t)\}$ (subscript $l$ for linear) calculated for the same strains leads to the pseudo-forces $\{P_{o}^{j}(t)\}$. The latter, which can be regarded as initial forces, are thus required to bring the linear solution into coincidence with the true nonlinear one. In a finite-element analysis, this calculation is performed for the stresses in the Gauss integration points. In the first pass through this part of the algorithm, when only specified exterior loads are processed (zeroth iteration), the pseudo-forces vanish. Formulating equilibrium leads to the pseudo-loads $\{Q^{j}(t)\}$. For a finite element, this is performed using the virtual-work equation, which will involve the strain–displacement matrix.

Steps 6–8:

After transforming the pseudo-loads to the frequency domain, the system's total displacements $\{u^{j}(\omega)\}$ are determined by backsubstitution (or multiplication with the transfer function). The superscript $t$ for total is omitted. These displacements transformed to the time domain $\{u^{j}(t)\}$ are the total displacements of the $j$th iteration and are used to calculate the total strains and linear internal forces within the structure $\{P_{l}^{j}(t)\}$. When the unbounded soil's interaction forces are needed in the iteration (as in the case of partial uplift of the basemat or separation of the base's sidewall), the linear part of these forces is calculated in the frequency domain (Eq. 4.3) and then transformed to the time domain.

## Step 9:

The total internal forces of the $j$th iteration are equal to the sum of the linear ones and the pseudo-forces. This step is only performed at the very end of the calculation or if the convergence criterion involves the total internal forces.

### 5.1.2 Convergence Criterion

The selected convergence criterion depends on the problem being solved. It can address the displacements, the out-of-balance loads, or the energy.

The change of the displacements between two consecutive iterations, expressed as a fraction of the final displacements, can be examined. With $\varepsilon_d$ denoting a displacement-convergence tolerance,

$$\frac{\|\{u^j(t)\} - \{u^{j-1}(t)\}\|}{\|\{u^j(t)\}\|} < \varepsilon_d \tag{5.4}$$

can be selected as the criterion which has to be satisfied for the whole time history. The Euclidean norm of a vector of order $n$, defined as

$$\|\{u\}\| = \sqrt{\sum_{i=1}^{n} u_i^2} \tag{5.5}$$

measures its "size." In Eq. 5.4, the final displacement, which should appear in the denominator but is unknown, is approximated by $\{u^j(t)\}$. All displacement components must have the same units.

The change of the pseudo-loads compared to the prescribed exterior loads can be chosen as the criterion, resulting in

$$\frac{\|\{Q^j(t)\} - \{Q^{j-1}(t)\}\|}{\|\{R(t)\}\|} < \varepsilon_p \tag{5.6}$$

Alternatively, the denominator of Eq. 5.6 can be replaced by $\|\{Q^1(t)\}\|$,—that is, the norm of the pseudo-loads in the first iteration.

The criterion can address both the displacements and the pseudo-loads in the form of energies. It can be formulated as

$$\frac{(\{u^j(t)\} - \{u^{j-1}(t)\})^T (\{Q^j(t)\} - \{Q^{j-1}(t)\})}{\{u^o(t)\}^T \{R(t)\}} < \varepsilon_e \tag{5.7}$$

Again, the denominator could be changed to $\{u^1(t)\}^T\{Q^1(t)\}$.

Finally, instead of a criterion which is satisfied for the whole time history as discussed above, it could be postulated to apply in the frequency domain.

### 5.1.3 Comments

The hybrid frequency–time-domain solution scheme is conceptionally simple and can easily be implemented in an existing nonlinear time-domain program or in a linear frequency-domain program.

The frequency-domain integration method is theoretically exact. The time step, which is determined by the response and not by the numerical integrator, can in many cases be chosen to be much larger than in time-domain methods, where, for an explicit method, stability constraints can dominate and where, for an implicit method, the size of the time step can be dictated by accuracy considerations to avoid excessive amplitude decay and period elongation.

When one knows that the nonlinearities will occur only in certain zones of the structure, the degrees of freedom in the remainder of the structure—which will remain linear—can be eliminated in the frequency domain, analogously to the procedure used in the static condensation process. The actual iterations (Steps 4 to 9 of Table 5-1), which are restricted to the zones of nonlinearity, are performed in a much smaller system.

Many modifications to the basic procedure described in Section 5.1.1 are possible, and in some cases even necessary, to arrive at a computationally attractive implementation [K4]. The formulation summarized in Table 5-1 works with the total load (prescribed exterior and pseudo) in each iteration. This total-displacement formulation allows initial approximations to be corrected more easily in later iterations than in alternative schemes where the structure is only loaded by the pseudo-loads or even their changes in the iterations. Many schemes developed for the constant-stiffness approach of the nonlinear finite-element analysis can be adopted, such as the use of over- and under-relaxation, resorting to artificial damping and applying the prescribed exterior load incrementally. The very efficient segmenting approach, whereby the total time interval is subdivided into segments, and for which convergence is achieved sequentially for each segment separately, is examined in Section 5.4.

Experience in applying the hybrid frequency–time domain procedure is limited and cannot yet be compared with the wealth of knowledge available for time-integration schemes.

## 5.2 MASS CONNECTED BY NONLINEAR SPRING TO SEMI-INFINITE ROD ON ELASTIC FOUNDATION

A simple example (Fig. 5-1) which can be interpreted as a nonlinear structure–soil system is used to illustrate the hybrid frequency–time-domain procedure. The nonlinear structure consists of a mass $m$ connected by a viscous damper, with a coefficient $c$, and a spring, with an elasto-plastic force-displacement relationship, to an elastic undamped semi-infinite rod on an elastic foundation that represents the unbounded linear soil. $k_s$ denotes the elastic spring constant and $F_y^+$ and $F_y^-$ the yielding values in tension and compression (Fig. 5-1b). The rod is characterized by $A$, $E$, $\rho$, and $k_g$, as introduced in Section

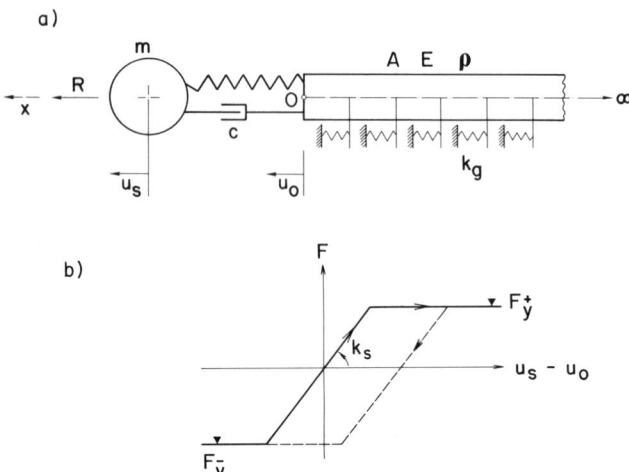

**Figure 5-1** Mass connected by nonlinear spring to semi-infinite rod on elastic foundation. a. Dynamic system. b. Elasto-plastic spring force-distortion relationship.

3.2.1. The mass point is subjected to the dynamic load

$$R(\bar{t}) = \frac{R_o}{2}\left[1 - \cos\left(2\pi \frac{\bar{t}}{\bar{t}_o}\right)\right] \qquad 0 < \bar{t} < \bar{t}_o$$
$$R(\bar{t}) = 0 \qquad \bar{t} > \bar{t}_o \qquad (5.8)$$

$\bar{t}$ is the dimensionless time (Eq. 3.65). $\bar{t}_o$ is selected as equal to 2. The displacements of the system with the two degrees of freedom are $u_s$ ($s$ for structure) and $u_o$. The following dimensionless ratios are selected: $\bar{k}_s = k_s/K = 1$, where $K$ is the static-stiffness coefficient of the semi-infinite rod (Eq. 3.79); $\bar{m} = m\kappa/(\rho A) = 0.5$; and $\bar{\zeta} = cc_l\kappa/(2K) = 0.3$. The ratios $F_y^+/R_o$ and $F_y^-/R_o$ are varied.

For the system shown in Fig. 5-1, a negative pseudo-force arises at the start of the first nonlinearity, which leads to a pseudo-load of the same magnitude acting on the mass in the positive $x$-direction and in the opposite direction in Point 0.

When working with a discrete Fourier transform, the total load is assumed to be periodic with a finite period. The pseudo-loads, in general, do not vanish at the end of the period. To allow the system's motion to be damped out at the end of the period, the pseudo-loads, after the time of interest for the response has passed, are gradually artificially reduced to zero, followed by the quiet zone up to the end of the (expanded) period (see Fig. 5-2b).

Selecting $F_y^+/R_o = 0.4$ and $F_y^-/R_o = \infty$ leads to a nonlinear behavior occurring only during the first excursion of the mass. Its displacement time-

Sec. 5.2    Mass Connected by Nonlinear Spring    219

histories, nondimensionalized by the static value $R_o/K$, $\bar{u}_s(\bar{t}) = u_s(\bar{t})/(R_o/K)$, corresponding to the initial phase (linear analysis with load $R(t)$), and after the first, second, and third iterations, are plotted in Fig. 5-2a. Convergence is achieved after three iterations. At the beginning, during the linear phase, the pseudo-force $P_o$ vanishes (Fig. 5-2b) and then remains constant until $\bar{t} = 25$ where it is gradually reduced to zero.

Also, when strong nonlinearities arise—as in the case $F_y^+/R_o = 0.1$, $F_y^-/R_o = 0$—the hybrid method convergences but after many more iterations. The responses after the 26th iteration up to the 30th iteration are plotted in Fig. 5-3. After 30 iterations, convergence is achieved for $\bar{t} < 25$. $\bar{u}_s(\bar{t})$ of Fig. 5-3a coincides with the result of the convolution-integral approach, using the dynamic-stiffness coefficient in the time domain, which is discussed in Chapter 6 (Fig. P 6-11b).

It is observed that the pseudo-force history (and thus the solution) corrects itself in a time progressing manner. The earlier part of the response-time history converges faster than the later one does.

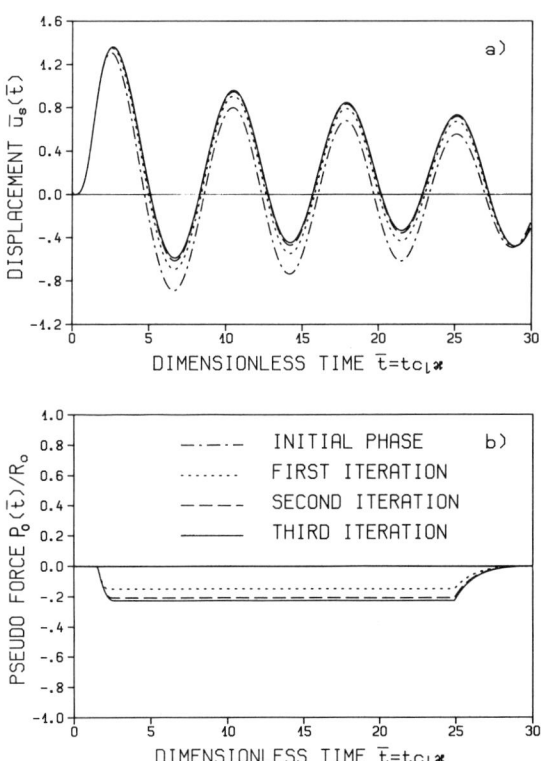

**Figure 5-2** $F_y^+/R_o = 0.4$, $F_y^-/R_o = \infty$. a. Displacement of mass. b. Pseudo-force.

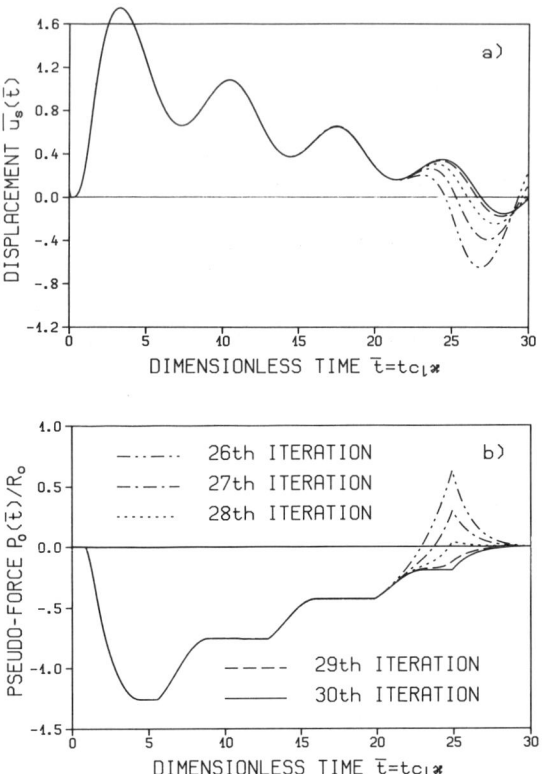

**Figure 5-3** $F_y^+/R_o = 0.1$, $F_y^-/R_o = 0$. a. Displacement of mass. b. Pseudo-force.

Finally, the computational effort is addressed. For the case with the weak nonlinearity, where convergence is achieved after three iterations (Fig. 5-2), the hybrid method is about twice as fast as the explicit algorithm which evaluates the convolution integral, discussed in Chapter 6. For the case with the strong nonlinearity with 30 iterations (Fig. 5-3), the hybrid method takes about three times longer than the explicit algorithm, whose running time is not affected by the extent of the nonlinearity.

## 5.3 STABILITY CRITERION FOR HARMONIC EXCITATION

### 5.3.1 One-Degree-of-Freedom System

**Scope of investigation.** It is appropriate to examine the stability criterion of a linear one-degree-of-freedom system for harmonic excitation, working exclusively in the frequency domain [W18]. The excitation with a

## Sec. 5.3  Stability Criterion for Harmonic Excitation

specific frequency ω acts infinitely long. In the derivation, an infinite series arises which leads to physical insight. The criterion for the harmonic excitation also forms the basis of that for a transient loading (Section 5.5).

The linear one-degree-of-freedom system of mass $m$, spring coefficient $k$, and damping coefficient $c$ is excited by the harmonic load $R \exp(i\omega t)$, with $R$ representing the amplitude and ω the frequency (Fig. 5-4a). The argument ω is deleted in this section. The displacement of the mass is denoted as $u \exp(i\omega t)$. The factor $\exp(i\omega t)$ is omitted in the following. The system in Fig. 5-4a corresponds to the actual nonlinear structure to be analyzed.

The exact (steady-state) solution is equal to

$$u = \frac{R}{k} \frac{1}{1 - \frac{\omega^2}{\omega_s^2} + 2i\zeta \frac{\omega}{\omega_s}} \tag{5.9}$$

with the actual system's natural frequency $\omega_s$ and damping ratio $\zeta$ specified as

$$\omega_s^2 = \frac{k}{m} \tag{5.10}$$

$$\zeta = \frac{c}{2m\,\omega_s} \tag{5.11}$$

The series of linear analyses of the hybrid frequency–time-domain method is performed with the pseudo-linear system shown in Fig. 5-4b. The spring

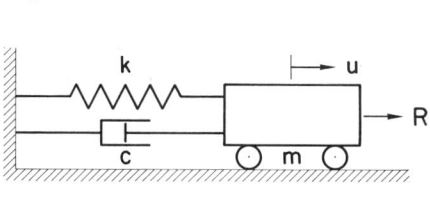

**Figure 5-4** One-degree-of-freedom system. a. Actual system. b. Pseudo-linear system.

coefficient $k_o$ differs from the one of the actual system in Fig. 5-4a while the mass and the damping coefficient are the same. The ratio $\alpha$ of the spring coefficient is defined as $k/k_o$. The amplitude of the pseudo-force in the spring equals $(k - k_o)u$, and the corresponding amplitude of the pseudo-load (which acts in addition to $R$) is equal to $-(k - k_o)u$. The steady-state response of the pseudo-system in the converged state is formulated as

$$u = \frac{R - (k - k_o)u}{k_o} \cdot \frac{1}{1 - \dfrac{\omega^2}{\omega_o^2} + 2i\zeta_o \dfrac{\omega}{\omega_o}} \tag{5.12}$$

with the corresponding values of the pseudo-system

$$\omega_o^2 = \frac{k_o}{m} \tag{5.13}$$

$$\zeta_o = \frac{c}{2m\omega_o} \tag{5.14}$$

Solving Eq. 5.12 for $u$ leads to the exact solution, Eq. 5.9, whereby the ratio $\alpha$ characterizing the pseudo-system drops out of the equation.

**Criterion.** Formulating Eq. 5.12 for the pseudo-system in the $j$th iteration using the known amplitude $u_{j-1}$ leads to

$$u_j = \left[ \frac{R}{k_o} + (1 - \alpha)u_{j-1} \right] \frac{1}{1 - \dfrac{\omega^2}{\omega_o^2} + 2i\zeta_o \dfrac{\omega}{\omega_o}} \tag{5.15}$$

Applying Eq. 5.15 for $j - 1, j - 2, \ldots, 1$ recursively and substituting in Eq. 5.15 results in

$$u_j = \frac{R}{k_o} \frac{1}{1 - \dfrac{\omega^2}{\omega_o^2} + 2i\zeta_o \dfrac{\omega}{\omega_o}} \left[ 1 + \frac{1 - \alpha}{1 - \dfrac{\omega^2}{\omega_o^2} + 2i\zeta_o \dfrac{\omega}{\omega_o}} \right.$$

$$\left. + \frac{(1 - \alpha)^2}{\left(1 - \dfrac{\omega^2}{\omega_o^2} + 2i\zeta_o \dfrac{\omega}{\omega_o}\right)^2} + \ldots + \frac{(1 - \alpha)^j}{\left(1 - \dfrac{\omega^2}{\omega_o^2} + 2i\zeta_o \dfrac{\omega}{\omega_o}\right)^j} \right] \tag{5.16}$$

The expression in brackets represents a geometric series with the factor

## Sec. 5.3  Stability Criterion for Harmonic Excitation

$(1 - \alpha)/(1 - \omega^2/\omega_o^2 + 2i\zeta_o\omega/\omega_o)$, whose sum can be substituted, leading to

$$u_j = \frac{R}{k_o} \frac{1}{1 - \frac{\omega^2}{\omega_o^2} + 2i\zeta_o \frac{\omega}{\omega_o}} \cdot \frac{1 - \left(\frac{1 - \alpha}{1 - \frac{\omega^2}{\omega_o^2} + 2i\zeta_o \frac{\omega}{\omega_o}}\right)^{j+1}}{1 - \frac{1 - \alpha}{1 - \frac{\omega^2}{\omega_o^2} + 2i\zeta_o \frac{\omega}{\omega_o}}}$$

$$= \frac{R}{k_o} \frac{1 - \left(\frac{1 - \alpha}{1 - \frac{\omega^2}{\omega_o^2} + 2i\zeta_o \frac{\omega}{\omega_o}}\right)^{j+1}}{\alpha - \frac{\omega^2}{\omega_o^2} + 2i\zeta_o \frac{\omega}{\omega_o}} \quad (5.17)$$

For an infinite geometric series ($j \to \infty$), convergence is reached ($u_j \to u$) if the magnitude of the factor is smaller than 1, resulting in

$$u = \lim_{j \to \infty} u_j = \frac{R}{k_o} \frac{1}{\alpha - \frac{\omega^2}{\omega_o^2} + 2i\zeta_o \frac{\omega}{\omega_o}} = \frac{R}{k\left(1 - \frac{\omega^2}{\omega_s^2} + 2i\zeta \frac{\omega}{\omega_s}\right)} \quad (5.18)$$

The right-hand side of Eq. 5.18, determined as the limiting displacement of a loading sequence acting on the pseudo-system represents the exact solution of the actual system (Eq. 5.9). Examining the criterion for stability further

$$\left| \frac{1 - \alpha}{\sqrt{\left(1 - \frac{\omega^2}{\omega_o^2}\right)^2 + \left(2\zeta_o \frac{\omega}{\omega_o}\right)^2}} \right| < 1 \quad (5.19)$$

applies. Expressed as a function of the pseudo-system's properties, $\alpha = k/k_o$ is bounded by

$$1 - \sqrt{\left(1 - \frac{\omega^2}{\omega_o^2}\right)^2 + \left(2\zeta_o \frac{\omega}{\omega_o}\right)^2} < \alpha < 1 + \sqrt{\left(1 - \frac{\omega^2}{\omega_o^2}\right)^2 + \left(2\zeta_o \frac{\omega}{\omega_o}\right)^2} \quad (5.20)$$

In Fig. 5-5, the two limits of $\alpha$ are plotted as a function of $\omega/\omega_o$ for the indicated $\zeta_o$. The domain of stability, where convergence is reached, is shown as hatched areas. For the static case $0 < \alpha < 2$ applies. This agrees with the well-known fact that the static solution diverges if the instantaneous slope of the force–displacement curve (the static-stiffness coefficient) corresponding to the exact solution is more than twice the initial (linear) slope of the

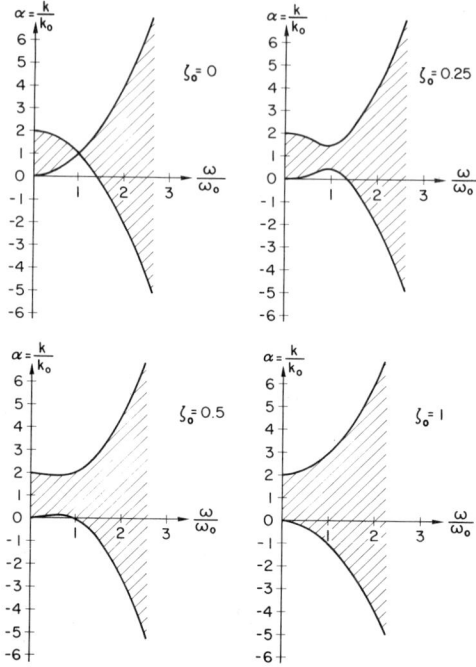

**Figure 5-5** Domain of convergence for harmonic excitation.

pseudo-system used in the initial-stress method. For increasing $\omega/\omega_o$, the permissible range for $\alpha$ is reduced up to the resonance of the pseudo-system ($\omega/\omega_o = 1$), whereby, for an undamped system ($\zeta_o = 0$), only $\alpha = 1$ leads to stability—that is, the actual system must be used. For larger $\omega/\omega_o$, the allowable range for $\alpha$ is increased again. As expected, if damping is present in the system, a larger domain of stability results.

For the undamped dynamic case, a physical interpretation exists for Eq. 5.20. As just mentioned, the ratio of the static-stiffness coefficients of the actual system and the pseudo-system must lie between 0 and 2 for convergence to occur in the static case. Analogously, postulating the same limits for the ratio of the dynamic-stiffness coefficients for harmonic excitation leads to

$$0 < \frac{k - \omega^2 m}{k_o - \omega^2 m} = \frac{\alpha - \dfrac{\omega^2}{\omega_o^2}}{1 - \dfrac{\omega^2}{\omega_o^2}} < 2 \tag{5.21}$$

For $\omega/\omega_o < 1$ and $>1$, Eq. 5.21 results in

$$\frac{\omega^2}{\omega_o^2} < \alpha < 2 - \frac{\omega^2}{\omega_o^2} \tag{5.22a}$$

and

$$2 - \frac{\omega^2}{\omega_o^2} < \alpha < \frac{\omega^2}{\omega_o^2} \tag{5.22b}$$

Sec. 5.3   Stability Criterion for Harmonic Excitation                         225

respectively. These two inequalities correspond to the two hatched areas shown in Fig. 5-5 for $\zeta_o = 0$.

A divergent situation is illustrated in Fig. 5-6, which is adapted from Ref. [S4]. The dynamic-stiffness coefficient of the actual system $k-\omega^2 m$ (solid line) is more than twice that of the pseudo-system $k_o-\omega^2 m$ (dashed line). The exact solution corresponds to the intersection of the applied load with the amplitude $R$ and the solid line with the slope $k-\omega^2 m$. The series of linear analyses, with the pseudo-system leading to the points $a, b, c, d, e, f, g, h$ . . . , diverges.

### 5.3.2 Multiple-Degree-of-Freedom System

The generalization to a system with multiple degrees of freedom proceeds as follows [D1]. The true instantaneous dynamic-stiffness matrix is denoted as $[S(\omega)]$, and the pseudo-linear dynamic-stiffness matrix is denoted as $[S_o(\omega)]$. For the $j$th iteration, the displacement amplitudes $\{u^j(\omega)\}$ are calculated, analogously to Eq. 5.2 of Table 5-1, as

$$\{u^j(\omega)\} = [S_o(\omega)]^{-1} (\{R(\omega)\} - ([S(\omega)] - [S_o(\omega)]) \{u^{j-1}(\omega)\}) \quad (5.23)$$

where $\{R(\omega)\}$ are the prescribed exterior load amplitudes, and the pseudo-forces are calculated as $([S(\omega)] - [S_o(\omega)]) \{u^{j-1}(\omega)\}$ (Eq. 5.1). The initial

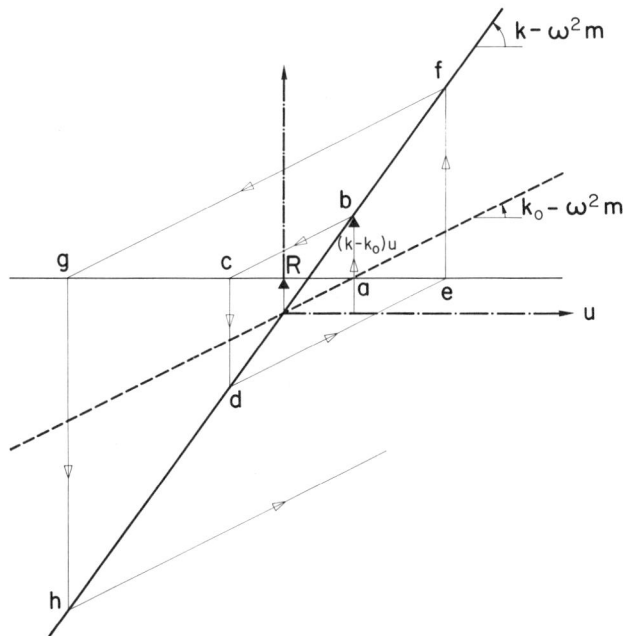

**Figure 5-6**   Divergence.

values for $j = 0$ are equal to

$$\{u^o(\omega)\} = [S_o(\omega)]^{-1} \{R(\omega)\} \tag{5.24}$$

Applying Eq. 5.23 recursively and substituting back in Eq. 5.23 results in

$$\{u^j(\omega)\} = ([I] + [A(\omega)] + [A(\omega)]^2 + \ldots + [A(\omega)]^j) [S_o(\omega)]^{-1} \{R(\omega)\} \tag{5.25}$$

where

$$[A(\omega)] = -[S_o(\omega)]^{-1} ([S(\omega)] - [S_o(\omega)]) = [I] - [S_o(\omega)]^{-1} [S(\omega)] \tag{5.26}$$

and with $[I]$ being the unit matrix. The sum of the geometric series appearing on the right-hand side of Eq. 5.25 can be derived by pre-multiplying Eq. 5.25 by $[A]$ and subtracting Eq. 5.25, leading to

$$\{u^j(\omega)\} = [S(\omega)]^{-1} [S_o(\omega)] ([I] - [A(\omega)]^{j+1}) [S_o(\omega)]^{-1} \{R(\omega)\} \tag{5.27}$$

The spectral decomposition of the nonsymmetric matrix $[A(\omega)]$ is introduced. Denoting the matrix of the eigenvectors as $[\Phi(\omega)]$ and placing the eigenvalues $\lambda_i(\omega)$ on the diagonal of $[\Lambda(\omega)]$ results in

$$[A(\omega)] = [\Phi(\omega)] [\Lambda(\omega)] [\Phi(\omega)]^{-1} \tag{5.28}$$

or

$$[A(\omega)]^{j+1} = [\Phi(\omega)] [\Lambda(\omega)]^{j+1} [\Phi(\omega)]^{-1} \tag{5.29}$$

With the spectral radius $\rho(\omega)$ of $[A(\omega)]$ defined as the largest absolute eigenvalue

$$\rho(\omega) = \max_i |\lambda_i(\omega)| \tag{5.30}$$

$[\Lambda(\omega)]^{j+1}$ will converge to zero for $j \to \infty$ if

$$\rho(\omega) < 1 \tag{5.31}$$

This transforms Eq. 5.27 for $j \to \infty$ to

$$\{u^j(\omega)\} = [S(\omega)]^{-1} \{R(\omega)\} \tag{5.32}$$

which, of course, is the correct result. The stability criterion is specified in Eq. 5.31. The smaller $\rho(\omega)$ is, the faster the convergence will be.

Applying this criterion to the one-degree-of-freedom system discussed in Section 5.3.1 leads to, with

$$S_o(\omega) = k_o + 2\zeta_o m\omega_o i\omega - \omega^2 m \tag{5.33a}$$

$$S(\omega) = k + 2\zeta m\omega_s i\omega - \omega^2 m \tag{5.33b}$$

Sec. 5.4  Properties of Convergence

to

$$\lambda = \frac{1 - \alpha}{\sqrt{\left(1 - \frac{\omega^2}{\omega_o^2}\right)^2 + \left(2\zeta_o \frac{\omega}{\omega_o}\right)^2}} \quad (5.34)$$

with

$$|\lambda| < 1 \quad (5.35)$$

This result is identical to the stability criterion expressed in Eq. 5.19.

The stability criterion expressed in Eq. 5.31 applies to a harmonic motion of infinite duration. It should not be enforced for all frequencies of a transient excitation, as this would be too strict (see Section 5.5).

## 5.4 PROPERTIES OF CONVERGENCE FOR ITERATIVE SCHEME WORKING WITH SEGMENTS OF TIME INTERVAL

### 5.4.1 Segmenting Procedure

The hybrid frequency–time-domain analysis does not need to achieve convergence by working with the total time interval (during which the response is to be evaluated) in each iteration. If the total time interval is used, one observes that the pseudo-force history (and thus the solution) corrects itself in a time progressing manner. This is already pointed out in Section 5.2 (Fig. 5-3). It is useless to perform iterations over the time interval's earlier portions after they have converged. The later portions of the pseudo-force history are evaluated in all iterations, but only in the final iterations do they begin to converge to the exact solution. If the total time interval is subdivided into segments, each consisting, in general, of several time steps, convergence can be reached for each segment separately, starting with the first and moving to the second only after reaching convergence in the first, and so on [K4]. Within a segment, the response converges for all times up to a specific time. New pseudo-forces for times larger than this specific time are then calculated only within this particular segment, and a new iteration is started. This procedure is repeated until the specific time coincides with the last time value of this segment, after which the next segment is investigated. In the limit, if the segment consists of only one time step, the initial-stress approach results, in which the pseudo-forces are calculated only for the current time step for which convergence must be achieved before processing the next time step.

The calculations performed show that it is of utmost importance not to update any response quantity or pseudo-force on that part of the time span which has previously converged—that is, for times less than the specific time just mentioned. Should such updates be performed, some very small changes in response would occur from one iteration to the next for times less than this specific time. These small changes affect the response of that part of the investigated segment which has not yet converged. This can cause divergence, especially when a true–false situation of the type encountered in contact problems occurs [D1].

In certain cases, to reach convergence for a transient excitation (Section 5.5), the use of sufficiently short segments is essential.

It must also be realized that the pseudo-linear system's equations of motion—not those of the true nonlinear system—are solved in the frequency domain. The pseudo-linear system's properties must therefore be used to obtain the natural frequencies and the values of critical damping, which are needed to establish the length of the time period to be used in the discrete form of the Fourier transform. In certain cases, a longer time period must be used (Section 5.5.3).

The following covers how the properties of convergence depend on the number of segments the total time interval is divided into (and thus on the number of time steps per segment). The extent by which the intermediate results depart from the exact (nonlinear) solution and the number of iterations are addressed. These aspects are discussed at first using a system with frequency-independent properties (Section 5.4.2) and then examining a rigid block with partial uplift (Section 5.4.3).

### 5.4.2 Nonlinear Two-Degree-of-Freedom System with Frequency-Independent Parameters

**Investigated system.** The investigated system with two degrees of freedom is shown in Fig. 5-7. A nonlinear but elastic spring with a linear coefficient $k$ and a yielding value $F_y$ is connected to two points with equal mass $m$. No hysteresis loop exists. This part of the system could represent a simple nonlinear structure. A spring with a coefficient $k$ and a damper with a coefficient $c$ (both coefficients being frequency-independent), coarsely modeling the soil, are attached to one of the mass points. The load $R$, consisting of a rounded triangular pulse with a peak value $R_o$ and a duration $t_o$, acts on the other mass point.

$$R(\bar{t}) = \frac{R_o}{2}\left[1 - \cos\left(2\pi \frac{\bar{t}}{\bar{t}_o}\right)\right], \quad 0 < \bar{t} < \bar{t}_o$$
$$R(\bar{t}) = 0, \quad \bar{t} > \bar{t}_o \quad (5.36)$$

$\bar{t}$ is the dimensionless time ($= t/\sqrt{m/k}$). The displacements of the mass points are denoted as $u_o$ and $u_s$.

## Sec. 5.4 Properties of Convergence

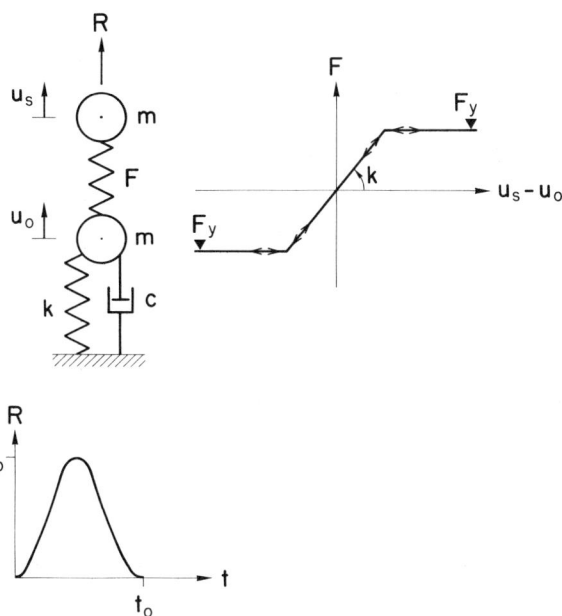

**Figure 5-7** Nonlinear two-degree-of-freedom system.

The following dimensionless quantities are used in the calculations: $c/\sqrt{km} = 0.2$, $F_y/R_o = 0.5$, $\bar{t}_o = 5$. For the sake of simplicity, the linear analyses with pseudo-loads normally performed in the frequency domain are carried out directly in the time domain.

**Rate of convergence.** In the linear analysis (in every iteration for a specific time segment of the pseudo-linear system excited by the applied load $R(t)$ and the pseudo-loads), an explicit integration—with $\beta = 0$, $\gamma = 0.5$ (Eqs. 2.60 and 2.62) with a dimensionless time step $\Delta t/\sqrt{m/k} = 0.4$—is used. The dimensionless total time interval during which the response is evaluated equals $t/\sqrt{m/k} = 60$, which results in a total of 150 time steps. Various calculations are performed, varying the number of time steps per segment (and thus also varying the number of segments of equal time interval that are used to represent the evaluated response's total time interval). In each segment, convergence is reached by performing the necessary number of iterations, whereby the following criterion must be satisfied for all time steps of the segment being processed (Eq. 5.4 applied to $u_s$ only)

$$\left| \frac{u_s^j - u_s^{j-1}}{u_s^j} \right| < 10^{-6} \tag{5.37}$$

In Fig. 5-8, the displacement $u_s$ (nondimensionalized with the factor $k/R_o$) is plotted as a function of the dimensionless time $\bar{t}$ for various numbers of time

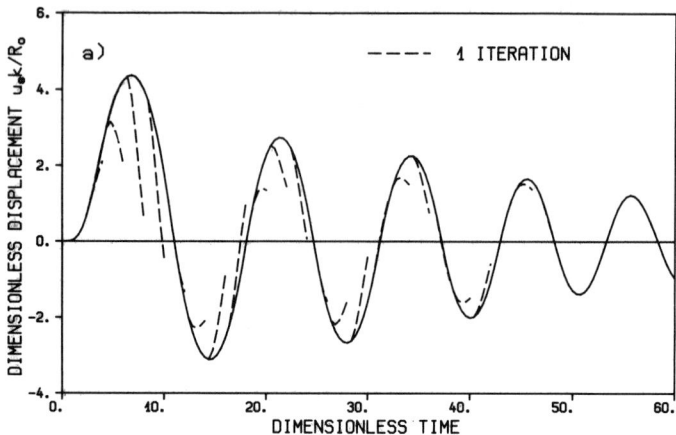

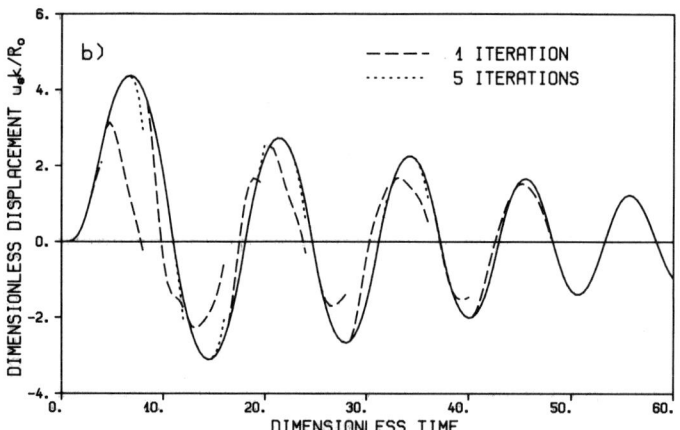

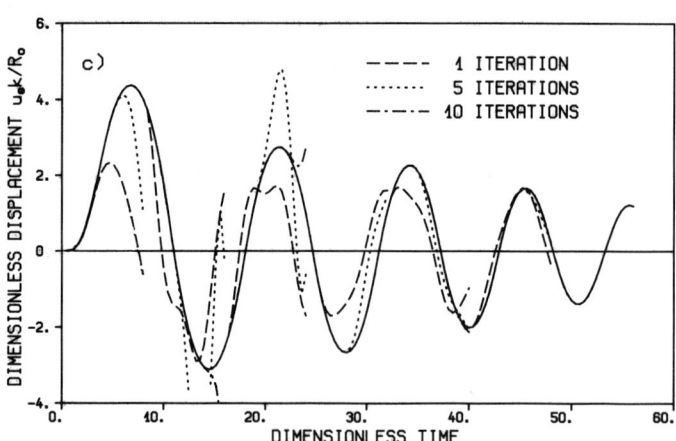

Sec. 5.4  Properties of Convergence 231

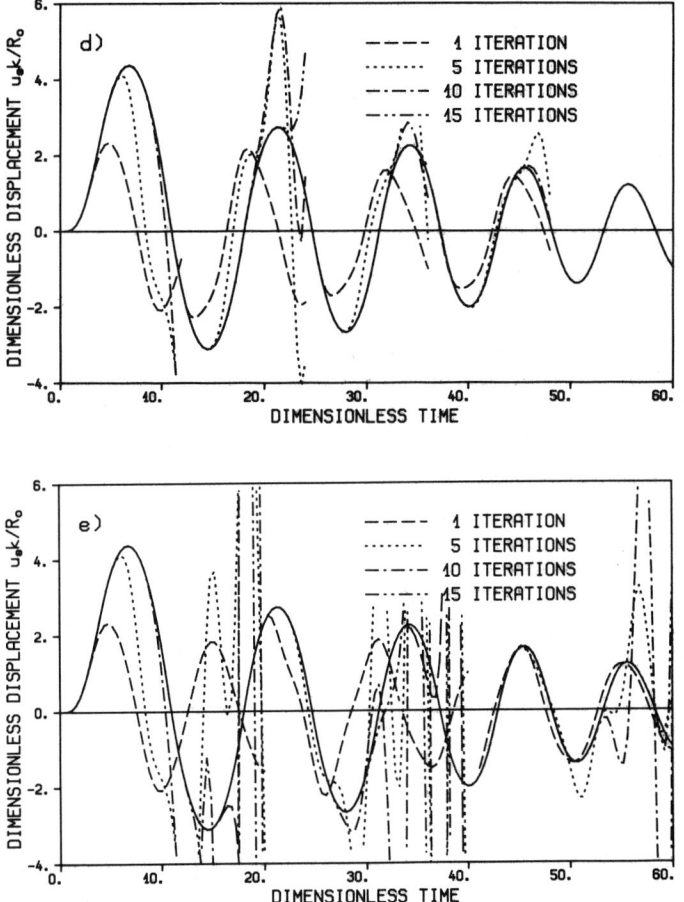

**Figure 5-8** Rate of convergence for varying number of time steps in each segment. a. 5 time steps. b. 10 time steps. c. 20 time steps. d. 30 time steps. e. 50 time steps.

segments. The solid line corresponds to the exact solution. Fig. 5-8a corresponds to 30 segments each with 5 time steps; Fig. 5-8b to 15 segments with 10 $\Delta t$; Fig. 5-8c to 7 segments with 20 $\Delta t$; Fig. 5-8d to 5 segments with 30 $\Delta t$; and Fig. 5-8e to 3 segments with 50 $\Delta t$.

Convergence is reached in all cases. This also applies to further cases (not shown) with segments of 100 $\Delta t$ and 150 $\Delta t$—the latter case corresponding to the iterations being performed over the evaluated response's total time interval. In this case of explicit time integration, the average number of iterations per segment is almost proportional to the number of time steps in the segment (for example, 90 for a segment of 150 $\Delta t$, 17 for a segment of 30 $\Delta t$). The intermediate results obtained after performing the indicated number of iterations in each segment are also shown. The larger the number of time steps per segment, the larger the deviations from the exact converged solution. While after a significant number of iterations very large oscillations still occur (for example, 15 in Fig. 5-8e), the solution then converges remarkably fast (for example, after 32, 34, and 19 iterations in the three segments of the case of Fig. 5-8e).

### 5.4.3 Rigid Block on Half-Space with Partial Uplift

**Investigated system.** As an example of a soil–structure-interaction analysis with abrupt nonlinearities occurring, a rigid block with individual footings, which can uplift, resting on the surface of an undamped half-space (Fig. 5-9) is examined [D1]. The four circular disk foundations of radius $a$ are placed at each corner of the rigid body of height $2h$, width $2b$, mass $m$ and mass moment of inertia $I$ referred to the center of mass. The half-space is characterized by the shear modulus $G$, shear-wave velocity $c_s$, and Poisson's ratio $v$. The following numerical values apply:

$h = 9\ m,\ b = 6\ m,\ a = 4\ m,\ m = 12.7 \cdot 10^6\ kg,$

$I = 6.4 \cdot 10^9\ kgm^2,\ G = 1.3 \cdot 10^9\ N/m^2,\ c_s = 750\ m/s,\ v = 1/3.$

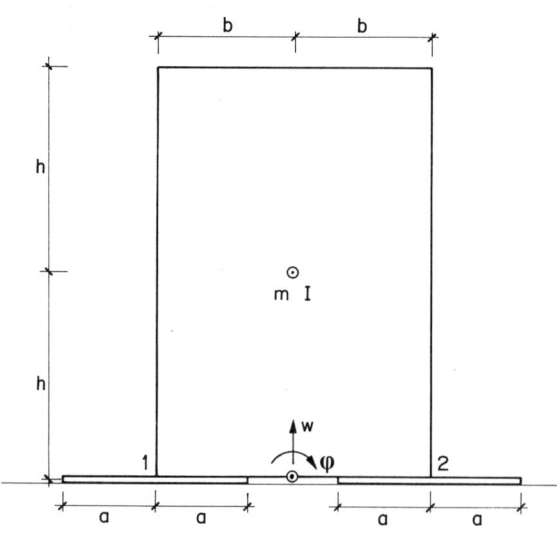

**Figure 5-9** Rigid block on circular footings with partial uplift.

## Sec. 5.4  Properties of Convergence

The idealized horizontal earthquake ground acceleration $\ddot{u}_g(t)$ acting during two seconds is selected as $g/10 \, [-3 \sin(8\pi t) + 9 \sin(24\pi t)]$, with $g = 9.81$ m/s².

Only the vertical and rocking motions $w(t)$ and $\varphi(t)$ of the block's bottom center are considered in the calculation, with the horizontal motion assumed identical to the ground motion.

The vertical dynamic-stiffness coefficient $S_v(a_o)$ of the individual disk footings is specified in closed form as

$$S_v(a_o) = K_v \left[ 1 - \frac{\mu_1 a_o^2}{1 + \frac{\mu_1^2}{\gamma_1^2} a_o^2} + i a_o \left( \frac{\mu_1}{\gamma_1} \frac{\mu_1 a_o^2}{1 + \frac{\mu_1^2}{\gamma_1^2} a_o^2} + \gamma_o \right) \right] \quad (5.38)$$

with

$$K_v = \frac{4Ga}{1 - \nu} \quad (5.39)$$

$$\gamma_o = 0.8 \quad (5.40a)$$

$$\gamma_1 = 0.29 \quad (5.40b)$$

$$\mu_1 = 0.35 \quad (5.40c)$$

$$a_o = \frac{\omega a}{c_s} \quad (5.41)$$

Equation 5.38 corresponds to the dynamic-stiffness coefficient of the discrete model (Eq. 2.34) whose values are specified in Table 2-2.

The natural frequency of the linear system associated with rocking is 3.9 Hz; while that associated with vertical motion equals 15.8 Hz; both are based on the static-stiffness coefficient $K$ of the individual footings. The corresponding ratios of critical damping are 0.07 and 0.28, respectively, calculated with the asymptotic value ($a_o \to \infty$) of the damper $C_v = K(\gamma_1 + \gamma_o)a/c_s$.

The logically selected pseudo-linear system is the one in which no uplift occurs. The equations of motion to be solved in the frequency domain at the jth iteration are then (corresponds to Eq. 5.2)

$$[-m\omega^2 + 4S_v(\omega)] w^j(\omega) = -W(\omega) - 2[Q_1^j(\omega) + Q_2^j(\omega)] \quad (5.42a)$$

$$[-(I + mh^2)\omega^2 + 4b^2 S_v(\omega) - mgh]\varphi^j(\omega)$$
$$= -mh\ddot{u}_g(\omega) - 2b[Q_1^j(\omega) - Q_2^j(\omega)] \quad (5.42b)$$

$$S_v(\omega)w_1^j(\omega) = S_v(\omega)[w^j(\omega) + b\varphi^j(\omega)] + Q_1^j(\omega) \quad (5.43a)$$

$$S_v(\omega)w_2^j(\omega) = S_v(\omega)[w^j(\omega) - b\varphi^j(\omega)] + Q_2^j(\omega) \quad (5.43b)$$

Equations 5.42a and 5.42b refer, respectively, to the block's vertical and rocking degrees of freedom. $Q_1^j(\omega)$ is the amplitude of the pseudo-load acting at a left corner of the block; while $Q_2^j(\omega)$ is that of the pseudo-load at a right corner. $W(\omega)$ is the Fourier transform of the weight $mg$ of the block. Equations 5.43a and 5.43b refer to the vertical degrees of freedom of the footings under the block's left and right corners, respectively. The motions are identified by the subscripts 1 and 2. They are equal to the corresponding motions of the corners of the block whenever these corners are in contact with the footings, but are different whenever uplift of these corners occurs. Four degrees of freedom must be introduced: two to describe the motion of the block $w^j(\omega)$, $\varphi^j(\omega)$ and two for the footings $w_1^j(\omega)$, $w_2^j(\omega)$. The left-hand side of Eq. 5.43a represents the amplitude of the total (nonlinear) interaction force $R_{1,nl}^j(\omega)$ equal to the sum of the amplitudes of the linear force $R_{1,l}^j(\omega) = S_v(\omega)[w^j(\omega) + b\varphi^j(\omega)]$ and of the pseudo-force $Q_1^j(\omega)$ (corresponds to Eq. 5.3). Equation 5.43b represents the analogous equation for the other corner.

The pseudo-forces, whose amplitudes appear for the $j$th iteration in Eqs. 5.42 and 5.43, are determined in the time domain based on the results of the ($j$-1)th iteration as (corresponds to Eq. 5.1, with $R_{1,nl}^{j-1}(t) = 0$ for uplift)

$$Q_1^j(t) = \min \{-R_{1,l}^{j-1}(t); 0\} \quad \text{when uplift} \qquad (5.44)$$
$$= 0 \quad \text{when contact}$$

and the analogous relationship for corner 2.

A corner is assumed either to lose contact from the supporting footing whenever the interaction force becomes positive (tension) or to gain contact whenever the upward footing displacement becomes larger than the upward corner displacement.

**Results.** A time period of 10.24 s with a time step of 0.01 s is selected. The time span of interest equals 3.2 s.

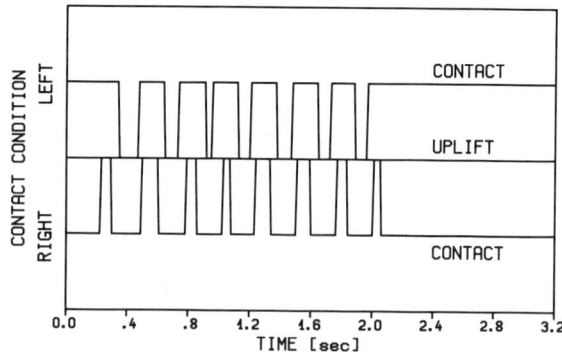

**Figure 5-10** Contact conditions.

Sec. 5.4    Properties of Convergence    235

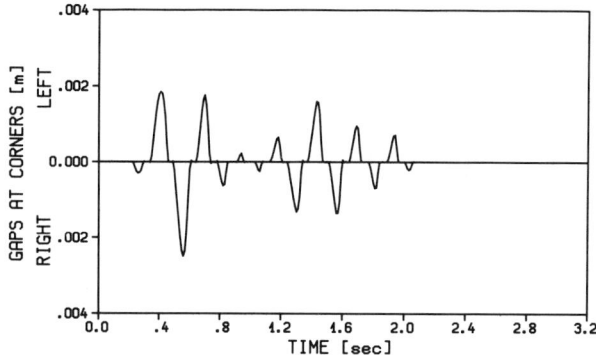

**Figure 5-11**  Vertical gaps between block and footings.

The high level of the nonlinearities occurring in this application is apparent in Fig. 5-10, where the contact conditions are plotted as a function of time. Further results include the time histories of the gaps at the corners (Fig. 5-11), of the (nonlinear) interaction forces (Fig. 5-12), and of the pseudo-forces (Fig. 5-13).

**Number of iterations.**   The number of segments into which the time interval of interest (3.2 s with 320 time steps) is divided is varied. The total number of iterations and the corresponding computer time are plotted versus the number of segments introduced in the calculation in Fig. 5-14. One first observes that the computer time is essentially proportional to the total number of iterations. Second, one sees that using many segments, each containing only a few time steps, wastes computer time. Using too few segments can cause instability. In this application, instability occurs when using a single segment of 320 time steps. This is discussed in depth in Section 5.5. Finally,

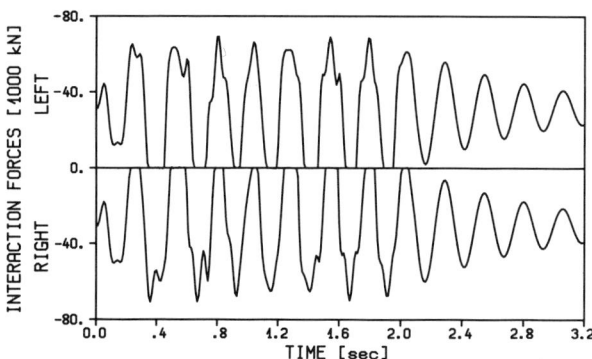

**Figure 5-12**  Total interaction forces.

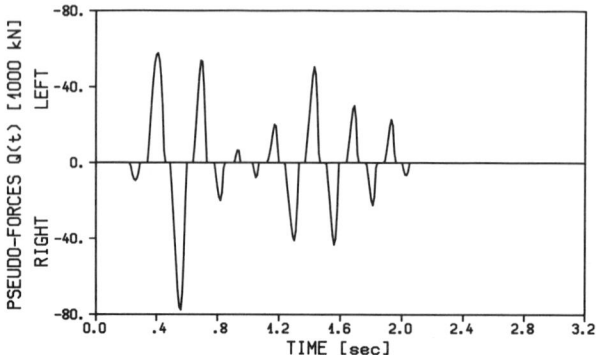

**Figure 5-13** Pseudo-forces.

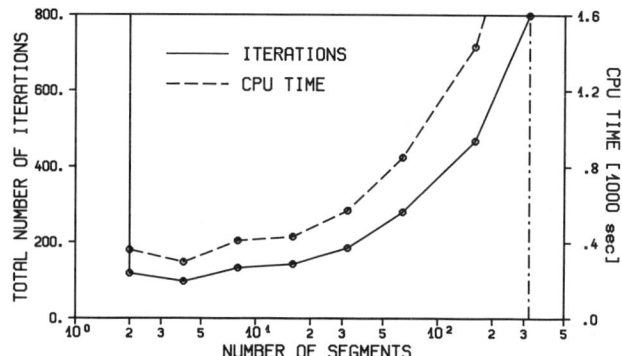

**Figure 5-14** Computational effort as function of number of segments.

one notes that the sensitivity of the total number of iterations to the number of segments is not dramatic when operating in a range relatively close to the optimum number of segments (equal to four in this application).

## 5.5 STABILITY CRITERION FOR TRANSIENT EXCITATION

### 5.5.1 Initial-Value Theorem

The stability criterion for a transient excitation is developed in Ref. [D1], on which Section 5.5 is based.

Based on the observation that the hybrid frequency–time-domain procedure converges in a time progressing manner (Section 5.2), one can expect that, in a favorable situation, convergence can be reached by treating the whole time span of interest at once. In the other extreme situation, con-

Sec. 5.5  Stability Criterion for Transient Excitation   237

vergence can only be attained by processing each time step independently and sequentially. Thus, a condition necessary to achieve the procedure's convergence is that the solution converges when each time step is considered separately. Working with one time step at a time basically corresponds to treating an initial-value problem for which the initial-value theorem applies. The latter states that the value at time $t = 0^+$ of a function of time $f(t)$ which is zero for $t < 0$ is given by

$$\lim_{t \to 0^+} f(t) = \lim_{i\omega \to \infty} [i\omega F(\omega)] \qquad (5.45)$$

where $F(\omega)$ is the Fourier transform of $f(t)$.

The discrete Fourier transform of a function, with the period $T$, and $N$ time steps of length $\Delta t$ ($\Delta t = T/N$), is defined with $t = n\Delta t$ ($n = 0, 1, \ldots, N - 1$) as

$$F(\omega) = \Delta t\, f(0) + \Delta t \sum_{n=1}^{N-1} f(n\Delta t)\exp(-i\omega n\Delta t) \qquad (5.46)$$

The highest frequency considered in the discrete form is the Nyquist frequency $\Omega = \pi N/T$. Evaluating $F(\omega)$ at $i\omega = \Omega$—that is, at $\omega = -i\Omega$—then results in

$$F(\omega = -i\Omega) = \Delta t\, f(0) + \Delta t \sum_{n=1}^{N-1} f(n\Delta t)\exp(-n\pi) \qquad (5.47)$$

$f(0)$ is approximately given by

$$f(0) \approx F(\omega = -i\Omega)/\Delta t \qquad (5.48)$$

Equation 5.48 is obtained by retaining only the first term of Eq. 5.47. It may be noted that the series $e^{-\pi} + e^{-2\pi} + \ldots + e^{-(N-1)\pi} \approx 0.045$, and thus the terms ignored remain small when compared to the first term, provided $f(t)$ does not increase with time.

### 5.5.2 Criterion

The considerations of Section 5.5.1, combined with those of Section 5.3.2, imply that, for a transient excitation, the condition referring to the harmonic case should not be formulated for all frequencies but only for the Nyquist frequency. Thus, a condition necessary to guarantee convergence of the hybrid frequency–time-domain procedure is that the spectral radius evaluated at $\omega = -i\Omega$ is less than unity. A series of calculations, to be presented in Section 5.5.3, confirms the validity of this statement. The analyses' results even suggest that this condition is sufficient to insure stability of the hybrid frequency–time-domain procedure. It will therefore be referred to as the *criterion of stability*. Mathematically, it is summarized as (Eqs.

5.31, 5.30, 5.26)

$$\rho(\omega = -i\Omega) < 1 \tag{5.49}$$

with

$$\rho(\omega) = \max_i |\lambda_i(\omega)| \tag{5.50}$$

where $\lambda_i(\omega)$ is the $i$th eigenvalue of

$$[A(\omega)] = [I] - [S_o(\omega)]^{-1}[S(\omega)] \tag{5.51}$$

It is important to observe that only the properties of the systems at the Nyquist frequency enter into this criterion and that the Nyquist frequency is related to the increment of time $\Delta t$ through

$$\Omega = \pi/\Delta t \tag{5.52}$$

It must also be noted that $[S(\omega)]$ refers to the numerical nonlinear system, which may differ significantly from the true physical one whenever, at some stage of the iterative procedure, the pseudo-loads force the system to operate in a nonlinear range that departs significantly from the converged one.

### 5.5.3 Numerical Validation

Various calculations are perfomed with the objective of numerically investigating the extent to which the hybrid frequency–time-domain procedure is sensitive to the value of $\rho(\omega = -i\Omega)$, and of confirming the validity of the criterion of stability derived in Section 5.5.2. For this investigation, the actual system, consisting of a one-degree-of-freedom oscillator of mass $m$, stiffness $k$, and damping $c$, is subjected to a force excitation $P(t)$. The system's response is obtained by use of the hybrid frequency–time-domain procedure, whereby the pseudo-linear system has mass $m_o$, stiffness $k_o$, and damping $c_o$. The equation of motion to be solved is thus

$$(-\omega^2 m_o + i\omega c_o + k_o)u^j(\omega) = P(\omega) + Q^j(\omega) \tag{5.53}$$

where $u^j(\omega)$, $P(\omega)$, and $Q^j(\omega)$ are the Fourier transforms of the displacement $u^j(t)$, of the exciting force $P(t)$, and of the pseudo-load $Q^j(t)$. The latter is obtained in the time domain through the equation

$$Q^j(t) = (m_o - m)\ddot{u}^{j-1}(t) + (c_o - c)\dot{u}^{j-1}(t) + (k_o - k)u^{j-1}(t) \tag{5.54}$$

In Eq. 5.54, the superscript $j$ refers to the $j$th iteration.

The properties of the original one-degree-of-freedom system are selected in consistent units as $m = 9$, $k = 5685$ (natural frequency $= 4$), and $c = 18.1$ (critical damping ratio $= .04$), with the force excitation $P(t)$ given by $P(t) = 100[-3\sin(8\pi t) + 9\sin(24\pi t)]$ for $0 < t < 2$ and by zero otherwise.

In a first set of calculations, the pseudo-linear system's properties are selected as $k_o = 5685$ and $c_o = 18.1$; while $m_o$ is varied as indicated in Table

## Sec. 5.5  Stability Criterion for Transient Excitation

5-2 to achieve a wide variation of $\rho(\omega = -i\Omega)$. Unless indicated otherwise, the period of time used in the discrete Fourier transformation is 10.24 (1024 time steps of length $\Delta t = .01$), and the time span for which the response is calculated is 3.2 (320 time steps). Longer time periods are used in some cases for the reason discussed in Section 5.4.1. The time span of 3.2 is divided into segments of equal lengths, whereby the segmenting approach is implemented as stated in Section 5.4.1. The minimum total number of iterations valid in the particular calculation performed is determined by varying the number of segments from one calculation to another.

The spectral radius $\rho(\omega = -i\Omega)$ is evaluated from the following equation, in which $\omega = -i\Omega$ is substituted ($\Omega = 100\pi$):

$$\rho(\omega) = \left| 1 - \frac{-\omega^2 m + i\omega c + k}{-\omega^2 m_o + i\omega c_o + k_o} \right| \qquad (5.55)$$

The minimum total number of iterations and the corresponding number of segments are presented in Table 5-2. The sign of $\rho$, as obtained by disregarding the absolute value in Eq. 5.55, is kept in the table for the sake of clarity.

The results obtained indicate that the optimum number of time segments depends on the value of the spectral radius $\rho(\omega = -i\Omega)$. The more the latter tends towards unity, the larger the number of time segments is. The dependence of the minimum total number of iterations on the value of the spectral radius $\rho(\omega = -i\Omega)$ is similar to that of the optimum number of segments. Typically, it becomes more difficult to reach convergence as $\rho(\omega = -i\Omega)$ tends towards unity. The procedure diverges for the value of $m_o$ equal 4.4, leading to $\rho(\omega = -i\Omega)$ larger than 1. For values of $m_o$ resulting in $\rho(\omega = -i\Omega)$ slightly below unity, time segments consisting of only one time step must be used to obtain convergence (for $m_o = 182$ and $m_o = 4.6$

**TABLE 5-2  Results for Pseudo-Linear System with $k_o = 5685$, $c_o = 18.1$**

| Mass $m_o$ | Spectral Radius $\rho(\omega = -i\Omega)$ | Optimum Number of Segments | Minimum Total Number of Iterations |
|---|---|---|---|
| 182 | .95 | 320 | 29840* |
| 45.5 | .80 | 16 | 766** |
| 15 | .40 | 4 | 83 |
| 9 | 0 | 1 | 2 |
| 6.4 | − .40 | 2 | 91 |
| 5 | − .78 | 8 | 963 |
| 4.6 | − .93 | 320 | 25488*** |
| 4.4 | −1.02 | | diverges |

*Extrapolated from 4 converged time steps; time period of 81.92
**Time period of 30.72
***Extrapolated from 20 converged time steps

TABLE 5-3 Results for Pseudo-Linear System with $k_o = 1895$, $c_o = 18.1$

| Mass $m_o$ | Spectral Radius $\rho(\omega = -i\Omega)$ | Optimum Number of Segments | Minimum Total Number of Iterations |
|---|---|---|---|
| 182 | .95 | 320 | 30528* |
| 45.5 | .80 | 20 | 944** |
| 15 | .39 | 10 | 184 |
| 9 | − .00 | 10 | 120 |
| 6.4 | − .42 | 10 | 327 |
| 5 | − .82 | 20 | 1671 |
| 4.6 | − .97 | 320 | 31378*** |
| 4.4 | −1.04 | | diverges |

*Extrapolated from 5 converged time steps; time period of 81.92
**Time period of 30.72
***Extrapolated from 18 converged time steps

in this application). These results are in total agreement with the criterion of stability of Eq. 5.49.

The results of a second set of calculations are presented in Table 5-3. The system differs from the previous one in the choice of $k_o$, which is set equal to 1895.

All previously described features remain the same. As seen from Fig. 5-15, in which the results of both sets of calculations are compared, the optimum number of segments and the minimum total number of iterations differ from one set to the other. This is caused by the change in $k_o$. In these calculations, it is assumed that convergence is reached whenever the change in pseudo-force from one iteration to the next is less than 1.

The static counterpart of the procedure is known to diverge in cases for which the spectral radius $\rho(\omega)$ evaluated at $\omega = 0$ is larger than 1 (Section 5.3.1). A further set of calculations is thus performed to see how the hybrid frequency–time-domain procedure performs in such cases. The same one-degree-of-freedom system is utilized, using $m_o = 9$ and $c_o = 18.1$ and varying $k_o$ as indicated in Table 5-4. Convergence is reached without difficulty in all cases, as expected from the criterion of stability and from the conclusions drawn from the results of the previous sets of calculations.

The results of the three sets of calculations thus fully sustain the validity of the criterion of stability. It must be noted that the criterion will be satisfied in most structural applications because it depends on the properties of the pseudo-linear system and the nonlinear system at the Nyquist frequency only. The Nyquist frequency is generally high enough (or easily increased by decreasing $\Delta t$) so $[S_o(\omega)]$ and $[S(\omega)]$ are dominated by the mass, which does not generally change from the actual system to the pseudo-linear one, so $[S_o(\omega = -i\Omega)] \approx [S(\omega = -i\Omega)]$, and thus $\rho(\omega = -i\Omega) \approx 0$.

### Sec. 5.5  Stability Criterion for Transient Excitation

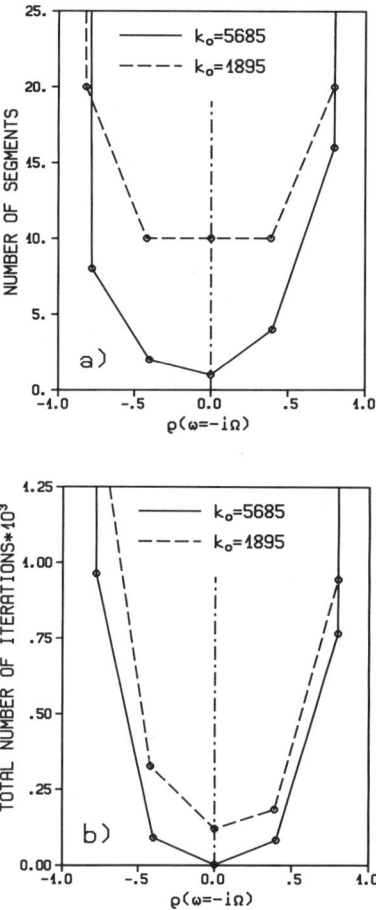

**Figure 5-15** Results of hybrid frequency–time-domain analyses of one-degree-of-freedom system.  a. Optimum number of segments versus spectral radius ($\omega = -i\Omega$).  b. Minimum total number of iterations versus spectral radius ($\omega = -i\Omega$).

**TABLE 5-4**  Results for Pseudo-Linear System with $m_o = 9$, $c_o = 18.1$

| Spring Coefficient $k_o$ | Spectral Radius | | Optimum Number of Segments | Minimum Total Number of Iterations |
|---|---|---|---|---|
| | $\rho(\omega = -i\Omega)$ | $\rho(\omega = 0)$ | | |
| 1137 | −.01 | −4.00 | 16 | 146 |
| 1895 | −.00 | −2.00 | 10 | 120 |
| 5685 | 0 | 0 | 1 | 2 |
| 11370 | .01 | .50 | 6 | 105 |
| 22740 | .02 | .75 | 15 | 235 |

## SUMMARY

1. The hybrid frequency–time-domain procedure bridges the gap between the time-domain solutions, used for the analyses of nonlinear systems, and the frequency-domain methods, used for the analyses of linear systems. In this procedure, a series of linear analyses is performed iteratively in the frequency domain. The nonlinearities are evaluated in the time domain, as in the standard time-integration schemes, and are introduced on the right-hand side of the equation of motion. By running through the response in time, the pseudo-forces, which are needed to bring the linear internal forces into coincidence with the true nonlinear ones corresponding to the same strains, are calculated. The pseudo-forces lead to pseudo-loads, which are added to the prescribed exterior loads in the linear analysis performed in the next iteration's frequency domain.
2. The hybrid frequency–time-domain procedure is conceptually simple and can easily be implemented in an existing computer program.
3. The hybrid frequency–time-domain procedure provides an accurate timewise solution without false damping or frequency distortion.
4. The scheme works in each iteration with the prescribed exterior loads and the pseudo-loads. This total-displacement procedure is more forgiving in the sense that initial approximations are corrected in later cycles, compared to methods which only process the changes of the pseudo-loads.
5. To achieve a periodic system with a finite period and zero initial conditions, the pseudo-loads, after the time of interest for the response has passed, are gradually artificially reduced to zero, followed by the quiet zone.
6. By eliminating the degrees of freedom in the structure's linear zones, it is possible to perform the iterations only over the locally nonlinear zones.
7. The hybrid frequency–time-domain procedure exhibits a forward time progressing convergence. The latter portion of the response time-history cannot converge until the initial portion has converged, because the pseudo-loads are calculated in the time domain.
8. Subdividing the total time interval into segments and achieving convergence sequentially for each segment in a time progressing manner significantly reduces the intermediate results' deviations from the exact solution and thus reduces the possibility of divergence and unstable behavior occurring. After convergence has been reached in the first part of a segment, the corresponding pseudo-loads are not changed.
9. The convergence criterion can, over the whole time history, address the change of the displacements between two consecutive iterations ex-

pressed as a fraction of the final displacements; address the change of the pseudo-loads compared to the prescribed exterior loads; or address the change of energies, formulated as the product of the changes of the displacements and of the pseudo-loads between two consecutive iterations, expressed as a fraction of the final energy.

10. The stability criterion for harmonic excitation of a multiple-degree-of-freedom system states that the spectral radius of a matrix, which is equal to the product of the inverse of the pseudo-linear system's dynamic-stiffness matrix and the difference between the actual nonlinear and linear ones, must be smaller than 1. Applying this stability criterion to a one-degree-of-freedom system excited by a harmonic force leads to the condition that convergence occurs when the ratio of the dynamic-stiffness coefficients of the actual system and of the pseudo-linear system lies between 0 and 2. This criterion is, however, too strict for a transient excitation. For this case, the criterion of stability requires that the same spectral radius, evaluated at a frequency equal to minus $i$ times the Nyquist frequency, must be less than 1. The latter is generally high enough so that the associated dynamic-stiffness matrices are dominated by the mass of the respective systems. The mass generally does not vary from the actual to the pseudo-linear system so that the spectral radius evaluated at the Nyquist frequency is generally close to zero. Such a value of the spectral radius always guarantees stability. A successful implementation of the procedure, however, requires the introduction of time segments. The number of segments does not strongly affect the total number of iterations and the necessary computer time, except when too many are used. Selecting too few time segments can cause divergence.

11. Additional parametric studies on a variety of soil–structure systems—also addressing other aspects—are essential to build up the necessary experience before the hybrid frequency–time-domain procedure can be applied with confidence.

# 6

# Substructure Method Using Dynamic Stiffness of Soil

In the substructure method of analyzing soil–structure interaction, the interaction horizon coincides with the (generalized) structure–soil interface (Section 1.3, Fig. 1-3a). The properties of the unbounded (linear) soil on the exterior are represented by a boundary condition in the form of a force–displacement relationship, which will be global in space and time. This relationship is expressed in the form of convolution integrals involving the dynamic-stiffness coefficients in the time domain with respect to the degrees of freedom of the nodes located on the structure–soil interface and the corresponding motions. Alternatively, Green's functions in the time domain can be used directly in the computational procedure, which is addressed in Chapter 7.

As the regular unbounded soil on the exterior of the interaction horizon up to infinity behaves linearly, this substructure can be analyzed in the frequency domain. Applying these well-known procedures results in the dynamic-stiffness coefficients in the frequency domain (Section 4.4), which must be transformed into the time domain before the total dynamic system with a nonlinear structure can be analyzed. Various alternatives to handle this substructure of the unbounded soil are possible. Instead of using stiffness coefficients, a formulation based on dynamic-flexibility coefficients can be applied, which are easier (at least conceptually) to transform from the frequency to the time domain. It is also feasible to calculate the unbounded soil's interaction forces in the frequency domain and then to transform these into the time domain; this, as will become apparent, corresponds to evaluating the convolution integrals in the frequency domain and thus avoids the difficulties associated with the transformation of the dynamic-stiffness coefficients.

## Sec. 6.1 Basic Equation of Motion

The basic equation of motion of soil–structure-interaction analysis in the time domain is derived in Section 6.1. Stiffness and flexibility formulations are examined. The dynamic-stiffness coefficients in the time domain are addressed in Section 6.2. Because in a practical case they are calculated as the Fourier transform of the corresponding values in the frequency domain, the behavior of the dynamic-stiffness coefficients for high frequencies is discussed in Section 6.3. The dynamic-flexibility coefficients in the time domain are treated in Section 6.4. The influence of incorporating hysteretic damping is described in Section 6.5. The discretization in the time domain as well as the corresponding computational procedure for an explicit and an implicit time-integration scheme of the total dynamic system are addressed in Section 6.6. For the unbounded soil's substructure, a stiffness and a flexibility formulation are possible. For the sake of illustration, a simple structure with and without a nonlinear base isolation is calculated in Section 6.7. The rigorous results are compared to those of an approximate analysis based on frequency-independent springs and dampers. The evaluation of the convolution integrals in the frequency domain is discussed in Section 6.8, whereby the Fourier transforms of the displacements are calculated recursively. An introduction to the recursive calculation of the convolution integrals in the time domain is presented in Section 6.9, which leads to a dramatic reduction of the computational effort. These two latter procedures are still in the formative stage.

## 6.1 BASIC EQUATION OF MOTION

### 6.1.1 Stiffness Formulation

To derive the equation of motion of the structure and the soil in the time domain, the dynamic system with a spatial discretization shown in Fig. 6-1 is examined. The loading is assumed to be introduced into the system via the soil, as is the case for seismic excitation. The system consists of a

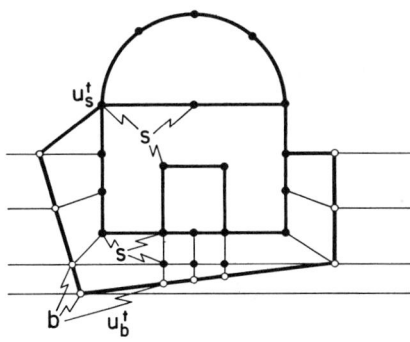

**Figure 6-1** Structure–soil system.

structure (and possibly of an irregular soil region adjacent to the structure), which can behave nonlinearly, and of a regular unbounded linear soil. The (generalized) structure–soil interface, with the nodes whose degrees of freedom are denoted by subscript $b$ (for base), separates these two regions. (In the following, it is implied that the word *structure* also includes the irregular soil region—that is, the expression *generalized* is omitted.) The subscript $s$ (for structure) denotes the degrees of freedom of the other nodes of the potentially nonlinear discretized system, which is one of the substructures. The other one consists of the unbounded soil with excavation (Fig. 6-2), which is indicated by the superscript $g$ (for ground). It is appropriate to work also with other subsystems of the soil. The unbounded soil without excavation— the free field—is denoted by $f$; and $e$ is used to designate the excavated soil.

The substructure ground's contribution to the equations of motion is determined as follows. The dynamic-stiffness matrix in the time domain of this unbounded soil for which the excavation is taken into account is denoted as $[S^g_{bb}(t)]$. Each column contains the forces at the nodes $b$ as a function of time required to produce a unit-impulse displacement at time zero—that is, a delta function $\delta(t)$ for that degree of freedom corresponding to the column. The vector of the displacements at the nodes $b$ in the same soil system $g$— for example, for seismic excitation—is denoted as the generalized scattered motion $\{u^g_b(t)\}$ (Fig. 6-2), which can be calculated from the free-field motion $\{u^f_b(t)\}$, as described in Section 6.1.3. Let $\{R_b(t)\}$ denote the vector of the interaction forces of the unbounded soil (system $g$) acting at the nodes $b$; and let $\{u^t_b(t)\}$ denote the vector of the dynamic system's total displacements. (A superscript is used only when confusion would otherwise arise.) $\{R_b\}$ is a function of the motion relative to the ground ($\{u^t_b\} - \{u^g_b\}$); as for the loading state specified by $\{u^g_b\}$, the surface that subsequently forms the structure–soil interface is a free surface. The contribution of the infinitesimal pulse ($\{u^t_b(\tau)\}$

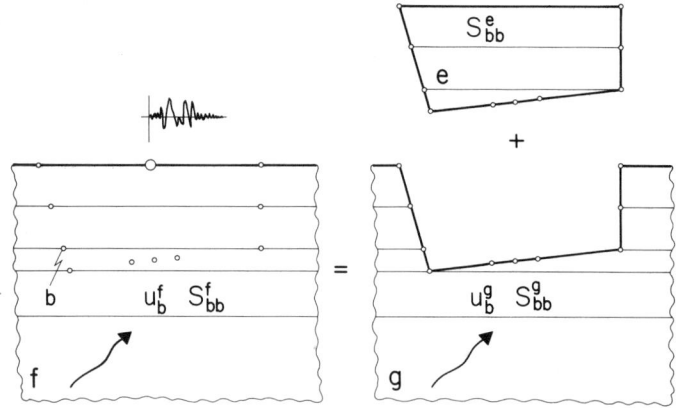

**Figure 6-2** Reference soil systems.

## Sec. 6.1  Basic Equation of Motion

$-\{u_b^g(\tau)\})\, d\tau$ acting at time $\tau$ to the interaction forces $\{dR_b(t)\}$ at time $t$ ($t > \tau$) depends on the dynamic stiffness evaluated for the time difference $t - \tau$, $[S_{bb}^g(t - \tau)]$, resulting in

$$\{dR_b(t)\} = [S_{bb}^g(t - \tau)]\, (\{u_b^t(\tau)\} - \{u_b^g(\tau)\})\, d\tau \qquad (6.1)$$

For the total time history, the ground's interaction forces $\{R_b(t)\}$ are equal to the convolution integral of the dynamic stiffness $[S_{bb}^g(t)]$ and the motion relative to the ground $\{u_b^t(t)\} - \{u_b^g(t)\}$

$$\{R_b(t)\} = \int_0^t [S_{bb}^g(t - \tau)]\, (\{u_b^t(\tau)\} - \{u_b^g(\tau)\})\, d\tau \qquad (6.2)$$

This is, of course, analogous to the familiar Duhamel integral introduced in elementary structural dynamics in the context of a flexibility formulation.

The structure's equation of motion is specified as

$$\begin{bmatrix} [M_{ss}] & [M_{sb}] \\ [M_{bs}] & [M_{bb}] \end{bmatrix} \begin{Bmatrix} \{\ddot{u}_s^t(t)\} \\ \{\ddot{u}_b^t(t)\} \end{Bmatrix} + \begin{Bmatrix} \{P_s(t)\} \\ \{P_b(t)\} \end{Bmatrix} = \begin{Bmatrix} \{0\} \\ -\{R_b(t)\} \end{Bmatrix} \qquad (6.3)$$

$[M]$ is the mass matrix and $\{P(t)\}$ the vector of the (nonlinear) internal forces of the structure at time $t$. As for an excitation introduced into the system via the soil, the nodes not in contact with the soil are not loaded, the right-hand side, corresponding to the degrees of freedom associated with $s$, is zero in a formulation in total displacements. If loads are applied directly to the structure, the corresponding loads $\{R_s(t)\}$ and $\{R_b^s(t)\}$ are added to the right-hand side of Eq. 6.3.

Substituting Eq. 6.2 into Eq. 6.3 leads to the basic equation of motion in total displacements of the substructure method [W11, W12].

$$\begin{bmatrix} [M_{ss}] & [M_{sb}] \\ [M_{bs}] & [M_{bb}] \end{bmatrix} \begin{Bmatrix} \{\ddot{u}_s^t(t)\} \\ \{\ddot{u}_b^t(t)\} \end{Bmatrix} + \begin{Bmatrix} \{P_s(t)\} \\ \{P_b(t)\} \end{Bmatrix}$$

$$+ \begin{Bmatrix} \{0\} \\ \int_0^t [S_{bb}^g(t - \tau)]\{u_b^t(\tau)\}\, d\tau \end{Bmatrix} = \begin{Bmatrix} \{0\} \\ \int_0^t [S_{bb}^g(t - \tau)]\, \{u_b^g(\tau)\}\, d\tau \end{Bmatrix} \qquad (6.4)$$

This equation, expressed in total motion, results in a simple procedure which can be used to calculate a general wave pattern consisting of inclined body waves and surface waves. The right-hand side representing the loads can be evaluated, once $\{u_b^g(t)\}$ is determined—that is, the motion in the nodes (which will subsequently lie on the structure–soil interface) of the ground with the excavation. In most cases, it is desirable to replace this generalized scattered motion by the free-field motion $\{u_b^f(t)\}$, which does not depend on the excavation (with the exception of the location of the nodes in which it is to be calculated). This is performed as follows:

The free-field system results when the excavated part of the soil is added to the soil with excavation. This also holds for the assembly process of the dynamic-stiffness matrices (Fig. 6-2).

$$[S_{bb}^g(t)] + [S_{bb}^e(t)] = [S_{bb}^f(t)] \tag{6.5}$$

By stipulating that the "structure" consists of the excavated part of the soil only, Eq. 6.4 can be formulated for this special case. No nodes with subscript $s$ exist. $\{u_b^t(t)\}$ equals $\{u_b^f(t)\}$ in this case.

$$[M_{bb}^e]\{\ddot{u}_b^f(t)\} + [K_{bb}^e]\{u_b^f(t)\} + \int_0^t [S_{bb}^g(t-\tau)]\{u_b^f(\tau)\}\,d\tau \tag{6.6}$$
$$= \int_0^t [S_{bb}^g(t-\tau)]\{u_b^g(\tau)\}\,d\tau$$

where the excavated soil's forces are expressed as the product of the static-stiffness matrix $[K_{bb}^e]$ and $\{u_b^f\}$. As for the linear excavated part,

$$[M_{bb}^e]\{\ddot{u}_b^f(t)\} + [K_{bb}^e]\{u_b^f(t)\} = \int_0^t [S_{bb}^e(t-\tau)]\{u_b^f(\tau)\}\,d\tau \tag{6.7}$$

applies, and Eq. 6.6 is reformulated after using Eq. 6.5 (specified for $t - \tau$) as

$$\int_0^t [S_{bb}^f(t-\tau)]\{u_b^f(\tau)\}\,d\tau = \int_0^t [S_{bb}^g(t-\tau)]\{u_b^g(\tau)\}\,d\tau \tag{6.8}$$

This equality of forces is quite a remarkable result in its own right. Substituting Eq. 6.8 in Eq. 6.4, the basic equation of motion, with the free-field motion determining the load vector, is specified as

$$\begin{bmatrix} [M_{ss}] & [M_{sb}] \\ [M_{bs}] & [M_{bb}] \end{bmatrix} \begin{Bmatrix} \{\ddot{u}_s^t(t)\} \\ \{\ddot{u}_b^t(t)\} \end{Bmatrix} + \begin{Bmatrix} \{P_s(t)\} \\ \{P_b(t)\} \end{Bmatrix}$$
$$+ \begin{Bmatrix} \{0\} \\ \int_0^t [S_{bb}^g(t-\tau)]\{u_b^t(\tau)\}\,d\tau \end{Bmatrix} = \begin{Bmatrix} \{0\} \\ \int_0^t [S_{bb}^f(t-\tau)]\{u_b^f(\tau)\}\,d\tau \end{Bmatrix} \tag{6.9}$$

Alternatively, using the free-field forces $\{P_b^f(t)\}$ (calculated from the corresponding surface tractions), which, taking equilibrium into consideration, are equal to $-\int_0^t [S_{bb}^e(t-\tau)]\{u_b^f(\tau)\}d\tau$, leads to the following modified right-hand side of the subvector in Eq. 6.9: $\int_0^t [S_{bb}^e(t-\tau)]\{u_b^f(\tau)\}\,d\tau - \{P_b^f(t)\}$. In this case, both the displacements and the surface tractions of the free field must be determined.

### 6.1.2 Flexibility Formulation

Alternatively, a flexibility formulation can be more appropriate to describe the unbounded soil's contribution to the equations of motion, together with the direct stiffness method for the structure (Sections 6.4, 6.6).

$[F^g_{bb}(t)]$ denotes the dynamic-flexibility matrix of the soil with excavation—that is, each column contains the displacements at nodes $b$ as a function of time resulting from a unit-impulse force applied at time zero in the direction of that degree of freedom corresponding to the column.

Analogous to Eq. 6.2, the displacement–interaction force relationship is formulated as

$$\{u^t_b(t)\} - \{u^g_b(t)\} = \int_0^t [F^g_{bb}(t - \tau)] \{R_b(\tau)\} \, d\tau \qquad (6.10)$$

The motion relative to the ground $\{u^t_b(t)\} - \{u^g_b(t)\}$ is thus equal to the convolution integral of the dynamic-flexibility $[F^g_{bb}(t)]$ and the soil's interaction forces $\{R_b(t)\}$. Equation 6.10 is used together with Eq. 6.3 in the computational procedure (Section 6.6).

A flexibility formulation for the unbounded soil which uses the free-field motion is derived in Problem 6.2.

### 6.1.3 Scattered Motion

To work with some forms of the basic equation of motion (Eqs. 6.4, 6.10), the generalized scattered motion $\{u^g_b(t)\}$ is needed. It can be expressed by the free-field motion $\{u^f_b(t)\}$ (Fig. 6-2). As this is a linear problem, it is appropriate to work in the frequency domain. The relationship corresponding to Eq. 6.8 equals (see also Eq. 4.8)

$$[S^f_{bb}(\omega)] \{u^f_b(\omega)\} = [S^g_{bb}(\omega)] \{u^g_b(\omega)\} \qquad (6.11)$$

The amplitudes of the generalized scattered motion in the frequency domain $\{u^g_b(\omega)\}$ follow from those of the free-field motion $\{u^f_b(\omega)\}$ as

$$\{u^g_b(\omega)\} = [S^g_{bb}(\omega)]^{-1} [S^f_{bb}(\omega)] \{u^f_b(\omega)\} \qquad (6.12)$$

## 6.2 DYNAMIC-STIFFNESS COEFFICIENT IN TIME DOMAIN

### 6.2.1 Calculation in Time Domain

The spherical cavity with symmetric waves, addressed in Sections 2.3 and 3.1.3, can be used to illustrate the procedure to calculate the dynamic-stiffness coefficient in the time domain and to discuss its vital aspects. For this case, it is possible to work in the time domain without introducing a transformation.

The radial displacement $u(r, t)$ as the solution of the wave equation is specified in Eq. 3.40. Deleting the term corresponding to the wave propagating in the negative radial direction and modifying the argument slightly to concisely satisfy the boundary condition on the wall of the cavity at $r = a$

leads to

$$u(r, t) = -\frac{f\left(t - \frac{r-a}{c_p}\right)}{r^2} - \frac{f'\left(t - \frac{r-a}{c_p}\right)}{rc_p} \qquad (6.13)$$

with $f' = df[t - (r - a)/c_p]/d[t - (r - a)/c_p]$. The initial conditions are specified as

$$u(r, t = 0) = 0 \qquad (6.14a)$$
$$\dot{u}(r, t = 0) = 0 \qquad (6.14b)$$

and the boundary condition corresponding to the definition of the dynamic-stiffness coefficient (Section 6.1.1) as

$$u(r = a, t) = \delta(t) \qquad (6.15)$$

$\delta(t)$ is the Dirac delta function.

Substituting Eq. 6.13 in Eq. 6.15 leads to the ordinary differential equation of first order

$$\frac{\dot{f}(t)}{ac_p} + \frac{f(t)}{a^2} = -\delta(t) \qquad (6.16)$$

where the prime used to indicate the differentiation with respect to the argument is replaced by the symbol of the time derivative.

The homogeneous equation

$$\dot{f}_h(t) + \frac{c_p}{a} f_h(t) = 0 \qquad (6.17)$$

leads to

$$f_h(t) = c_1 \exp\left(-\frac{c_p}{a} t\right) \qquad (6.18)$$

with the integration constant $c_1$. The particular solution equals

$$f_p(t) = -ac_p \exp\left(-\frac{c_p}{a} t\right) \qquad t \geq 0 \qquad (6.19a)$$

or introducing the Heaviside function $H(t)$ ($=0$ for $t < 0$, $= 1$ for $t \geq 0$)

$$f_p(t) = -ac_p \exp\left(-\frac{c_p}{a} t\right) H(t) \qquad (6.19b)$$

The differentiation of $f_p(t)$ also leads to a term with a Dirac function [$\delta(t) = \dot{H}(t)$]

$$\dot{f}_p(t) = c_p^2 \exp\left(-\frac{c_p}{a} t\right) H(t) - ac_p \exp\left(-\frac{c_p}{a} t\right) \delta(t) \qquad (6.20)$$

### Sec. 6.2  Dynamic-Stiffness Coefficient in Time Domain

The total solution $f[t - (r - a)/c_p]$ equals

$$f\left(t - \frac{r-a}{c_p}\right)$$
$$= \exp\left[-\frac{c_p}{a}\left(t - \frac{r-a}{c_p}\right)\right]\left[c_1 - ac_p H\left(t - \frac{r-a}{c_p}\right)\right] \quad (6.21)$$

Substituting Eq. 6.21 in Eq. 6.13 results in

$$u(r, t) = \exp\left[-\frac{c_p}{a}\left(t - \frac{r-a}{c_p}\right)\right]\left[\frac{a}{r}\delta\left(t - \frac{r-a}{c_p}\right)\right.$$
$$\left. + \left(-\frac{c_p}{r} + \frac{ac_p}{r^2}\right)H\left(t - \frac{r-a}{c_p}\right) + c_1\left(\frac{1}{ar} - \frac{1}{r^2}\right)\right] \quad (6.22)$$

As the pulse propagates in the radial direction with the velocity $c_p$, $u(r, t)$ must vanish for $t < (r - a)/c_p$. This condition leads to $c_1 = 0$—that is, only the particular solution is needed (Eq. 6.18). Thus, the final displacement is

$$u(r, t) = \exp\left[-\frac{c_p}{a}\left(t - \frac{r-a}{c_p}\right)\right]\left[\frac{a}{r}\delta\left(t - \frac{r-a}{c_p}\right)\right.$$
$$\left. + \frac{c_p}{r}\left(-1 + \frac{a}{r}\right)H\left(t - \frac{r-a}{c_p}\right)\right] \quad (6.23)$$

It is instructive to study the time history $u(t)$ at a specific location $r$ ($> a$). As already mentioned, for $t < (r - a)/c_p$, the displacement vanishes. For $t = (r - a)/c_p$, the Dirac pulse with a value $a/r$ appears; and for $t \geq (r - a)/c_p$, $u(t)$ changes sign and diminishes with time.

The radial stress $\sigma_r(r, t)$ follows from Hooke's law and, using the strain-displacement relationship (Eq. 2.16), is

$$\sigma_r(r, t) = G\left[\frac{c_p^2}{c_s^2}u(r, t)_{,r} + \left(\frac{c_p^2}{c_s^2} - 2\right)\frac{2}{r}u(r, t)\right] \quad (6.24)$$

The dynamic-stiffness coefficient $S(t)$ is equal to the pressure at $r = a$ $[-\sigma_r(r = a, t)]$

$$S(t) = 4\frac{G}{a}\left[\left(1 - \frac{c_p^2}{4c_s^2}\right)\delta(t) + \frac{c_p^2}{4c_s^2}\frac{a}{c_p}\dot{\delta}(t)\right.$$
$$\left. + \frac{c_p^2}{4c_s^2}\frac{c_p}{a}\exp\left(-\frac{c_p}{a}t\right)\right] \quad t \geq 0 \quad (6.25)$$

The first two terms on the right-hand side are singular, and the third one with the exponential function is the regular part. For $t < 0$, $S(t) = 0$.

Analogous to $S(\omega)$, which is formulated as a function of the dimensionless frequency $a_o = \omega\, a/c_p$ (Eq. 2.12), it is convenient to introduce the dimensionless time $\bar{t}$ in $S(t)$

$$\bar{t} = t\frac{c_p}{a} \tag{6.26}$$

Using

$$\delta(t) = \frac{c_p}{a}\delta(\bar{t}) \tag{6.27a}$$

$$\dot{\delta}(t) = \frac{c_p^2}{a^2}\frac{d\delta(\bar{t})}{d\bar{t}} \tag{6.27b}$$

and substituting the static-stiffness coefficient $K = 4G/a$ transforms Eq. 6.25 to

$$S(\bar{t}) = K\frac{c_p}{a}\left[\left(1 - \frac{c_p^2}{4c_s^2}\right)\delta(\bar{t}) + \frac{c_p^2}{4c_s^2}\frac{d\delta(\bar{t})}{d\bar{t}} + \frac{c_p^2}{4c_s^2}\exp(-\bar{t})\right] \tag{6.28}$$

The value $a/c_p\, S(\bar{t})/K = a/c_p\, \bar{S}(\bar{t})$ is a dimensionless variable which is a function of $\bar{t}$. A similar relationship exists in the frequency domain $[S(a_o)/K = \bar{S}(a_o)]$. Note that in the time domain the dynamic-stiffness coefficient divided by the static coefficient must be multiplied by $a/c_p$ for a dimensionless quantity to be generated.

On the right-hand side of Eq. 6.28, the third term, representing the regular part, is plotted for $v = 1/3$ ($c_p/c_s = 2$) in Fig. 6-3.

To be able to evaluate the physical significance of the singular terms with $\delta(t)$ and $\dot{\delta}(t)$ in $S(t)$, the force–displacement relationship in the form of the convolution integral is formulated. Analogous to Eq. 6.2, the pressure

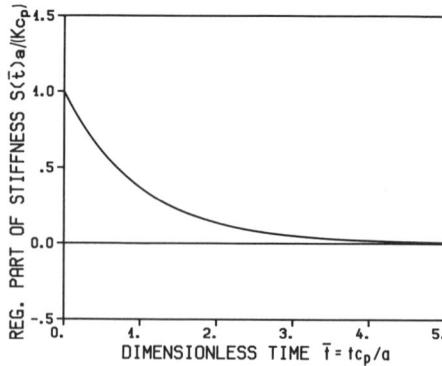

Figure 6-3 Regular part of dynamic-stiffness coefficient in time domain, spherical cavity.

Sec. 6.2  Dynamic-Stiffness Coefficient in Time Domain    253

$p(t)$ follows as

$$p(t) = \int_o^t S(t - \tau) u(\tau) \, d\tau \tag{6.29}$$

Substituting Eq. 6.25 into Eq. 6.29 results in

$$p(t) = K\left[\left(1 - \frac{c_p^2}{4c_s^2}\right) u(t) + \frac{c_p^2}{4c_s^2} \frac{a}{c_p} \dot{u}(t) \right. \\ \left. + \frac{c_p^2}{4c_s^2} \frac{c_p}{a} \int_o^t \exp\left[-\frac{c_p}{a}(t - \tau)\right] u(\tau) \, d\tau\right] \tag{6.30}$$

The force–displacement relationship at time $t$ consists of three parts: a spring with a dimensionless coefficient $1 - c_p^2/(4c_s^2)$ arising from $\delta(t)$; a damper with a coefficient $c_p a/(4c_s^2)$ occurring from $\dot{\delta}(t)$; and the convolution integral. The first two parts involve the motion at $t$ only, while the third processes the displacements from zero up to $t$.

In general, the dynamic-stiffness coefficient cannot be determined working only in the time domain, because the differential equation cannot be solved without applying a transformation into the frequency domain. It is possible to transform the dynamic-stiffness coefficient directly from the frequency to the time domain. This procedure is examined in the next section.

### 6.2.2 Transformation from Frequency to Time Domain

**Fourier-transform pair.** By definition (Section 6.1.1), the dynamic-stiffness coefficient in the time domain $S(t)$ is equal to the force as a function of time $R(t)$ that produces a displacement $u(t)$ of unit impulse $\delta(t)$ (Dirac delta function). Transforming this displacement into the frequency domain

$$u(\omega) = \int_{-\infty}^{+\infty} u(t) \exp(-i\omega t) \, dt = \int_{-\infty}^{+\infty} \delta(t) \exp(-i\omega t) \, dt = 1 \tag{6.31}$$

formulating the force–displacement relationship in the frequency domain

$$R(\omega) = S(\omega) u(\omega) = S(\omega) \tag{6.32}$$

and applying the inverse transformation, leads to

$$R(t) = \frac{1}{2\pi} \int_{-\infty}^{+\infty} R(\omega) \exp(i\omega t) \, d\omega = \frac{1}{2\pi} \int_{-\infty}^{+\infty} S(\omega) \exp(i\omega t) \, d\omega \tag{6.33}$$

which means, as $S(t) = R(t)$,

$$S(t) = \frac{1}{2\pi} \int_{-\infty}^{+\infty} S(\omega) \exp(i\omega t) \, d\omega \tag{6.34}$$

that is, $S(t)$ and $S(\omega)$ form a Fourier-transform pair.

Using dimensionless quantities, with $a$ denoting a characteristic length of the basemat,

$$a_o = \frac{\omega a}{c_s} \tag{6.35}$$

$$\bar{t} = t\frac{c_s}{a} \tag{6.36}$$

the Fourier transformation of the dynamic-stiffness coefficient equals

$$S(\bar{t}) = \frac{c_s}{2\pi a} \int_{-\infty}^{+\infty} S(a_o) \exp(ia_o \bar{t}) \, da_o \tag{6.37}$$

Based on the Fourier integral theorem the existence of $S(t)$ is guaranteed when $S(\omega)$ appearing in Eq. 6.34 is absolutely integrable over the $\omega$-axis—that is

$$\int_{-\infty}^{+\infty} |S(\omega)| \, d\omega < \infty \tag{6.38}$$

Due to the presence of radiation damping occurring in an unbounded domain, the values of the peaks of $S(\omega)$ multiplied by $d\omega$ remain finite, even in a system without material damping. However, for $\omega \to \infty$, $S(\omega)$ does not converge to zero. As proven in Section 6.3, the dynamic-stiffness coefficient for the undamped soil with excavation (system ground) in the frequency domain converges for $\omega \to \infty$ to a constant and an imaginary linear term

$$\lim_{\omega \to \infty} S(\omega) = K(k + i\omega c) \tag{6.39}$$

Splitting $S(\omega)$ into this limit—which is the singular part—and the remaining regular part allows the Fourier transformation to be performed. The regular part, which decays towards zero for $\omega \to \infty$, can, at least in principle, be evaluated. The transformation of the singular part equals

$$\frac{K}{2\pi} \int_{-\infty}^{+\infty} (k + i\omega c) \exp(i\omega t) \, d\omega = K[k\delta(t) + c\dot{\delta}(t)] \tag{6.40}$$

As an example, the spherical cavity is again addressed, using dimensionless variables. The dynamic-stiffness coefficient in the frequency domain, specified in Eqs. 2.25 and 2.26, is repeated here for easy reference ($a_o = \omega a/c_p$)

$$S(a_o) = K[k(a_o) + ia_o c(a_o)] \tag{6.41}$$

$$k(a_o) = 1 - \frac{\frac{c_p^2}{4c_s^2} a_o^2}{1 + a_o^2} \tag{6.42a}$$

$$c(a_o) = \frac{c_p^2}{4c_s^2} \frac{a_o^2}{1 + a_o^2} \tag{6.42b}$$

## Sec. 6.2  Dynamic-Stiffness Coefficient in Time Domain

The asymptotic values of $k(a_o)$ and $c(a_o)$ for $a_o \to \infty$ are equal to

$$\lim_{a_o \to \infty} k = 1 - \frac{c_p^2}{4c_s^2} \qquad (6.43a)$$

$$\lim_{a_o \to \infty} c = \frac{c_p^2}{4c_s^2} \qquad (6.43b)$$

Splitting $S(a_o)$ into the singular part corresponding to these limits and the remaining regular part leads to

$$S(a_o) = K\left[1 - \frac{c_p^2}{4c_s^2} + ia_o \frac{c_p^2}{4c_s^2} + \frac{c_p^2}{4c_s^2}\frac{1}{1+ia_o}\right] \qquad (6.44)$$

Substituting Eq. 6.44 into Eq. 6.37 (with $c_s$ replaced by $c_p$) and performing the transformation leads to [W17]

$$\bar{S}(\bar{t}) = K\frac{c_p}{a}\left[\left(1 - \frac{c_p^2}{4c_s^2}\right)\delta(\bar{t}) + \frac{c_p^2}{4c_s^2}\frac{d\delta(\bar{t})}{d\bar{t}} + \frac{c_p^2}{4c_s^2}\exp(-\bar{t})\right] \qquad (6.45)$$

which agrees with the corresponding expression working only in the time domain (Eq. 6.28).

It is worth mentioning that the asymptotic value of the imaginary part of $S(a_o)$—$4G/a \cdot i\omega a/c_p \cdot c_p^2/(4c_s^2)$—is equal to $\rho c_p i\omega$; that is, the standard damping coefficient is recovered (Section 3.5).

**Initial characteristics of regular part and corresponding high-frequency behavior.** Based on the initial-value theorem, the asymptotic behavior for $ia_o \to \infty$ of the regular part in the frequency domain $S_r(ia_o = \infty)$ determines the initial characteristics of the regular part in the time domain—that is, the value $\bar{S}_r(\bar{t} = 0^+)$ and the slope $d\bar{S}_r(\bar{t} = 0^+)/d\bar{t}$.

$$\lim_{\bar{t} \to 0^+}\left[\frac{a}{c_s}\bar{S}_r(\bar{t})\right] = \lim_{ia_o \to \infty}[ia_o S_r(a_o)] \qquad (6.46a)$$

$$\lim_{\bar{t} \to 0^+}\left[\frac{a}{c_s}\frac{d\bar{S}_r(\bar{t})}{d\bar{t}}\right] = \lim_{ia_o \to \infty}\left[-a_o^2 S_r(a_o) - ia_o \frac{a}{c_s}\bar{S}_r(\bar{t} = 0^+)\right] \qquad (6.46b)$$

Applying these equations to the spherical cavity, with $S_r(a_o)$ specified in the third term of Eq. 6.44 and replacing $c_s$ in Eq. 6.46 by $c_p$, leads to

$$\frac{a}{c_p}\bar{S}_r(\bar{t} = 0^+) = K\frac{c_p^2}{4c_s^2} \qquad (6.47a)$$

$$\frac{a}{c_p}\frac{d\bar{S}_r(\bar{t} = 0^+)}{d\bar{t}} = -K\frac{c_p^2}{4c_s^2} \qquad (6.47b)$$

which are the same values as calculated from Eq. 6.45.

**Layered half-space.** In the frequency domain, analytical solutions in closed form do not exist for the dynamic-stiffness coefficients for more realistic cases, such as a basemat resting on a layered half-space. In principle, the transformation from the frequency to the time domain must be performed numerically. The asymptotic value of $S_{bb}^g(\omega)$ for $\omega \to \infty$, consisting of a spring and a damper for the undamped system ground (Eq. 6.39), can be determined for quite general cases (see Section 6.3.2). For certain simple cases, asymptotic expressions of $S_{bb}^g(\omega)$ applicable for large $\omega$ can be derived. The corresponding transformation from the frequency to the time domain can then be performed analytically for the range above a specific frequency. Below this frequency, $S_{bb}^g(\omega)$ is calculated numerically (see Problem 6.8). For a layered half-space, $S_{bb}^g(\omega)$ exhibits strong oscillations up to a large frequency, which leads to a fine mesh and thus to a significant computational effort. Experience shows that at least six points are needed to accurately represent the wave length $2\pi c_s/\omega$. The maximum length of an element (boundary or finite element) thus equals $\pi c_s/(3\omega)$. If, on a line running across the basemat of length $2a$ (for example, the diameter), $m$ elements are chosen, the dimensionless frequency $a_o = \omega a/c_s$ up to the value $\pi m/6$ is adequately modeled using this criterion. As an alternative, the convolution integral can be evaluated in the frequency domain which avoids the Fourier transformation of $S_{bb}^g(\omega)$ (Section 6.8).

Summarizing, the following transformation for the dynamic-stiffness matrix of the undamped soil, taking the excavation into account (system $g$, Fig. 6-2), can thus be performed.

$$[S_{bb}^g(\omega)] = [k_{bb}^g] + i\omega [c_{bb}^g] + [S_{r,bb}^g(\omega)] \tag{6.48}$$

$$[S_{bb}^g(t)] = [k_{bb}^g]\delta(t) + [c_{bb}^g]\dot{\delta}(t) + [S_{r,bb}^g(t)] \tag{6.49}$$

where $[k_{bb}^g]$ and $[c_{bb}^g]$ denote the asymptotic values for $\omega \to \infty$ of the constant and linear terms, respectively. The subscript $r$ stands for the regular part. The force–displacement relationship is formulated as

$$\{R_b(t)\} = [k_{bb}^g]\{u_b(t)\} + [c_{bb}^g]\{\dot{u}_b(t)\} + \int_o^t [S_{r,bb}^g(t - \tau)]\{u_b(\tau)\}\, d\tau \tag{6.50}$$

which corresponds for $\{u_b(t)\} = \{u_b^t(t)\} - \{u_b^g(t)\}$ to Eq. 6.2.

The transformation for the soil's undamped excavated part, which is a bounded system, is straightforward.

$$[S_{bb}^e(\omega)] = -\omega^2 [M_{bb}^e] + [K_{bb}^e] \tag{6.51}$$

$$[S_{bb}^e(t)] = [M_{bb}^e]\ddot{\delta}(t) + [K_{bb}^e]\delta(t) \tag{6.52}$$

The corresponding force–displacement relationship equals

$$\{P_b^e(t)\} = [M_{bb}^e]\{\ddot{u}_b(t)\} + [K_{bb}^e]\{u_b(t)\} \tag{6.53}$$

Sec. 6.2  Dynamic-Stiffness Coefficient in Time Domain

Finally, adding the contributions of the system ground and system excavation leads to the corresponding relationships for the system free field.

**Vertical dynamic-stiffness coefficient of disk on damped half-space.** Dynamic-stiffness coefficients in the time domain, even for a homogeneous half-space, have hardly been calculated. As an example, the vertical stiffness coefficient of a disk resting on a damped half-space is determined through the inverse Fourier transformation [W12]. Material damping can be described using the three-parameter Kelvin model (Fig. 6-4b), which does exhibit instantaneous elasticity, in contrast to the two-parameter model, also called the Voigt model (Fig. 6-4a). Fig. 6-4 corresponds to a one-dimensional stress-strain state.

For the Voigt model, the complex material constant $E^*$ (Young's modulus) is defined as

$$E^* = E(1 + 2\bar{\zeta}\omega i) = E(1 + 2\zeta\, a_o i) \tag{6.54}$$

with the damping ratio $\bar{\zeta} = c/(2k)$ and the corresponding dimensionless value $\zeta = \bar{\zeta} c_s/a$ as in Eq. 2.45.

For the three-parameter model, the flexibility relationship equals

$$\frac{1}{k_1} + \frac{1}{k_2 + i\omega c} \tag{6.55}$$

and its inverse, the stiffness relationship

$$\frac{k_1[(k_1 + k_2)k_2 + \omega^2 c^2 + i\omega k_1 c]}{(k_1 + k_2)^2 + \omega^2 c^2} \tag{6.56}$$

Dividing Eq. 6.56 by the static value ($\omega = 0$), which corresponds to the elastic material constant $E$, leads to the factor with which the latter is multiplied to define the viscoelastic material constant. Assuming the two spring constants to be equal ($k_1 = k_2$), and with $\bar{\zeta} = c/(2k)$, results in the complex material constant $E^*$

$$E^* = E\frac{1 + 2\bar{\zeta}\omega i}{1 + \bar{\zeta}\omega i} = E\frac{1 + 2\zeta a_o i}{1 + \zeta a_o i} \tag{6.57}$$

In three-dimensional elasticity, $G$ and $\lambda + 2G$ are multiplied by the factors specified in Eqs. 6.54 and 6.57.

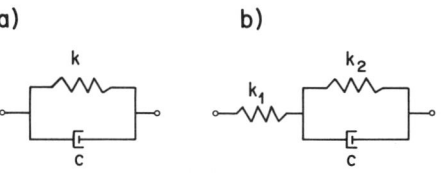

**Figure 6-4** Linearly viscoelastic material models. a. Voigt model. b. Three-parameter model.

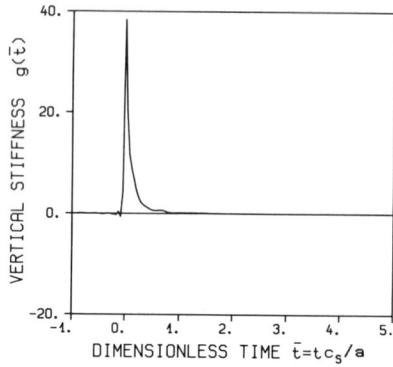

**Figure 6-5** Regular part of vertical dynamic-stiffness coefficient in time domain, half-space.

To calculate the dynamic-stiffness coefficient in the frequency domain in the vertical direction, the disk of radius $a$ is discretized using 25 annular rings of equal width (boundary elements). Over each annular ring the (unknown) soil reaction is assumed to be constant. Poisson's ratio $\nu$ equals 1/3, and the three-parameter model with $\zeta = 0.1$ is used. An asymptotic high-frequency expansion for the dynamic-stiffness coefficient is specified in Ref. [B2]. Substituting the complex elastic constants (analogous to Eq. 6.57) into the dimensionalized form of this asymptotic value leads, for the vertical direction, to

$$\lim_{a_o \to \infty} S(a_o) = \frac{4G\,a}{1 - \nu}\left[1.482\,a_o i + \left(0.786 - \frac{0.371}{\zeta}\right)\right] \quad (6.58)$$

with $a_o = \omega a/c_s$. This equation does not apply for $\zeta \to 0$, as $1/(\zeta a_o) \ll 1$ is assumed in the derivation. This singular part can be interpreted in the time domain as a viscous damper and a spring. The remaining regular part is transformed numerically back into the time domain. The dynamic-stiffness coefficient in the time domain $S(\bar{t})$, with $\bar{t} = tc_s/a$, equals

$$S(\bar{t}) = \frac{4Ga}{1-\nu}\frac{c_s}{a}\left[\left(0.786 - \frac{0.371}{\zeta}\right)\delta(\bar{t}) + 1.482\frac{d\delta(\bar{t})}{d\bar{t}} + g(\bar{t})\right] \quad (6.59)$$

where $g(\bar{t})$ corresponds to the transform of the regular part. $g(\bar{t})$ is plotted in Fig. 6-5.

## 6.3 HIGH-FREQUENCY BEHAVIOR OF WAVES FROM VIBRATING SOURCE

### 6.3.1 Increased Directionality

To be able to perform the inverse Fourier transform of the dynamic-stiffness coefficient from the frequency to the time domain, the asymptotic value for the frequency approaching infinity must be determined. This limit,

### Sec. 6.3 High-Frequency Behavior of Waves from Vibrating Source

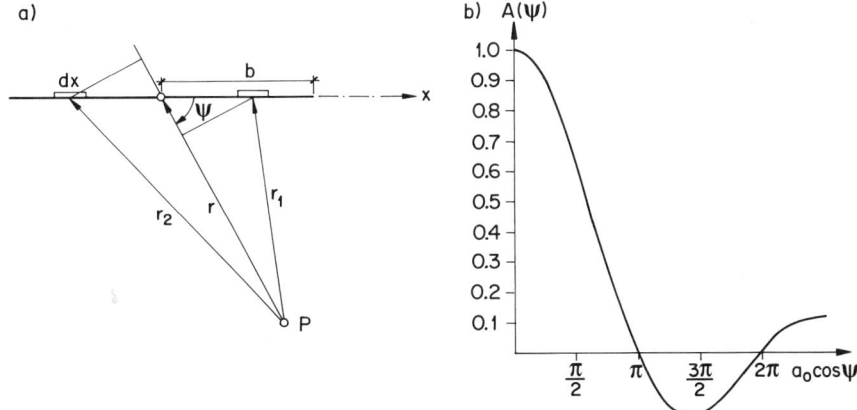

**Figure 6-6** Directionality of waves as function of frequency. a. Crude model. b. Angular dependence.

consisting of a constant real term and an imaginary term which is linear in frequency (Section 6.2), forms the singular part of the dynamic-stiffness coefficient of the undamped soil, taking the excavation into consideration (system ground).

The wave pattern arising from a vibrating source for increasing frequency is first addressed. A very crude model based on Ref. [G2] is used to illustrate the waves' increased directionality, which occurs when the frequency is increased. A strip of width $2b$ resting on the surface of a "half-plane" is assumed to radiate a single type of wave away from the line of contact with a continuous distribution of infinitesimal sources of constant intensity (Fig. 6-6a). The wave's amplitude at a point $P$ of the half-plane is examined. The point $P$, at a distance $r$ from the strip's center and at an angle of incidence $\psi$, is assumed to be reasonably far away from the vibrating strip. This allows the distances $r_1$ and $r_2$ from two radiating sources lying symmetrically at equal distance $x$ from the center to be expressed as

$$r_1 = r - x \cos\psi \quad (6.60a)$$

$$r_2 = r + x \cos\psi \quad (6\text{-}60b)$$

With the foregoing simplification, the contribution of these two symmetric infinitesimal sources to the wave amplitude $dA(r, \psi)$ at $P$ equals

$$dA(r, \psi) = \exp[i(\omega t - kr_1)]dx + \exp[i(\omega t - kr_2)]dx \quad (6.61)$$

or, using Eq. 6.60,

$$dA(r, \psi) = \exp[i(\omega t - kr)][\exp(ikx \cos\psi) + \exp(-ikx \cos\psi)]dx \quad (6.62)$$
$$= 2 \exp[i(\omega t - kr)]\cos(kx \cos\psi)dx$$

$k$ denotes the wave number ($=\omega/c$, with the wave velocity $c$).

After integration over the continuous source distribution,

$$A(r, \psi) = 2 \exp[i(\omega t - kr)] \int_{-b}^{+b} \cos\left(\frac{\omega}{c} \cos\psi \, x\right) dx \tag{6.63}$$

$$= 4 \exp[i(\omega t - kr)] \frac{c}{\omega \cos\psi} \sin\left(\frac{\omega b}{c} \cos\psi\right)$$

follows. Introducing the dimensionless frequency $a_o = \omega b/c$

$$A(r, \psi) = 4b \exp[i(\omega t - kr)] A(\psi) \tag{6.64}$$

results, with the angular dependence $A(\psi)$ specified as

$$A(\psi) = \frac{\sin(a_o \cos\psi)}{a_o \cos\psi} \tag{6.65}$$

$A(\psi)$ is plotted in Fig. 6-6b. For a constant $A(\psi)$, the product $a_o \cos\psi$ must remain constant. For increasing frequency $a_o$, the $\cos\psi$ must decrease and thus $\psi$ will increase. Significant spreading will occur at low frequencies, but at high frequencies a strong directionality will arise. In the limit for $a_o \to \infty$, $\cos\psi$ will be zero, resulting in $\psi = 90°$ and thus in wave propagation perpendicular to the vibrating surface—the strip.

The fact that most of the waves emitted from a vibrating source propagate through only a portion of the semi-infinite medium is caused by the so-called destructive interference. The arriving waves' phase differences are such that the waves tend to cancel each other out.

### 6.3.2 Dynamic-Stiffness Coefficient for Infinite Frequency

**Plane vibrating surface.** To discuss the asymptotic value of the dynamic-stiffness coefficient for infinite frequency, the undamped half-plane, with an arbitrary displacement amplitude $v(x)$ enforced on the surface in the out-of-plane direction $y$, is addressed first (Fig. 6-7). The corresponding load amplitude is denoted as $q(x)$. As $\omega$ is assumed to be infinite, this argument is omitted. $v(x)$ can be expanded in the horizontal direction $x$ into a Fourier integral with terms $\exp(-ikx)$, $k$ being the wave number (Section 4.2.2). The following transformations apply:

$$v(k) = \frac{1}{2\pi} \int_{-\infty}^{+\infty} v(x) \exp(ikx) dx \tag{6.66a}$$

$$v(x) = \int_{-\infty}^{+\infty} v(k) \exp(-ikx) dk \tag{6.66b}$$

The corresponding amplitude in the $k$-domain $q(k)$ is calculated applying

Sec. 6.3   High-Frequency Behavior of Waves from Vibrating Source   261

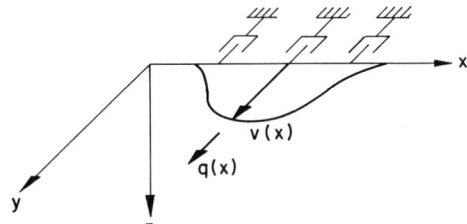

Figure 6-7 Surface of half-plane with out-of-plane loading.

the direct-stiffness approach. The dynamic-stiffness coefficient $S_{\text{SH}}(k, \omega)$ equals (Eq. 4.23)

$$S_{\text{SH}}(k, \omega) = iktG \tag{6.67}$$

with the wave number in the $z$-direction $kt$ specified in Eq. 3.193, as

$$kt = \sqrt{\frac{\omega^2}{c_s^2} - k^2} \tag{6.68}$$

In the limit $\omega \to \infty$ with $\omega > k$, $kt$ equals $\omega/c_s$, leading with $G = \rho c_s^2$ to

$$S_{\text{SH}}(k, \omega = \infty) = S_{\text{SH}} = i\omega\rho c_s \tag{6.69}$$

$S_{\text{SH}}$ is thus independent of $k$.
The load amplitude $q(k)$ follows as

$$q(k) = S_{\text{SH}} v(k) \tag{6.70}$$

The inverse Fourier transform of Eq. 6.70 results in

$$q(x) = S_{\text{SH}} v(x) \tag{6.71}$$

using Eq. 6.66b and the corresponding relationship for $q$. Substituting Eq. 6.69 leads to

$$q(x) = \rho c_s i\omega v(x) = \rho c_s \dot{v}(x) \tag{6.72}$$

This is the exact transmitting boundary condition for the one-dimensional wave propagation of the SH-waves in the direction $z$ (normal to the vibrating surface) with the velocity $c_s$ ( Eqs. 3.12, 3.109). The asymptotic value of the dynamic-stiffness coefficient for $\omega \to \infty$ thus consists of an imaginary linear term in $\omega$ with a constant $\rho c_s$. The corresponding force–velocity relationship $q(x)dx - \dot{v}(x)$ can be interpreted as a damper with a coefficient $\rho c_s\, dx$, which is independent of neighboring points (Fig. 6-7).

This concept can be generalized in various respects as follows: First, the in-plane case, with amplitudes of the displacements prescribed in the horizontal $x$-direction $u(x)$ and in the vertical $z$-direction $w(x)$ on the surface of a half-plane, is addressed. The corresponding load amplitudes are denoted as $p(x)$ and $r(x)$. For $\omega \to \infty$ and $\omega > k$, the dynamic-stiffness matrix $[S_{\text{P-SV}}]$ equals (Eq. 4.30)

$$[S_{\text{P-SV}}] = iG\frac{\omega}{c_s}\begin{bmatrix} 1 & 0 \\ 0 & \dfrac{c_p}{c_s} \end{bmatrix} \tag{6.73}$$

The horizontal and vertical directions are thus uncoupled. The first element on the diagonal $S_{\text{SV}}$ related to the horizontal degree of freedom and the second one $S_{\text{P}}$ related to the vertical one are associated with vertically propagating (that is, perpendicular to the vibrating surface) SV-waves with velocity $c_s$ and P-waves with velocity $c_p$, respectively.

$$S_{\text{SV}} = i\omega\rho c_s \tag{6.74a}$$

$$S_{\text{P}} = i\omega\rho c_p \tag{6.74b}$$

The force–displacement relationships in the $k$-domain

$$p(k) = i\omega\rho c_s u(k) \tag{6.75a}$$

$$r(k) = i\omega\rho c_p w(k) \tag{6.75b}$$

are transformed back to the time domain

$$p(x) = \rho c_s i\omega u(x) = \rho c_s \dot{u}(x) \tag{6.76a}$$

$$r(x) = \rho c_p i\omega w(x) = \rho c_p \dot{w}(x) \tag{6.76b}$$

Second, addressing the three-dimensional case (in cylindrical coordinates), the loading is expanded, in a Fourier series, in the circumferential direction, and the Fourier transformation is replaced by the Bessel transformation in the radial direction involving the wave number. As the same dynamic-stiffness coefficients apply for cylindrical coordinates as for plane waves expressed in Cartesian coordinates (Section 4.2.4), the derivation is analogous.

Third, if some material damping of the hysteretic type is present, the results remain unchanged when a layer of finite depth instead of a half-space is examined. The dynamic-stiffness matrix of a layer of depth $d$ for the out-of-plane motion equals (Eq. 4.22)

$$[S_{\text{SH}}] = \frac{ktG}{\sin ktd}\begin{bmatrix} \cos ktd & -1 \\ -1 & \cos ktd \end{bmatrix} \tag{6.77}$$

As just shown, $kt$ converges for $\omega \to \infty$ to $\omega/c_s^*$ with

$$c_s^* = c_s\sqrt{1 + 2\zeta i} \sim c_s(1 + \zeta i) \tag{6.78}$$

leading to

$$kt \approx \frac{\omega}{c_s}(1 - \zeta i) \tag{6.79}$$

## Sec. 6.3 High-Frequency Behavior of Waves from Vibrating Source

The imaginary part of $kt$—$Im(kt) = -\omega\zeta/c_s$—is thus negative. The factor appearing in the first element of $[S_{SH}]$ (Eq. 6.77) is expanded, using straightforward algebra, as

$$\frac{\cos[Re(ktd) + i\,Im(ktd)]}{\sin[Re(ktd) + i\,Im(ktd)]} = \frac{\sin[Re(ktd)]\cos[Re(ktd)](1 - \tanh^2[Im(ktd)])}{\sin^2[Re(ktd)] + \cos^2[Re(ktd)]\tanh^2[Im(ktd)]}$$

$$-i\frac{\tanh[Im(ktd)]}{\sin^2[Re(ktd)] + \cos^2[Re(ktd)]\tanh^2[Im(ktd)]} \quad (6.80)$$

As the $\tanh[Im(ktd)] = \tanh(-\omega\zeta d/c_s)$ with a negative argument converges for $\omega \to \infty$ to the value $-1$, the real and imaginary parts converge to 0 and $+i$, respectively. This also occurs for an infinitesimally small hysteretic damping ratio $\zeta$. The first element of $[S_{SH}]$ (Eq. 6.77) thus converges to $iktG^* = i\omega\rho c_s(1 + \zeta i)$, which, for an infinitesimally small $\zeta$, equals $i\omega\rho c_s$. This is the same result as for the half-space (Eq. 6.69).

In summary, a vibrating source will lead in the limit of an infinitely high frequency to a one-dimensional wave propagation in the direction perpendicular to the vibrating plane. The two tangential components correspond to shear waves, and the normal component corresponds to a dilatational wave.

When this undamped one-dimensional wave propagation corresponds to that present in a prismatic bar (as for a vibrating horizontal plane), the dynamic-stiffness coefficient is equal to an imaginary linear term in the frequency, with a coefficient equal to the product of the area, the mass density, and the corresponding wave velocity. For a layered system, the properties of the material adjacent to the vibrating source—that is, the structure–soil interface—are used. For instance, for a rigid disk of radius $a$ resting on the surface of a layered half-space, the asymptotic values of the dynamic-stiffness coefficients are determined as follows:

For the horizontal direction $\pi a^2 \rho c_s i\omega = \pi G a i a_o$
For the vertical direction $\pi a^2 \rho c_p i\omega = \pi G \sqrt{2(1-v)/(1-2v)}\, a i a_o$
For the rocking motion $\pi a^4/4 \rho c_p i\omega = \pi/4\, G\sqrt{2(1-v)/(1-2v)}\, a^3 i a_o$
For the torsional motion $\pi a^4/2 \rho c_s i\omega = \pi/2 G a^3 i a_o$
Where $a_o = \omega a/c_s$ and with the material properties of the top layer.

**Curved vibrating surface.** When the vibrating surface is curved, a one-dimensional wave propagation in the direction perpendicular to the surface still occurs for $\omega \to \infty$. The asymptotic value of the undamped system's dynamic-stiffness coefficient will consist of an imaginary linear term in $\omega$ with the same value as for a flat vibrating surface (see preceding passage), corresponding to a damper, and of a constant, corresponding to a spring. To determine the latter, it is, in general, necessary to calculate the corresponding dynamic-stiffness coefficient and to determine its asymptotic value. This situation occurs, for example, for the spherical cavity (Eq. 6.43).

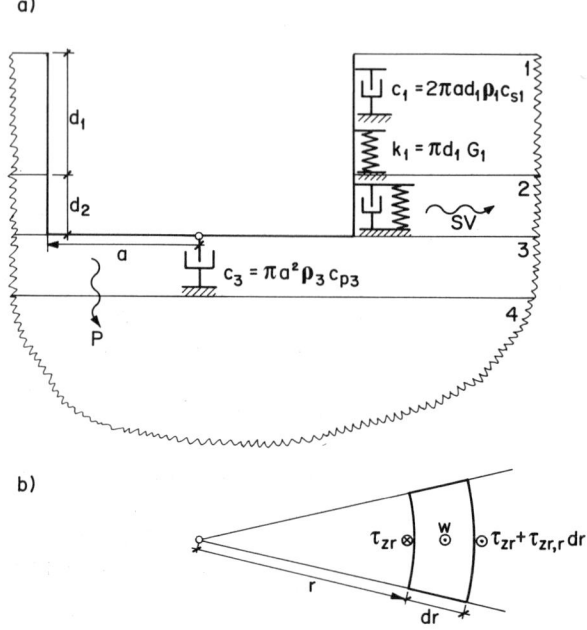

**Figure 6-8** Asymptotic dynamic-stiffness coefficient in vertical direction for embedded cylinder. a. System. b. Infinitesimal element.

Another example is addressed next (Fig. 6-8a). The asymptotic value of the dynamic-stiffness coefficient of a rigid cylinder embedded in an undamped layered half-space in the vertical direction is calculated. The contributions of the circular basemat and of the vertical wall—for each cylindrical segment of the latter—can be determined separately. In the area of contact, the basemat's vertical motion will generate P-waves propagating vertically, leading to the stiffness coefficient just mentioned ($\pi a^2 \rho_3 c_{p3} i\omega$, determined with the soil properties of the layer 3 adjacent to the basemat). The vertical motion $w$ of a segment of the cylindrical wall will result in SV-waves propagating radially in a cylindrical coordinate system, leading to a shear stress $\tau_{zr}$ (Fig. 6-8b). The dynamic-stiffness coefficient is derived as follows: Substituting the stress–displacement relationship

$$\tau_{zr}(\omega) = G w_{,r}(\omega) \tag{6.81}$$

into the equilibrium equation

$$\tau_{zr,r}(\omega) + \frac{1}{r} \tau_{zr}(\omega) + \rho \omega^2 w(\omega) = 0 \tag{6.82}$$

### Sec. 6.3 High-Frequency Behavior of Waves from Vibrating Source

leads to the equation of motion

$$G\left[w(\omega)_{,rr} + \frac{1}{r}w(\omega)_{,r}\right] + \rho\omega^2 w(\omega) = 0 \tag{6.83}$$

With $a_o = \omega a/c_s$, the solution of this Bessel equation of order zero with the parameter $a_o/a$ is equal to

$$w(a_o) = c_1 J_o\left(\frac{a_o}{a}r\right) + c_2 Y_o\left(\frac{a_o}{a}r\right) \tag{6.84}$$

Introducing zeroth-order Hankel functions of the first and second kind defined as

$$H_o^{(1)}\left(\frac{a_o}{a}r\right) = J_o\left(\frac{a_o}{a}r\right) + iY_o\left(\frac{a_o}{a}r\right) \tag{6.85a}$$

$$H_o^{(2)}\left(\frac{a_o}{a}r\right) = J_o\left(\frac{a_o}{a}r\right) - iY_o\left(\frac{a_o}{a}r\right) \tag{6.85b}$$

leads to

$$w(a_o) = d_1 H_o^{(1)}\left(\frac{a_o}{a}r\right) + d_2 H_o^{(2)}\left(\frac{a_o}{a}r\right) \tag{6.86}$$

The radiation condition demands that only outgoing waves arise, propagating in the positive $r$-direction. For large arguments

$$H_o^{(1)}\left(\frac{a_o}{a}r\right) \sim \sqrt{\frac{2a}{\pi a_o r}} \exp\left[i\left(\frac{a_o}{a}r - \frac{\pi}{4}\right)\right] \tag{6.87a}$$

$$H_o^{(2)}\left(\frac{a_o}{a}r\right) \sim \sqrt{\frac{2a}{\pi a_o r}} \exp\left[-i\left(\frac{a_o}{a}r - \frac{\pi}{4}\right)\right] \tag{6.87b}$$

As $H_o^{(1)}(a_o r/a)$ corresponds with a time factor $\exp(+i\omega t)$ to an incoming wave, $d_1 = 0$. For a prescribed displacement amplitude $w_o(a_o)$ at $r = a$

$$d_2 = \frac{w_o(a_o)}{H_o^{(2)}(a_o)} \tag{6.88}$$

The solution thus equals

$$w(a_o) = \frac{H_o^{(2)}\left(\frac{a_o}{a}r\right)}{H_o^{(2)}(a_o)} w_o(a_o) \tag{6.89}$$

Substituting into the stress–displacement relationship (Eq. 6.81) yields $\tau_{zr}(r = a, a_o)$

$$\tau_{zr}(r = a, a_o) = -\frac{G}{a} a_o \frac{H_1^{(2)}(a_o)}{H_0^{(2)}(a_o)} w_o(a_o) \quad (6.90)$$

Defining the dynamic-stiffness coefficient of the cylindrical segment of circumference $2\pi a$ and height $d$ as

$$S(a_o) = -\frac{2\pi a d \tau_{zr}(r = a, a_o)}{w_o(a_o)} \quad (6.91)$$

leads to

$$S(a_o) = 2\pi d G a_o \frac{H_1^{(2)}(a_o)}{H_0^{(2)}(a_o)} \quad (6.92)$$

Using the asymptotic expansion for $a_o \to \infty$

$$\frac{H_1^{(2)}(a_o)}{H_0^{(2)}(a_o)} \sim i + \frac{1}{2a_o} \quad (6.93)$$

results in the asymptotic value of $S(a_o)$ for $a_o \to \infty$

$$\lim_{a_o \to \infty} S(a_o) = 2\pi d G \left( a_o i + \frac{1}{2} \right) = 2\pi a d \rho c_s i \omega + \pi d G \quad (6.94)$$

Besides the familiar damper with a coefficient $2\pi a d \rho c_s$, a spring with a constant $\pi d G$ results (Fig. 6.8a).

Summarizing, for the general case of an inclined curved vibrating surface, the damper coefficient per unit area will be $\rho c_p$ in the perpendicular direction and $\rho c_s$ in the two tangential directions. The imaginary linear term in $\omega$ of the dynamic-stiffness coefficient's asymptotic value can then be easily determined. However, to calculate the constant term (and thus the spring coefficient), the one-dimensional wave equation in the perpendicular direction must be solved.

The asymptotic value of the dynamic-stiffness matrix consisting of discrete uncoupled dampers and, if present, of uncoupled springs with frequency-independent coefficients can be used to model the unbounded soil for a high-frequency excitation, such as occurs in shock problems.

**Material damping.** Finally, the asymptotic value of the dynamic-stiffness coefficient for $\omega \to \infty$ with material damping is addressed. The case of hysteretic damping is examined in Section 6.5. For the three-parameter Kelvin model (Eq. 6.57), the asymptotic value consists of a constant and an imaginary linear term in the frequency (Eq. 6.58 and Problem 6.8), which can be interpreted as a spring and a damper. For the Voigt model, this no

Sec. 6.4   Dynamic-Flexibility Coefficient in Time Domain        267

longer applies, as is examined in the following using the spherical cavity for illustration.

Substituting the complex elastic constants, whose definition is analogous to that in Eq. 6.54, into the equation of the limit of the dynamic-stiffness coefficient in the frequency domain (Eq. 2.24) leads to

$$\lim_{a_o \to \infty} S(a_o) = 4\frac{G}{a}(1 + 2\zeta a_o i)\left[1 - \frac{c_p^2}{4c_s^2} + i\frac{a_o}{\sqrt{1 + 2\zeta a_o i}}\frac{c_p^2}{4c_s^2}\right] \quad (6.95)$$

or

$$\lim_{a_o \to \infty} S(a_o) = 4\frac{G}{a}\left[\frac{\sqrt{2}\,c_p^2}{4\,c_s^2}\sqrt{\zeta}\,(a_o i)^{3/2} + 2\zeta\left(1 - \frac{c_p^2}{4c_s^2}\right)a_o i \right. \\ \left. + \frac{\sqrt{2}\,c_p^2}{16c_s^2}\frac{1}{\sqrt{\zeta}}(a_o i)^{1/2} + 1 - \frac{c_p^2}{4c_s^2}\right] \quad (6.96)$$

This singular part consisting of four terms is more complicated. It is discussed in depth in Problem 6.8.

## 6.4 DYNAMIC-FLEXIBILITY COEFFICIENT IN TIME DOMAIN

### 6.4.1 Calculation in Time Domain

The dynamic-stiffness coefficient, which appears in the convolution integral describing the interaction force–displacement relationship, contains significant high-frequency components leading to the singular part. The difficulties in determining these components can be avoided when a flexibility formulation is selected. The dynamic-flexibility coefficient in the frequency domain, being the inverse of the dynamic-stiffness coefficient, converges to zero for the frequency approaching infinity. Such influence functions (Green's functions) in the form of flexibilities are widely used in elasto-dynamics.

Again, it is possible, without introducing a transformation, to derive the dynamic-flexibility coefficient for the spherical cavity with symmetric waves, working in the time domain. The radial displacement $u(r, t)$ specified in Eq. 6.13 is repeated here for easy reference:

$$u(r, t) = -\frac{f\left(t - \dfrac{r-a}{c_p}\right)}{r^2} - \frac{f'\left(t - \dfrac{r-a}{c_p}\right)}{rc_p} \quad (6.97)$$

The initial conditions are specified in Eq. 6.14 and the boundary condition, corresponding to the dynamic-flexibility coefficient's definition (Section 6.1.2),

is specified as
$$p(t) = \delta(t) \tag{6.98}$$
where $p(t)$ denotes the pressure on the wall of the cavity at $r = a$, and $\delta(t)$ is the Dirac delta function.

The pressure is equal to the negative radial stress
$$p(t) = -\sigma_r(r = a, t) \tag{6.99}$$
with the radial stress–displacement relationship specified in Eq. 6.24. Substituting Eq. 6.97 into the latter leads to the following ordinary differential equation of second order for $f(t)$:
$$\ddot{f}(t) + \alpha \dot{f}(t) + \alpha \frac{c_p}{a} f(t) = -\frac{a}{\rho} \delta(t) \tag{6.100}$$
where the prime used to indicate differentiation with respect to the argument is replaced by the symbol of the time derivative and with
$$\alpha = 4 \frac{c_s^2}{a c_p} \tag{6.101}$$

Equation 6.100 corresponds to the equation of motion of a one-degree-of-freedom system with viscous damping. The unit-impulse-response function $h(t)$ is the solution of
$$m\ddot{h}(t) + c\dot{h}(t) + kh(t) = \delta(t) \tag{6.102}$$
and is equal to the damped vibration for an initial velocity $\delta(t)\, dt/m = 1/m$
$$h(t) = \frac{1}{\omega_d m} \exp(-\omega_o \zeta t) \sin\omega_d t \tag{6.103}$$
with the natural frequency $\omega_o$, the damping ratio $\zeta$, and the damped frequency $\omega_d$
$$\omega_o = \sqrt{k/m} \tag{6.104}$$
$$\zeta = \frac{c}{2\sqrt{km}} \tag{6.105}$$
$$\omega_d = \omega_o \sqrt{1 - \zeta^2} \tag{6.106}$$
$f(t)$ follows from Eq. 6.103, comparing Eq. 6.102 to Eq. 6.100,
$$f(t) = -\frac{2a}{\rho \beta} \exp\left(-\frac{\alpha}{2} t\right) \sin\left(\frac{\beta}{2} t\right) \tag{6.107}$$
where
$$\beta = \alpha \sqrt{\frac{c_p^2}{c_s^2} - 1} \tag{6.108}$$

## Sec. 6.4  Dynamic-Flexibility Coefficient in Time Domain

Substituting Eq. 6.107 into Eq. 6.97 leads to

$$u(r, t) = \frac{a}{\rho}\left[\frac{1}{c_p r}\cos\left[\frac{\beta}{2}\left(t - \frac{r-a}{c_p}\right)\right]\right.$$

$$\left. + \frac{2}{\beta}\left(\frac{1}{r^2} - \frac{\alpha}{2c_p r}\right)\sin\left[\frac{\beta}{2}\left(t - \frac{r-a}{c_p}\right)\right]\right]\exp\left[-\frac{\alpha}{2}\left(t - \frac{r-a}{c_p}\right)\right],$$

$$t > \frac{r-a}{c_p} \quad (6.109)$$

The dynamic-flexibility coefficient $F(t)$ is equal to $u(r = a, t)$.

$$F(t) = \left[\frac{1}{\rho c_p}\cos\left(\frac{\beta}{2}t\right) + \left(\frac{2}{a\rho\beta} - \frac{\alpha}{\rho\beta c_p}\right)\sin\left(\frac{\beta}{2}t\right)\right]\exp\left(-\frac{\alpha}{2}t\right) \quad (6.110)$$

or

$$F(t) = \frac{1}{4}\frac{a}{G}\left[\alpha\cos\left(\frac{\beta}{2}t\right) - \frac{1}{2}\frac{\alpha^2}{\beta}\left(2 - \frac{c_p^2}{c_s^2}\right)\sin\left(\frac{\beta}{2}t\right)\right]\exp\left(-\frac{\alpha}{2}t\right),$$

$$t \geq 0 \quad (6.111)$$

For $t < 0$, $F(t) = 0$.

Introducing the dimensionless time $\bar{t}$ defined in Eq. 6.26 and the static-flexibility coefficient $1/K = a/(4G)$ leads to

$$F(\bar{t}) = \frac{1}{K}\frac{c_p}{a}\left[4\frac{c_s^2}{c_p^2}\cos\left(2\frac{c_s^2}{c_p^2}\sqrt{\frac{c_p^2}{c_s^2} - 1}\,\bar{t}\right)\right.$$

$$\left. + 2\left(1 - 2\frac{c_s^2}{c_p^2}\right)\sqrt{\frac{1}{c_p^2/c_s^2 - 1}}\sin\left(2\frac{c_s^2}{c_p^2}\sqrt{\frac{c_p^2}{c_s^2} - 1}\,\bar{t}\right)\right]\exp\left(-2\frac{c_s^2}{c_p^2}\bar{t}\right) \quad (6.112)$$

The value $a/c_p\, F(\bar{t})\, K$ is a dimensionless variable which is a function of $\bar{t}$. It is plotted for $\nu = 1/3$ ($c_p/c_s = 2$) in Fig. 6-9.

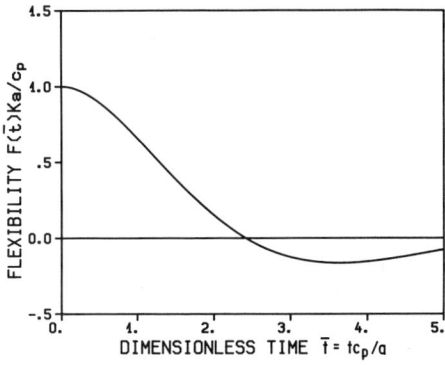

**Figure 6-9** Dynamic-flexibility coefficient in time domain, spherical cavity.

### 6.4.2 Transformation from Frequency to Time Domain

**Fourier-transform pair.** As in the case of the stiffness coefficient, the dynamic-flexibility coefficient in the time domain is generally determined directly through a Fourier transformation from the corresponding value in the frequency domain. It follows, analogous to the derivation in Section 6.2.2, that $F(t)$ and $F(\omega)$ form a Fourier-transform pair.

$$F(t) = \frac{1}{2\pi} \int_{-\infty}^{+\infty} F(\omega) \exp(i\omega t) \, d\omega \tag{6.113}$$

Using dimensionless quantities as defined in Eqs. 6.35 and 6.36,

$$F(\bar{t}) = \frac{c_s}{a} \frac{1}{2\pi} \int_{-\infty}^{+\infty} F(a_o) \exp(ia_o\bar{t}) \, da_o \tag{6.114}$$

applies.

The inversion of the dynamic-stiffness matrix in the frequency domain $[S(\omega)]$ leads to the dynamic-flexibility matrix $[F(\omega)]$

$$[F(\omega)] = [S(\omega)]^{-1} \tag{6.115}$$

The individual elements of $[F(\omega)]$ are absolutely integrable over the $\omega$-axis (see Eq. 6.38). They converge for $\omega \to \infty$ to zero. No singular part exists for a dynamic-flexibility coefficient.

As an example, the spherical cavity is examined using dimensionless quantities. Taking the inverse of $S(a_o)$, specified in Eqs. 6.41 and 6.42, results with $a_o = \omega a/c_p$ in

$$F(a_o) = \frac{1}{K} \frac{a_o - i}{a_o + i\left(\frac{1}{4}\frac{c_p^2}{c_s^2} a_o^2 - 1\right)} \tag{6.116}$$

The real and imaginary parts of $F(a_o)$, which both decay strongly for large $\omega$'s, are plotted in Fig. 6-10 for $\nu = 1/3$.

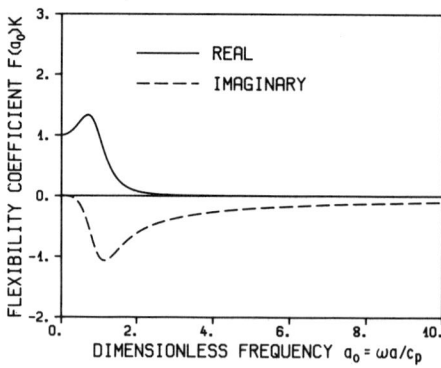

**Figure 6-10** Dynamic-flexibility coefficient in frequency domain, spherical cavity.

Sec. 6.4    Dynamic-Flexibility Coefficient in Time Domain    271

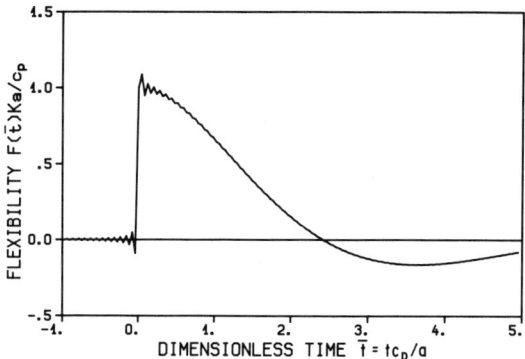

**Figure 6-11** Dynamic-flexibility coefficient in time domain, by numerical transform.

Substituting Eq. 6.116 into Eq. 6.114 (with $c_p$ replacing $c_s$) and performing the transformation (Ref. [C1, Nos. 448.8 and 449]) leads to Eq. 6.112 which is derived working only in the time domain.

In more complicated cases, the transformation from the frequency to the time domain must be performed numerically. The standard Fast Fourier Transform can be applied for the flexibility coefficient. This is shown in Fig. 6-11 for the spherical cavity, selecting a frequency span extending to $a_o = 80$ and divided into 256 increments $\Delta a_o$. An improvement results if the easily derivable asymptotic expression

$$\lim_{a_o \to \infty} F(a_o) = \frac{1}{4} \frac{a}{G} \left[ -4i \frac{c_s^2}{c_p^2} \frac{1}{a_o} - 4 \left( 1 - 4 \frac{c_s^2}{c_p^2} \right) \frac{c_s^2}{c_p^2} \left( \frac{1}{a_o} \right)^2 \right] \quad (6.117)$$

is transformed analytically above the frequency $a_o = 80$. Adding this analytical part to the numerical one leads to a result which coincides with the exact curve of Fig. 6-9 (comparison not shown).

**Initial characteristics and corresponding high-frequency behavior.** As for the dynamic-stiffness coefficient's regular part (Eq. 6.46), the high-frequency asymptotic behavior of the dynamic-flexibility coefficient in the frequency domain $F(a_o \to \infty)$ determines the initial characteristics of the corresponding quantity in the time domain—that is, the value $F(\bar{t} = 0^+)$ and the slope $dF(\bar{t} = 0^+)/d\bar{t}$.

$$\lim_{\bar{t} \to 0^+} \left[ \frac{a}{c_s} F(\bar{t}) \right] = \lim_{ia_o \to \infty} [ia_o F(a_o)] \quad (6.118a)$$

$$\lim_{\bar{t} \to 0^+} \left[ \frac{a}{c_s} \frac{dF(\bar{t})}{d\bar{t}} \right] = \lim_{ia_o \to \infty} \left[ -a_o^2 F(a_o) - ia_o \frac{a}{c_s} F(\bar{t} = 0^+) \right] \quad (6.118b)$$

Applying these relationships to the spherical cavity, with $F(a_o)$ specified in

**272**  Substructure Method Using Dynamic Stiffness of Soil  Chap. 6

Equation 6.116 and replacing $c_s$ by $c_p$, leads to

$$\frac{a}{c_p} F(\bar{t} = 0^+) = \frac{1}{K} 4 \frac{c_s^2}{c_p^2} \tag{6.119a}$$

$$\frac{a}{c_p} \frac{dF(\bar{t} = 0^+)}{d\bar{t}} = \frac{1}{K} 4 \frac{c_s^2}{c_p^2} \left(1 - 4 \frac{c_s^2}{c_p^2}\right) \tag{6.119b}$$

which agrees with the values determined from Eq. 6.112.

It follows from Eq. 6.119a that $F(\bar{t} = 0^+)$ equals $1/(\rho c_p)$, which is equal to the inverse of the asymptotic damping coefficient for $a_o \to \infty$ (Eq. 6.43b).

**Initial value.** This initial value $F(t = 0^+)$ can be easily verified by applying the law of conservation of momentum. Denoting the applied pressure as $p(t)$,

$$p(t) = \rho \Delta r \ddot{u}(t = 0) \tag{6.120}$$

holds, with $\Delta r = c_p \Delta t$. Substituting the Dirac delta function for $p(t)$ in Eq. 6.120 and integrating with respect to time leads to

$$1 = \rho c_p \Delta t \dot{u}(t = 0^+) = \rho c_p u(t = 0^+) \tag{6.121}$$

or

$$F(t = 0^+) = u(t = 0^+) = \frac{1}{\rho c_p} \tag{6.122}$$

The increase of the area in the radial direction has no effect for $\bar{t} = 0^+$.

This result can be generalized for a curved surface. The initial dynamic-flexibility coefficient for a unit-impulse force per unit area is equal to $1/(\rho c_p)$ in the direction perpendicular to the surface and equal to $1/(\rho c_s)$ in the two tangential directions. As for $t = 0^+$, the wave has not yet propagated to neighboring nodes, and no coupling of the initial values exists—that is the initial dynamic-flexibility matrix is diagonal or has a small band width. These properties correspond, of course, to those of the asymptotic imaginary part of the dynamic-stiffness coefficient for $\omega \to \infty$, represented by the damper (see Section 6.3.2, Curved Vibrating Surface).

**Layered half-space.** The rigid circular disk of radius $a$ resting on the surface of a half-space is addressed. As is well established, the coupling term of the horizontal and rocking degrees of freedom can be disregarded, thus allowing each of the dynamic-stiffness matrix's terms to be analyzed independently from the others. Applying the boundary-element method to calculate the dynamic-stiffness coefficient $S(a_o)$, the disk is discretized using 25 annular rings of equal width. Over each annular ring, the (unknown) soil reaction is assumed to be constant for the vertical degree of freedom and to

## Sec. 6.4 Dynamic-Flexibility Coefficient in Time Domain

vary as the first symmetric Fourier term circumferentially for the rigid disk's other two degrees of freedom in the horizontal direction.

It is appropriate to use nondimensional dynamic-stiffness coefficients $\bar{S}(a_o)$ that result from $S(a_o)$ through division by the corresponding static value (horizontal: $8Ga/(2 - v)$; vertical: $4Ga/(1 - v)$; rocking: $8Ga^3/(3(1 - v))$; $G$ = shear modulus; $v$ = Poisson's ratio). The Fourier transform of $\bar{F}(a_o)$ = $1/\bar{S}(a_o)$ leads to the dynamic-flexibility coefficient in the time domain $\bar{F}(\bar{t})$ (Eq. 6.114).

Material damping is introduced using the Voigt and the three-parameter Kelvin models described in Section 6.2.2.

The integrations are performed numerically up to $a_o = 16$. Above this frequency, the expansions valid for large $a_o$ are integrated analytically [B2]. For the undamped case, these values, which are based on an approximate theory, are as follows for $v = 1/3$:

Horizontal: $\quad \bar{F}(a_o) \approx 1.528 \, (a_o i)^{-1} + 1.119 \, a_o^{-2}$ \quad (6.123a)

Vertical: $\quad \bar{F}(a_o) \approx 0.955 \, (a_o i)^{-1} + 0.358 \, a_o^{-2}$ \quad (6.123b)

Rocking: $\quad \bar{F}(a_o) \approx 2.546 \, (a_o i)^{-1} + 1.904 \, a_o^{-2}$ \quad (6.123c)

The first terms on the right-hand sides of Eqs. 6.123a and 6.123b, 6.123c correspond to vertically propagating S- and P-waves, respectively (Eq. 6.76).

The dynamic-flexibility coefficients in the time domain are plotted versus $\bar{t} = tc_s/a$ in Fig. 6-12 [W12]. In addition to the elastic case, the results are shown for the three-parameter model (Eq. 6.57) and Voigt model (Eq. 6.54) with a damping factor $\zeta = 0.10$. From a practical point of view, the numerical and analytical integrations lead to $\bar{F}(\bar{t}) = 0$ for $\bar{t} < 0$. The expansion of the dynamic-flexibility coefficient in the frequency domain for large $a_o$ (Equation 6.123) determines the corresponding value in the time domain for $\bar{t} = 0^+$ (Eq. 6.118a). This leads to $\bar{F}(\bar{t} = 0^+)a/c_s$ equal to 1.528, 0.955, and 2.546 for the horizontal, vertical, and rocking degrees of freedom, respectively. It also follows from Eq. 6.123 that the values for $[d\bar{F}(\bar{t} = 0^+)/d\bar{t}] \, a/c_s$ are equal to $-1.119$, $-0.358$, and $-1.904$ for the three degrees of freedom (Eq. 6.118b).

The initial values can also be verified as follows: For instance, for the rocking degree of freedom, the applied moment $M(t)$ can be formulated, analogously to Eq. 6.120, as

$$M(t) = I_o \rho c_p \Delta t \ddot{\varphi} \, (t = 0) = I_o \rho c_p \dot{\varphi} \, (t = 0) \quad (6.124)$$

with the moment of inertia $I_o \, (= \pi a^4/4)$. The time integral of Eq. 6.124 results in

$$F(t = 0^+) = \varphi(t = 0^+) = \frac{1}{I_o \rho c_p} \quad (6.125)$$

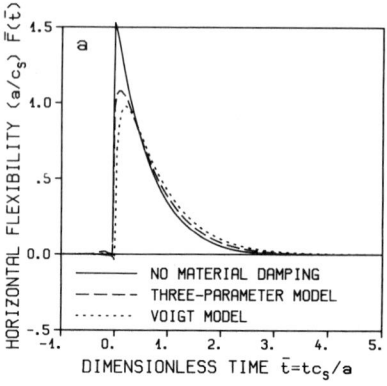
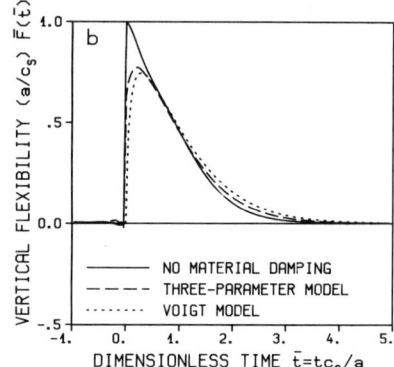
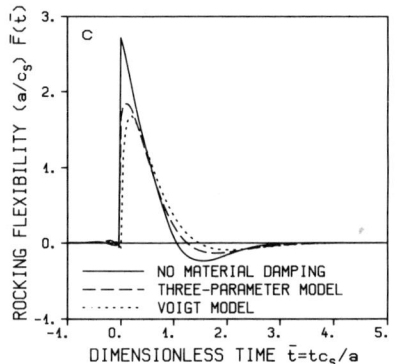

**Figure 6-12** Dynamic-flexibility coefficient in time domain, half-space.
a. Horizontal. b. Vertical.
c. Rocking

Normalizing with the static value $3(1 - \nu)/(8Ga^3)$, Eq. 6.125 is transformed to

$$\overline{F}(\bar{t} = 0^+)\frac{a}{c_s} = \frac{32}{3\pi(1 - \nu)}\frac{c_s}{c_p} \quad (6.126)$$

For $\nu = 1/3$, the right-hand side of Eq. 6.126 equals 2.546. For the horizontal and vertical degrees of freedom, $\overline{F}(\bar{t} = 0^+)a/c_s$ equals $8/[\pi(2 - \nu)]$ and $4c_s/[\pi(1 - \nu)c_p]$, respectively.

The dynamic-flexibility coefficients in the time domain for the rigid circular disk of radius $a$ resting on the surface of a layer built-in at its base are discussed next. This system and the half-space examined before represent the two limiting cases of a site. The layer's depth is selected equal to the radius. Poisson's ratio = 1/3. The static-flexibility coefficients of the half-space (and not of the layer) are used to "non-dimensionalize" the built-in layer's corresponding dynamic-flexibility coefficients, which are shown in Fig. 6-13 [W12]. The models used for material damping and the selected damping

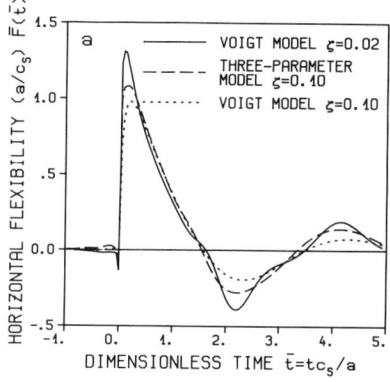
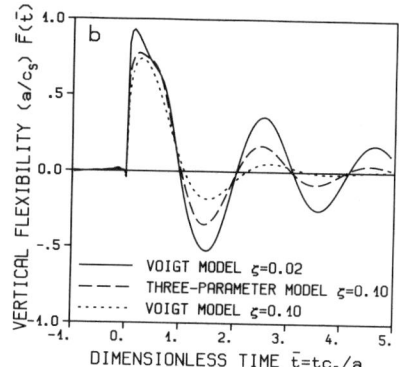
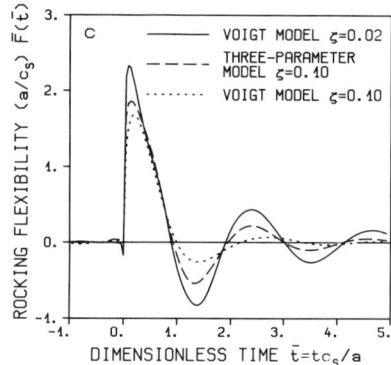

**Figure 6-13** Dynamic-flexibility coefficient in time domain, layer built-in.
a. Horizontal. b. Vertical.
c. Rocking.

factors are specified in the figure. The reflections of the waves at the base of the layer lead to strong oscillations in $\bar{F}(\bar{t})$. The early-time behavior of the half-space's and of the layer's flexibility are identical for the same material model and damping factor; in particular the maxima are the same.

In an actual nonlinear analysis, the dynamic-stiffness coefficient is often approximated with a spring and a viscous damper whose coefficients are frequency-independent (see Sections 3.6 and 3.11). For instance, for the rocking degree of freedom of a disk on an undamped half-space, the following equation applies:

$$\bar{S}(a_o) = k + ia_o c \qquad (6.127)$$

with $k = 1$ and $c = 0.15$. The corresponding dynamic-flexibility coefficient $\bar{F}(\bar{t})$, determined from Eq. 6.114 with $\bar{F}(a_o) = 1/\bar{S}(a_o)$, is compared to the rigorous value of Fig. 6-12c in Fig. 6-14. As $c = 0.15$ does not correspond to the asymptotic value of $\bar{S}(a_o)$ (inverse of Eq. 6.123c, $a_o i/2.546$), the value differs significantly for $\bar{t} = 0^+$.

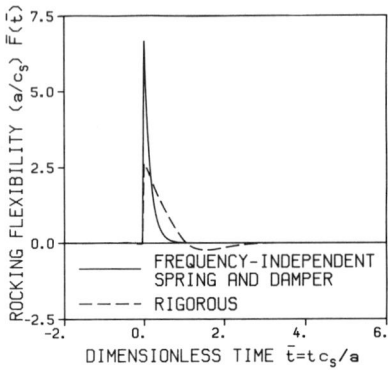

**Figure 6-14** Dynamic-flexibility coefficient in time domain, half-space, no material damping, rocking.

## 6.5 HYSTERETIC DAMPING

### 6.5.1 Non-Causal Behavior

The material damping occurring in the soil and the structure mainly involves a frictional loss of energy. The damping forces, which are independent of frequency, are proportional to the displacements, but are in phase with the velocities ($\sim K|u|\dot{u}/|\dot{u}|$). This represents (linear) hysteretic damping, also called linear structural damping. The constitutive law can be directly incorporated into a nonlinear step-by-step time-domain analysis.

In a linear analysis in the frequency domain, in which the energy loss per cycle is addressed, the two Lamé constants are multiplied by $[1 + 2\zeta i \, \text{sign}(\omega)]$, where, as $\omega$ can also become negative, the sign of $\omega$ [denoted as $\text{sign}(\omega)$] is taken into account. $\zeta$ is the damping ratio. Difficulties arise in transforming this relation back to the time domain.

This is easily verified for the single spring with stiffness $K$ whose corresponding stiffness coefficient for hysteretic damping is formulated as

$$S(\omega) = K\left[1 + 2\zeta i \, \text{sign}(\omega)\right] \tag{6.128}$$

Its inverse Fourier transform equals [B3]

$$S(t) = K\left[\delta(t) + \frac{2\zeta}{\pi t}\right] \tag{6.129}$$

As Eq. 6.129 also applies for $t < 0$, a response prior to the application of the excitation ($t = 0$) arises, which is called *non-causal behavior*. However, this precursor response [C5], which is physically unrealizable, is small, as is shown in the next section.

The corresponding expression for the flexibility coefficient

$$F(\omega) = \frac{1}{K}\frac{1}{1 + 4\zeta^2}\left[1 - 2\zeta i \, \text{sign}(\omega)\right] \tag{6.130}$$

Sec. 6.5  Hysteretic Damping

is transformed to

$$F(t) = \frac{1}{K} \frac{1}{1 + 4\zeta^2} \left[ \delta(t) - \frac{2\zeta}{\pi t} \right] \qquad (6.131)$$

which again also applies for $t < 0$.

### 6.5.2 Spherical Cavity with Symmetric Waves

Returning to the cavity embedded in the infinite space, the dynamic-stiffness coefficient's behavior is addressed first. Replacing the two Lamé constants by their complex counterparts, which is equivalent to replacing $G$, $c_p$, and $c_s$ by $G[1 + 2\zeta i \operatorname{sign}(a_o)]$, $c_p[1 + 2\zeta i \operatorname{sign}(a_o)]^{1/2}$ and $c_s[1 + 2\zeta i \operatorname{sign}(a_o)]^{1/2}$ in Eqs. 6.41 and 6.42, leads to the damped system's dynamic-stiffness coefficient.

$$S(a_o) = K[1 + 2\zeta i \operatorname{sign}(a_o)] \left\{ 1 - \frac{1}{4}\frac{c_p^2}{c_s^2} + i a_o [1 + 2\zeta i \operatorname{sign}(a_o)]^{-1/2} \frac{1}{4}\frac{c_p^2}{c_s^2} \right\}$$

$$+ K[1 + 2\zeta i \operatorname{sign}(a_o)] \left\{ \frac{\frac{1}{4}\frac{c_p^2}{c_s^2} - i a_o [1 + 2\zeta i \operatorname{sign}(a_o)]^{-1/2} \frac{1}{4}\frac{c_p^2}{c_s^2}}{1 + a_o^2 [1 + 2\zeta i \operatorname{sign}(a_o)]^{-1}} \right\} \qquad (6.132)$$

The first term represents the singular part, which must be transformed analytically. The second term is regular and is transformed numerically.

The transform of the singular part is [W17]

$$K \frac{c_p}{a} \left\{ \left(1 - \frac{c_p^2}{4c_s^2}\right) \left[ \delta(\bar{t}) - \frac{2\zeta}{\pi \bar{t}} \right] + \frac{c_p^2}{4c_s^2} \left[ \frac{Re + 2\zeta \operatorname{Im}}{Re^2 + \operatorname{Im}^2} \frac{d\delta(\bar{t})}{d\bar{t}} \right. \right.$$

$$\left. \left. + \frac{2\zeta Re - \operatorname{Im}}{Re^2 + \operatorname{Im}^2} \frac{1}{\pi \bar{t}^2} \right] \right\} \qquad (6.133)$$

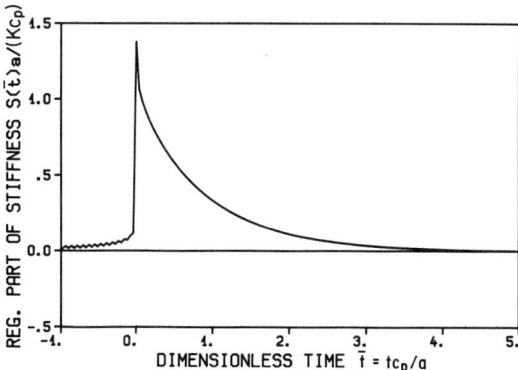

**Figure 6-15** Dynamic-stiffness coefficient in time domain with hysteretic damping, spherical cavity.

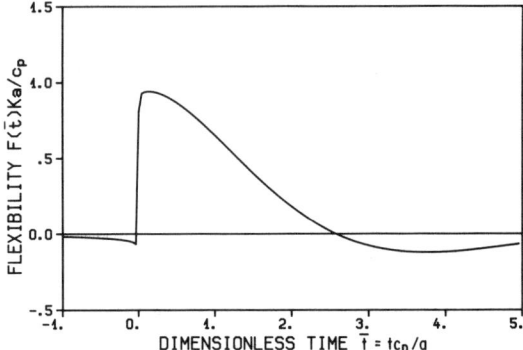

**Figure 6-16** Dynamic-flexibility coefficient in time domain with hysteretic damping, spherical cavity.

with $Re$ and $Im$ denoting the real and imaginary parts of $\sqrt{1 + 2\zeta\, i\, \text{sign}(a_o)}$, respectively.

The regular part is plotted for $\zeta = 0.05$ in Fig. 6-15, whereby the asymptotic expansion for large frequencies is transformed partially analytically, as described next for the flexibility coefficient. A small non-causal behavior is observed.

Proceeding analogously, the dynamic-flexibility coefficient with hysteretic damping can be formulated. The asymptotic expansion $\lim_{a_o \to \infty} F(a_o)$ follows from Eq. 6.117. Performing the integration numerically up to $a_o = 80$, using the same number of increments $\Delta a_o$ as in the undamped case and performing an analytical transform above $a_o = 80$, leads to the dynamic-flexibility coefficient, which is plotted in Fig. 6-16 for $\zeta = 0.05$. Although the non-causal behavior is clearly visible for $\bar{t} < 0$, its influence on any result should in general be small.

The non-causal behavior can also be verified directly in the frequency domain. The Hilbert transform relation between the real and imaginary parts is not satisfied in this case [B3].

## 6.6 COMPUTATIONAL PROCEDURE

### 6.6.1 Newmark Time-Integration Scheme

As one of the many possible ways to perform the time integration, the step-by-step procedure based on the Newmark method is selected to discuss the computational procedure.

The formulation expressed in the total motion (superscript $t$) for the $n$th

### Sec. 6.6 Computational Procedure

step leading from time $(n-1)\Delta t$, where all variables are known (subscript $n-1$) to time $n\Delta t$ (subscript $n$), is based on the following expressions:

$$\{u^t\}_n = \{u^t\}_{n-1} + \Delta t \{\dot{u}^t\}_{n-1} + \left(\frac{1}{2} - \beta\right) \Delta t^2 \{\ddot{u}^t\}_{n-1} + \beta \Delta t^2 \{\ddot{u}^t\}_n \quad (6.134\text{a})$$

$$\{\dot{u}^t\}_n = \{\dot{u}^t\}_{n-1} + (1 - \gamma)\Delta t\{\ddot{u}^t\}_{n-1} + \gamma \Delta t \{\ddot{u}^t\}_n \quad (6.134\text{b})$$

The parameters $\beta$ and $\gamma$ determine the variation of the displacements, velocities, and accelerations within each time interval and thus the accuracy, the stability, and the cost of the calculation. For instance, as an example of an implicit algorithm, $\beta = 0.25$ and $\gamma = 0.5$ lead to the constant-average acceleration method, which is unconditionally stable. Together with the equations of motion satisfied at the discrete time station $n\Delta t$, Eq. 6.134 forms a system of equations with unknown displacements $\{u^t\}_n$, velocities $\{\dot{u}^t\}_n$, and accelerations $\{\ddot{u}^t\}_n$ at time $n\Delta t$.

Both the explicit and implicit integration schemes can be based on Eq. 6.134, whereby, besides the choice of the values $\beta$ and $\gamma$, the implementation differs. The methods as applied to a nonlinear structure are well known and will not be repeated here. In the following, the discussion of the explicit and implicit methods will thus be limited to the contribution of the unbounded soil's interaction forces, which leads to convolution integrals. The stiffness and the flexibility formulations are addressed.

#### 6.6.2 Explicit Algorithm

**Predictor-corrector scheme.** As already described in Eqs. 2.60 to 2.62, predictor values for the displacements $\{\tilde{u}^t\}_n$ and the velocities $\{\tilde{\dot{u}}^t\}_n$ (a tilde denotes these quantities)

$$\{\tilde{u}^t\}_n = \{u^t\}_{n-1} + \Delta t\{\dot{u}^t\}_{n-1} + \left(\frac{1}{2} - \beta\right) \Delta t^2 \{\ddot{u}^t\}_{n-1} \quad (6.135\text{a})$$

$$\{\tilde{\dot{u}}^t\}_n = \{\dot{u}^t\}_{n-1} + (1 - \gamma) \Delta t \{\ddot{u}^t\}_{n-1} \quad (6.135\text{b})$$

are determined first. All internal forces of the structure–soil system—including the interaction forces—are calculated with these predicted values. Formulating equilibrium then leads to the accelerations $\{\ddot{u}^t\}_n$, which finally allows the predicted values to be corrected as follows:

$$\{u^t\}_n = \{\tilde{u}^t\}_n + \beta \Delta t^2 \{\ddot{u}^t\}_n \quad (6.136\text{a})$$

$$\{\dot{u}^t\}_n = \{\tilde{\dot{u}}^t\}_n + \gamma \Delta t \{\ddot{u}^t\}_n \quad (6.136\text{b})$$

Substituting Eq. 6.135 into Eq. 6.136 leads to the original Newmark expressions (Eq. 6.134).

For $\beta = 0$ (and $\gamma = 0.5$), the predicted displacements $\{\tilde{u}^t\}_n$ are the final ones $\{u^t\}_n$.

**Stiffness formulation.** The procedure to calculate the interaction forces of the unbound soil $\{R_b\}_n$ is explained based on the equation with the dynamic-stiffness matrix of the system ground $[S_{bb}^g(t)]$ (Eq. 6.2, leading to the basic equation of motion, Eq. 6.4). Decomposing into the singular and regular parts results in Eq. 6.50, which for the time station $n\Delta t$, is formulated as

$$\{R_b\}_n = [k_{bb}^g](\{\bar{u}_b^t\}_n - \{u_b^g\}_n) + [c_{bb}^g](\{\dot{\bar{u}}_b^t\}_n - \{\dot{u}_b^g\}_n)$$

$$+ \int_o^{(n-1)\Delta t} [S_{r,bb}^g(t - \tau)] (\{u_b^t(\tau)\} - \{u_b^g(\tau)\}) \, d\tau \quad (6.137)$$

$$+ \int_{(n-1)\Delta t}^{n\Delta t} [S_{r,bb}^g(t - \tau)] (\{\bar{u}_b^t(\tau)\} - \{u_b^g(\tau)\}) \, d\tau$$

Because for the $n$th time step the (known) predicted values appear on the right-hand side, the latter can be calculated. It can be appropriate when evaluating the convolution integral to assume a linear variation of the displacements over the $n$th time step ($0 < \tau' < \Delta t$) (and, analogously, over the previous time steps)

$$\{\bar{u}_b^t(\tau')\}_n - \{u_b^g(\tau')\}_n$$

$$= \left(1 - \frac{\tau'}{\Delta t}\right)(\{u_b^t\}_{n-1} - \{u_b^g\}_{n-1}) + \frac{\tau'}{\Delta t}(\{\bar{u}_b^t\}_n - \{u_b^g\}_n) \quad (6.138)$$

The convolution integral is then formulated as

$$\int_o^{(n-1)\Delta t} [S_{r,bb}^g(t - \tau)] (\{u_b^t(\tau)\} - \{u_b^g(\tau)\}) \, d\tau$$

$$+ \int_{(n-1)\Delta t}^{n\Delta t} [S_{r,bb}^g(t - \tau)] (\{\bar{u}_b^t(\tau)\} - \{u_b^g(\tau)\}) \, d\tau$$

$$= \sum_{i=1}^{n-1} [S_{r,bb}^g]_{n-i} (\{u_b^t\}_i - \{u_b^g\}_i) \quad (6.139)$$

$$+ [S_{r,bb}^g]_o (\{\bar{u}_b^t\}_n - \{u_b^g\}_n)$$

with

$$[S_{r,bb}^g]_{n-i} = \int_o^{\Delta t} \frac{\tau'}{\Delta t} [S_{r,bb}^g((n + 1 - i)\Delta t - \tau')] \, d\tau'$$

$$+ \int_o^{\Delta t} \left(1 - \frac{\tau'}{\Delta t}\right) [S_{r,bb}^g((n - i)\Delta t - \tau')] \, d\tau' \quad (6.140a)$$

$$[S_{r,bb}^g]_o = \int_o^{\Delta t} \frac{\tau'}{\Delta t} [S_{r,bb}^g(\Delta t - \tau')] \, d\tau' \quad (6.140b)$$

## Sec. 6.6  Computational Procedure

For a lumped-mass matrix, the accelerations follow from the basic equation of motion (Eq. 6.4) as

$$\{\ddot{u}_b^t\}_n = -[M_{bb}]^{-1}(\{P_b\}_n + \{R_b\}_n) \tag{6.141}$$

where $\{P_b\}_n$ are the internal forces of the structure acting at the base. In the $n$th time step, all previous displacements $\{u_b^t\}_i$ ($i = 1, \ldots, n-1$), which have to be stored during the analysis, are processed.

**Flexibility formulation.** The displacement–interaction force relationship is specified in Eq. 6.10, which is formulated for the station $n\Delta t$ as

$$\{u_b^t\}_n - \{u_b^g\}_n = \int_0^{n\Delta t} [F_{bb}^g(t-\tau)]\{R_b(\tau)\}\, d\tau \tag{6.142}$$

Again, the predicted displacements $\{\tilde{u}_b^t\}_n$ appear, which are known (Eq. 6.135).

Assuming a linear temporal variation of the concentrated interaction forces over the $i$th time step ($0 < \tau' < \Delta t$)

$$\{R_b(\tau')\}_i = \left(1 - \frac{\tau'}{\Delta t}\right)\{R_b\}_{i-1} + \frac{\tau'}{\Delta t}\{R_b\}_i \tag{6.143}$$

Eq. 6.142 leads to

$$\{\tilde{u}_b^t\}_n - \{u_b^g\}_n = \sum_{i=1}^{n-1}[F_{bb}^g]_{n-i}\{R_b\}_i + [F_{bb}^g]_o\{R_b\}_n \tag{6.144}$$

with

$$[F_{bb}^g]_{n-i} = \int_o^{\Delta t}\frac{\tau'}{\Delta t}[F_{bb}^g((n+1-i)\Delta t - \tau')]d\tau' \tag{6.145a}$$

$$+ \int_o^{\Delta t}\left(1 - \frac{\tau'}{\Delta t}\right)[F_{bb}^g((n-i)\Delta t - \tau')]d\tau'$$

$$[F_{bb}^g]_o = \int_o^{\Delta t}\frac{\tau'}{\Delta t}[F_{bb}^g(\Delta t - \tau')]d\tau' \tag{6.145b}$$

In the algorithm $\{R_b\}_i$ for $i = 1, 2, \ldots, n-1$ are known ($\{R_b\}_o$ vanishes). Equation 6.144 can then be solved for the unknown interaction forces $\{R_b\}_n$ [W 12].

$$\{R_b\}_n = [F_{bb}^g]_o^{-1}\left(\{\tilde{u}_b^t\}_n - \{u_b^g\}_n - \sum_{i=1}^{n-1}[F_{bb}^g]_{n-i}\{R_b\}_i\right) \tag{6.146}$$

No inversion must be performed, as $[F_{bb}^g]_o$ is a diagonal matrix (see Section 6.4.2, Initial Value). $[F_{bb}^g]_o^{-1}$ is identified as the instantaneous dynamic-stiffness matrix ($= [S_{bb}^g]_o$).

The accelerations $\{\ddot{u}_b^t\}_n$ again follow from Eq. 6.141.

In the flexibility formulation, all interaction forces $\{R_b\}_i$ ($i = 1, \ldots, n - 1$) must be stored during the analysis.

### 6.6.3 Implicit Algorithm

**Stiffness formulation.** The interaction forces as specified in Eq. 6.50 are formulated for the time station $n\Delta t$ as

$$\{R_b\}_n = [k_{bb}^g]\,(\{u_b^t\}_n - \{u_b^g\}_n) + [c_{bb}^g]\,(\{\dot{u}_b^t\}_n - \{\dot{u}_b^g\}_n) \tag{6.147}$$
$$+ \int_o^{n\Delta t} [S_{r,bb}^g(t - \tau)]\,(\{u_b^t(\tau)\} - \{u_b^g(\tau)\})\,d\tau$$

To be able to express $\{R_b\}_n$ as a function of only one unknown at time station $n\Delta t$—for example, $\{u_b^t\}_n$—Eq. 6.134a is solved for $\{\ddot{u}_b^t\}_n$ in terms of $\{u_b^t\}_n$. Substituted in Eq. 6.134b leads to $\{\dot{u}_b^t\}_n$ as a function of $\{u_b^t\}_n$. To be able to evaluate the convolution integral in Eq. 6.147, the variation of $\{u_b^t(\tau')\}$ over each time interval must be introduced, which should be consistent with the choice of $\beta$ and $\gamma$. This would involve, besides the displacements $\{u_b^t\}_i$, also the velocities $\{\dot{u}_b^t\}_i$ and the accelerations $\{\ddot{u}_b^t\}_i$ ($i = 1, \ldots, n - 1$). To avoid the corresponding storage and calculational efforts, a linear variation over the $i$th time step is assumed ($0 < \tau' < \Delta t$), as in Eq. 6.138:

$$\{u_b^t(\tau')\}_i - \{u_b^g(\tau')\}_i$$
$$= \left(1 - \frac{\tau'}{\Delta t}\right)(\{u_b^t\}_{i-1} - \{u_b^g\}_{i-1}) + \frac{\tau'}{\Delta t}(\{u_b^t\}_i - \{u_b^g\}_i) \tag{6.148}$$

The convolution integral is written as

$$\int_o^{n\Delta t} [S_{r,bb}^g(t - \tau)]\,(\{u_b^t(\tau)\} - \{u_b^g(\tau)\})\,d\tau$$
$$= \sum_{i=1}^n [S_{r,bb}^g]_{n-i}(\{u_b^t\}_i - \{u_b^g\}_i) \tag{6.149}$$

with $[S_{r,bb}^g]_{n-i}$ defined for $i = 1, \ldots, n - 1$ in Eq. 6.140a and for $i = n$ in Eq. 6.140b.

This allows $\{R_b\}_n$ (Eq. 6.147) to be rewritten as

$$\{R_b\}_n = [k_{bb}^g]\,(\{u_b^t\}_n - \{u_b^g\}_n) + [c_{bb}^g]\left(\frac{\gamma}{\beta\Delta t}\{u_b^t\}_n\right.$$
$$\left. - \frac{\gamma}{\beta\Delta t}\{u_b^t\}_{n-1} + \left(1 - \frac{\gamma}{\beta}\right)\{\dot{u}_b^t\}_{n-1} + \left(1 - \frac{\gamma}{2\beta}\right)\Delta t\{\ddot{u}_b^t\}_{n-1} - \{\dot{u}_b^g\}_n\right)$$
$$+ [S_{r,bb}^g]_o(\{u_b^t\}_n - \{u_b^g\}_n) + \sum_{i=1}^{n-1}[S_{r,bb}^g]_{n-i}(\{u_b^t\}_i - \{u_b^g\}_i) \tag{6.150}$$

## Sec. 6.6  Computational Procedure

These interaction forces of the soil will affect the total structure–soil system's basic equation of motion (Eq. 6.4) as follows: The contribution to the "effective" stiffness matrix (appearing as the coefficient matrix on the left-hand side of the equations with the unknown $\{u_b^t\}_n$) equals

$$[k_{bb}^g] + \frac{\gamma}{\beta \Delta t} [c_{bb}^g] + [S_{r,bb}^g]_o \qquad (6.151a)$$

while the contribution to the "effective" load vector on the right-hand side is written as

$$\frac{\gamma}{\beta \Delta t} [c_{bb}^g] \{u_b^t\}_{n-1} + \left(\frac{\gamma}{\beta} - 1\right) [c_{bb}^g] \{\dot{u}_b^t\}_{n-1} + \left(\frac{\gamma}{2\beta} - 1\right) \Delta t \, [c_{bb}^g] \{\ddot{u}_b^t\}_{n-1}$$

$$+ ([k_{bb}^g] + [S_{r,bb}^g]_o) \{u_b^g\}_n + [c_{bb}^g] \{\dot{u}_b^g\}_n \qquad (6.151b)$$

$$+ \sum_{i=1}^{n-1} [S_{r,bb}^g]_{n-i} (\{u_b^g\}_i - \{u_b^t\}_i)$$

For a nonlinear structure, iterations in each time step must be performed. Using an incremental formulation,

$$\{u_b^j\}_n = \{u_b^{j-1}\}_n + \{\Delta u_b^j\}_n \qquad (6.152)$$

applies for the $j$th iteration in the $n$th time step. The interaction forces can be written incrementally as (Eq. 6.147)

$$\{R_b^j\}_n = \{R_b^{j-1}\}_n + [k_{bb}^g]\{\Delta u_b^j\}_n + [c_{bb}^g]\{\Delta \dot{u}_b^j\}_n$$

$$+ \int_{(n-1)\Delta t}^{n\Delta t} [S_{r,bb}^g(t - \tau)]\{\Delta u_b^j(\tau)\}_n \, d\tau \qquad (6.153)$$

whereby the integral will only involve $[S_{r,bb}^g]_o$ (Eq. 6.140b). The discretization, working with the increment $\{\Delta u_b^j\}_n$ as unknown, proceeds as before.

**Flexibility formulation.** The displacement–interaction force relationship specified in Eq. 6.10 is formulated for the time station $n\Delta t$

$$\{u_b^t\}_n - \{u_b^g\}_n = \int_o^{n\Delta t} [F_{bb}^g(t - \tau)] \{R_b(\tau)\} \, d\tau \qquad (6.154)$$

Assuming a linear temporal variation for $\{R_b(\tau)\}$ as in Eq. 6.143, with the definition specified in Eq. 6.145, and solving for $\{R_b\}_n$ as in Eq. 6.146 leads to

$$\{R_b\}_n = [F_{bb}^g]_o^{-1} \left(\{u_b^t\}_n - \{u_b^g\}_n - \sum_{i=1}^{n-1} [F_{bb}^g]_{n-i} \{R_b\}_i\right) \qquad (6.155)$$

The contribution of these interaction forces of the soil to the total structure–soil system's "effective" stiffness matrix consists of $[F_{bb}^g]_o^{-1} = [S_{bb}^g]_o$, which is the instantaneous dynamic-stiffness matrix containing the singular and regular parts. This is consistent with the result shown in Eq. 6.151a.

The term $[F_{bb}^g]_o^{-1} (\sum_{i=1}^{n-1} [F_{bb}^g]_{n-i}\{R_b\}_i + \{u_{b}^g\}_n)$ forms the contribution of the "effective" load vector which arises from the soil.

## 6.7 STRUCTURE WITH BASE ISOLATION

As an example, the simple structure with mass $m$ shown in Fig. 6-17 with and without base isolation, is investigated. $3/4\ m$ is concentrated at the height $h$. The mass moment of inertia associated with the rocking degree of freedom at the rigid basemat of radius $a$ equals $1/4\ a^2 m$. For a structure with base isolation, half of the mass at the basemat $(= m/4)$ is assigned to each part of it. The fixed-base frequency and the damping ratio of the structure without the base isolation equal 4 Hz and 0.07, respectively. The nonlinear isolation mechanism, acting in the horizontal direction and located between the upper and lower basemats, consists of Neoprene pads, resulting in a frequency of 1 Hz, and friction plates with a coefficient of 0.17. The following parameters apply: $a/c_s = 0.06$ s, $h/a = 1.5$, $m/(\rho a^3) = 3$ ($\rho$ = mass density of soil). Horizontal and vertical artificial time-histories which follow the U.S. NRC response spectra [U2], both normalized to 0.21g, are used. The soil is modeled as a half-space without material damping and also as a layer built-in at its base, with the depth equal to the radius, using a Voigt model with $\zeta = 0.02$. Poisson's ratio $\nu = 1/3$.

Calculations are performed not only with the rigorous procedure based on the convolution integral, which uses the flexibility formulation (dynamic-flexibility coefficients, Figs. 6-12 and 6-13) but also approximately, using (frequency-independent) springs and viscous dampers to represent the soil. The static-spring coefficient and the damping coefficient for high frequency are used. As is visible from Table 6-1 for the (linear) structure without base isolation, the approximate method works better for the half-space than for the layer, because in the latter case, the dynamic-stiffness coefficients depend strongly on the frequency. Another choice of the frequency-independent coefficients of the springs and dampers—for example, evaluated at the fun-

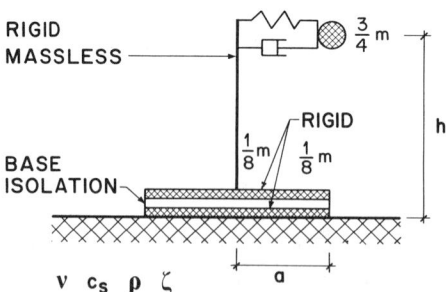

**Figure 6-17** Model of investigated structure.

**TABLE 6-1  Maximum Total Acceleration [g]**

| | Without Base Isolation | | | | With Base Isolation | |
|---|---|---|---|---|---|---|
| | Half-Space | | Layer | | Half-Space | |
| | Convolution Integral | Spring and Damper | Convolution Integral | Spring and Damper | Convolution Integral | Spring and Damper |
| Horizontal at Top | 0.415 | 0.360 | 0.860 | 0.511 | 0.189 | 0.188 |
| Horizontal at Upper Basemat | 0.245 | 0.234 | 0.240 | 0.247 | 0.251 | 0.223 |
| Horizontal at Lower Basemat | 0.245 | 0.234 | 0.240 | 0.247 | 0.271 | 0.284 |
| Vertical | 0.273 | 0.262 | 0.498 | 0.315 | 0.273 | 0.262 |

damental frequencies—would result in a better agreement. For the nonlinear case, the two procedures lead to similar results because the soil's influence is diminished for a structure with base isolation.

## 6.8 RECURSIVE EVALUATION OF CONVOLUTION INTEGRAL IN FREQUENCY DOMAIN

### 6.8.1 Introductory Remarks

It is desirable to avoid the transformation of the linear unbounded soil's dynamic-stiffness (-flexibility) matrix from the frequency domain to the time domain. In such a scheme, the interaction forces at the end of the $n$th time step—that is, at time $t = n\Delta t$ (subscript $n$)—are calculated in the frequency domain using $[S^g_{bb}(\omega)]$ as follows:

$$\{R_b(\omega)\}_n = [S^g_{bb}(\omega)] (\{u^t_b(\omega)\}_n - \{u^g_b(\omega)\}_n) \qquad (6.156)$$

where $\{u^t_b(\omega)\}_n$ denotes the Fourier transform of the total motion from time $t = 0$ to $t = n\Delta t$. Applying the inverse Fourier transform to $\{R_b(\omega)\}_n$ (which must only be evaluated at time $t = n\Delta t$) will then lead to the interaction forces $\{R_b\}_n$. The details of this transformation are discussed in Section 6.8.3. Full use can be made of the existing computer programs to calculate $[S^g_{bb}(\omega)]$, and only a minor additional programming effort is required to implement this procedure. However, in this procedure the computational effort is still high, because for each time step a Fourier transformation using the Fast Fourier Transform (FFT) is necessary ($\{u^t_b(t)\}_n \rightarrow \{u^t_b(\omega)\}_n$). The high storage requirements are also unchanged.

In addition to this concept of evaluating the interaction forces in the frequency domain, the amplitudes of the displacements in the frequency domain $\{u^t_b(\omega)\}_n$ can be calculated recursively using only the amplitudes at the previous time step $\{u^t_b(\omega)\}_{n-1}$ and the displacements at time $t = n\Delta t$, $\{u^t_b\}_n$, and time $t = (n-1)\Delta t$, $\{u^t_b\}_{n-1}$. The Fourier transformation is thus avoided. So, to determine the amplitudes at time $n\Delta t$, only the values at time $(n-1)\Delta t$ and not at earlier times are needed. The recursive evaluation is, in principle, rigorous. This scheme will lead to a significant reduction in the computational effort and storage requirement when interpolation in the frequency domain is applied. This procedure is described in detail in Section 6.8.4.

As an illustrative example, the spherical cavity, with symmetric waves, which has mass attached to its wall, and is embedded in a full infinite space, is examined for an earthquake excitation in Section 6.8.2. The number of operations is addressed in Section 6.8.5.

Section 6.8 is based on Refs. [M4] and [M5].

### Sec. 6.8  Recursive Evaluation of Convolution Integral in Frequency Domain

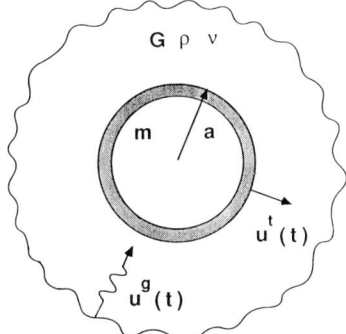

**Figure 6-18** Spherical cavity with mass embedded in infinite space.

#### 6.8.2 Spherical Cavity with Mass with Symmetric Waves

For the sake of illustration, a simple one-degree-of-freedom system is analyzed for earthquake excitation. A spherical cavity of radius $a$ is embedded in infinite space with shear modulus $G$, density $\rho$, and Poisson's ratio $\nu$. A constant distributed mass $m$ (per unit area) is attached to the cavity's wall (Fig. 6-18). The earthquake motion $u^g(t)$—assumed to act radially in this spherically symmetric system—consists of a 10-second artificial accelerogram normalized to 1.0 g, which closely follows the U.S. NRC response spectrum for 7% damping [U2]. The time-histories of the accelerations, velocities, and displacements are shown in Fig. 6-19. The system shown in Fig. 6-18 can be regarded as a simple problem of seismic soil–structure interaction, with the mass representing the structure, and the full space surrounding the cavity representing the soil. Although the system is linear, the procedures to be developed can be applied just as well to a nonlinear case.

The basic equation of motion in the time domain (Eq. 6.4) is formulated as

$$m\ddot{u}^t(t) + R(t) = 0 \tag{6.157}$$

where the interaction force $R(t)$ of the full space acting on the cavity's wall is specified in the stiffness formulation (Eq. 6.2) as

$$R(t) = \int_0^t S(t - \tau)(u^t(\tau) - u^g(\tau))\,d\tau \tag{6.158}$$

and in the flexibility formulation (Eq. 6.10) as

$$u^t(t) - u^g(t) = \int_0^t F(t - \tau)R(\tau)\,d\tau \tag{6.159}$$

The displacement of the mass is denoted as $u^t(t)$. Where appropriate, the subscripts $b$ and $g$ are dropped for the sake of simplicity. The dynamic-stiffness coefficients of the full space surrounding the cavity in the frequency

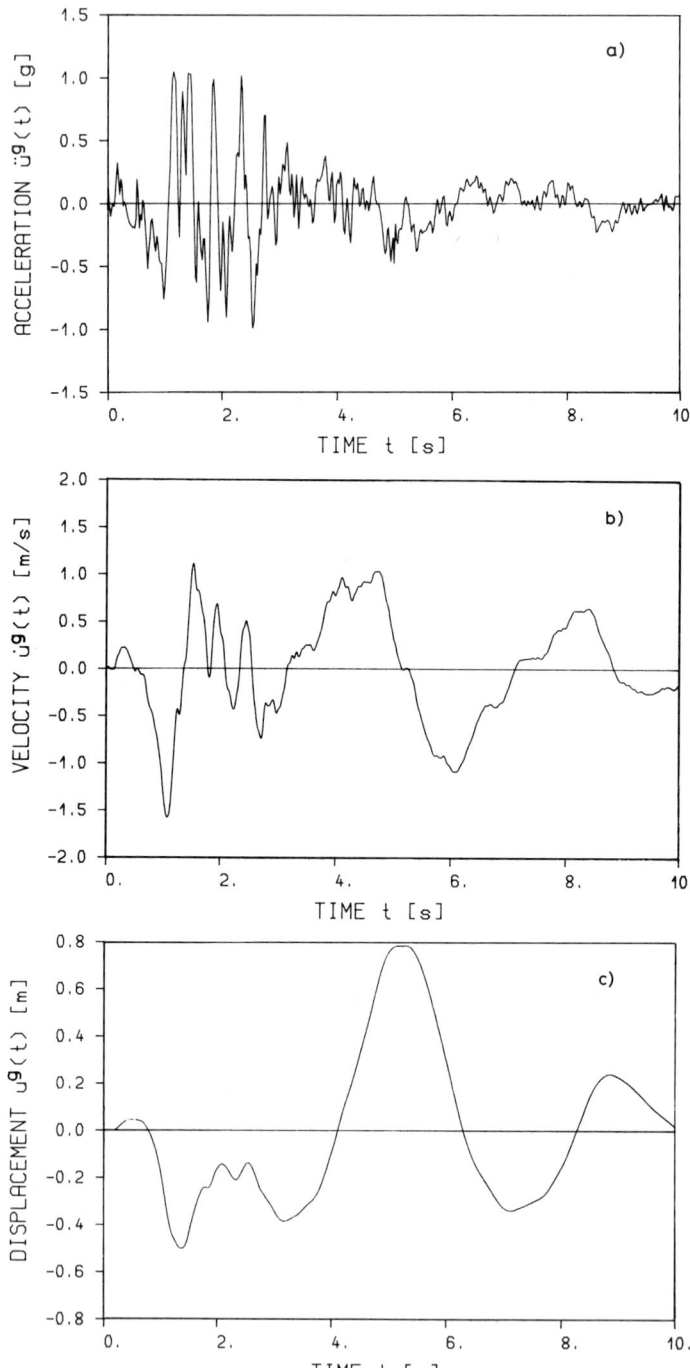

**Figure 6-19** Artificial earthquake time-history.

Sec. 6.8  Recursive Evaluation of Convolution Integral in Frequency Domain   289

and time domains are specified in Eqs. 6.44 and 6.45; the corresponding dynamic-flexibility coefficients in Eqs. 6.116 and 6.112.

All analyses are performed for $c_p/a = 15 \text{ s}^{-1}$, $\nu = 0$ ($c_p^2/c_s^2 = 2$), and $m = a\rho$.

Evaluating the convolution integral in either a stiffness or a flexibility formulation and using an implicit or an explicit time integration scheme as described in Section 6.6 leads to the acceleration time history of the mass $\ddot{u}^t(t)$, shown in Fig. 6-20. $\Delta t$ is selected as 0.005 s for the explicit and 0.04 s for the implicit scheme.

This spherical cavity embedded in an infinite space exhibits most properties of a homogeneous half-space. To check the behavior for the other limiting case of a site—a layer built-in at its base—a semi-infinite rod resting on an elastic foundation is examined in Problem 6.13.

### 6.8.3 Successive Fourier Transformations

**Fourier series.** The evaluation of the unbounded soil's interaction forces in the frequency domain, based on successive Fourier transformations, proceeds as follows.

The interaction forces at time $n\Delta t$ $\{R_b\}_n$ can be determined by working in the frequency domain, using a Fourier series representation of the periodic response of period $T$. The dynamic system's response is known up to time $(n - 1)\Delta t$. Defining the relative motion as

$$\{u_b(t)\} = \{u_b^t(t)\} - \{u_b^g(t)\} \qquad (6.160)$$

the following equations apply:

$$\{u_b(\omega_j)\}_n = \int_0^T \{u_b(\tau)\} \exp(-i\omega_j \tau) \, d\tau \qquad (6.161)$$

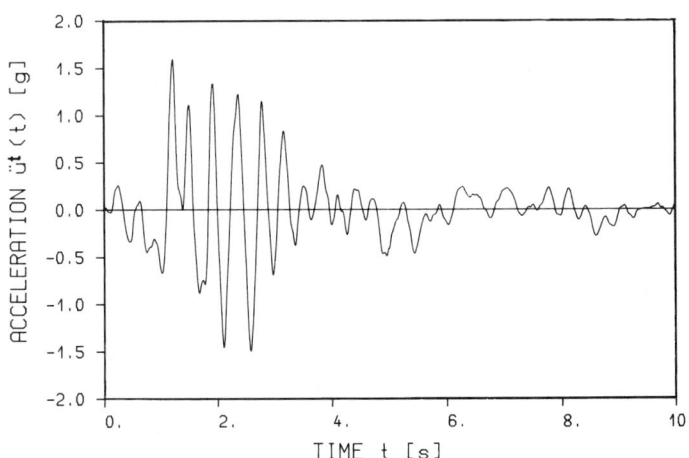

**Figure 6-20** Acceleration-time-history of mass.

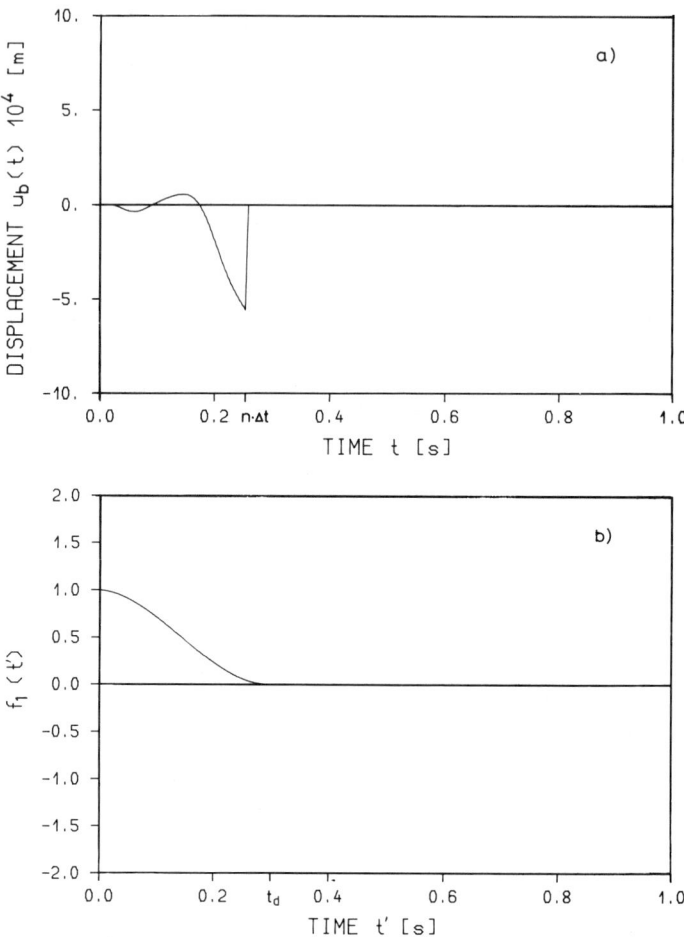

**Figure 6-21** Suppression of unrealistic high-frequency components. a. Unprocessed time-history. b. Polynomial to construct decay function (unit displacement). c. Polynomial to construct decay function (unit slope). d. Augmented displacement time-history.

$$\{R_b(\omega_j)\}_n = [S_{bb}^g(\omega_j)] \{u_b(\omega_j)\}_n \tag{6.162}$$

$$\{R_b\}_n = \frac{1}{T} \sum_{j=-\infty}^{+\infty} \{R_b(\omega_j)\}_n \exp(i\omega_j t) \tag{6.163}$$

In Eq. 6.161, $\{u_b(\tau)\}$ denotes the time history of the displacements from $\tau = 0$ to time $\tau = n\Delta t$, with those at $n\Delta t$ being predicted by the corresponding values at time $(n - 1)\Delta t$ in an explicit time integration scheme (Section 6.6.2).

The period $T$ includes an interval of zero load. This so-called quiet zone allows the free vibration components of the system's response to be damped out when the load's new period starts.

### Sec. 6.8  Recursive Evaluation of Convolution Integral in Frequency Domain    291

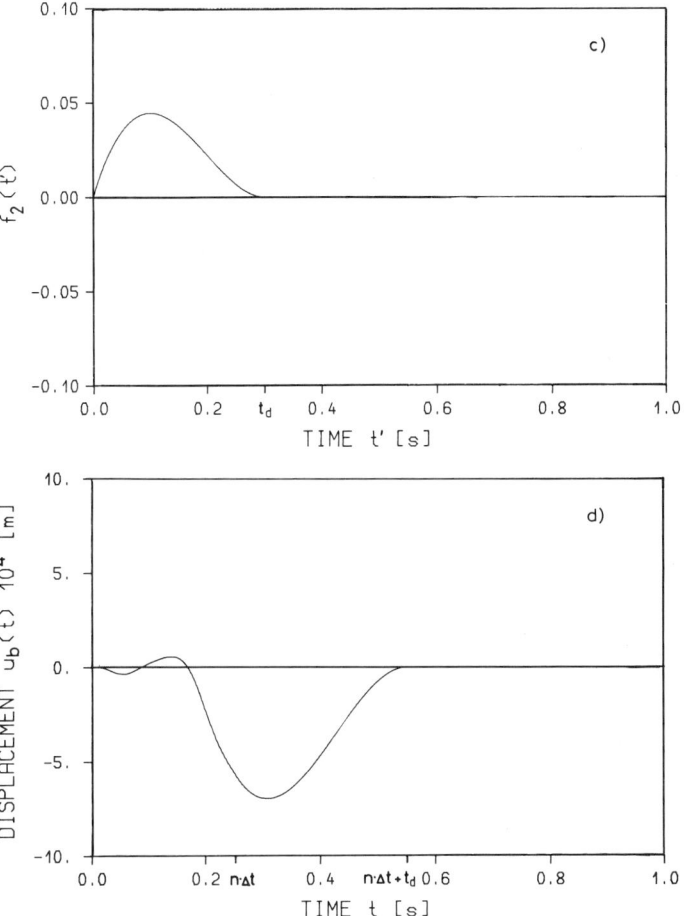

**Figure 6-21**  Continued

**Discrete Fourier transform.** The period $T$ is divided into $N^T$ increments of $\Delta t$. The frequency span is also divided into the same number of increments $\Delta \omega$, where $\Delta \omega = 2\pi/T$. With $T = N^T \Delta t$ and $\omega_j = j\Delta \omega$, Eq. 6.161 is formulated as a finite sum of discrete terms:

$$\{u_b(\omega_j)\}_n = \Delta t \sum_{k=0}^{N^T-1} \{u_b(t)\}_k \exp\left(-2\pi i \frac{jk}{N^T}\right) \quad (6.164)$$

This operation is performed with the Fast Fourier Transform algorithm. Equation 6.163 is evaluated only for $n\Delta t$ and thus represents a simple summation.

In the computational procedure to determine $\{u_b(\omega_j)\}_n$ (Eq. 6.164), $\{u_b(t)\}$ is specified (known or predicted) up to time $n\Delta t$ (Fig. 6-21a). In the range from $n\Delta t$ to $T$, it cannot be predicted accurately—only estimated. Assuming zero values for $\tau > t = n\Delta t$ would introduce a strong discontinuity which would lead to large amplitudes $\{u_b(\omega_j)\}_n$ in the higher-frequency range which are not actually present in the problem to be solved. This is illustrated in Fig. 6-22a, where the magnitude of the amplitudes of the displacement in the frequency domain is plotted as a function of $\omega$. As the dynamic-stiffness matrix $[S_{bb}^g(\omega_j)]$ (regular and singular parts) does not decay for increasing $\omega_j$, large amplitudes $\{R_b(\omega_j)\}_n$ would also arise in the higher-frequency range. To avoid this phenomenon, decay functions, which avoid discontinuities in the values and slopes at time $n\Delta t$, are added to $\{u_b(t)\}$ in the range $n\Delta t$ to $n\Delta t + t_d$, where $t_d$ is the decay time. The decay functions $f(t')$ ($t' = t - n\Delta t$) are constructed from the polynomials $f_1(t')$ and $f_2(t')$, with a unit value and a unit slope at $t' = 0$, respectively, and zero value and slope at $t' = t_d$.

$$f_1(t') = \frac{2}{t_d^3} t'^3 - \frac{3}{t_d^2} t'^2 + 1 \qquad (6.165a)$$

$$f_2(t') = \frac{t'^3}{t_d^2} - \frac{2}{t_d} t'^2 + t' \qquad (6.165b)$$

The decay functions are thus formulated as

$$\{u_b(t')\} = \{u_b\}_n f_1(t') + \{\dot{u}_b\}_n f_2(t') \qquad (6.166)$$

These functions are shown in Figs. 6-21b and 6-21c, and the augmented displacement time-history, which is actually used in Eq. 6.164, is shown in Fig. 6-21d. $|u_b(\omega)_n|$, using the augmented displacement time-history, is shown in Fig. 6-22b.

As the evaluation of Eq. 6.162 is computationally expensive, it is advantageous to calculate it, not for all $\omega_j$, but only for $N^\omega \ll N^T/2$.

**Spherical cavity.** A quiet zone of approximately 10 s is added to the time-history of the earthquake shown in Fig. 6-19 resulting in a period $T = 20.48$ s. Selecting $\Delta t = 0.005$ s results in $N^T = 4096$ and $\Delta\omega = 0.3068$ s$^{-1}$; the maximum frequency represented (Nyquist frequency, Eq. 5.52) equals $1/(2\Delta t) = 100$ Hz.

The interaction force $R(t)$ is evaluated using the discretized form of Eqs. 6.161 to 6.163. In this explicit stiffness formulation, the total dynamic-stiffness coefficient $S(\omega)$ (Eq. 6.44), including the singular part, is used for the calculation. As discussed further in Section 6.8.4, the decay time $t_d$ is chosen as 0.3 s. The successive Fourier transformations lead to an acceleration-time-history which, from a practical point of view, is identical to that of the "rigorous" solution determined in the time domain—shown in Fig. 6-20 (comparison not shown). Fig. 6-21 refers to the unprocessed and the augmented time histories at time $t = 0.25$ s ($n = 50$). Fig. 6-22 actually applies to the displacement's amplitude at time $t = 0.25$ s, $|u(\omega)_{n=50}|$.

Sec. 6.8    Recursive Evaluation of Convolution Integral in Frequency Domain        293

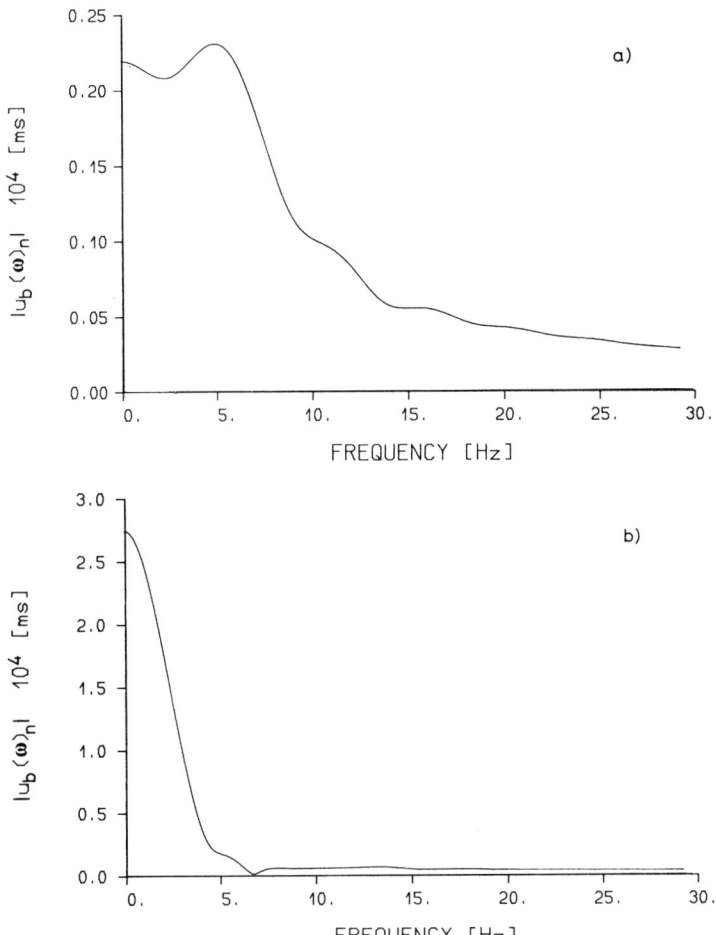

**Figure 6-22**  Amplitudes of displacements in frequency domain a. Unprocessed time-history. b. Augmented time-history.

Selecting $N^\omega = 800$—that is, taking all frequencies only up to 39.06 Hz—leads to results which do not differ from selecting $N^\omega = 2048$.

### 6.8.4 Recursive Procedure

**Stiffness formulation with explicit time integration.** At first, an equation for the interaction forces in the time domain, shown as a function of the dynamic-stiffness coefficients in the frequency domain and the time-history of the displacements, is obtained. In Eq. 6.163, the interaction forces $\{R_b\}_n$ at time $t = n\Delta t$ are expressed as a sum of the corresponding complex amplitudes in the frequency domain $\{R_b(\omega_j)\}_n$, which are related by Eq. 6.162

to the corresponding amplitudes of the displacements in the frequency domain $\{u_b(\omega_j)\}_n$. The latter vector, specified in Eq. 6.161, is reformulated as follows:

$$\{u_b(\omega_j)\}_n = \int_0^t \{u_b(\tau)\} \exp(-i\omega_j \tau)\, d\tau + \int_t^T \{u_b(\tau)\} \exp(-i\omega_j \tau)\, d\tau \quad (6.167)$$

Substituting Eq. 6.167 into Eq. 6.162 and the resulting equation into Eq. 6.163 leads to

$$\{R_b\}_n = \frac{1}{T} \sum_{j=-\infty}^{+\infty} [S_{bb}^g(\omega_j)] \left( \int_0^t \{u_b(\tau)\} \exp(-i\omega_j \tau)\, d\tau \right) \exp(i\omega_j t)$$

$$+ \frac{1}{T} \sum_{j=-\infty}^{+\infty} [S_{bb}^g(\omega_j)] \left( \int_t^T \{u_b(\tau)\} \exp(-i\omega_j \tau)\, d\tau \right) \exp(i\omega_j t) \quad (6.168)$$

Changing the order of integration and summation, the second term on the right-hand side can be rewritten as

$$\frac{1}{T} \sum_{j=-\infty}^{+\infty} [S_{bb}^g(\omega_j)] \left( \int_t^T \{u_b(\tau)\} \exp(-i\omega_j \tau)\, d\tau \right) \exp(i\omega_j t)$$

$$= \frac{1}{T} \int_t^T \sum_{j=-\infty}^{+\infty} [S_{bb}^g(\omega_j)] \exp(i\omega_j(t-\tau))\{u_b(\tau)\}\, d\tau \quad (6.169)$$

Analogous to Eq. 6.34,

$$[S_{bb}^g(t-\tau)] = \frac{1}{T} \sum_{j=-\infty}^{+\infty} [S_{bb}^g(\omega_j)] \exp(i\omega_j(t-\tau)) \quad (6.170)$$

applies, which is substituted into Eq. 6.169 resulting in

$$\frac{1}{T} \sum_{j=-\infty}^{+\infty} [S_{bb}^g(\omega_j)] \left( \int_t^T \{u_b(\tau)\} \exp(-i\omega_j \tau)\, d\tau \right) \exp(i\omega_j t)$$

$$= \int_t^T [S_{bb}^g(t-\tau)] \{u_b(\tau)\}\, d\tau \quad (6.171)$$

This term vanishes as $[S_{bb}^g(t-\tau)]$ equals zero for $t - \tau < 0$ (causal behavior). Thus, Eq. 6.168 consists only of the first term

$$\{R_b\}_n = \frac{1}{T} \sum_{j=-\infty}^{+\infty} [S_{bb}^g(\omega_j)] \left( \int_0^t \{u_b(\tau)\} \exp(-i\omega_j \tau)\, d\tau \right) \exp(i\omega_j t) \quad (6.172)$$

As expected, to determine the reaction forces at time $t$ $\{R_b\}_n$, the time-history $\{u_b(\tau)\}$ for $0 < \tau < t$ only is needed.

The integral appearing in Eq. 6.172 can be evaluated recursively for each $\omega_j$ as follows:

$$\int_0^{t=n\Delta t} \{u_b(\tau)\} \exp(-i\omega_j \tau)\, d\tau = \int_0^{(n-1)\Delta t} \{u_b(\tau)\} \exp(-i\omega_j \tau)\, d\tau$$

$$+ \int_{(n-1)\Delta t}^{n\Delta t} \{\tilde{u}_b(\tau)\} \exp(-i\omega_j \tau)\, d\tau \quad (6.173)$$

### Sec. 6.8  Recursive Evaluation of Convolution Integral in Frequency Domain

where the relative predicted displacements equal

$$\{\bar{u}_b(t)\} = \{\bar{u}_b^t(t)\} - \{u_b^g(t)\} \tag{6.174}$$

Assuming a linear variation of the displacements $\{\bar{u}_b(\tau)\}$ over the $n$th time step as in Eq. 6.138, the second term on the right-hand side of Eq. 6.173 can be rewritten as

$$\int_{(n-1)\Delta t}^{n\Delta t} \{\bar{u}_b(\tau)\} \exp(-i\omega_j \tau) \, d\tau = c_1(\omega_j)\{\bar{u}_b\}_{n-1} + c_2(\omega_j)\{\bar{u}_b\}_n \tag{6.175}$$

where

$$c_1(\omega_j) = \left(-\frac{\exp(-i\omega_j \Delta t)}{\Delta t \omega_j^2} - \frac{i}{\omega_j} + \frac{1}{\Delta t \omega_j^2}\right) \exp(-i\omega_j(n-1)\Delta t) \tag{6.176a}$$

$$c_2(\omega_j) = \left(\exp(-i\omega_j \Delta t)\frac{i\omega_j \Delta t + 1}{\Delta t \omega_j^2} - \frac{1}{\Delta t \omega_j^2}\right) \exp(-i\omega_j(n-1)\Delta t) \tag{6.176b}$$

The integral up to $t = n\Delta t$ for each frequency $\omega_j$ present on the left-hand side of Eq. 6.173 is thus equal to the sum of the same integral for the previous time step $(n-1)\Delta t$ and the right-hand side of Eq. 6.175, which involves only the displacements in the time domain at $(n-1)\Delta t$ and the predicted values at $t = n\Delta t$. In the practical evaluation of $\{R_b\}_n$ (Eq. 6.172), it is important to realize that $[S_{bb}^g(\omega_j)]$ (singular and regular parts) does not decay for high frequencies. The product of $[S_{bb}^g(\omega_j)]$ and the integral $\int_0^t \{u_b(\tau)\} \exp(-i\omega_j \tau) \, d\tau$ representing the interaction forces' Fourier transform will thus exhibit high-frequency components, which are unrealistic, for example, for a seismic excitation. To avoid this phenomenon, the time-history $\{u_b(\tau)\}$ is augmented by the decay functions introduced in Section 6.8.3 for $t < \tau < t + t_d$. This leads to an approximation. The integral appearing in Eq. 6.172 is thus actually evaluated for $0 < \tau < t + t_d$. Equation 6.172, when using the total dynamic-stiffness matrix $[S_{bb}^g(\omega_j)]$, thus represents only a formal result.

Closed form solutions exist for the additional term involving the decay functions.

$$\int_0^{t_d} [\{\bar{u}_b\}_n f_1(t') + \{\dot{\bar{u}}_b\}_n f_2(t')] \exp(-i\omega_j t') \, dt'$$

$$= \{\bar{u}_b\}_n \left(-\frac{1}{\omega_j^4 t_d^3}\right) [6 \exp(-i\omega_j t_d)(i\omega_j t_d + 2) + i\omega_j^3 t_d^3 + 6i\omega_j t_d - 12]$$

$$+ \{\dot{\bar{u}}_b\}_n \left(-\frac{1}{\omega_j^4 t_d^2}\right) [2 \exp(-i\omega_j t_d)(i\omega_j t_d + 3) + \omega_j^2 t_d^2 + 4i\omega_j t_d - 6]$$

$$\tag{6.177}$$

It should be noted that, when evaluating the first term on the right-hand side of Eq. 6.173, the effects of the previous time step's decay function are not included.

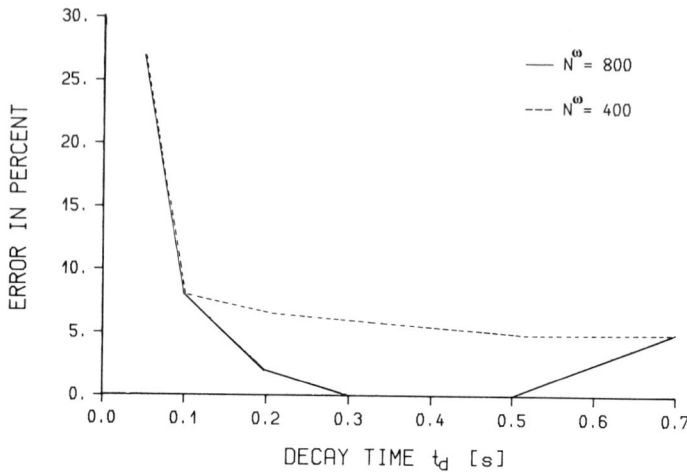

**Figure 6-23** Influence of duration of decay function on accuracy for the convolution integral with total dynamic stiffness.

If the soil's dynamic-stiffness matrix is decomposed into the regular and singular parts, Eq. 6.172 is rewritten as

$$\{R_b\}_n = \frac{1}{T} \sum_{j=-\infty}^{+\infty} [S^g_{r,bb}(\omega_j)] \left( \int_o^t \{u_b(\tau)\} \exp(-i\omega_j\tau)\, d\tau \right) \exp(i\omega_j t)$$
$$+ [k^g_{bb}]\{\bar{u}_b\}_n + [c^g_{bb}]\{\dot{\bar{u}}_b\}_n \qquad (6.178)$$

In this case, no decay functions need be introduced.

The illustrative example of the spherical cavity is calculated with the stiffness formulation using the explicit time integration with $\Delta t = 0.005$ s. First, the soil's total dynamic-stiffness coefficient appears in the convolution integral (Eq. 6.172). A decay function is introduced (Eqs. 6.165 and 6.166). The number of frequencies $N^\omega$ and the decay function's duration are varied. In Fig. 6-23, the maximum error occurring anywhere in the time-history of the acceleration of the mass $\ddot{u}^t(t)$ is plotted versus $t_d$ for two $N^\omega$'s. For instance, for $t_d = 0.3$ s and $N^\omega = 800$ (39.06 Hz), a perfect agreement results from a practical point of view. For smaller $t_d$'s, the high-frequency behavior leads to a significant error. For $t_d = 0.7$ s, the low-frequency content is overemphasized, again leading to unacceptable results. For $N^\omega = 400$, convergence is not reached.

Second, as an intermediate case between those described by Eqs. 6.172 and 6.178, only the damper (which is easy to determine) is removed from the convolution integral, leading to a singular term $c\dot{u}^t$. The maximum error in $\ddot{u}^t(t)$ is plotted versus $t_d$ for $N^\omega = 400$ in Fig. 6-24. As expected, the accuracy is improved.

Third, evaluating the convolution integral with the dynamic-stiffness coefficient's regular part—that is, applying Eq. 6.178—leads to highly ac-

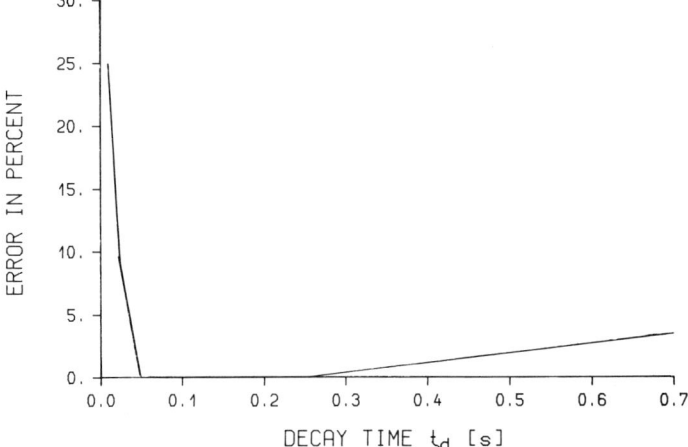

**Figure 6-24** Influence of duration of decay function on accuracy for the convolution integral with damper separated from total dynamic stiffness.

curate results, as is visible from Fig. 6-25 ($N^\omega = 400$). For realistic $N^\omega$'s such as $N^\omega = 400$, no decay function is actually needed.

**Flexibility formulation with explicit time integration.** To evaluate the convolution integral involving the interaction forces recursively in the frequency domain, Eq. 6.142 is split into two integrals as follows:

$$\{\bar{u}_b^t\}_n - \{u_b^g\}_n = \int_o^{(n-1)\Delta t} [F_{bb}^g(t-\tau)] \{R_b(\tau)\} \, d\tau$$
$$+ \int_{(n-1)\Delta t}^{n\Delta t} [F_{bb}^g(t-\tau)] \{R_b(\tau)\} \, d\tau \qquad (6.179)$$

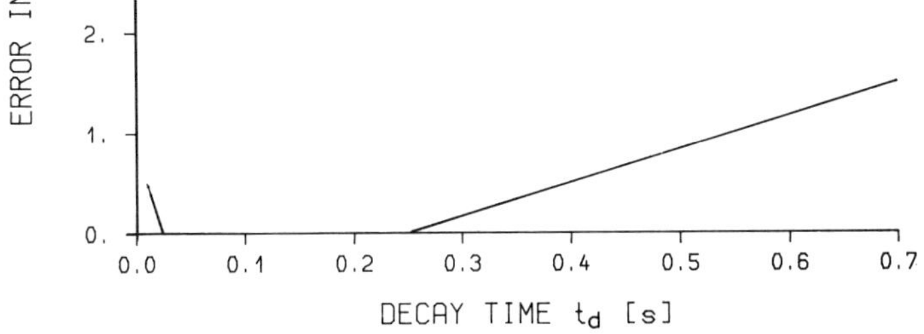

**Figure 6-25** Influence of duration of decay function on accuracy for the convolution integral with regular part of dynamic stiffness.

Analogous to Eq. 6.172, but integrating only up to $(n - 1)\Delta t$, the first integral on the right-hand side is equal to

$$\int_0^{(n-1)\Delta t} [F_{bb}^g(t - \tau)]\{R_b(\tau)\} d\tau$$

$$= \frac{1}{T} \sum_{j=-\infty}^{+\infty} [F_{bb}^g(\omega_j)] \left( \int_0^{(n-1)\Delta t} \{R_b(\tau)\} \exp(-i\omega_j \tau) d\tau \right) \exp(i\omega_j t) \quad (6.180)$$

The integral $\int_0^{(n-1)\Delta t} \{R_b(\tau)\} \exp(-i\omega_j \tau) d\tau$, appearing on the right-hand side of Eq. 6.180, is known, as it has been evaluated recursively at the end of the previous time step (see Eq. 6.184). After substituting Eq. 6.143 into the second integral on the right-hand side of Eq. 6.179, the following results:

$$\int_{(n-1)\Delta t}^{n\Delta t} [F_{bb}^g(t - \tau)]\{R_b(\tau)\} d\tau = [F_{bb}^g]_1^* \{R_b\}_{n-1} + [F_{bb}^g]_o \{R_b\}_n \quad (6.181)$$

where

$$[F_{bb}^g]_1^* = \int_0^{\Delta t} \left(1 - \frac{\tau'}{\Delta t}\right) [F_{bb}^g(\Delta t - \tau')] d\tau' \quad (6.182)$$

and $[F_{bb}^g]_o$ is defined in Eq. 6.145b. Substituting Eqs. 6.180 and 6.181 into Eq. 6.179 and solving for $\{R_b\}_n$ results in

$$\{R_b\}_n = [F_{bb}^g]_o^{-1} \Bigg( \{\bar{u}_b^t\}_n - \{u_b^g\}_n - [F_{bb}^g]_1^* \{R_b\}_{n-1}$$

$$- \frac{1}{T} \sum_{j=-\infty}^{+\infty} [F_{bb}^g(\omega_j)] \left( \int_0^{(n-1)\Delta t} \{R_b(\tau)\} \exp(-i\omega_j \tau) d\tau \right) \exp(i\omega_j t) \Bigg)$$

$$(6.183)$$

After determining $\{R_b\}_n$, the recursive integral is updated as a preparation for the operations in the next time step:

$$\int_0^{t=n\Delta t} \{R_b(\tau)\} \exp(-i\omega_j \tau) d\tau = \int_0^{(n-1)\Delta t} \{R_b(\tau)\} \exp(-i\omega_j \tau) d\tau$$

$$+ \int_{(n-1)\Delta t}^{n\Delta t} \{R_b(\tau)\} \exp(-i\omega_j \tau) d\tau \quad (6.184)$$

$$= \int_0^{(n-1)\Delta t} \{R_b(\tau)\} \exp(-i\omega_j \tau) d\tau + c_1(\omega_j) \{R_b\}_{n-1} + c_2(\omega_j) \{R_b\}_n$$

Selecting $\Delta t = 0.005$ s, $N^\omega = 400$ and applying Eq. 6.183 leads to an acceleration-time–history of the mass on the spherical cavity's wall, which from a practical point of view coincides with that of Fig. 6-20.

### Sec. 6.8  Recursive Evaluation of Convolution Integral in Frequency Domain

**Stiffness formulation with implicit time integration.** Again, the convolution integral appearing in the equation for the interaction forces (Eq. 6.147) is split up into two integrals:

$$\{R_b\}_n = [k_{bb}^g](\{u_b^t\}_n - \{u_b^g\}_n) + [c_{bb}^g](\{\dot{u}_b^t\}_n - \{\dot{u}_b^g\}_n)$$

$$+ \int_0^{(n-1)\Delta t} [S_{r,bb}^g(t-\tau)](\{u_b^t(\tau)\} - \{u_b^g(\tau)\})\,d\tau \quad (6.185)$$

$$+ \int_{(n-1)\Delta t}^{n\Delta t} [S_{r,bb}^g(t-\tau)](\{u_b^t(\tau)\} - \{u_b^g(\tau)\})\,d\tau$$

The first integral is transformed to

$$\int_0^{(n-1)\Delta t} [S_{r,bb}^g(t-\tau)](\{u_b^t(\tau)\} - \{u_b^g(\tau)\})\,d\tau$$

$$= \frac{1}{T}\sum_{j=-\infty}^{+\infty} [S_{r,bb}^g(\omega_j)] \left( \int_0^{(n-1)\Delta t} (\{u_b^t(\tau)\} - \{u_b^g(\tau)\}) \exp(-i\omega_j\tau)\,d\tau \right) \exp(i\omega_j t) \quad (6.186)$$

The integral appearing on the right-hand side of Eq. 6.186—$\int_0^{(n-1)\Delta t} (\{u_b^t(\tau)\} - \{u_b^g(\tau)\}) \exp(-i\omega_j\tau)\,d\tau$—is known, as it has been evaluated recursively at the end of the previous time step (see Eq. 6.191).

The second integral on the right-hand side of Eq. 6.185 leads to

$$\int_{(n-1)\Delta t}^{n\Delta t} [S_{r,bb}^g(t-\tau)](\{u_b^t(\tau)\} - \{u_b^g(\tau)\})\,d\tau \quad (6.187)$$

$$= [S_{r,bb}^g]_1^*(\{u_b^t\}_{n-1} - \{u_b^g\}_{n-1}) + [S_{r,bb}^g]_0(\{u_b^t\}_n - \{u_b^g\}_n)$$

where

$$[S_{r,bb}^g]_1^* = \int_0^{\Delta t} \left(1 - \frac{\tau'}{\Delta t}\right) [S_{r,bb}^g(\Delta t - \tau')]\,d\tau' \quad (6.188)$$

and $[S_{r,bb}^g]_0$ is defined in Eq. 6.140b. Equation 6.150 can thus be formulated as

$$\{R_b\}_n = [k_{bb}^g]\left(\{u_b^t\}_n - \{u_b^g\}_n\right) + [c_{bb}^g]\left(\frac{\gamma}{\beta\Delta t}\{u_b^t\}_n \right.$$

$$\left. - \frac{\gamma}{\beta\Delta t}\{u_b^t\}_{n-1} + \left(1 - \frac{\gamma}{\beta}\right)\{\dot{u}_b^t\}_{n-1} + \left(1 - \frac{\gamma}{2\beta}\right)\Delta t\{\ddot{u}_b^t\}_{n-1} - \{\dot{u}_b^g\}_n\right)$$

$$+ [S_{r,bb}^g]_0\left(\{u_b^t\}_n - \{u_b^g\}_n\right) + [S_{r,bb}^g]_1^*\left(\{u_b^t\}_{n-1} - \{u_b^g\}_{n-1}\right)$$

$$+ \frac{1}{T}\sum_{j=-\infty}^{+\infty} [S_{r,bb}^g(\omega_j)] \left( \int_0^{(n-1)\Delta t} (\{u_b^t(\tau)\} - \{u_b^g(\tau)\}) \exp(-i\omega_j\tau)\,d\tau \right) \exp(i\omega_j t)$$

$$(6.189)$$

The contribution to the "effective" stiffness matrix of the structure–soil system is specified in Eq. 6.151a, while the contribution to the "effective" load vector is

$$\frac{\gamma}{\beta \Delta t} [c_{bb}^g]\{u_b^t\}_{n-1} + \left(\frac{\gamma}{\beta} - 1\right)[c_{bb}^g]\{\dot{u}_b^t\}_{n-1}$$

$$+ \left(\frac{\gamma}{2\beta} - 1\right)\Delta t [c_{bb}^g]\{\ddot{u}_b^t\}_{n-1} + ([k_{bb}^g] + [S_{r,bb}^g]_o)\{u_b^g\}_n$$

$$+ [c_{bb}^g]\{\dot{u}_b^g\}_n - [S_{r,bb}^g]_1^* (\{u_b^t\}_{n-1} - \{u_b^g\}_{n-1})$$

$$- \frac{1}{T} \sum_{j=-\infty}^{+\infty} [S_{r,bb}^g(\omega_j)] \left( \int_o^{(n-1)\Delta t} (\{u_b^t(\tau)\} - \{u_b^g(\tau)\}) \exp(-i\omega_j \tau) \, d\tau \right) \exp(i\omega_j t)$$

(6.190)

Analogous to Eq. 6.184, the integral is evaluated recursively in preparation for the next time step:

$$\int_o^{t=n\Delta t} (\{u_b^t(\tau)\} - \{u_b^g(\tau)\}) \exp(-i\omega_j \tau) \, d\tau$$

$$= \int_o^{(n-1)\Delta t} (\{u_b^t(\tau)\} - \{u_b^g(\tau)\}) \exp(-i\omega_j \tau) \, d\tau + c_1(\omega_j)\{u_b\}_{n-1} + c_2(\omega_j)\{u_b\}_n$$

(6.191)

Applying Eq. 6.189 with $\Delta t = 0.04$ s and $N^\omega = 400$ leads to highly accurate results for the spherical cavity.

**Flexibility formulation with implicit time integration.** Equation 6.183 still applies. Dropping tildes leads to

$$\{R_b\}_n = [F_{bb}^g]_o^{-1} \left( \{u_b^t\}_n - \{u_b^g\}_n - [F_{bb}^g]_1^* \{R_b\}_{n-1} \right.$$

$$\left. - \frac{1}{T} \sum_{j=-\infty}^{+\infty} [F_{bb}^g(\omega_j)] \left( \int_o^{(n-1)\Delta t} \{R_b(\tau)\} \exp(-i\omega_j \tau) \, d\tau \right) \exp(i\omega_j t) \right)$$

(6.192)

The contribution of these interaction forces to the "effective" stiffness matrix of the total structure–soil system is the same as explained in Section 6.6.3 (Flexibility Formulation). The contribution to the "effective" load vector now equals

$$[F_{bb}^g]_o^{-1} \left( \frac{1}{T} \sum_{j=-\infty}^{+\infty} [F_{bb}^g(\omega_j)] \left( \int_o^{(n-1)\Delta t} \{R_b(\tau)\} \exp(-i\omega_j \tau) \, d\tau \right) \exp(i\omega_j t) \right.$$

$$\left. + [F_{bb}^g]_1^* \{R_b\}_{n-1} + \{u_b^g\}_n \right)$$

(6.193)

The integral in Eq. 6.193 has been evaluated recursively at the end of the previous time step and thus is known (Eq. 6.184). After determining $\{u_b^t\}_n$ by solving the discretized system of equations, the interaction forces $\{R_b\}_n$ follow from Eq. 6.192.

Based on Eq. 6.192 with $\Delta t = 0.04$ s and $N^\omega = 400$, the result shown in Fig. 6-20 is recalculated.

### 6.8.5 Number of Operations

The number of operations is governed by the evaluation of the convolution integrals in the various methods. The other operations can be disregarded in a comparison. $N^s$ denotes the number of boundary elements introduced in the spatial discretization (with 3 degrees of freedom per element). The temporal discretization is characterized by the number of time steps $N^t$. $N^T$ denotes the number of time increments into which the total period used for the Fourier transform is divided, and $N^\omega$ denotes the number of frequencies.

The direct time-domain evaluation of the convolution integral (Section 6.6) requires

$$\frac{(N^t)^2}{2} (3N^s)^2 \tag{6.194}$$

operations. The successive evaluation in the frequency domain using Fast Fourier Transforms (Eqs. 6.161 to 6.163) leads to

$$N^t \left[ N^T (\log_2 N^T) \, 3N^s + N^\omega (3N^s)^2 + N^\omega \, 3N^s \right] \tag{6.195}$$

operations. As in the case of the recursive evaluation of the convolution integral in the frequency domain (Section 6.8.4), no Fourier transformations are performed anymore, and the first term in the number of operations specified in Eq. 6.195 is reduced significantly, leading to the following total number of operations:

$$N^t \left[ N^\omega \, 3N^s + N^\omega (3N^s)^2 + N^\omega \, 3N^s \right] \tag{6.196}$$

When the recursive evaluation is performed for all frequencies, that is $N^\omega = N^t/2$, no reduction of the computational effort results when compared with the direct time-domain evaluation. This is easily verified by substituting $N^\omega = N^t/2$ in Eq. 6.196 which leads essentially to Eq. 6.194. Note also, that the recursive evaluation in the frequency domain involves complex variables compared to real variables in the direct calculation in the time domain.

The recursive evaluation allows $\Delta\omega$ to be chosen larger than $2\pi/T$. This will reduce $N^\omega$ significantly without changing the maximum frequency represented. Such interpolation in the frequency domain is often used in the standard complex response procedure (Chapter 4). Compared to the direct evaluation of the convolution integrals in the time domain (Eq. 6.194), the

**TABLE 6-2  Number of Operations for Evaluation of Convolution Integrals**

| Direct Time-Domain Evaluation | Successive Evaluation in Frequency Domain Using FFTs | Recursive Evaluation in Frequency Domain |
|---|---|---|
| $6.5 \cdot 10^{10}$ | $4.2 \cdot 10^{10}$ | $4.5 \cdot 10^{9}$ |

proposed method is much more efficient for a small number of frequencies $N^\omega$. As an example, an earthquake time-history of 30 s duration with a $\Delta t = 0.005$ s is addressed, resulting in $N^t = 6000$ and $N^T = 8192$ ($T = 40.96$s). $N^\omega$ is selected as 200, and $\Delta\omega = 4 \times 2\pi/T = 0.6208 \text{ s}^{-1}$, leading to a maximum frequency of 19.76 Hz. No decay function is introduced. For 20 three-dimensional boundary elements ($N^s = 20$), the number of operations for the three procedures is compared in Table 6-2.

## 6.9 RECURSIVE EVALUATION OF CONVOLUTION INTEGRAL IN TIME DOMAIN

### 6.9.1 Introductory Remarks

Instead of evaluating the convolution integral recursively in the frequency domain (Section 6.8), which is, in principle, a rigorous procedure, the same concept can also be applied in the time domain. This recursive evaluation in the time domain, which is, in general, only approximate but highly accurate, leads to a larger reduction in the computational effort than working recursively in the frequency domain with using interpolation in the frequency domain. The resulting number of operations is similar to that of the standard direct (non-recursive) procedure of determining the complex response in the frequency domain (Chapter 4), which is only applicable to linear systems. The recursive evaluation of the convolution integrals corresponding to the interaction forces of the unbounded soil in the time domain at each time step thus makes the time-domain analysis using the substructure method computationally competitive with the corresponding direct frequency-domain procedure (see Section 6.9.5).

For instance, in the stiffness formulation of the recursive evaluation, the interaction force at a specific time $t$ is calculated from the most recent interaction forces and the most recent past displacements. Thus, the procedure does not correspond to a truncation of the convolution integral that only retains recent past displacements (see Section 7.7.4). Processing both most recent past displacements and interaction forces is equivalent to evaluating an approximate dynamic-stiffness coefficient in the time domain directly, i.e. from zero to time $t$. This approximate dynamic-stiffness coefficient, which has to be stable, must always be compared to the exact one to be able to judge the achieved accuracy of the recursive evaluation. From

## Sec. 6.9  Recursive Evaluation of Convolution Integral in Time Domain

the definition of the convolution integral it follows that a good agreement of the dynamic-stiffness coefficients must exist over the whole time history of the excitation in the range of the non-negligible stiffness coefficient. From the corresponding plots of the dynamic-stiffness coefficients (and flexibility coefficients) for the half-space (Figs. 6-5, 6-3, 6-12) and for the layer built-in at its base (Figs. 6-13, P6-7, P6-9), the non-negligible range extends approximately up to the dimensionless time $\bar{t}$ equal to 4 and 10, respectively. The corresponding times $t = a/c_s \bar{t}$ with a typical value $a/c_s = 0.1$ are equal to 0.4 s and 1 s, respectively. It follows that for the majority of transient excitations, in particular for seismic loading, the dynamic-stiffness coefficient must be adequately approximated over its whole range. The recursive formulation also leads to a significant reduction of the storage requirements.

The recursive evaluation of the convolution integral in the time domain can be applied to the stiffness and flexibility formulations using the explicit and implicit algorithms (Section 6.6). The procedure is developed based on the stiffness formulation as described typically in Eq. 6.147 (see Problems 6.15, 6.16, and 6.17 for the flexibility formulations). The singular part consisting of the first two terms on the right-hand side need not to be addressed. The regular part of the interaction forces $\{R_{r,b}\}_n$ at time $t = n\Delta t$ is equal to

$$\{R_{r,b}\}_n = \int_o^{n\Delta t} [S^g_{r,bb}(t - \tau)](\{u^t_b(\tau)\} - \{u^g_b(\tau)\}) \, d\tau \quad (6.197)$$

The subscripts $r$, $b$ and the superscripts $t$, $g$ are dropped for the sake of conciseness in the following. The expression "regular part" of the interaction force is also replaced by the word "force."

In the recursive evaluation, the forces $\{R\}_n$ at time $t = n\Delta t$ are computed from the $n$th displacements $\{u\}_n$ and the $M$ and $L$ past values of the forces and displacements, respectively. To evaluate a typical convolution integral appearing in Eq. 6.197 written as

$$\{R\}_n = \int_o^t [S(t - \tau)] \{u(\tau)\} \, d\tau \quad (6.198)$$

the recursive formulation leads to

$$\{R\}_n = \sum_{i=1}^M [a]_i \{R\}_{n-i} + \Delta t \sum_{i=0}^L [b]_i \{u\}_{n-i} \quad (6.199)$$

$[a]_i$ and $[b]_i$ are matrices to be determined.

The choice of a recursive equation is, in general, not unique and many possibilities exist. Two options that work with the dynamic-stiffness coefficients in the time domain $[S(t)]$ are addressed. The first, called the *impulse-invariant method* [V6], sets the approximate dynamic-stiffness coefficients corresponding to the recursive formulation equal to the exact ones in a certain time range (Section 6.9.2). This results in a system of equations with the unknowns $[a]_i$ and $[b]_i$. In the second procedure, *the segment approach* [M6], the dynamic-stiffness coefficients in the time domain are interpolated piecewise (Section 6.9.3). Applying the so-called $z$-transformation then results in an explicit recursive equation without solving a system of equations. Other concepts are developed in Problems 6.17 and 6.18. In Section 6.9.4, the recursive relationship is developed directly from the dynamic-stiffness coefficients in the frequency domain, whereby the latter are approximated as ratios of polynomials. The number of operations of the various procedures is addressed in Section 6.9.5. The rod on an elastic foundation is analyzed recursively for a short duration load and as the model for the unbounded soil in a simple nonlinear soil-structure-interaction analysis for seismic excitation in Section 6.9.6.

This Section is based on Ref. [W20], which contains additional information.

### 6.9.2 Impulse-Invariant Method

**System of equations for recursive coefficients.** The $M + L + 1$ unknown matrices $[a]_i$, $[b]_i$ in Eq. 6.199 are calculated by solving a system of equations that are established by equating the rigorous and recursive formulations for a discretized unit impulse displacement at the first $M + L + 1$ time stations. As the displacements are assumed to vary linearly over the time steps (as described in Eq. 6.148), the unit-impulse displacement for each degree of freedom is applied in the form of a triangle (zero at $t = 0$, increasing linearly up to $t = \Delta t$ where the value equals $1/\Delta t$ and then decreasing linearly up to $t = 2\Delta t$). The displacements over the first and second time steps are thus specified as

$$\{u(\tau)\}_1 = \frac{\tau}{\Delta t}\left\{\frac{1}{\Delta t}\right\} \tag{6.200a}$$

$$\{u(\tau)\}_2 = \left(1 - \frac{\tau}{\Delta t}\right)\left\{\frac{1}{\Delta t}\right\} \tag{6.200b}$$

$\tau$ is measured from the beginning of the corresponding time step ($0 \leq \tau \leq \Delta t$). The vector $\{1/\Delta t\}$ contains the displacements $\{u\}_1$ at $n = 1$ ($t = \Delta t$).

Substituting Eq. 6.200 into the rigorous expression for the forces (Eq. 6.198) leads to

$$\{R\}_n = [S]_{n-1}\left\{\frac{1}{\Delta t}\right\} \tag{6.201}$$

where for $n > 1$

$$[S]_{n-1} = \int_o^{\Delta t} [S(n\Delta t - \tau)]\frac{\tau}{\Delta t}d\tau + \int_o^{\Delta t} [S((n-1)\Delta t - \tau)]\left(1 - \frac{\tau}{\Delta t}\right)d\tau$$

$$\tag{6.202a}$$

Sec. 6.9  Recursive Evaluation of Convolution Integral in Time Domain

and for $n = 1$

$$[S]_o = \int_o^{\Delta t} [S(\Delta t - \tau)] \frac{\tau}{\Delta t} d\tau \qquad (6.202b)$$

hold. Equation 6.202 is identical to Eq. 6.140. Equation 6.201 (formulated for $n - i$) is substituted on the right-hand side of Eq. 6.199, which results in the recursive expression for the forces as a function of $\{1/\Delta t\}$ (all $\{u\}_{n-i}$ are zero with the exception of $n = i + 1$ for which it equals $\{1/\Delta t\}$). This transformed recursive equation for $\{R\}_n$ is formulated for the first $M + L + 1$ steps and is equated to the rigorous one (Eq. 6.201). This leads for $N$ degrees of freedom to the following system of equations

$$\begin{bmatrix} [I]^{(L+1)N \times (L+1)N} & [G]^{(L+1)N \times MN} \\ [O]^{MN \times (L+1)N} & [H]^{MN \times MN} \end{bmatrix} \cdot \begin{bmatrix} \Delta t[B]^{(L+1)N \times N} \\ [A]^{MN \times N} \end{bmatrix} = \begin{bmatrix} [P]^{(L+1)N \times N} \\ [Q]^{MN \times N} \end{bmatrix}$$

(6.203)

where

$$[G] = \begin{bmatrix} [O] & [O] & \vdots & [O] \\ [S]_o & [O] & \vdots & [O] \\ [S]_1 & [S]_o & \vdots & [O] \\ \cdot & \cdot & & \\ \cdot & \cdot & & \\ \cdot & \cdot & & \\ [S]_{L-1} & [S]_{L-2} & \vdots & [S]_{L-M} \end{bmatrix} \qquad (6.204a)$$

$$[H] = \begin{bmatrix} [S]_L & [S]_{L-1} & \vdots & [S]_{L-M+1} \\ [S]_{L+1} & [S]_L & \vdots & [S]_{L-M+2} \\ [S]_{L+2} & [S]_{L+1} & \vdots & [S]_{L-M+3} \\ \cdot & \cdot & & \cdot \\ \cdot & \cdot & & \cdot \\ \cdot & \cdot & & \cdot \\ [S]_{L+M-1} & [S]_{L+M-2} & \vdots & [S]_L \end{bmatrix} \qquad (6.204b)$$

$$[B] = \begin{bmatrix} [b]_o^T \\ [b]_1^T \\ [b]_2^T \\ \cdot \\ \cdot \\ \cdot \\ [b]_L^T \end{bmatrix} \quad (6.204c) \qquad [A] = \begin{bmatrix} [a]_1^T \\ [a]_2^T \\ [a]_3^T \\ \cdot \\ \cdot \\ \cdot \\ [a]_M^T \end{bmatrix} \quad (6.204d)$$

$$[P] = \begin{bmatrix} [S]_o^T \\ [S]_1^T \\ [S]_2^T \\ \vdots \\ \vdots \\ [S]_L^T \end{bmatrix} \quad (6.204e) \qquad [Q] = \begin{bmatrix} [S]_{L+1}^T \\ [S]_{L+2}^T \\ [S]_{L+3}^T \\ \vdots \\ \vdots \\ [S]_{L+M}^T \end{bmatrix} \quad (6.204f)$$

and with $[I]$ and $[O]$ denoting the unit and zero matrices. The order of the matrices is indicated with superscripts in Eq. 6.203. Due to the special structure of Eq. 6.203, $[A]$ containing the recursive coefficients of the forces is determined independently of the influence of $[B]$ as

$$[A] = [H]^{-1}[Q] \qquad (6.205)$$

$[B]$ then follows as

$$[B] = \frac{1}{\Delta t}([P] - [G][A]) \qquad (6.206)$$

**Special cases.** For $M = L = 1$, Eq. 6.203 with Eq. 6.204 leads to

$$[a]_1 = [S]_2[S]_1^{-1} \qquad (6.207a)$$

$$[b]_o = \frac{1}{\Delta t}[S]_o \qquad (6.207b)$$

$$[b]_1 = \frac{1}{\Delta t}([S]_1 - [S]_2[S]_1^{-1}[S]_o) \qquad (6.207c)$$

In principle, the integrals appearing in Eq. 6.202 can be calculated exactly. From a practical point of view, it can be appropriate to assume that the dynamic-stiffness matrices vary linearly over the $n$th time step

$$[S(\tau)]_n = \left(1 - \frac{\tau}{\Delta t}\right)[S((n-1)\Delta t)] + \frac{\tau}{\Delta t}[S(n\Delta t)] \qquad (6.208)$$

For $M = L = 1$ and for a one-degree-of-freedom system ($N = 1$), the recursive formulation leads to

$$R_n = \frac{S(\Delta t) + 4S(2\Delta t) + S(3\Delta t)}{S(0) + 4S(\Delta t) + S(2\Delta t)} R_{n-1} + \frac{2S(0) + S(\Delta t)}{6} u_n \Delta t + \frac{1}{6}$$
$$\times \left[ S(0) + 4S(\Delta t) + S(2\Delta t) - \frac{S(\Delta t) + 4S(2\Delta t) + S(3\Delta t)}{S(0) + 4S(\Delta t) + S(2\Delta t)}(2S(0) + S(\Delta t)) \right] u_{n-1} \Delta t$$

$$(6.209)$$

### Sec. 6.9 Recursive Evaluation of Convolution Integral in Time Domain

**Optimum agreement of dynamic-stiffness coefficients.** As described in Eq. 6.203, the recursive coefficients follow from equating the dynamic-stiffness coefficients of the exact and recursive formulations at the first $L + M + 1$ time stations. In particular, $[a]_i$ are calculated by enforcing these discretized unit-impulse functions at all time stations between $L + 2$ and $L + M + 1$, that is, at $M$ consecutive time stations. In general, the dynamic-stiffness coefficients at later times will differ. A better overall agreement of the dynamic-stiffness coefficients in the range necessary to evaluate accurately the convolution integral can be achieved by increasing the data set used to calculate $[a]_i$, whereby least-square procedures could be applied. It is possible to construct a rectangular matrix by equating the dynamic-stiffness coefficients at $K$ time stations $(K > M)$, with $(K + L + 1)\Delta t$ representing the total range where a good agreement is necessary for acceptable results. Point collocation can, therefore, be applied by selecting those time stations where the absolute value of the pivot is the largest and then performing a partial inversion. The procedure can best be illustrated with a one-degree-of-freedom system. To process the first unknown, the element with the largest absolute value in the first column of the matrix of order $K \times M$ is selected as the pivot and a partial inversion is performed. Then the second unknown is addressed by applying the same criterion for the pivot to all elements in the second column with the exception of that corresponding to the row that was used for the partial inversion of the first unknown. After performing the corresponding partial inversion, the third unknown is examined, etc. At all time stations corresponding to the rows where a pivot is selected, the two dynamic-stiffness coefficients will agree. After determining the $[b]_i$, the recursive formulation (Eq. 6.199) can be applied to the discretized unit-impulse leading to the approximate dynamic-stiffness coefficients that will, in general, differ at all other time stations from the exact ones.

As an example, the dynamic-stiffness coefficient of the semi-infinite rod on an elastic foundation is addressed. $S(t)$ is specified in Problem 6.9. The non-dimensional regular part $\overline{S}(\bar{t})/(c_l \kappa)$ appearing in the convolution integral equals $J_1(\bar{t})/\bar{t}$ (Fig. P6-9). This exact coefficient is also plotted as a dashed line in Figs. 6-26 and 6-27.

The recursive equation (Eq. 6.199) for $M = L = 3$ formulated in dimensionless variables equals

$$\overline{R}_n = a_1 \overline{R}_{n-1} + a_2 \overline{R}_{n-2} + a_3 \overline{R}_{n-3} + b_0 \overline{u}_n \Delta \bar{t}$$
$$+ b_1 \overline{u}_{n-1} \Delta \bar{t} + b_2 \overline{u}_{n-2} \Delta \bar{t} + b_3 \overline{u}_{n-3} \Delta \bar{t} \quad (6.210)$$

$\Delta \bar{t}$ is selected as 0.1. For $K = 3$—at all the time stations between $\bar{t} = 0.5$ and 0.7 the dynamic-stiffness coefficients are set equal to determine $a_1$ to $a_3$—the following coefficients result

$$a_1 = 2.98626 \qquad a_2 = -2.97757 \qquad a_3 = 0.99128$$

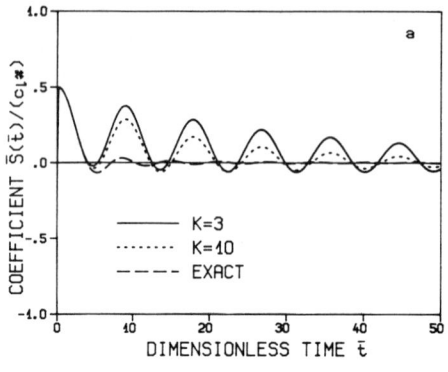

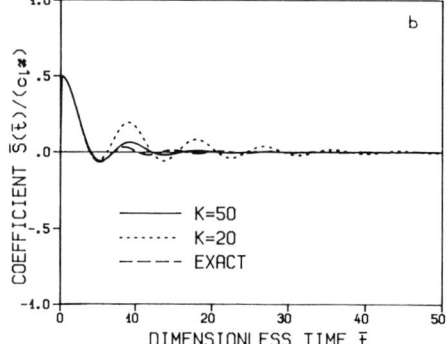

**Figure 6-26** Comparison of regular part of dynamic-stiffness coefficient for increasing data.

The corresponding dynamic-stiffness coefficient shown as a solid line in Fig. 6-26a deviates significantly from the exact value. For $K = 50$, increasing the potential data set to include all time stations between $\bar{t} = 0.5$ and $5.4$ (whereby the time stations 0.5, 2.7, and 4.9 are selected in the algorithm) leads to

$$a_1 = 2.92621 \qquad a_2 = -2.85924 \qquad a_3 = 0.93285$$

As can be seen from the corresponding solid line in Fig. 6-26b, a dramatic improvement results. The computational effort in evaluating the convolution integral is unaffected by increasing the available potential data set. For $K = 10$ (Fig. 6-26a) and $K = 20$ (Fig. 6-26b) the dynamic-stiffness coefficients lie between the values for $K = 3$ and 50.

In general, the number $M$ of recursive coefficients for the forces in Eq. 6.199 is selected before the coefficients $[a]_i$ are calculated. $M$ cannot, however, be chosen to be too large. For instance for $M = L = K = 15$ and for $\Delta t = 0.1$ an instable behavior occurs, as is visible for the dynamic-stiffness coefficient at $\bar{t} = 27$ in Fig. 6-27. This arises as in the partial inversion

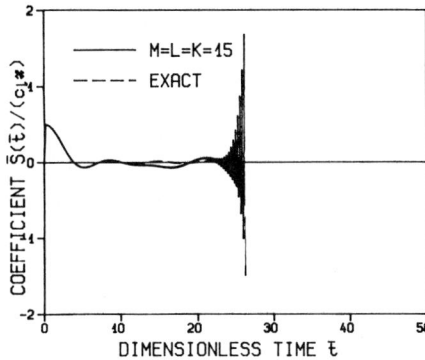

**Figure 6-27** Instable behavior of regular part of dynamic-stiffness coefficient.

process described above, the pivots become very small. The pivots corresponding to $a_i$ in ascending order ($i = 1, 2, \ldots, 15$) are equal to $0.4 \cdot 10^{-1}$, $0.1 \cdot 10^{-2}$, $-0.8 \cdot 10^{-5}$, $0.2 \cdot 10^{-7}$, $-0.8 \cdot 10^{-10}$, $0.7 \cdot 10^{-13}$, and the rest are all $10^{-13}$ or $10^{-14}$. It follows that accurate results can be expected for $M = 4$ or 5, which is confirmed for an actual earthquake excitation in Section 6.9.6. By limiting the pivot to a reasonable size ($10^{-10}$ in this case), the number $M$ can thus be determined in the algorithm and the danger of inaccurate results can be avoided.

Numerical experimentation shows that $L$ should be chosen to be equal to $M$, which is expected intuitively. Increasing $M$ (and $L$) leads to more accurate results. The same tendency occurs for larger $\Delta t$ (if not in contradiction with other constraints concerning the time step). When the dynamic-stiffness coefficients in the time domain are calculated numerically by applying the Fast Fourier Transform, zig-zagging of $[S(t)]$ for small $t$ can arise, when the high-frequency contribution is neglected. This can cause divergence of the solution for small $M$ and $L$.

**Exact recursive equation.** When the dynamic-stiffness coefficient in the frequency domain is a ratio of two polynomials, the appropriately chosen recursive equation leads to the rigorous solution. Factors appearing in the numerator and denominator must be canceled before determining $M$ and $L$. $M$ is equal to the degree of the polynomial in $\omega$ (or to be more precise, in $i\omega$) of the denominator, and $L$ is equal to that of the numerator. When a linear variation of the displacement in each time step is assumed, $L$ is increased by one, that is, $L - 1$ is equal to the degree of the numerator (see Eq. 6.226, and compare Eq. 6.227 to Eq. 6.225). The proof of this remarkable result, which involves $z$-transforms, follows as a by-product of the procedure addressed in Section 6.9.4, which determines the recursive equation directly from the dynamic-stiffness coefficient in the frequency domain—from the frequency-response function.

### 6.9.3 Segment Approach Based on z-Transform

**Concept.** While the impulse-invariant method provides a clear physical insight into the procedure that is used to derive the recursive equation, it does suffer from two disadvantages. First, by equating the approximate dynamic-stiffness coefficient corresponding to the recursive equation to the exact one at only a few distinct time stations, no control over the agreement of the two stiffness coefficients at other time stations exists before the recursive coefficients are determined. Second, a system of equations has to be solved to calculate the recursive coefficients, which for a multiple-degree-of-freedom system especially is a formidable task. The segment approach avoids these two disadvantages, but the storage requirement is somewhat larger than in the impulse-invariant method (see Section 6.9.5).

In the segment approach, the dynamic-stiffness coefficients in the time domain $[S(t)]$ (unit-impulse functions) are divided into (time) segments (Fig. 6-28). In each segment, each coefficient of $[S(t)]$ is approximated by the product of a polynomial and an exponential function with a negative exponent. By selecting many segments, a large range, in the limit the whole range, of $[S(t)]$ can be approximated, whereby the deviation from the exact values can be influenced and is known before the recursive coefficients are determined. It is also possible to determine the recursive equation corresponding to this approximation rigorously without solving a system of equations, which also avoids any potential numerical problems. Such an explicit formulation is especially important for the case of a system with many degrees of freedom. It turns out that by selecting the same type of interpolation functions for all degrees of freedom and for all segments results in the same diagonal $[a]_i$-matrix, which is independent of the segment. This property further reduces the computational effort. The $[b]_i$-matrix of Eq. 6.199 will, in general, be fully populated.

The recursive equation is derived concisely using the so-called z-transformation. Only fundamentals of this procedure, which is related to the discrete Fourier transformation, are needed to be able to follow the deriva-

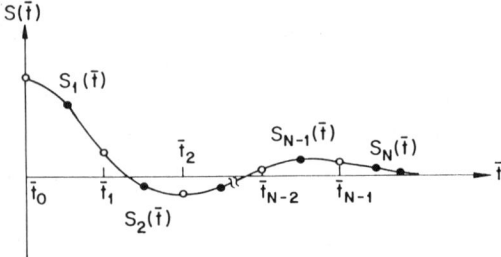

**Figure 6-28** Subdivision of regular part of dynamic-stiffness coefficient into segments.

### Sec. 6.9 Recursive Evaluation of Convolution Integral in Time Domain

tion. The reader not familiar with the $z$-transformation can either consult elementary text books such as [B3, 01] or disregard the derivation, addressing directly the final recursive equation.

**Derivation.** $[S(\bar{t})]$ is approximated in each segment using second-degree polynomials as follows. The subdivision into $N$ segments for a dynamic-stiffness coefficient $S(\bar{t})$ is illustrated in Fig. 6-28.

$$[S(\bar{t})] = \begin{matrix} ([c_o]_1 + [c_1]_1 \bar{t} + [c_2]_1 \bar{t}^2) \exp(-\bar{t}) & \bar{t}_o = 0 < \bar{t} < \bar{t}_1 \\ ([c_o]_2 + [c_1]_2 \bar{t} + [c_2]_2 \bar{t}^2) \exp(-\bar{t}) & \bar{t}_1 < \bar{t} < \bar{t}_2 \\ ([c_o]_i + [c_1]_i \bar{t} + [c_2]_i \bar{t}^2) \exp(-\bar{t}) & \bar{t}_{i-1} < \bar{t} < \bar{t}_i \\ ([c_o]_N + [c_1]_N \bar{t} + [c_2]_N \bar{t}^2) \exp(-\bar{t}) & \bar{t}_{N-1} < \bar{t} < \bar{t}_N = \infty \end{matrix}$$

(6.211)

$\bar{t}$ is the dimensionless time (Eq. 6.36) with $\bar{t}_i$ denoting the time at the end of the $i$th segment, $N$ is the number of segments, whereby the $N$th segment extends to infinity (the number of bounded segments is thus $N - 1$), $[c_j]_i$ are the coefficients of the $i$th segment ($j = 0, 1, 2$) determined from the values at the beginning, center and end of each segment with the exception of the $N$th one where besides the beginning, two arbitrary time stations are selected. Equation 6.211 can be written by introducing Heaviside step functions $H$ as

$$[S(\bar{t})] = \sum_{i=1}^{N} ([c_o]_i + [c_1]_i \bar{t} + [c_2]_i \bar{t}^2) \exp(-\bar{t})(H(\bar{t} - \bar{t}_{i-1}) - H(\bar{t} - \bar{t}_i))$$

(6.212)

or

$$[S(\bar{t})] = ([c_o]_1 + [c_1]_1 \bar{t} + [c_2]_1 \bar{t}^2) \exp(-\bar{t}) H(\bar{t})$$

$$+ \sum_{i=2}^{N} ([c_o]_i - [c_o]_{i-1} + ([c_1]_i - [c_1]_{i-1}) \bar{t} \quad (6.213)$$

$$+ ([c_2]_i - [c_2]_{i-1}) \bar{t}^2) \exp(-\bar{t}) H(\bar{t} - \bar{t}_{i-1})$$

The first term on the right-hand side of Eq. 6.213 can be incorporated into the sum defining $[c_j]_o = 0$. The sum then starts at $i = 1$. Expressing the polynomial and exponential functions in $\bar{t} - \bar{t}_{i-1}$ instead of in $\bar{t}$ results in

$$[S(\bar{t})] = \sum_{i=1}^{N} \exp(-\bar{t}_{i-1})([d_o]_i + [d_1]_i (\bar{t} - \bar{t}_{i-1}) + [d_2]_i (\bar{t} - \bar{t}_{i-1})^2)$$

$$\times \exp[-(\bar{t} - \bar{t}_{i-1})] H(\bar{t} - \bar{t}_{i-1})$$

(6.214)

with

$$[d_o]_i = [c_o]_i - [c_o]_{i-1} + ([c_1]_i - [c_1]_{i-1}) \bar{t}_{i-1} + ([c_2]_i - [c_2]_{i-1}) \bar{t}_{i-1}^2 \quad (6.215a)$$

$$[d_1]_i = [c_1]_i - [c_1]_{i-1} + 2([c_2]_i - [c_2]_{i-1})\bar{t}_{i-1} \qquad (6.215b)$$

$$[d_2]_i = [c_2]_i - [c_2]_{i-1} \qquad (6.215c)$$

To derive the corresponding recursive equation for the total system corresponding to Eq. 6.214, it is sufficient to examine one element in one term of the sum. Shifting the sequence by $\bar{t}_{i-1}$ and disregarding a constant factor, the latter is formulated as

$$h(\bar{t}) = (d_o + d_1\bar{t} + d_2\bar{t}^2)\exp(-\bar{t}) \qquad (6.216)$$

To be able to apply the $z$-transformation, Eq. 6.216 is discretized with the values selected at the middle of the time steps, resulting in

$$h(n) = \left(d_o + d_1\left(\frac{2n+1}{2}\right)\Delta\bar{t} + d_2\left(\frac{2n+1}{2}\right)^2\Delta\bar{t}^2\right)\exp\left(-\frac{2n+1}{2}\Delta\bar{t}\right)$$

$$n = 0, 1, \ldots \qquad (6.217)$$

The right-sided $z$-transform equals

$$H(z) = \sum_{n=0}^{\infty} h(n) z^{-n} \qquad (6.218)$$

The infinite sums corresponding to $d_o$, $d_1$, and $d_2$ are evaluated separately.

$$H(z) = H_o(z) + H_1(z) + H_2(z) \qquad (6.219)$$

where

$$H_o(z) = \sum_{n=0}^{\infty} d_o \exp\left(-\frac{2n+1}{2}\Delta\bar{t}\right) z^{-n} = d_o \exp\left(-\frac{\Delta\bar{t}}{2}\right) \sum_{n=0}^{\infty} \exp(-n\Delta\bar{t}) z^{-n}$$

$$= d_o \frac{\exp(-\Delta\bar{t}/2)}{1 - \exp(-\Delta\bar{t})z^{-1}}$$

$$(6.220)$$

$$H_1(z) = \sum_{n=0}^{\infty} d_1 \frac{2n+1}{2}\Delta\bar{t} \exp\left(-\frac{2n+1}{2}\Delta\bar{t}\right) z^{-n}$$

$$= d_1 \frac{\Delta\bar{t}}{2}\exp\left(-\frac{\Delta\bar{t}}{2}\right)\sum_{n=0}^{\infty}\exp(-n\Delta\bar{t})z^{-n} + d_1\Delta\bar{t}\exp\left(-\frac{\Delta\bar{t}}{2}\right)\sum_{n=0}^{\infty}n\exp(-n\Delta\bar{t})z^{-n}$$

$$= d_1 \frac{\frac{\Delta\bar{t}}{2}\exp\left(-\frac{\Delta\bar{t}}{2}\right)(1 + \exp(-\Delta\bar{t})z^{-1})}{(1 - \exp(-\Delta\bar{t})z^{-1})^2}$$

$$(6.221)$$

Sec. 6.9  Recursive Evaluation of Convolution Integral in Time Domain

$$H_2(z) = \sum_{n=0}^{\infty} d_2 \left(\frac{2n+1}{2}\Delta\bar{t}\right)^2 \exp\left(-\frac{2n+1}{2}\Delta\bar{t}\right) z^{-n}$$

$$= d_2 \frac{\Delta\bar{t}^2 \exp\left(-\frac{\Delta\bar{t}}{2}\right)\left(\frac{1}{4} + \frac{3}{2}\exp(-\Delta\bar{t})z^{-1} + \frac{1}{4}\exp(-2\Delta\bar{t})z^{-2}\right)}{(1 - \exp(-\Delta\bar{t})z^{-1})^3}$$

(6.222)

This leads to

$$H(z) = \frac{\left(d_o + \frac{\Delta\bar{t}}{2}d_1 + \frac{\Delta\bar{t}^2}{4}d_2\right)\exp\left(-\frac{\Delta\bar{t}}{2}\right)}{1 - 3\exp(-\Delta\bar{t})z^{-1} + 3\exp(-2\Delta\bar{t})z^{-2} - \exp(-3\Delta\bar{t})z^{-3}}$$

$$+ \frac{\left(-2d_o + \frac{3\Delta\bar{t}^2}{2}d_2\right)\exp\left(-\frac{3\Delta\bar{t}}{2}\right)z^{-1} + \left(d_o - \frac{\Delta\bar{t}}{2}d_1 + \frac{\Delta\bar{t}^2}{4}d_2\right)\exp\left(-\frac{5\Delta\bar{t}}{2}\right)z^{-2}}{1 - 3\exp(-\Delta\bar{t})z^{-1} + 3\exp(-2\Delta\bar{t})z^{-2} - \exp(-3\Delta\bar{t})z^{-3}}$$

(6.223)

Formulating the force-displacement relationship in the $z$-domain

$$P(z) = H(z)\,U(z) \tag{6.224}$$

and performing the inverse $z$-transformation by identifying the coefficients of the terms involving $z^{-n}$ in the power series in the numerator and denominator results in

$$P_n = 3\exp(-\Delta\bar{t})\,P_{n-1} - 3\exp(-2\Delta\bar{t})\,P_{n-2}$$

$$+ \exp(-3\Delta\bar{t})\,P_{n-3} + \left(d_o + \frac{\Delta\bar{t}}{2}d_1 + \frac{\Delta\bar{t}^2}{4}d_2\right)\exp\left(-\frac{\Delta\bar{t}}{2}\right) U(n)$$

$$+ \left(-2d_o + \frac{3\Delta\bar{t}^2}{2}d_2\right)\exp\left(-\frac{3\Delta\bar{t}}{2}\right) U(n-1)$$

$$+ \left(d_o - \frac{\Delta\bar{t}}{2}d_1 + \frac{\Delta\bar{t}^2}{4}d_2\right)\exp\left(-\frac{5\Delta\bar{t}}{2}\right) U(n-2)$$

(6.225)

For a piecewise linear variation of the displacements over each time step

$$U(n) = \frac{\Delta\bar{t}}{2}(u_{n-1} + u_n) \tag{6.226}$$

is substituted, leading to

$$P_n = 3\exp(-\Delta \bar{t}) P_{n-1} - 3\exp(-2\Delta \bar{t}) P_{n-2} + \exp(-3\Delta \bar{t}) P_{n-3}$$

$$+ \left(\frac{d_o}{2} + \frac{\Delta \bar{t}}{4} d_1 + \frac{\Delta \bar{t}^2}{8} d_2\right) \exp\left(-\frac{\Delta \bar{t}}{2}\right) u_n \Delta \bar{t}$$

$$+ \left(\frac{d_o}{2} - d_o \exp(-\Delta \bar{t}) + \frac{\Delta \bar{t}}{4} d_1 + \frac{\Delta \bar{t}^2}{8} d_2 + \frac{3}{4}\Delta \bar{t}^2 d_2 \exp(-\Delta \bar{t})\right) \exp\left(-\frac{\Delta \bar{t}}{2}\right) u_{n-1} \Delta \bar{t}$$

$$+ \left(-d_o + \frac{d_o}{2} \exp(-\Delta \bar{t}) - \frac{\Delta \bar{t}}{4} d_1 \exp(-\Delta \bar{t}) + \frac{3}{4}\Delta \bar{t}^2 d_2 + \frac{\Delta \bar{t}^2}{8} d_2 \exp(-\Delta \bar{t})\right)$$

$$\times \exp\left(-\frac{3\Delta \bar{t}}{2}\right) u_{n-2} \Delta \bar{t} + \left(\frac{d_o}{2} - \frac{\Delta \bar{t}}{4} d_1 + \frac{\Delta \bar{t}^2}{8} d_2\right) \exp\left(-\frac{5\Delta \bar{t}}{2}\right) u_{n-3} \Delta \bar{t}$$

(6.227)

The terms appearing in Eq. 6.214 can be constructed from Eq. 6.216 by replacing $\bar{t}$ by $\bar{t} - \bar{t}_{i-1}$, which corresponds to shifting the sequence $h(n)$ of Eq. 6.217 by $-\bar{t}_{i-1}/\Delta \bar{t} = -n_{i-1}$. The corresponding z-transform of $h(n - n_{i-1})$ equals $z^{-n_{i-1}} H(z)$ with $H(z)$ specified in Eq. 6.223. The displacements (input) associated with the numerator of the inverse z-transform are thus shifted by $-n_{i-1}$. Taking the factor $\exp(-\bar{t}_{i-1})$ also into account transforms Eq. 6.214 to the final recursive equation

$$\{R\}_n = 3\exp(-\Delta \bar{t}) [I] \{R\}_{n-1} - 3\exp(-2\Delta \bar{t}) [I] \{R\}_{n-2}$$

$$+ \exp(-3\Delta \bar{t}) [I] \{R\}_{n-3} + \Delta \bar{t} \sum_{i=1}^{N} \exp(-\bar{t}_{i-1}) \sum_{j=0}^{3} [b_i]_j \{u\}_{n-j-n_{i-1}} \quad (6.228)$$

where

$$[b_i]_0 = \left(\frac{[d_o]_i}{2} + \frac{\Delta \bar{t}}{4}[d_1]_i + \frac{\Delta \bar{t}^2}{8}[d_2]_i\right) \exp\left(-\frac{\Delta \bar{t}}{2}\right) \quad (6.229a)$$

$$[b_i]_1 = \left(\frac{[d_o]_i}{2} - [d_o]_i \exp(-\Delta \bar{t}) + \frac{\Delta \bar{t}}{4}[d_1]_i + \frac{\Delta \bar{t}^2}{8}[d_2]_i\right.$$

$$\left. + \frac{3}{4}\Delta \bar{t}^2 [d_2]_i \exp(-\Delta \bar{t})\right) \exp\left(-\frac{\Delta \bar{t}}{2}\right) \quad (6.229b)$$

$$[b_i]_2 = \left(-[d_o]_i + \frac{[d_o]_i}{2}\exp(-\Delta \bar{t}) - \frac{\Delta \bar{t}}{4}[d_1]_i \exp(-\Delta \bar{t}) + \frac{3}{4}\Delta \bar{t}^2 [d_2]_i\right.$$

$$\left. + \frac{\Delta \bar{t}^2}{8}[d_2]_i \exp(-\Delta \bar{t})\right) \exp\left(-\frac{3\Delta \bar{t}}{2}\right) \quad (6.229c)$$

### Sec. 6.9  Recursive Evaluation of Convolution Integral in Time Domain

$$[b_i]_3 = \left( \frac{[d_o]_i}{2} - \frac{\overline{\Delta t}}{4}[d_1]_i + \frac{\overline{\Delta t}^2}{8}[d_2]_i \right) \exp\left( -\frac{5\overline{\Delta t}}{2} \right) \quad (6.229d)$$

As the recursive coefficients of the forces—3 $\exp(-\overline{\Delta t})$, etc.—are independent of the coefficients $d_o$, $d_1$, $d_2$ of the polynomials, the coefficients of the contributions of the various degrees of freedom to a specific degree of freedom are the same, resulting in a diagonal matrix for the recursive coefficients of the forces, that is, 3 $\exp(-\overline{\Delta t})[I]$, etc.

As in the derivation analytical $z$-transforms are used, the recursive equation (Eq. 6.228) is rigorous for the piecewise interpolation function expressed in Eq. 6.211. Instead of using second-order polynomials, other interpolation schemes whose $z$-transforms can be calculated analytically could be selected.

Finally, the region of convergence of the $z$-transform is addressed. The poles of $H(z)$, which correspond to the roots of the denominator (Eq. 6.223), appear at $1 - \exp(-\overline{\Delta t})z^{-1} = 0$, that is, at $z = \exp(-\overline{\Delta t})$. As this value is always smaller than 1, the series is absolutely convergent.

Note, that the recursive evaluation in the time domain based on the segment approach and the recursive evaluation in the frequency domain calculated at selected frequencies use interpolation in their respective domains.

#### 6.9.4 Recursive Equation Directly from Dynamic-Stiffness Coefficient in Frequency Domain

**Derivation.** It is an attractive procedure to derive the recursive equation starting from the dynamic-stiffness coefficient in the frequency domain (frequency-response function). Each coefficient of the dynamic-stiffness matrix $[S(\omega)]$ is treated separately.

The dimensionless forms of frequency $a_o$ and time $\bar{t}$ are used in the derivation. Using a curve-fitting technique based on the least square method (see below), the regular part $S(a_o)$ is approximated as a ratio of two polynomials in $ia_o$

$$S(a_o) \simeq S(ia_o) = \frac{P(ia_o)}{Q(ia_o)}$$

$$= K \frac{1 + p_1(ia_o) + p_2(ia_o)^2 + \ldots + p_{M-1}(ia_o)^{M-1}}{1 + q_1(ia_o) + q_2(ia_o)^2 + \ldots + q_M(ia_o)^M} \quad (6.230)$$

$K$ denotes the static-stiffness coefficient. The two constant coefficients of $P$ and $Q$ are both 1. The order of the numerator is selected as one less than that of the denominator. The coefficients $p_i$, $q_i$ are real.

The ratio of the polynomials can be expressed in a partial-fraction expansion of the form

$$\sum_{i=1}^{M} \frac{A_i}{ia_o - s_i} \tag{6.231}$$

where $s_i$ are the roots of $Q(ia_o)$, that is, the poles of $S(ia_o)$, and the $A_i$'s are the residues at the poles

$$A_i = (ia_o - s_i) S(ia_o)|_{ia_o = s_i} \tag{6.232}$$

Equation 6.231 applies only for first-order poles. If $S(ia_o)$ has multiple-order poles, the corresponding term (for a pole at $s_j$ of order $l$) is replaced by

$$\sum_{i=1}^{l} \frac{A_i}{(ia_o - s_j)^i} \tag{6.233}$$

with

$$A_i = \frac{1}{(l-i)!} \frac{d^{l-i}}{dia_o^{l-i}} [ia_o - s_j]^l S(ia_o)|_{ia_o = s_j} \tag{6.234}$$

As multiple poles do not modify the derivation significantly, they are not addressed any further. See also Eqs. 6.216 to 6.225 for expressions of the z-transform in the presence of multiple poles.

The Fourier transform of Eq. 6.231 equals

$$S(\bar{t}) = \sum_{i=1}^{M} A_i \exp(s_i \bar{t}) \quad (\bar{t} \geq 0) \tag{6.235}$$

Discretizing $S(\bar{t})$ at the middle of the time step ($n = 0, 1, \ldots$)

$$S(n) = \sum_{i=1}^{M} A_i \exp\left[s_i \left(\frac{2n+1}{2}\right) \Delta \bar{t}\right] \tag{6.236}$$

and applying the right-sided z-transformation results in (see also Eq. 6.220)

$$S(z) = \sum_{i=1}^{M} \sum_{n=0}^{\infty} S(n) z^{-n} = \sum_{i=1}^{M} \frac{A_i \exp(s_i \Delta \bar{t}/2)}{1 - \exp(s_i \Delta \bar{t}) z^{-1}} \tag{6.237}$$

The term $\exp(s_i \Delta \bar{t})$ is the pole of $S(z)$. For stability, $|\exp(s_i \Delta \bar{t})| < 1$ is required. Equation 6.237 can be written in the form

$$S(z) = \frac{b_0 + b_1 z^{-1} + b_2 z^{-2} + \ldots + b_{M-1} z^{-(M-1)}}{1 - a_1 z^{-1} - a_2 z^{-2} - \ldots - a_M z^{-M}} \tag{6.238}$$

Performing the inverse z-transformation by identifying the coefficients of the terms involving $z^{-n}$ in the numerator and in the denominator leads to the recursive equation

$$R_n = \sum_{i=1}^{M} a_i R_{n-i} + \sum_{i=0}^{M-1} b_i U(n-i) \tag{6.239}$$

Substituting for a piecewise linear variation of the displacements over each time step

$$U(n) = \frac{\overline{\Delta t}}{2}(u_{n-1} + u_n) \qquad (6.240)$$

in Eq. 6.239 leads to the final recursive equation in the so-called direct form.

Alternatively, the Eq. 6.237 can be processed in the so-called parallel form as follows, leading to a simple explicit expression. Addressing each term identified by $i$ separately and performing the corresponding inverse $z$-transformation results in

$$R_n^i = \exp(s_i \overline{\Delta t}) R_{n-1}^i + A_i \exp\left(s_i \frac{\overline{\Delta t}}{2}\right) U(n) \qquad (6.241)$$

where $R_n^i$ will either be real or appear in complex conjugate pairs. Substituting Eq. 6.240 transforms Eq. 6.241 to

$$R_n^i = \exp(s_i \overline{\Delta t}) R_{n-1}^i + \frac{\overline{\Delta t}}{2} \exp\left(s_i \frac{\overline{\Delta t}}{2}\right)(u_n + u_{n-1}) \qquad (6.242)$$

The total (real) force $R_n$ follows from

$$R_n = \sum_{i=1}^{M} R_n^i \qquad (6.243)$$

The direct (Eq. 6.239) and parallel forms (Eq. 6.241) derived from the same $z$-transform (Eq. 6.237) correspond to the same recursive equation. For infinite bit precision of the word length, the same numerical results will arise. For a finite word length of the recursive coefficients and of the variables, which depends on the hardware and the software implemented, differences will exist. Especially when the poles of $S(z)$, $\exp(s_i \Delta t)$, are close to the unit circle, a sensitivity problem exists for the direct method.

The only approximation introduced in this procedure consists of replacing $S(a_o)$ by a ratio of two polynomials. The recursive equation in the form of Eq. 6.239 is a rigorous expression for the ratio of the two polynomials, as only analytical $z$-transformations are performed in its derivation. The above derivation serves thus also as the proof referred to at the end of Section 6.9.2.

**Least-square method.** $2M - 1$ unknown coefficients $p_i$, $q_i$ in Eq. 6.230 have to be determined using a curve-fitting technique with a complex $S(a_o)$ based on the least-square method. Denoting the exact (complex) value of the dynamic-stiffness coefficient discretized in the $j$th data point at $a_{oj}$ as $S(a_{oj})$, the square of the following Euclidean norm $\epsilon$ is made a minimum.

$$\epsilon^2 = \|S(a_{oj})Q(ia_{oj}) - P(ia_{oj})\|^2 \qquad (6.244)$$

where

$$\epsilon^2 = \sum_{j=1}^{J} |S(a_{oj})Q(ia_{oj}) - P(ia_{oj})|^2$$

$$= \sum_{j=1}^{J} (S(a_{oj})Q(ia_{oj}) - P(ia_{oj}))(S^*(a_{oj})Q^*(ia_{oj}) - P^*(ia_{oj}))$$

(6.245)

$J$ is the number of data points (frequencies) used in the curve-fitting process. An asterisk as superscript denotes the conjugate complex value. $\epsilon^2$ is a real quadratic form in $p_i$, $q_i$. For a minimum of $\epsilon^2$,

$$\frac{\partial \epsilon^2}{\partial p_i} = 0 \quad i = 1, 2, \ldots M - 1 \quad (6.246a)$$

$$\frac{\partial \epsilon^2}{\partial q_i} = 0 \quad i = 1, 2, \ldots M \quad (6.246b)$$

applies, which leads to a system of $2M - 1$ linear equations with real coefficients for the $2M - 1$ real unknowns $p_i$, $q_i$.

**Example**

The dynamic-stiffness coefficient of the semi-infinite rod on an elastic foundation derived in Section 3.2.4 is discussed for the sake of illustration. This is a stringent test, as the real and imaginary parts of the total stiffness coefficient are zero above and below the cutoff frequency, respectively. The regular part (see Problem 6.9), non-dimensionalized with its static value, $i(\sqrt{a_o^2 - 1} - a_o)$, is plotted as a dashed line in Figs. 6-29 and 6-31.

$M$ is first selected equal to 8; $S(a_o)$ is approximated as a ratio of a 7th degree and a 8th degree polynomial (Eq. 6.230). Fifty equally spaced data points are used for the curve fitting, covering the frequency range up to $a_o = 3$ ($J = 50$, $\Delta a_o = 0.06$). Applying the least-square method results in coefficients $p_i$, $q_i$ of the same sign and order of magnitude. As can be seen from the comparison of the real and imaginary parts of the dynamic-stiffness coefficients in the frequency domain in Fig. 6-29, the agreement is excellent. The curves coincide from a practical point of view. This also applies for the regular part of the dynamic-stiffness coefficients in the time domain (Fig. 6-30).

The first two poles $s_i$ (Eq. 6.231) are real, the remaining six are complex conjugates. The real part is negative, as is necessary for a stable system (argument of exponential function in Eq. 6.235). The poles of the z-transform $\exp(s_i \Delta \bar{t})$ lie just slightly inside the unit circle in the z-domain (for example, for $\Delta \bar{t} = 0.1$, $\exp(s_1 \Delta \bar{t}) = 0.99995$, $\exp(s_7 \Delta \bar{t}) = 0.99054 + i \cdot 0.09809$, calculated with double precision), which still guarantees stability.

Selecting $M = 5$ leads to less accurate, but still acceptable results. The comparison for the regular part of the dynamic-stiffness coefficient is shown in the frequency and time domains in Figs. 6-31 and 6-32, keeping the other parameters of the least-square method unchanged.

Sec. 6.9  Recursive Evaluation of Convolution Integral in Time Domain        319

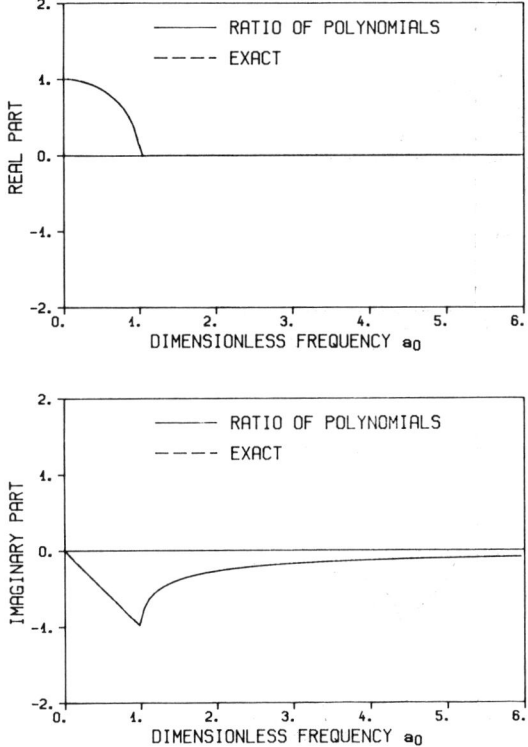

**Figure 6-29** Regular part of dynamic-stiffness coefficient in frequency domain using ratio of polynomials (8th degree in denominator).

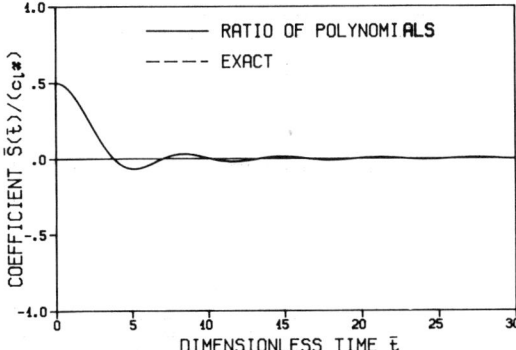

**Figure 6-30** Regular part of dynamic-stiffness coefficient in time domain using ratio of polynomials (8th degree in denominator).

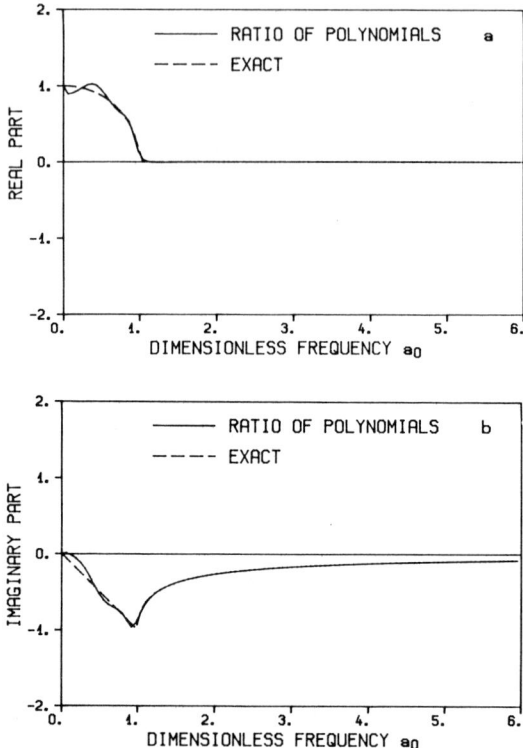

**Figure 6-31** Regular part of dynamic-stiffness coefficient in frequency domain using ratio of polynomials (5th degree in denominator).

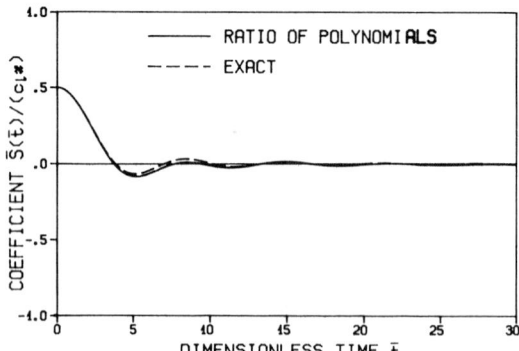

**Figure 6-32** Regular part of dynamic-stiffness coefficient in time domain using ratio of polynomials (5th degree in denominator).

Finally, the recursive equation determined by the impulse-invariant method with $M = L = K = 15$, which leads to instable behavior (Fig. 6-27), is addressed further. Starting from this recursive equation (such as Eq. 6.239) it is possible to construct the corresponding $z$-transform (Eq. 6.238), which is a ratio of two polynomials in $z^{-1}$. The roots of the polynomials in the numerator and in the denominator can be determined. Some roots are identical in the numerator and denominator. After canceling the corresponding factors (in this case 7), a $z$-transform associated with $M = L = 8$ results. The corresponding recursive equation does not exhibit instable behavior.

### 6.9.5 Number of Operations

The number of operations in evaluating the convolution integrals in the various methods is determined as follows. $N^t$ denotes the number of time steps, $N^s$ the number of boundary elements introduced in the spatial discretization (with 3 degrees of freedom per element).

The computational effort of the direct evaluation of the convolution integral (Section 6.6) is discussed in Section 6.8.5 (Eq. 6.194).

The impulse-invariant method (Eq. 6.199) leads to the following number of operations (direct form)

$$N^t(M + L + 1)(3N^s)^2 \qquad (6.247)$$

In the segment approach with $N$ segments (of which $N - 1$ are bounded) and $n_i$ time steps in the $i$th segment (Eq. 6.228)

$$N^t 3(3N^s) + \sum_{i=1}^{N-1} 4(N^t - n_{i-1})(3N^s)^2 \qquad (6.248)$$

operations result. If it is assumed that the number of time steps ($= t_1/\Delta t$) in each segment is the same, this expression can be formulated as

$$N^t 3(3N^s) + [N^t 4(N-1) - 2(N-1)(N-2) t_1/\Delta t](3N^s)^2 \qquad (6.249)$$

As an example, the same case is examined as for the recursive evaluation in the frequency domain in Section 6.8.5. An earthquake time history of 30 s duration with a $\Delta t = 0.005$ s is addressed, resulting in $N^t = 6000$. For twenty 3-dimensional boundary elements ($N^s = 20$) the number of operations for the three methods is compared in Table 6-3. For the segment approach $N$ and $t_1/\Delta t$ are varied as indicated. For one segment, a dramatic reduction of almost three orders of magnitude occurs. The storage requirement ($= (t_{N-1}/\Delta t + 7)3N^s$) of the segment approach is reduced compared to that of the direct evaluation, but is larger than that of the impulse-invariant method ($= (M + L + 1)3N^s$), which also leads to a significant reduction in the number of operations. This recursive evaluation in the time domain is much more efficient than that in the frequency domain shown in Table 6-2. The resulting efficiency depends on the type of problem to be solved.

Finally, the number of operations for the standard direct (non-recursive)

**TABLE 6-3** Number of Operations for Evaluation of Convolution Integrals

| | Impulse-Invariant Method $M = L$ = 3 | | Segment Approach | | | |
|---|---|---|---|---|---|---|
| Direct Evaluation | | $N - 1$ \ $\frac{t_1}{\Delta t}$ | 60 (0.01 $N^t$) | 150 (0.025 $N^t$) | 300 (0.05 $N^t$) | 600 (0.1 $N^t$) |
| $6.5 \cdot 10^{10}$ | $1.5 \cdot 10^8$ | 1 | $8.7 \cdot 10^7$ | $8.7 \cdot 10^7$ | $8.7 \cdot 10^7$ | $8.7 \cdot 10^7$ |
| | | 2 | $1.73 \cdot 10^8$ | $1.72 \cdot 10^8$ | $1.70 \cdot 10^8$ | $1.65 \cdot 10^8$ |
| | | 3 | $2.58 \cdot 10^8$ | $2.54 \cdot 10^8$ | $2.47 \cdot 10^8$ | $2.34 \cdot 10^8$ |
| | | 4 | $3.42 \cdot 10^8$ | $3.34 \cdot 10^8$ | $3.21 \cdot 10^8$ | $2.95 \cdot 10^8$ |

procedure working in the frequency domain is discussed. The Fourier transformation from $\{u(t)\}$ to $\{u(\omega)\}$, the calculation of $\{R(\omega)\}$ and the inverse transformation resulting in $\{R(t)\}$ lead for the contribution of the soil to

$$N^t \ln(2N^t) \, 4(3N^s) + N^t \, 4(3N^s)^2 \tag{6.250}$$

operations. It is assumed that the Fast Fourier Transform is applied and that these operations are performed for $N^t$ frequencies. For the example ($N^t = 6000$, $N^s = 20$), $10^8$ operations result, which is about the same as for the recursive evaluation in the time domain.

### 6.9.6 Semi-Infinite Rod on Elastic Foundation

**Benchmark problem.** The benchmark problem used to examine the performance of the approximate local boundary conditions in the direct method (Chapter 3) is addressed again. The undamped rod with area $A$, modulus of elasticity $E$, mass density $\rho$ and spring coefficient per unit length $k_g$ is subjected to a prescribed support movement at point 0 (Fig. 3-4a)

$$u_o(\bar{t}) = \frac{u_o}{2}\left[1 - \cos\left(2\pi \frac{\bar{t}}{\bar{t}_o}\right)\right], \, 0 < \bar{t} < \bar{t}_o$$

$$u_o(\bar{t}) = 0, \, \bar{t} > \bar{t}_o \tag{6.251}$$

The corresponding dimensionless time $\bar{t}_o = t_o c_l \kappa$ is selected as equal to 2. $c_l$ and $\kappa$ are specified in Eqs. 3.56b and 3.56a.

Using the results of Problems 6.9 and 6.11, the dimensionless reaction force $\bar{P}_o(\bar{t}) = P_o(\bar{t})/(Ku_o)$ equals

$$\bar{P}_o(\bar{t}) = \frac{d\bar{u}_o(\bar{t})}{d\bar{t}} + \int_o^{\bar{t}} \frac{1}{\bar{t} - \bar{\tau}} J_1(\bar{t} - \bar{\tau}) \bar{u}_o(\bar{\tau}) \, d\bar{\tau} \tag{6.252}$$

with $\bar{u}_o(\bar{t}) = u_o(\bar{t})/u_o$ and with the static-stiffness coefficient $K$ (Eq. 3.79). The second term on the right-hand side of Eq. 6.252 represents the regular part of the reaction force, consisting of a convolution integral of the regular part of the dynamic-stiffness coefficient $J_1(\bar{t})/\bar{t}$ and of the displacement $\bar{u}_o(\bar{t})$.

Sec. 6.9    Recursive Evaluation of Convolution Integral in Time Domain          323

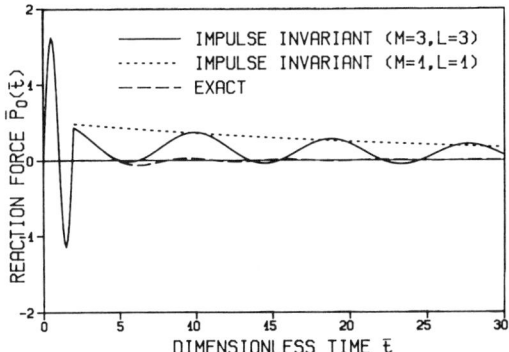

**Figure 6-33**  Varying $M$ and $L$ of impulse-invariant method ($\bar{\Delta t} = 0.1$).

$\bar{P}_o(\bar{t})$ determined directly from Eq. 6.252 is plotted as a dashed line in the following figures.

The impulse-invariant method is examined first, whereby $K$ is selected equal to $M$—the data set used to calculate the recursive coefficients $a_i$ is the smallest possible. With $\bar{\Delta t} = 0.1$, the results are plotted for $M = L = 1$ and $M = L = 3$ in Fig. 6-33. As expected, selecting a larger number for $M$ and $L$ in the impulse-invariant method improves the accuracy. A better agreement also occurs when for a fixed $M = L = 3$ a larger $\bar{\Delta t}$ is used, as is visible from the comparison for $\bar{\Delta t} = 0.1$ and 0.4 in Fig. 6-34. The result of the impulse-invariant method with $M = L = 3$ is compared to that of the segment approach with only one bounded segment and with $\bar{t}_1 = 1.4$, which leads to the same computational effort, in Fig. 6-35 ($\bar{\Delta t} = 0.2$). As the segment approach adds a decaying tail for the range $\bar{t} > 1.4$, the accuracy is superior to that of the impulse-invariant method. Both incorporate the dynamic-stiffness coefficient up to $\bar{t} = 1.4$. As expected, increasing the number of segments improves the accuracy, as can be seen from the results for 1 and 4 bounded segments in Fig. 6-36 ($\bar{\Delta t} = 0.1$, $\bar{t}_1 = 1.4$, $\bar{t}_4 = 5.6$).

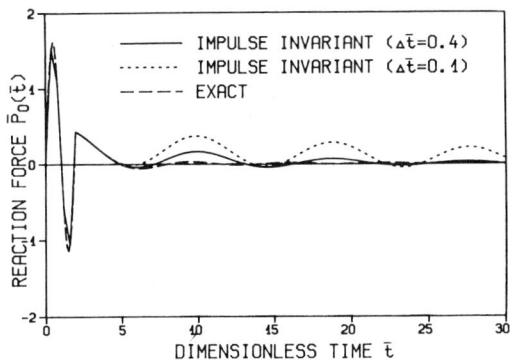

**Figure 6-34**  Varying $\bar{\Delta t}$ of impulse-invariant method ($M = L = 3$).

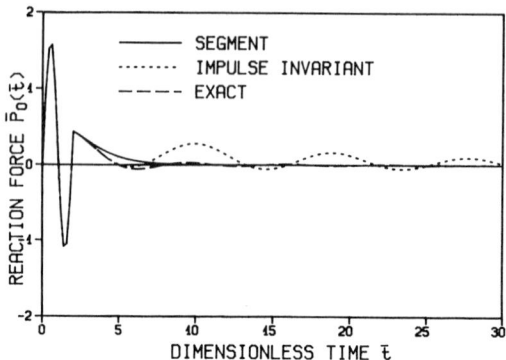

**Figure 6-35** Comparison of segment approach and impulse-invariant method ($\Delta \bar{t} = 0.2$).

**Nonlinear structure.** A simple nonlinear structure-soil system (Fig. 6-37a), which is similar to the example examined in Section 5.2, is analyzed for seismic excitation. The nonlinear structure consists of two masses $m_s$ and $m_o$ which are connected by a viscous damper with a coefficient $c$ and a spring with an elasto-plastic force-displacement relationship (Fig. 6-37b). $k_s$ denotes the elastic spring constant and $F_y$ the yielding value in tension and compression. The soil is modeled as a semi-infinite rod on elastic foundation and is characterized by $A$, $E$, $\rho$ and $k_g$.

The following dimensionless coefficients are selected: $\bar{k}_s = k_s/K = 1$, where $K$ is the static-stiffness coefficient of the semi-infinite rod (Eq. 3.79), $\bar{m} = m_s\kappa/(\rho A) = 0.5$, $m_o/m_s = 0.5$, $\bar{\zeta} = c\, c_l\kappa/(2K) = 0.3$ and $F_y/(k_s u_o^g) = 1/6$ ($u_o^g$ = max. value). Strong nonlinearities occur, as the yielding value turns out to be 0.26 times the corresponding maximum linear spring force. The total displacements of the system with two degrees of freedom are denoted as $u_s^t$ and $u_o^t$. The 10 second velocity and displacement time histories

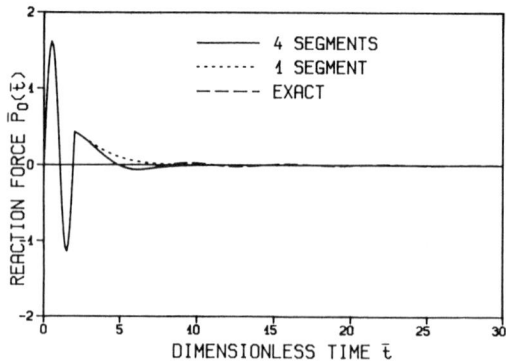

**Figure 6-36** Varying number of segments ($\Delta \bar{t} = 0.1$).

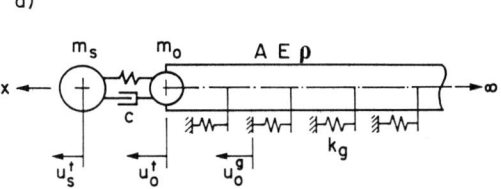

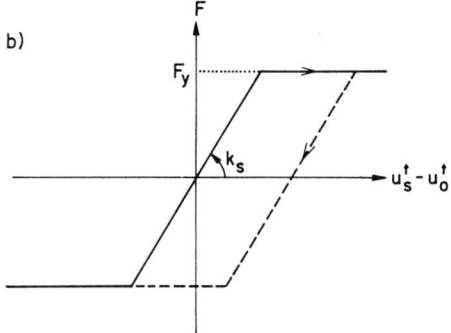

**Figure 6-37** Mass connected by nonlinear spring to semi-infinite rod on elastic foundation. a. Dynamic system. b. Elasto-plastic spring force-distortion relationship.

$u_o^g$ and $u_o^g$ are plotted versus the dimensionless time $\bar{t} = tc_l\kappa$ in Fig. 6-38.

The exact solution is calculated based on an explicit time integration procedure ($\Delta \bar{t} = 0.1$) with a direct evaluation of the convolution integral as described in Problem 6.11 for a similar case. The corresponding result is presented as a dashed line in the following figures.

The displacement of the mass $u_s^t$ is plotted in Fig. 6-39, whereby the convolution integral is calculated recursively with the impulse-invariant method ($\Delta \bar{t} = 0.1$). $M = L = K = 3$ and $M = L = 3, K = 50$ (Figs. 6-39a and b) lead to unacceptable results. The corresponding approximate dynamic-stiffness coefficients are compared to the exact values in Figs. 6-26a and b, respectively. Selecting $M = L = 5, K = 100$ leads to an accurate solution (Fig. 6-39c).

In Fig. 6-40 the segment approach is applied, varying the number of bounded segments as indicated. The length of a segment equals $\bar{t}_1 = 2$ and $\Delta \bar{t} = 0.1$.

Finally, the procedure that approximates the dynamic-stiffness coefficient in the frequency domain with a ratio of two polynomials is examined. As for the dynamic-stiffness coefficients (Figs. 6-29, 6-30 for $M = 8$, Figs. 6-31, 6-32 for $M = 5$), the 8th degree polynomial in the denominator leads to (slightly) more accurate results than the 5th degree one does (Fig. 6-41). $\Delta \bar{t}$ equals 0.1.

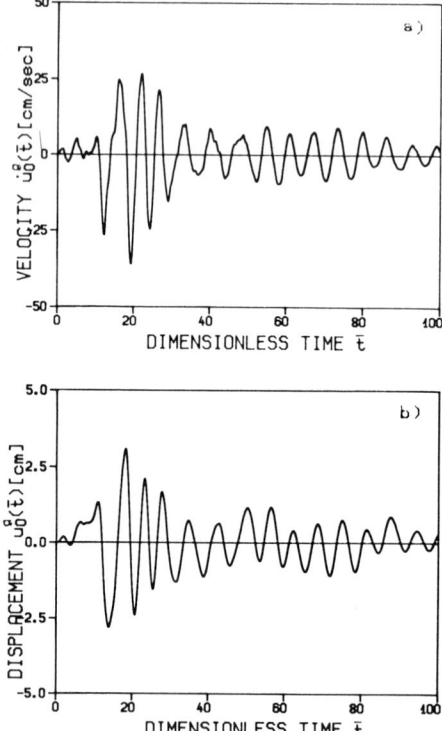

**Figure 6-38** Earthquake time history.

## SUMMARY

1. The dynamic-stiffness coefficient in the time domain is defined as follows: the force as a function of time required to produce a unit-impulse displacement at time zero. The dynamic-flexibility coefficient in the time domain is equal to the displacement as a function of time caused by a unit-impulse force applied at time zero.
2. The basic equation of motion for the substructure method of analyzing a nonlinear structure's interaction—including an irregular soil region—with the linear unbounded soil is formulated in the time domain. The total displacements are associated with the nodes within the structure and on the structure–soil interface. The unbounded soil's contribution, as expressed by the interaction forces acting at the nodes on the structure–soil interface, involves convolution integrals of the dynamic-stiffness coefficients of the unbounded soil with excavation (system ground) in the time domain and the corresponding motion relative to the generalized scattered input one. The load acting at the nodes on the structure–soil interface thus consists of the convolution integrals of the dy-

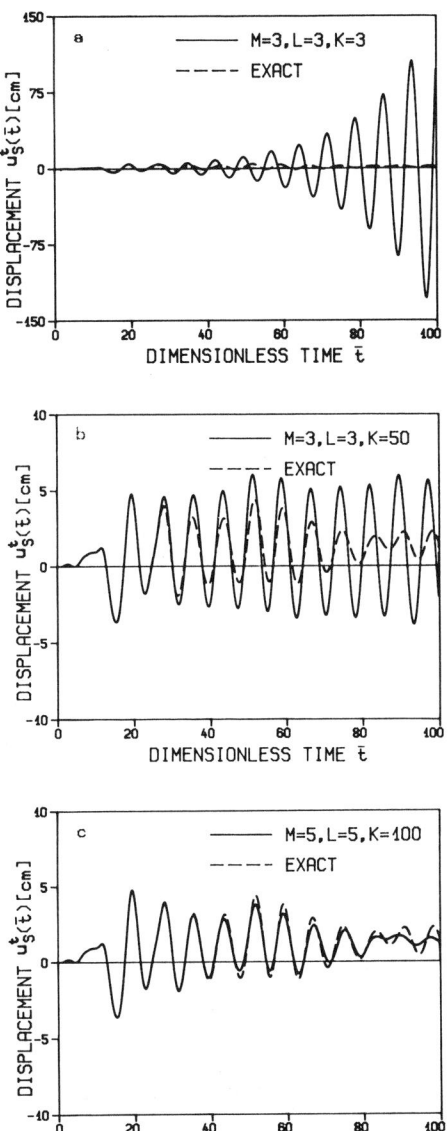

**Figure 6-39** Displacement of mass varying $M$, $L$, $K$ in impulse-invariant method ($\Delta \bar{t} = 0.1$).

namic-stiffness coefficients of the soil with excavation and the generalized scattered motion or, as an alternative, integrals of the dynamic-stiffness coefficients of the continuous soil (free field) and the free-field motion. The contribution of the unbounded soil with excavation, expressed as the motion in the nodes on the structure–soil interface relative to the generalized scattered motion, can also be formulated as the convolution

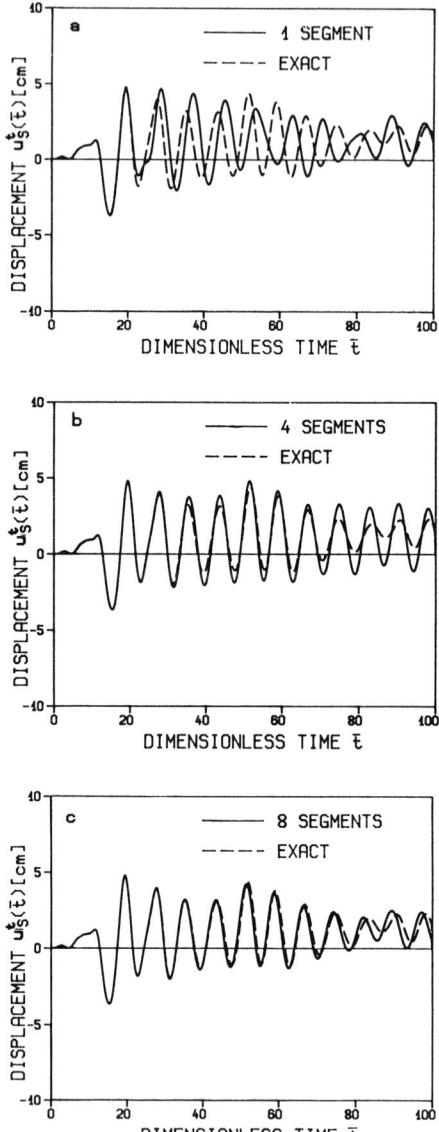

**Figure 6-40** Displacement of mass varying number of segments in segment approach ($\Delta \bar{t} = 0.1$).

integrals of the dynamic-flexibility coefficients of the soil with excavation in the time domain and the interaction forces.

3. Also for nonlinear soil–structure-interaction analysis, the loading environment (arising, for example, from seismic excitation) is defined by the free-field motion determined on the interface between the discretized and unbounded regions—that is, on the structure–soil interface—

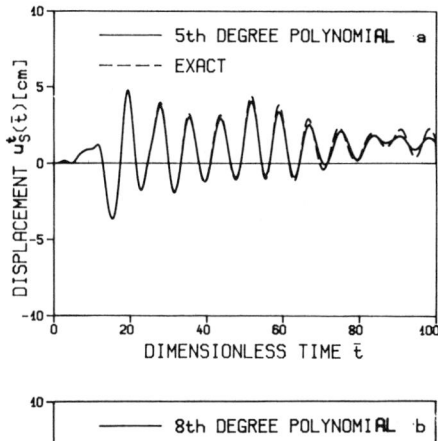
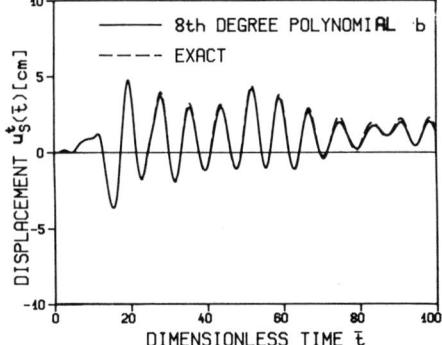

**Figure 6-41** Displacement of mass varying degree of polynomial in denominator of method based on ratio of polynomials in frequency domain ($\Delta \bar{t} = 0.1$).

based on a linear analysis. The corresponding generalized scattered motion can easily be calculated working in the frequency domain.

4. The dynamic-stiffness coefficient in the time domain is calculated, in general, as the Fourier transform of the corresponding value in the frequency domain. The dynamic-stiffness coefficient in the frequency domain is decomposed into the singular part and the remaining regular part. The latter is transformed numerically—possibly analytically above a certain frequency, if a high-frequency analytical expression exists. The initial value and initial slope of the dynamic-stiffness coefficient's regular part can be determined from its high-frequency behavior.

5. The singular part of the dynamic-stiffness coefficient is equal to its asymptotic value for infinite frequency. This limit consists of a real constant and an imaginary term, which is linear in frequency for the undamped soil with excavation (system ground). Its Fourier transform equals a Dirac delta function and its time derivative, which leads in the interaction force–displacement relationship to a spring and a damper, respectively.

6. An increased directionality of the waves occurs when the vibrating surface's frequency is increased. In the limit of an infinitely large frequency, a one-dimensional wave propagation in the direction perpendicular to the vibrating curved surface takes place. The two tangential components correspond to shear waves, and the normal component corresponds to a dilatational wave. The asymptotic value of the dynamic-stiffness coefficient of the undamped soil (system ground) per unit area consists of an imaginary linear term in frequency with a coefficient equal to the product of the mass density and the corresponding wave velocity, determined with the properties of the soil adjacent to the vibrating surface (corresponding to a damper). If the vibrating surface is curved, a constant term (corresponding to a spring) also occurs in the limit of the dynamic-stiffness coefficient, determined by solving the one-dimensional wave propagation problem. No coupling of the neighboring nodes' asymptotic values exists.

   For a bounded system, such as the excavated soil, the asymptotic value also contains a term consisting of the mass multiplied by the square of the frequency. Its Fourier transform equals the second time derivative of the Dirac delta function, which leads to a mass term in the force–displacement relationship. The system free field of the soil also has such a term.

7. Introducing material damping, modeled as a three-parameter Kelvin model, again leads to terms that can be interpreted as a spring and a damper. Voigt damping, however, results in more complicated singular terms, which, for the interaction force–displacement relationship, results in a convolution integral involving the velocity and the acceleration.

8. As the dynamic-flexibility coefficient in the frequency domain—being the inverse of the stiffness coefficient—converges to zero for the frequency approaching infinity, the corresponding coefficient in the time domain, determined as the Fourier transform, is simpler to calculate, because no singular part exists. The flexibility coefficient's initial value and initial slope can be determined from its high-frequency behavior.

9. For a general curved vibrating surface, the initial dynamic-flexibility coefficients per unit area are equal to the inverse of the product of the mass density and the shear-wave and dilatational-wave velocities for the two tangential and the perpendicular directions, respectively, calculated with the properties of the soil adjacent to the vibrating surface. No coupling of the neighboring nodes' initial values exists. These features correspond to those of the asymptotic dynamic-stiffness coefficient for the frequency approaching infinity.

10. The dynamic-flexibility coefficients in the time domain are calculated for a rigid circular disk resting on the surface of a half-space and on the

surface of a layer built-in at its base. The three-parameter Kelvin and the Voigt models are used to represent material damping.

11. Hysteretic damping, routinely used in the frequency domain, leads to non-causal behavior—that is, to a response prior to the excitation's application—which is physically unrealizable. This effect can, however, be ignored from a practical point of view.

12. In the computational procedure, the structure's properties, including those of the irregular soil, are formulated with the direct-stiffness method. The unbounded soil's contribution to the equations of motion can be established using either a stiffness or a flexibility formulation. In the former and latter cases, all previous displacements and interaction forces, respectively, at the nodes lying on the structure–soil interface must be stored and processed for the calculation of the convolution integrals in each time step. In both cases, the explicit algorithm with a predictor-corrector scheme and the implicit algorithm, both based on the Newmark method, can be applied.

13. The linear unbounded soil's interaction forces can be calculated recursively in the frequency domain. To evaluate the displacement amplitudes in the frequency domain at a specific time, only the corresponding amplitudes at the previous time station and the displacements at the present and the previous time stations are needed. The computational procedures, working with stiffness and flexibility formulations in conjunction with explicit and implicit time integration schemes, are derived.

The recursive evaluation in the frequency domain is, in principle, rigorous and does thus not represent an approximation. This is the case for all the flexibility formulations and for the stiffness formulations working with the regular part of the dynamic-stiffness coefficients, such as with an implicit time integration scheme. For the stiffness formulation with an explicit time integration working with the total dynamic-stiffness coefficients, decay functions must be introduced to suppress the unrealistic high-frequency content. This leads to an approximation.

In contrast to the calculation of the interaction forces in the time domain as convolution integrals, only the familiar dynamic-stiffness (or -flexibility) coefficients in the frequency domain are needed. No transformation of these coefficients into the time domain is thus required, which, for the case of the dynamic-stiffness coefficient, would require a decomposition into a regular and a singular part. The unbounded soil is thus treated as a linear substructure and can be calculated in the frequency domain, as is customary for a linear soil–structure interaction problem.

Because a recursive evaluation is used, no Fast Fourier Transforms are required in the algorithm to calculate a transient.

The recursive evaluation in the frequency domain also leads to the following advantages:

    a. Significant reduction of up to one order of magnitude in the number of operations resulting from the avoidance of Fast Fourier Transforms and from using a larger frequency increment (interpolation in frequency domain).

    b. Drastic decrease in the storage requirements, as no time-histories of the displacements (or interaction forces) need be stored.

14. The interaction forces of the linear unbounded soil can also be calculated recursively in the time domain. In this procedure, the forces at a specific time are computed from the displacements at the same time and from the most recent forces and most recent past displacements. This recursive evaluation of the convolution integrals in the time domain applies to the stiffness and flexibility formulations. It is, in principle, only approximate. Its accuracy can be improved by including more terms. When the dynamic-stiffness coefficients can be expressed as the ratios of two polynomials in frequency, the appropriately chosen recursive equations are exact.

    Three possibilities of choosing a recursive equation are discussed.

    a. The impulse-invariant method, where the unknown recursive coefficients are calculated by solving a system of equations that are established by equating the rigorous and recursive formulations for a discretized unit impulse displacement. After determining the coefficients, the approximate dynamic-stiffness coefficients in the time domain corresponding to the recursive formulation have to be compared over the applicable time range with the exact ones. The recursive formulation corresponds to the direct evaluation of the convolution integrals using these approximate dynamic-stiffness coefficients.

    b. In the segment approach, the dynamic-stiffness coefficients in the time domain are interpolated piecewise. Applying the $z$-transformation analytically then results in an explicit recursive equation without solving a system of equations.

    c. By approximating the dynamic-stiffness coefficients as the ratios of two polynomials in frequency using a curve-fitting technique based on the least-square method, the recursive coefficients can be determined explicitly starting directly in the frequency domain.

The recursive evaluation of the convolution integrals in the time domain leads to a dramatic reduction in the computational effort up to two and three orders of magnitude and in the storage requirement. This makes the time-domain analysis using the substructure method computationally competitive with the corresponding direct (non-recursive) frequency-domain procedure of determining the complex response which is, however, only applicable to a linear (total) system.

## PROBLEMS

**6.1** The general case of nonlinear soil–structure-interaction analysis cannot yet be solved. The unbounded soil, and thus also the free field, must remain linear. Various approximate procedures, however, are possible. One of them is discussed in the following, using a simple one-dimensional model for illustration and working essentially in the frequency domain.

A rigid body with mass $m_s$ (the "structure") rests on "soil," which is represented by a nonlinear spring and a mass $m_g$ (Fig. P 6-1a). The spring force equals $ku^3$, where $k$ is a constant, and $u$ is the distortion. The end of the spring which is not attached to $m_g$ is excited by the displacement $u_g(t) = u_g \sin \omega t$. Select $m_s/m_g = 1.7$ and $4\omega^2(m_s + m_g)/(3ku_g^2) = 1.35$.

The direct "rigorous" nonlinear solution is determined first. The equation of motion, with $u_s(t)$ denoting the total displacement of $m_s$, equals

$$(m_s + m_g)\ddot{u}_s(t) + k[u_s(t) - u_g(t)]^3 = 0$$

This nonlinear differential equation in $u_s(t)$ is solved (approximately) with the Ritz averaging method. Assuming $u_s(t) = u_s \sin \omega t$ and setting the average value of the virtual work equal to zero results in

$$\int_0^{2\pi/\omega} [-\omega^2(m_s + m_g) u_s \sin \omega t + k(u_s - u_g)^3 \sin^3 \omega t] \sin \omega t \, dt = 0$$

which leads with $u_f(t) = u_f \sin \omega t$ to

$$(u_f - u_g)^3 - \frac{4\omega^2}{3k} m_g u_f = 0$$

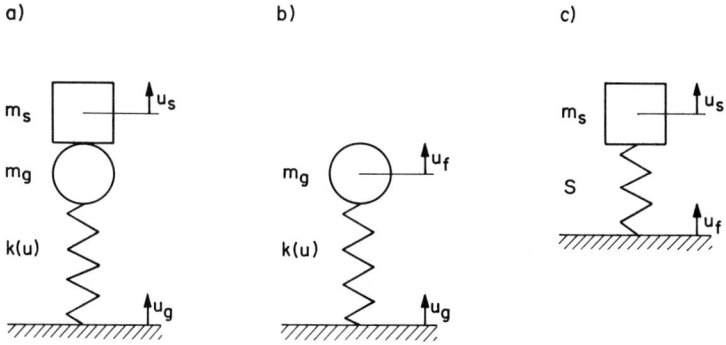

**Figure P6-1** Nonlinear unbounded soil. a. Total system. b. Free field. c. Linearized substructures.

or

$$(u_s - u_g)^3 - \frac{4\omega^2}{3k}(m_s + m_g)u_s = 0$$

with the solution $u_s = 2.5 u_g$.

The "free-field" response is calculated next (Fig. P 6-1b). The equation of motion, with $u_f(t)$ denoting the displacement of $m_g$, equals

$$m_g \ddot{u}_f(t) + k[u_f(t) - u_g(t)]^3 = 0$$

which leads with $u_f(t) = u_f \sin \omega t$ to

$$(u_f - u_g)^3 - \frac{4\omega^2}{3k} m_g u_f = 0$$

with the solution $u_f = 2u_g$.

Find an approximate solution for the rigid body's displacement, based on the substructure method using the tangential stiffness of the soil evaluated at the free-field response (Fig. P 6-1c).

*Solution*

Force–displacement relationship of soil:

$$P = -\omega^2 m_g u + k(u - u_g)^3$$

Tangent:

$$\frac{dP}{du} = -\omega^2 m_g + 3k(u - u_g)^2$$

Stiffness $S$ of soil:

$$S = -\omega^2 m_g + 3k(u_f - u_g)^2$$

Basic equation of motion:

$$m_s \ddot{u}_s(t) + S u_s(t) = S u_f(t)$$

$$(-\omega^2 m_s + S)u_s = S u_f$$

$$u_s = \frac{1}{1 - \dfrac{\omega^2 m_s}{-\omega^2 m_g + 3k(u_f - u_g)^2}} u_f$$

$$u_s = 2.64 u_g$$

**6.2** Derive a flexibility formulation for the unbounded soil based on the free-field motion $\{u_b^f(t)\}$ (and not on $\{u_b^g(t)\}$ as in Eq. 6.10), which can be applied together with the direct stiffness method for the appropriately modified structure.

*Solution*

Introducing the interaction forces with respect to the unbounded free-field soil $\{R_b^f(t)\}$, the displacement–force relationship of the system free-field equals

$$\{u_b^t(t)\} - \{u_b^e(t)\} = \int_0^t [F_{bb}^f(t - \tau)] \{R_b^f(\tau)\} \, d\tau$$

Subtracting the identity of the excavated soil

$$[M_{bb}^e] \{\ddot{u}_b^t(t)\} + [K_{bb}^e] \{u_b^t(t)\} = \{R_b^e(t)\}$$

with $\{R_b^e(t)\}$ denoting the forces of the linear excavated soil for the total motion, from the structure's equation of motion (Eq. 6.3) and noting that $\{R_b^f(t)\} = \{R_b(t)\} + \{R_b^e(t)\}$ leads to

$$\begin{bmatrix} [M_{ss}] & \vdots & [M_{sb}] \\ [M_{bs}] & \vdots & [M_{bb}] - [M_{bb}^e] \end{bmatrix} \begin{Bmatrix} \{\ddot{u}_s^t(t)\} \\ \{\ddot{u}_b^t(t)\} \end{Bmatrix} + \begin{Bmatrix} \{P_s(t)\} \\ \{P_b(t)\} - [K_{bb}^e]\{u_b^t(t)\} \end{Bmatrix} = \begin{Bmatrix} \{0\} \\ -\{R_b^f(t)\} \end{Bmatrix}$$

This represents the equation of motion of a discretized system consisting of the (nonlinear) structure and, in the embedded region, of the difference between the structure and the linear soil.

**6.3** Determine, for the conical rod in shear (Section 3.1.2), the dynamic-stiffness coefficient in the time domain at $x = l$, solving the wave equation in the time domain (without a transformation into the frequency domain). Specify the force–displacement relationship.

*Solution*

Displacement $w$ for the wave propagation in the positive $x$-direction (Eq. 3.26):

$$w(x, t) = \frac{l}{x} \delta\left(t - \frac{x - l}{c_s}\right)$$

Shear-force $Q$ (Eq. 3.22):

$$Q = GA w_{,x} = GA \left[-\frac{1}{l}\delta(t) - \frac{1}{c_s}\dot{\delta}(t)\right]$$

Dynamic-stiffness coefficient $S(t)$:

$$S(t) = -Q = \frac{GA}{l}\delta(t) + A\rho c_s \dot{\delta}(t)$$

Force–displacement relationship:

$$R(t) = \frac{GA}{l} w(t) + A\rho c_s \dot{w}(t)$$

No convolution integral!

**6.4** Using dimensionless coefficients, determine the dynamic-stiffness coefficient in the time domain of the discrete model shown in Fig. 2-4.

*Solution*

Eq. 2.34:

$$S(a_o) = K\left[1 - \frac{\mu_1 a_o^2}{1 + \frac{\mu_1^2}{\gamma_1^2} a_o^2} - \mu_o a_o^2 + ia_o\left(\frac{\mu_1}{\gamma_1}\frac{\mu_1 a_o^2}{1 + \frac{\mu_1^2}{\gamma_1^2} a_o^2} + \gamma_o\right)\right]$$

$$\lim_{a_o \to \infty} S(a_o) = K\left[1 - \frac{\gamma_1^2}{\mu_1} - \mu_o a_o^2 + ia_o(\gamma_1 + \gamma_o)\right]$$

$$S(a_o) = \lim_{a_o \to \infty} S(a_o) + K\frac{\gamma_1^2}{\mu_1}\frac{1}{1 + i\frac{\mu_1 a_o}{\gamma_1}}$$

$$S(\bar{t}) = K\frac{c_s}{a}\left[\left(1 - \frac{\gamma_1^2}{\mu_1}\right)\delta(\bar{t}) + \mu_o\frac{d^2\delta(\bar{t})}{d\bar{t}^2} + (\gamma_1 + \gamma_o)\frac{d\delta(\bar{t})}{d\bar{t}} + \frac{\gamma_1^3}{\mu_1^2}\exp\left(-\frac{\gamma_1}{\mu_1}\bar{t}\right)\right]$$

**6.5** Determine the dynamic-stiffness coefficient in the time domain $S(\bar{t})$ of the semi-infinite rod with exponentially increasing area (Fig. P 3-5a)

$$A(x) = A_o \exp\left(\frac{x}{f}\right)$$

by calculating the Fourier transform of $S(a_o)$ [W11].

*Solution*

The displacement amplitude in the frequency domain $u(x, a_o)$ for a value = 1 at $x = 0$, after suppressing the incoming wave, is specified in Problem 3.5 as

$$u(x, a_o) = \exp\left[-\frac{x}{2f}(1 + \sqrt{1 - 4a_o^2})\right]$$

with

$$a_o = \frac{\omega f}{c_l}$$

The dynamic-stiffness coefficient $S(a_o)$ is defined as

$$S(a_o) = -EA_o u_{,x}(x = 0, a_o)$$

resulting in

$$S(a_o) = \frac{EA_o}{f}\frac{1}{2}(1 + \sqrt{1 - 4a_o^2})$$

$S(a_o)$ is nondimensionalized with the static value $K = EA_o/f$, and the nondimensional form is split into its real and imaginary parts

$$S(a_o) = K[k(a_o) + ia_o c(a_o)]$$

resulting, for the regions below and above the cutoff frequency, in

$$a_o \leq 0.5 \quad k = \frac{1}{2}(1 + \sqrt{1 - 4a_o^2})$$

$$c = 0$$

$$a_o \geq 0.5 \quad k = \frac{1}{2}$$

$$c = \sqrt{1 - \frac{1}{4a_o^2}}$$

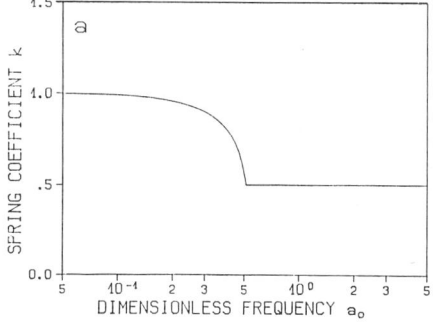

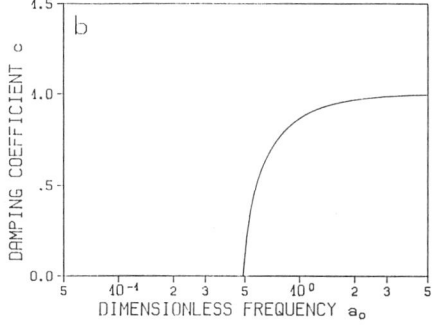

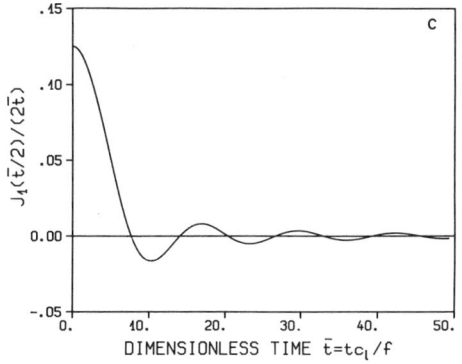

**Figure P6-5** Semi-infinite rod with exponentially increasing area. a. and b. Dynamic-stiffness coefficient in frequency domain. c. Regular part of dynamic-stiffness coefficient in time domain.

The spring coefficient $k$ and the damping coefficient $c$ are plotted as a function of $a_o$ in Fig. P 6-5a and b.

The inverse Fourier transform in dimensionless variables is equal to (Eq. 6.37)

$$S(\bar{t}) = \frac{c_l}{2\pi f} \int_{-\infty}^{+\infty} S(a_o) \exp(ia_o \bar{t}) \, da_o$$

with

$$\bar{t} = \frac{c_l}{f} t$$

The asymptotic value of $S(a_o)$ for $a_o \to \infty$ follows as

$$\lim_{a_o \to \infty} S(a_o) = K \left( \frac{1}{2} + ia_o \right)$$

Decomposing $S(a_o)$ into this singular part and the remaining regular one leads to

$$S(\bar{t})$$
$$= \frac{Kc_l}{f} \left[ \frac{1}{2\pi} \int_{-\infty}^{+\infty} \left( \frac{1}{2} + ia_o \right) \exp(ia_o \bar{t}) \, da_o + \frac{1}{2\pi} \int_{-\infty}^{+\infty} \left( -ia_o + \frac{\sqrt{1 - 4a_o^2}}{2} \right) \exp(ia_o \bar{t}) \, da_o \right]$$

which results in

$$S(\bar{t}) = \frac{Kc_l}{f} \left[ \frac{1}{2} \delta(\bar{t}) + \frac{d\delta(\bar{t})}{d\bar{t}} + \frac{1}{2\bar{t}} J_1\left(\frac{\bar{t}}{2}\right) \right] \quad \bar{t} \geq 0$$

The regular part is transformed using the Fourier integral tables of Ref. [C1, No. 556.1]. $J_1$ is the Bessel function of the first kind and of the first order. For $\bar{t} < 0$, $S(\bar{t}) = 0$. The transformed regular part $J_1(\bar{t}/2)/(2\bar{t})$ is plotted as a function of $\bar{t}$ in Fig. P 6-5c. For $\bar{t} = 0^+$, the value equals 0.125, which can be checked using Eq. 6.46a.

$$\frac{f}{c_l K} S_r(\bar{t} = 0^+) = \lim_{ia_o \to \infty} \left[ a_o^2 + \frac{ia_o \sqrt{1 - 4a_o^2}}{2} \right] = \frac{1}{8}$$

**6.6** The undamped semi-infinite rod with exponentially increasing area, examined in Problems 3.5 and 6.5, is subjected to a prescribed support movement at Point 0 (Fig. P3-5a):

$$u_o(\bar{t}) = \frac{u_o}{2} \left[ 1 - \cos\left(2\pi \frac{\bar{t}}{\bar{t}_o}\right) \right] = u_o \bar{u}_o(\bar{t}) \quad 0 < \bar{t} < \bar{t}_o$$

$$u_o(\bar{t}) = 0 \quad \bar{t} > \bar{t}_o$$

The corresponding dimensionless time $\bar{t}_o = t_o c_l/f$ is selected as $= 2$.
Plot the dimensionless reaction force $\bar{P}_o(\bar{t}) = P_o(\bar{t})/(Ku_o)$, separating the contributions of the spring, the damper, and the convolution integral.

## Result

Force–displacement relationship (Problem 6.5)

$$P_o(t) = K\left[\frac{u_o(t)}{2} + \frac{f}{c_l}\dot{u}_o(t) + \frac{1}{2}\int_o^t \frac{1}{t-\tau}J_1\left(\frac{c_l}{2f}(t-\tau)\right)u_o(\tau)\,d\tau\right]$$

or in dimensionless quantities

$$\bar{P}_o(\bar{t}) = \frac{\bar{u}_o(\bar{t})}{2} + \frac{d\bar{u}_o(\bar{t})}{d\bar{t}} + \frac{1}{2}\int_o^{\bar{t}} \frac{1}{\bar{t}-\bar{\tau}}J_1\left(\frac{\bar{t}-\bar{\tau}}{2}\right)\bar{u}_o(\bar{\tau})\,d\bar{\tau}$$

For $\bar{t} < \bar{t}_o$ (where the support movement is non-zero), the damper supplies the largest contribution, followed by the spring and the convolution integral (Fig. P 6-6).

**6.7** In Section 3.9.3, the dynamic-stiffness matrix in the frequency domain, corresponding to a parabolic enforced out-of-plane displacement of a layer built-in at its base of depth $d$ (Fig. 3-27), is determined. Calculate the dynamic-stiffness in the time domain of the first coefficient.

## Solution

Dynamic-stiffness coefficient in frequency domain (Eq. 3.207):

$$S(a_o) = \frac{16\,G}{\pi^3}\sum_{j=1}^{\infty}\frac{1}{(2j-1)^3}\left[3 - \frac{(-1)^{j+1}8}{(2j-1)\pi}\right]^2\sqrt{1 - \frac{4a_o^2}{(2j-1)^2\pi^2}}$$

with $a_o = \omega\,d/c_s$.

Static-stiffness coefficient $K$:

$$K = \sum_{j=1}^{\infty} K_j$$

$$K_j = \frac{16\,G}{\pi^3}\frac{1}{(2j-1)^3}\left[3 - \frac{(-1)^{j+1}8}{(2j-1)\pi}\right]^2$$

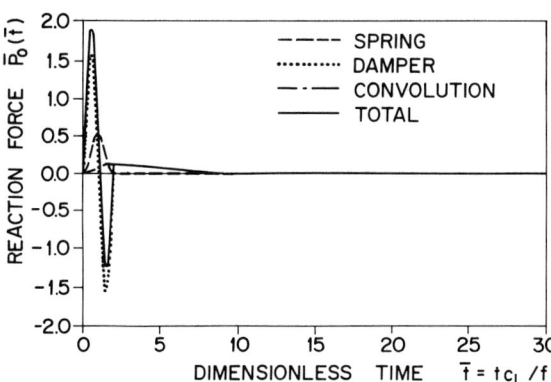

**Figure P6-6** Contribution of singular and regular parts to reaction force.

Singular part of $S(a_o)$:

$$\lim_{a_o \to \infty} S(a_o) = \sum_{j=1}^{\infty} K_j \frac{2}{(2j-1)\pi} a_o i$$

Regular part of $S(a_o)$:

$$S_r(a_o) = \sum_{j=1}^{\infty} K_j \frac{2}{(2j-1)\pi} \left( \sqrt{\frac{(2j-1)^2 \pi^2}{4} - a_o^2} - ia_o \right)$$

Dynamic-stiffness coefficient in time domain, with $\bar{t} = tc_s/d$ ([C1, No. 556.1]):

$$S(\bar{t}) = \frac{c_s}{d} \sum_{j=1}^{\infty} K_j \frac{2}{(2j-1)\pi} \left( \frac{d\delta(\bar{t})}{d\bar{t}} + \frac{(2j-1)\pi}{2\bar{t}} J_1 \left[ \frac{(2j-1)\pi \bar{t}}{2} \right] \right)$$

Dimensionless representation of $S_r(\bar{t})$:

$$\frac{d}{c_s} \frac{S_r(\bar{t})}{K} = \frac{\sum_{j=1}^{\infty} \frac{1}{(2j-1)^3} \left[ 3 - \frac{(-1)^{j+1} 8}{(2j-1)\pi} \right]^2 J_1 \left[ \frac{(2j-1)\pi}{2} \bar{t} \right]}{\bar{t} \sum_{j=1}^{n} \frac{1}{(2j-1)^3} \left[ 3 - \frac{(-1)^{j+1} 8}{(2j-1)\pi} \right]^2}$$

The function $S_r(\bar{t}) \, d/(Kc_s)$ is plotted in Fig. P 6-7.

**6.8** The dynamic-stiffness coefficients in the frequency and time domains of the undamped rod with exponentially increasing area are derived in Problem 6.5. For two damping models with $\zeta = 0.2$, the three-parameter model (Eq. 6.57) and the Voigt model (Eq. 6.54), determine the dynamic-stiffness coefficients in the frequency domain using the correspondence principle. Calculate asymptotic expansions applicable to large frequencies and in particular the singular part. For the three-parameter model, transform the remaining part of the asymptotic expansion analytically and the rest of the regular expansion numerically to calculate the dynamic-stiffness coefficient in the time domain. For the Voigt model, discuss the singular part's existence. Use in-

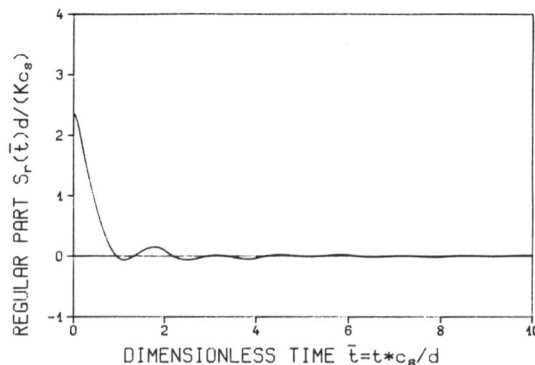

**Figure P6-7** Regular part of dynamic-stiffness coefficient in time domain.

tegration by parts to derive tractable expressions for the force–displacement relationship with convolution integrals involving accelerations and velocities [W12].

*Solution*
*Three-Parameter Model* (Fig. 6-4b)

Replacing $E$ by

$$E^* = E \frac{1 + 2\zeta a_o i}{1 + \zeta a_o i}$$

in

$$S(a_o) = \frac{EA_o}{f} \frac{1}{2} (1 + \sqrt{1 - 4a_o^2})$$

leads to

$$S(a_o) = K \frac{1}{2} \frac{1 + 2\zeta a_o i}{1 + \zeta a_o i} \left[ 1 + \sqrt{1 - 4a_o^2 \frac{1 + \zeta a_o i}{1 + 2\zeta a_o i}} \right]$$

Using $K = EA_o/f$ to nondimensionalize $S(a_o)$ as

$$S(a_o) = K [k(a_o) + i a_o\, c(a_o)]$$

determines the spring coefficient $k(a_o)$ and the damping coefficient $c(a_o)$ (Fig. P 6-8a and b for $\zeta = 0.2$).

For large $a_o$, the asymptotic expansion equals

$$S(a_o) \approx K \left[ \sqrt{2}\, a_o i + \left(1 - \frac{\sqrt{2}}{4\zeta}\right) - \left(\frac{1}{2\zeta} - \frac{\sqrt{2}\,(7/8 + \zeta^2)}{4\zeta^2}\right)(a_o i)^{-1} \right]$$

The expansion is not applicable for $\zeta \to 0$, as $1/(\zeta a_o) \ll 1$ is assumed in its derivation.

$S(a_o)$ is decomposed into the regular and singular terms. The regular part (exact expression minus the first two terms of the asymptotic expression) is numerically transformed up to a particular frequency using the Fast Fourier Transform. Above this frequency, the third term of the asymptotic expansion is integrated analytically. The singular part can be interpreted as a viscous damper and a spring. The resulting dynamic-stiffness coefficient in the time domain equals

$$S(\bar{t}) = K \frac{c_1}{f} \left[ \left(1 - \frac{\sqrt{2}}{4\zeta}\right) \delta(\bar{t}) + \sqrt{2}\, \frac{d\delta(\bar{t})}{d\bar{t}} + g(\bar{t}) \right]$$

where $Kc_1/f\, g(\bar{t})$ denotes the transform of the entire regular part. $g(\bar{t})$ is plotted in Fig. P 6-8c. Although a numerical integration is performed, $g(\bar{t}) = 0$ for $\bar{t} < 0$. The value of $g(\bar{t}) = 0^+$ follows from the third term of the asymptotic expansion as (Eq. 6.46a)

$$\frac{f}{c_1 K} S_r(\bar{t} = 0^+) = \lim_{i a_o \to \infty} \left[ -\frac{1}{2\zeta} + \frac{\sqrt{2}\,(7/8 + \zeta^2)}{4\zeta^2} \right] = 5.586$$

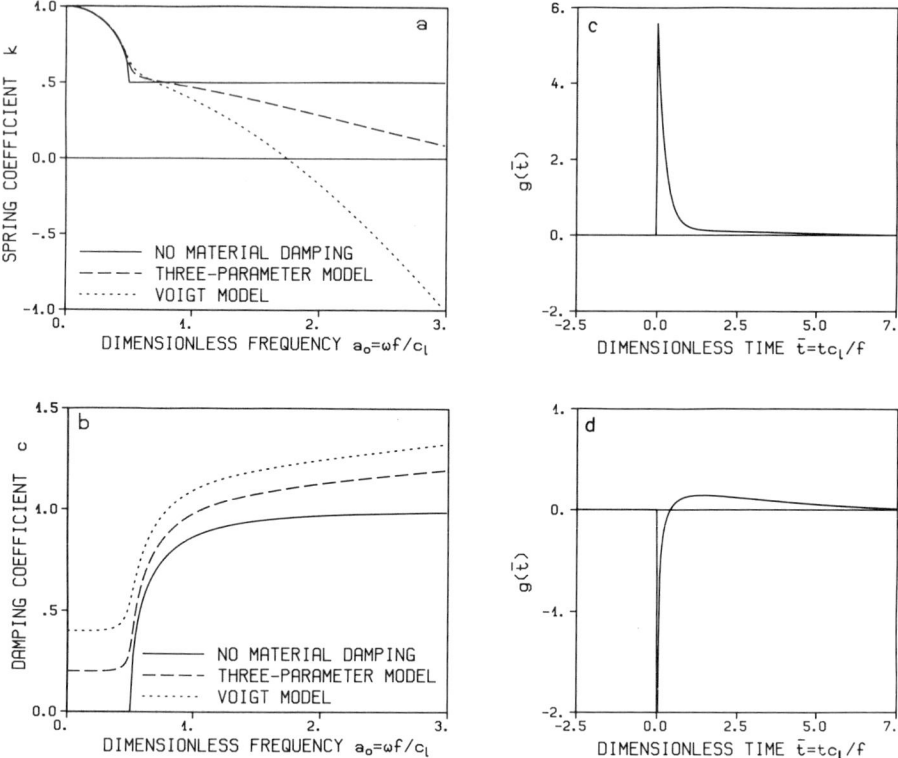

**Figure P6-8** Semi-infinite rod with exponentially increasing area with damping. a. and b. Dynamic-stiffness coefficient in frequency domain. c. Regular part of dynamic-stiffness coefficient in time domain, three-parameter model. d. Regular part of dynamic-stiffness coefficient in time domain, Voigt model.

*Voigt Model* (Fig. 6-4a)

With $E^* = E(1 + 2\zeta a_o i)$, the correspondence principle results in

$$S(a_o) = K\frac{1}{2}(1 + 2\zeta a_o i)\left[1 + \sqrt{1 - \frac{4a_o^2}{1 + 2\zeta a_o i}}\right]$$

The corresponding spring and damping coefficients are plotted in Fig. P 6-8a and b. For large $a_o$, the asymptotic expansion is equal to

$$S(a_o) \approx K\left[\sqrt{2\zeta}(a_o i)^{3/2} + \zeta a_o i + \frac{1 + \zeta^2}{2\sqrt{2\zeta}}(a_o i)^{1/2} + \frac{1}{2} + \frac{\sqrt{2\zeta}}{32}\left(6 - \frac{1}{\zeta^2} - \zeta^2\right)(a_o i)^{-1/2}\right]$$

The singular part consisting of the first four terms is of a more complicated form. It can be shown that the inverse transform equals

$$S(\bar{t}) = K\frac{c_l}{f}\left[\sqrt{\frac{2\zeta}{\pi}}\frac{3}{4}\bar{t}^{-5/2} + \zeta\frac{d\delta(\bar{t})}{dt} - \frac{1 + \zeta^2}{4\sqrt{2\zeta\pi}}\bar{t}^{-3/2} + \frac{1}{2}\delta(\bar{t}) + g(\bar{t})\right]$$

where $g(\bar{t})\, c_l/f$ represents the transform of the regular part shown for $\zeta = 0.2$ in Fig. 6-8d. Because the inverse transform of the first term of the regular part $[\sim(a_o i)^{-1/2}]$ is proportional to $\bar{t}^{-1/2}$, $g(\bar{t}=0^+)$ is infinite. However, the corresponding convolution integral exists. The second and fourth terms can again be interpreted as a viscous damper and a spring, respectively. The first and third terms exhibit strong singularities at $\bar{t} = 0$, whereby the transformation is not valid for $\bar{t} = 0$. As a consequence, the convolution integral (with the displacement) as appearing in Eq. 6.50 cannot be performed. However, applying integration by parts leads to tractable expressions with convolution integrals involving acceleration and velocities instead of displacements. For instance, the contribution of the third term (disregarding a constant) to the convolution integral equals

$$\int_o^t (t-\tau)^{-3/2} u(\tau)\, d\tau = 2(t-\tau)^{-1/2} u(\tau)\Big|_{\tau=0}^t - 2\int_o^t (t-\tau)^{-1/2} \dot{u}(\tau)\, d\tau$$

$$= -2\int_o^t (t-\tau)^{-1/2} \dot{u}(\tau)\, d\tau$$

The force–displacement relationship for the Voigt model then equals

$$R(t) = K\left[ \sqrt{\frac{2\zeta}{\pi}}\left(\frac{f}{c_l}\right)^{3/2}\int_o^t (t-\tau)^{-1/2} \ddot{u}(\tau)\, d\tau + \frac{\zeta f}{c_l}\dot{u}(t)\right.$$

$$\left. + \frac{1+\zeta^2}{2\sqrt{2\zeta\pi}}\sqrt{\frac{f}{c_l}}\int_o^t (t-\tau)^{-1/2} \dot{u}(\tau)\, d\tau + \frac{1}{2}u(t) + \int_o^t g(t-\tau)\, u(\tau)\, d\tau\right]$$

The first and third terms of this equation can be also easily derived by introducing accelerations and velocities instead of displacements in the frequency domain. This reduces the power of $a_o i$ of the first and third terms of $S(a_o)$ in the convolution product (appearing in the equation corresponding to Eq. 6.50 in the frequency domain) to powers of $-1/2$.

**6.9** For the semi-infinite rod resting on an elastic foundation examined in Section 3.2 (Fig. 3-4a), derive the dynamic-stiffness coefficient in the time domain $S(\bar{t})$, the force–displacement relationship, and the dynamic-flexibility coefficient in the time domain $F(\bar{t})$.

*Solution*

Dynamic-stiffness coefficient in frequency domain with $a_o = \omega/(c_l\kappa)$ (Eqs. 3.80, 3.81):

$$S(a_o) = Ki\sqrt{a_o^2 - 1}$$

Fourier transformation with $\bar{t} = tc_l\kappa$ (Eq. 6.37):

$$S(\bar{t}) = \frac{c_l\kappa}{2\pi}\int_{-\infty}^{+\infty} S(a_o)\exp(ia_o\bar{t})\, da_o$$

$$\lim_{a_o\to\infty} S(a_o) = Kia_o$$

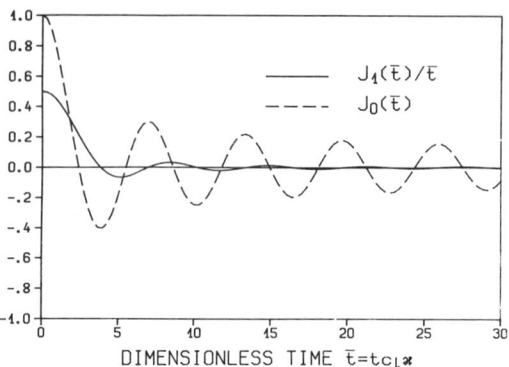

**Figure P6-9** Bessel functions appearing in dynamic-stiffness and -flexibility coefficients in time domain.

Decomposition of $S(a_o)$ into singular and remaining regular parts:

$$S(\bar{t}) = Kc_l\kappa \int_{-\infty}^{+\infty} \left[\frac{ia_o}{2\pi} + \frac{i}{2\pi}(\sqrt{a_o^2 - 1} - a_o)\right] \exp(ia_o\bar{t})\, da_o$$

The first term within the bracket is equal to $d\delta(\bar{t})/d\bar{t}$. The second term can be shown to result in $J_1(\bar{t})/\bar{t}$, where $J_1$ is the Bessel function of the first kind and of the first order (Ref. [C1, No. 556.1]).

This leads to the dynamic-stiffness coefficient in time domain ($\bar{t} \geq 0$):

$$S(\bar{t}) = Kc_l\kappa \left[\frac{d\delta(\bar{t})}{d\bar{t}} + \frac{1}{\bar{t}}J_1(\bar{t})\right]$$

The (transformed) regular part $J_1(\bar{t})/\bar{t}$ is plotted as a function of $\bar{t}$ in Fig. P 6-9.

Using

$$\frac{d\delta(\bar{t})}{d\bar{t}} = \frac{1}{c_l^2\kappa^2}\dot{\delta}(t)$$

and the convolution integral (Eq. 6.2)

$$P_o(t) = \int_0^t S(t - \tau)\, u_o(\tau)\, d\tau$$

leads to the force–displacement relationship

$$P_o(t) = K\left[\sqrt{\frac{A\rho}{k_g}}\, \dot{u}_o(t) + \int_0^t \frac{1}{t-\tau} J_1\left(\sqrt{\frac{k_g}{A\rho}}(t-\tau)\right) u_o(\tau)\, d\tau\right]$$

Dynamic-flexibility coefficient in frequency domain

$$F(a_o) = \frac{1}{S(a_o)} = -\frac{1}{K}\frac{i}{\sqrt{a_o^2 - 1}}$$

Fourier transformation (Eq. 6.114)

$$F(\bar{t}) = \frac{c_l \kappa}{2\pi} \int_{-\infty}^{+\infty} F(a_o) \exp(ia_o \bar{t}) \, da_o$$

resulting in (Ref. [C1, No. 557])

$$F(\bar{t}) = \frac{c_l \kappa}{K} J_o(\bar{t})$$

$J_o(\bar{t})$ decays less strongly than the regular part of the dynamic-stiffness coefficient $J_1(\bar{t})/\bar{t}$ (Fig. P 6-9).

**6.10** In Problem 6.8, the dynamic-stiffness coefficients in the frequency and time domains of the semi-infinite rod with exponentially increasing area are derived for the three-parameter model (Eq. 6.57) and for the Voigt model (Eq. 6.54). Determine the corresponding (dimensionless) dynamic-flexibility coefficients in the time domain $\bar{F}(\bar{t})$ for the two damped cases with $\zeta = 0.2$ and for the elastic one. Use asymptotic expansions which are valid for large $a_o$ and which can be transformed analytically. Derive the initial characteristics for the three cases using the high-frequency behavior and, for the elastic rod, basing them on the law of conservation of momentum. Assume further that the elastic rod's dynamic-stiffness coefficient is approximated, with the static coefficient, by a spring, and, with the coefficient corresponding to high frequency, by a damper and calculate the corresponding dynamic-flexibility coefficient in the time domain.

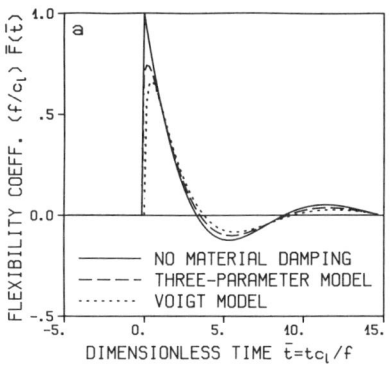

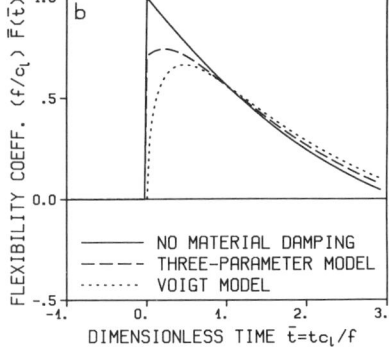

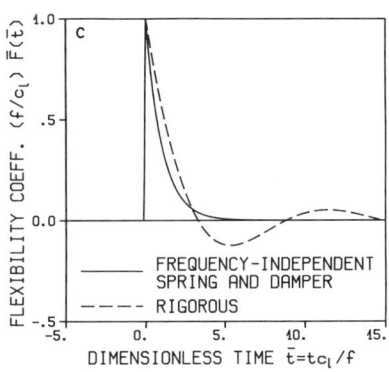

**Figure P6-10** Dynamic-flexibility coefficient in time domain, semi-infinite rod with exponentially increasing area. a. $-5 \leq \bar{t} \leq 15$. b. $-1 \leq \bar{t} \leq 3$. c. Comparison with spring and damper with frequency-independent coefficients.

*Solution*

Fourier transformation (Eq. 6.114):

$$\overline{F}(\bar{t}) = K \frac{c_l}{f} \frac{1}{2\pi} \int_{-\infty}^{+\infty} \frac{1}{\overline{S}(a_o)} \exp(i a_o \bar{t}) \, da_o$$

Asymptotic expansion, valid for large $a_o$:

Elastic: $\quad \overline{F}(a_o) \approx (a_o i)^{-1} + \frac{1}{2} a_o^{-2}$

Three-parameter model: $\quad \overline{F}(a_o) \approx \frac{1}{\sqrt{2}} (a_o i)^{-1} + \left(\frac{1}{2} - \frac{\sqrt{2}}{8\zeta}\right) a_o^{-2}$

Voigt model: $\quad \overline{F}(a_o) \approx \frac{1}{\sqrt{2\zeta}} (a_o i)^{-3/2} + \frac{1}{2} a_o^{-2}$

Plot of $\overline{F}(\bar{t})$ multiplied by $f/c_l$ in Fig. P6-10a and b.
Initial characteristics based on high-frequency behavior as just specified (Eq. 6.118):

Elastic: $\quad \dfrac{f}{c_l} \overline{F}(\bar{t} = 0^+) = 1$

$\dfrac{f}{c_l} \dfrac{d\overline{F}(\bar{t} = 0^+)}{d\bar{t}} = -\dfrac{1}{2}$

Three-parameter model: $\quad \dfrac{f}{c_l} \overline{F}(\bar{t} = 0^+) = \dfrac{1}{\sqrt{2}}$

$\dfrac{f}{c_l} \dfrac{d\overline{F}(\bar{t} = 0^+)}{d\bar{t}} = -\left(\dfrac{1}{2} - \dfrac{\sqrt{2}}{8\zeta}\right) = 0.384$

Voigt model: $\quad \dfrac{f}{c_l} \overline{F}(\bar{t} = 0^+) = 0$

$\dfrac{f}{c_l} \dfrac{d\overline{F}(\bar{t} = 0^+)}{d\bar{t}} = \infty$

Law of conservation of momentum (Eq. 6.122):

$$F(\bar{t} = 0^+) = \frac{1}{A_o \rho c_l}$$

$$\overline{F}(\bar{t} = 0^+) = K F(\bar{t} = 0^+) = \frac{c_l}{f}$$

Approximation:

$$\overline{S}(a_o) = k + i a_o c$$

With $k = 1$ and $c = 1$:

$$\overline{F}(\bar{t}) = \frac{c_l}{f} \frac{1}{2\pi} \int_{-\infty}^{+\infty} \frac{1}{k + i a_o c} \exp(i a_o \bar{t}) \, da_o$$

Plot of $\bar{F}(\bar{t})$ multiplied by $f/c_l$ in Fig. P 6-10c and comparison with rigorous value from Fig. P 6-10a.

**6.11** Solve the nonlinear system, consisting of a mass $m$ connected, by a viscous damper with a coefficient $c$ and a spring with an elasto-plastic force–displacement relationship, to an elastic semi-infinite rod on an elastic foundation (Fig. 5-1), with an explicit algorithm using the stiffness formulation. The problem with the corresponding parameters is defined at the beginning of Section 5.2. Select $F_y^+/R_o = 0.1$ and $F_y^-/R_o = 0$, which corresponds to a strong nonlinearity. Use the Newmark algorithm with $\beta = 0$, $\gamma = 0.5$, and with a time step $\Delta \bar{t} = 0.02$. Plot the (nondimensionalized) displacement of the mass $\bar{u}_s(\bar{t}) = u_s(t) \, K/R_o$ and the spring force $F(\bar{t})/R_o$.

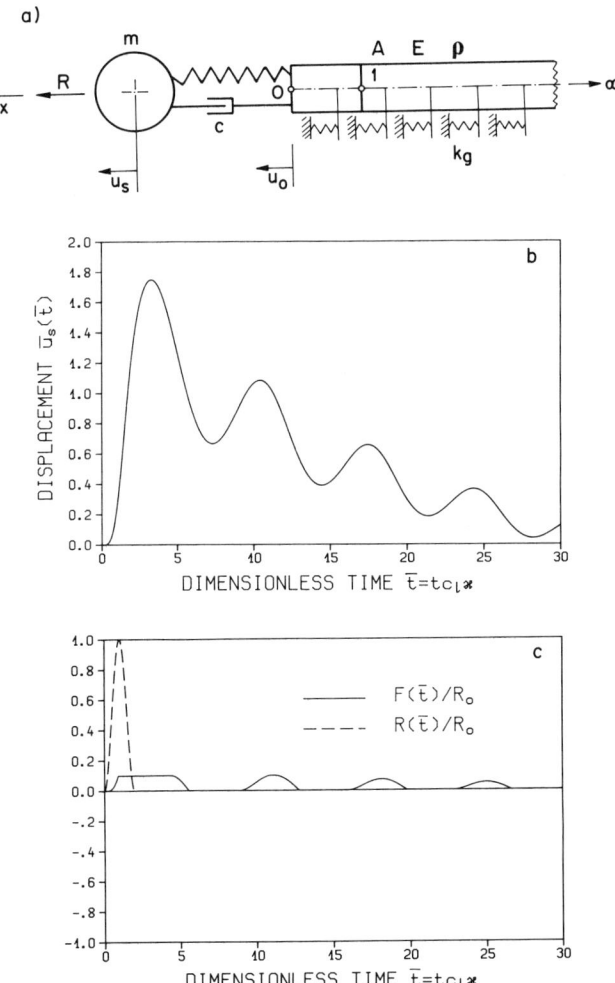

**Figure P6-11** Mass connected by nonlinear spring to semi-infinite rod on elastic foundation. a. Dynamic system. b. Displacement of mass. c. Spring force.

*Solution*

To be able to use the standard explicit algorithm (Eq. 6.141), a mass must be present at Point 0 (Fig. P 6-11a). To achieve this, a finite rod element is placed next to this point on the semi-infinite rod. This increases the number of degrees of freedom to 3 (additional Node 1). The finite rod's static-stiffness matrix is specified in Eq. 3.87. The interaction force–displacement relationship of the semi-infinite system (which is used to formulate equilibrium in Node 1) is formulated in Problem 6.9 as

$$P_1(t) = K\left[\sqrt{\frac{A\rho}{k_g}}\,\dot{u}_1(t) + \int_o^t \frac{1}{t-\tau} J_1\left(\sqrt{\frac{k_g}{A\rho}}(t-\tau)\right) u_1(\tau)\, d\tau\right]$$

From the plots of $\bar{u}_s(t)$ and $F(t)/R_o$ shown in Figs. P 6-11b and c, the strong nonlinear behavior is visible.

**6.12** The concepts developed in this text can be applied to other media. In particular, it is possible to analyze fluid–structure interaction in the time domain analogously to soil–structure interaction.

A simple situation is shown in Fig. P 6-12a. A dam consists of a vertical rigid wall. The reservoir of constant depth $d$ extends to infinity and rests on rigid rock. The inviscid water's motion is irrotational and of small amplitude. In the following, the dynamic-stiffness coefficient of the reservoir in the time domain $S(t)$—that is, the horizontal resultant force exerted by the wall on the water is calculated as a function of time $t$, $R(t)$, which leads to a unit-impulse horizontal displacement along the wall at $x = 0$ at time $t = 0$, $u(t) = \delta(t)$, [C2, W2].

The wave-equation expressed in the hydrodynamic pressure $p(x, z, t)$ is derived as follows: With $u$ and $w$ denoting the horizontal and vertical displacements and $\rho$ the mass density, the equations of equilibrium are equal to

$$p_{,x} + \rho \ddot{u} = 0$$

$$p_{,z} + \rho \ddot{w} = 0$$

The pressure–displacement relationship is formulated as

$$p = -K(u_{,x} + w_{,z})$$

with the bulk modulus $K$. This leads to the wave equation

$$p_{,xx} + p_{,zz} - \frac{\ddot{p}}{c^2} = 0$$

with the wave velocity (acoustic speed of sound) $c$

$$c = \sqrt{K/\rho}$$

The conditions on the finite boundaries expressed in $p$ are as follows. Examining an infinitesimal element along the wall $x = 0$ leads to

$$x = 0: \quad p_{,x} = -\rho \ddot{u}$$

whereby $\ddot{u}$ is known (as a function of $u$ at $x = 0$). Ignoring waves at the free surface

## Chap. 6 Problems

$z = 0$, the boundary condition equals

$$z = 0: \quad p = 0$$

Along the bottom of the reservoir $z = d$, $\ddot{w}$ vanishes, resulting in

$$z = d: \quad p_{,z} = 0$$

Derive the dynamic-stiffness coefficient in the frequency domain $S(a_o)$ with the dimensionless frequency $a_o = \omega d/c$. Specialize for the incompressible case ($K = \infty$). Divide by the bulk modulus $K$ and decompose the result into a real and an imaginary part, which can be interpreted as a spring with a coefficient $k_o$ and a damper with a dimensionless coefficient $c_o$. Plot $k_o$ and $c_o$ as a function of $a_o$. Analytically transform the dynamic-stiffness coefficient into the time domain $\bar{S}(t)$ with the dimensionless time $\bar{t} = tc/d$. Formulate the force–displacement relationship $R(t) - u(t)$.

### Solution

The steps are analogous to those leading to the dynamic-stiffness coefficients of the layer built-in at its base for out-of-plane motion (Section 3.9.3). However, important differences do exist, because the unknown is a pressure in contrast to a displacement. The boundary conditions at the free surface and at the bottom are reversed. The boundary condition along the wall involves the unknown variable's derivative and not the variable itself, and the dynamic-stiffness coefficient follows from the variable itself (the pressure) and not from its derivative (the shear stress). These latter differences lead to a dynamic-stiffness coefficient of the fluid with certain properties not encountered in that of the soil.

Wave equation in frequency domain:

$$p(\omega)_{,xx} + p(\omega)_{,zz} + \frac{\omega^2}{c^2} p(\omega) = 0$$

Solution analogous to Eq. 3.195, after enforcing radiation condition (outgoing pressure wave only):

$$p(\omega) = [A \exp(iktz) + B \exp(-iktz)] \exp(-ikx)$$

with wave number $k = \omega/c$ and (Eq. 3.194):

$$t = i\sqrt{1 - \frac{\omega^2}{c^2 k^2}}$$

Boundary condition at $z = 0$: $p(\omega) = 0$:

$$B = -A$$

Boundary condition at $z = d$: $p(\omega)_{,z} = 0$:
Characteristic equation $\cos ktd = 0$
Satisfied for discrete values $j = 1, 2, \ldots, \infty$:

$$k_j t_j d = \frac{(2j - 1)\pi}{2}$$

**350**      Substructure Method Using Dynamic Stiffness of Soil      Chap. 6

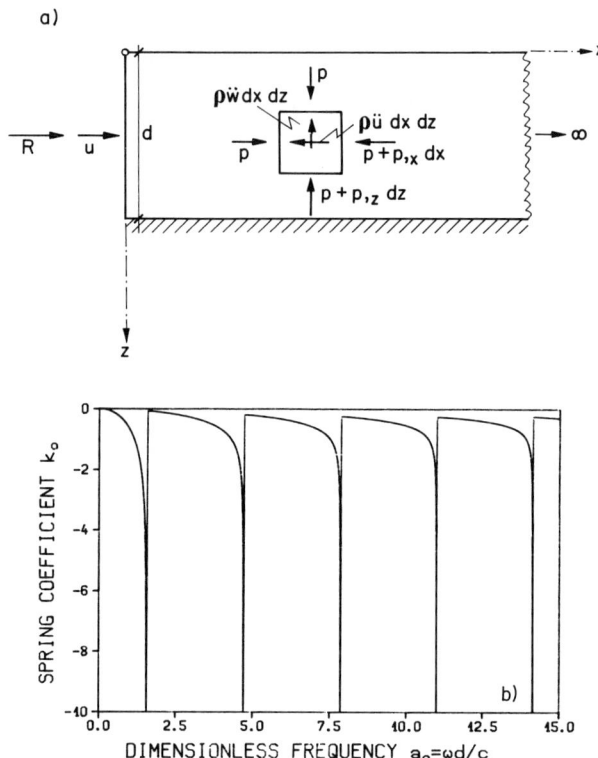

**Figure P6-12** Dam-water interaction. a. Rigid wall with semi-infinite reservoir. b. and c. Dynamic-stiffness coefficient in frequency domain. d. Regular part of dynamic-stiffness coefficient in time domain.

Solution: $p(\omega) = i \sum_{j=1}^{\infty} A_j \sin k_j t_j z \, \exp(-ik_j x)$

Boundary condition at $x = 0$: $p(\omega)_{,x} = \rho\omega^2 u(\omega)$:

$$\sum_{j=1}^{\infty} A_j k_j \sin k_j t_j z = \rho\omega^2 u(\omega)$$

Fourier series expansion with terms $\sin k_j t_j z$:

$$\rho\omega^2 u(\omega) = \left( \sum_{j=1}^{\infty} \frac{4}{(2j-1)\pi} \sin k_j t_j z \right) \rho\omega^2 u(\omega)$$

$$A_j = \frac{4}{(2j-1)\pi} \frac{1}{k_j} \rho\omega^2 u(\omega)$$

Interaction force $R(\omega) = \int_o^d p(\omega) \, dz$ at $x = 0$:

$$R(\omega) = i \left( \sum_{j=1}^{\infty} \frac{8}{(2j-1)^2 \pi^2} \frac{1}{k_j} \right) d\rho\omega^2 u(\omega)$$

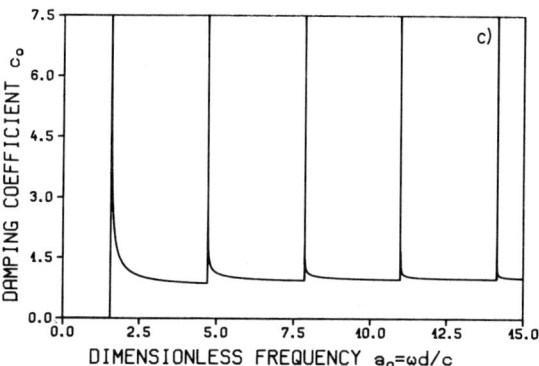

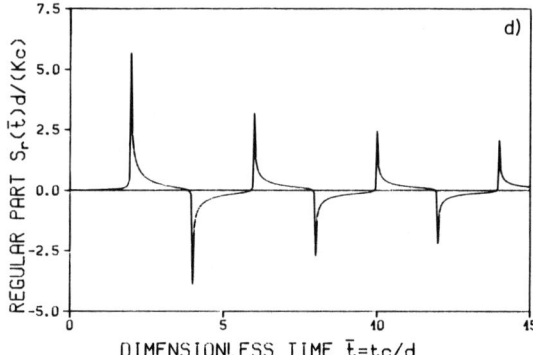

**Figure P6-12** Continued

with (Eq. 3.206): $k_j = -i(2j-1)\dfrac{\pi}{2}\sqrt{1 - \dfrac{4a_o^2}{(2j-1)^2\pi^2}}\dfrac{1}{d}$

$$R(a_o) = 16K \sum_{j=1}^{\infty} \frac{1}{(2j-1)^3\pi^3} \frac{1}{\sqrt{1 - \dfrac{4a_o^2}{(2j-1)^2\pi^2}}} [-a_o^2\, u(a_o)]$$

$-a_o^2 u(a_o)$ represents the amplitude of the acceleration in dimensionless quantities. The square root appears in the denominator and not in the numerator, as in Eq. 3.207.

Incompressible fluid ($K = \infty \rightarrow a_o = 0$):

$$R(\omega) = 16d^2\rho \sum_{j=1}^{\infty} \frac{1}{(2j-1)^3\pi^3} [-\omega^2 u(\omega)]$$

$$R(\omega) = m\ddot{u}(\omega)$$

with the frequency-independent mass:

$$m = \frac{16}{\pi^3} d^2\, \rho \sum_{j=1}^{\infty} \frac{1}{(2j-1)^3} = 0.543\, d^2\rho$$

Dynamic-stiffness coefficient in frequency domain (compressible fluid):

$$S(a_o) = -16K a_o^2 \sum_{j=1}^{\infty} \frac{1}{(2j-1)^2 \pi^2 \sqrt{(2j-1)^2\pi^2 - 4a_o^2}}$$

Spring and damping coefficients:

$$\frac{S(a_o)}{K} = k_o(a_o) + ia_o c_o(a_o)$$

$k_o(a_o)$ and $c_o(a_o)$ are plotted versus $a_o$ in Fig. P 6-12b and c. Below the cutoff frequency $a_o = \pi/2$, which is equal to the reservoir's fundamental frequency, $c_o$ vanishes. Discontinuities involving infinite values of $c_o$ and $k_o$ exist at all natural frequencies $a_o = (2j-1)\pi/2$.

Fourier transformation (Eq. 6.37):

$$\bar{S}(\bar{t}) = \frac{c}{2\pi d} \int_{-\infty}^{+\infty} S(a_o) \exp(ia_o \bar{t})\, da_o$$

Asymptotic value:

$$\lim_{a_o \to \infty} S(a_o) = 8K \sum_{j=1}^{\infty} \frac{1}{(2j-1)^2 \pi^2} i a_o$$

Decomposition in singular and remaining regular parts:

$$S(a_o) = 8K \sum_{j=1}^{\infty} \frac{1}{(2j-1)^2\pi^2} \left( ia_o - \frac{a_o^2}{\sqrt{\frac{(2j-1)^2\pi^2}{4} - a_o^2}} - ia_o \right)$$

$$= 8K \sum_{j=1}^{\infty} \frac{1}{(2j-1)^2\pi^2} \left( ia_o - \frac{\frac{(2j-1)^2\pi^2}{4}}{\sqrt{\frac{(2j-1)^2\pi^2}{4} - a_o^2}} + \sqrt{\frac{(2j-1)^2\pi^2}{4} - a_o^2} - ia_o \right)$$

The Fourier transformation of the first term within the parenthesis equals $d\bar{\delta}(\bar{t})/d\bar{t}$; that of the second term is specified in Ref. [C1, No. 557); and that of the third and fourth terms combined is specified in Ref. [C1, No. 556.1]:

$$\bar{S}(\bar{t}) = 8K \frac{c}{d} \sum_{j=1}^{\infty} \frac{1}{(2j-1)^2\pi^2} \left( \frac{d\bar{\delta}(\bar{t})}{d\bar{t}} - \frac{(2j-1)^2\pi^2}{4} J_o\left[\frac{(2j-1)\pi}{2}\bar{t}\right] \right.$$
$$\left. + \frac{(2j-1)\pi}{2\bar{t}} J_1\left[\frac{(2j-1)\pi}{2}\bar{t}\right] \right)$$

The regular part $S_r(\bar{t})$, consisting of the second and third term, is plotted in dimensionless form in Fig. P 6-12d (using 40 terms in the series).

Force–displacement relationship (Eq. 6.29):

$$R(t) = \int_o^t S(t - \tau) u(\tau) d\tau$$

$$R(t) = 8K \sum_{j=1}^{\infty} \left\{ \frac{1}{(2j-1)^2 \pi^2} \frac{d}{c} \ddot{u}(t) - \frac{c}{4d} \int_o^t J_o \left[ \frac{(2j-1)\pi}{2} \frac{c}{d}(t - \tau) \right] u(\tau) d\tau \right.$$

$$\left. + \frac{1}{2} \int_o^t \frac{1}{(2j-1)\pi} \frac{1}{t - \tau} J_1 \left[ \frac{(2j-1)\pi}{2} \frac{c}{d}(t - \tau) \right] u(\tau) d\tau \right\}$$

Incompressible fluid:

$$R(t) = m \ddot{u}(t)$$

with $m = 0.543 \, d^2 \, \rho$.

**6.13** The semi-infinite rod on an elastic foundation examined in Section 3.2 exhibits a cutoff frequency, and the wave pattern is dispersive. The area of the rod is denoted by $A$, the modulus of elasticity by $E$, the mass density by $\rho$, and the static spring stiffness per unit length of the elastic foundation by $k_g$ (Fig. 3-4a). The rod is subjected to a prescribed support movement at point 0.

$$u_0(\bar{t}) = \frac{u_0}{2} \left[ 1 - \cos(2\pi \frac{\bar{t}}{\bar{t}_0}) \right] \quad 0 < \bar{t} < \bar{t}_0$$

$$u_0(\bar{t}) = 0 \quad \bar{t} > \bar{t}_0$$

$u_0(\bar{t})$ denotes the axial displacement, and $\bar{t}$ denotes the dimensionless time $\bar{t} = tc_I \kappa$. $\bar{t}_0$ equals 2.

Evaluate the convolution integral for the reaction force recursively in the frequency domain [M4]. Select the dimensionless period $\bar{T} = 40.96$ and $\Delta \bar{t} = 0.02$. Calculate the reaction force, first using the total dynamic-stiffness coefficient (singular and regular parts) with $N^\omega = 400$ and a decay function with $\bar{t}_d = 1.2$, and then using the regular part only with $N^\omega = 40$ and no decay function.

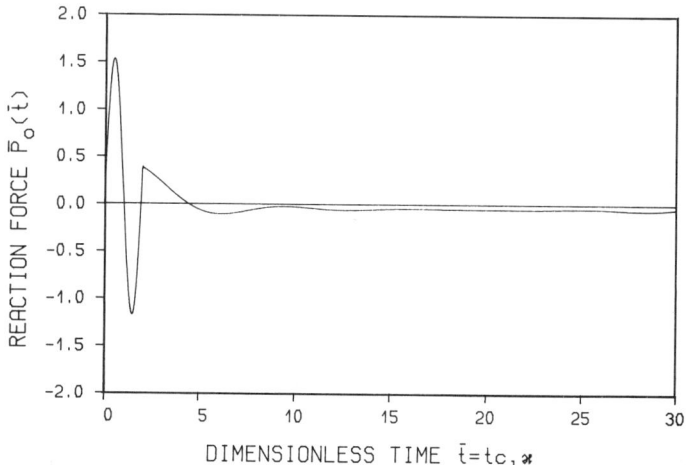

**Figure P6-13** Reaction force calculated recursively in frequency domain.

## Solution

The dimensionless reaction force $\bar{P}_o(\bar{t}) = P_o(t)/(Ku_o)$ is evaluated first based on Eq. 6.172 with Eq. 6.177, and then on Eq. 6.178. From a practical point of view, the results are identical (Fig. P 6-13) and agree well with the solution using Eq. 3.85 (Fig. 3-8b).

**6.14** Determine the recursive equation for the spherical cavity of radius $a$, shear modulus $G$, shear-wave and dilatational velocities $c_s$ and $c_p$ in non-dimensional form. Assume the displacements are discretized as impulses at the center of the time steps. Apply the impulse-invariant method that leads to a system of equations to be solved and the z-transform technique that directly results in the explicit expression for the recursive equation.

## Solution

As the regular part $S(a_o)$ is a ratio of two polynomials of degree 0 in the numerator and of degree 1 in the denominator (Eq. 6.44), the recursive equation with $M = 1$, $L = 0$ is exact.

$$R_n = a_1 R_{n-1} + b_o U_n$$

with the impulse $U_n\left(= u\left[\left(n - \frac{1}{2}\right)\Delta\bar{t}\right]\Delta\bar{t}\right)$

$$S(\bar{t}) = K \frac{c_p^2}{4c_s^2} \frac{c_p}{a} \exp(-\bar{t}) \qquad \text{(Eq. 6.28)}$$

with $K = 4G/a$ and $\bar{t} = tc_p/a$

Impulse-invariant method.
Unit-impulse displacement discretized at center of first time step: $u(\Delta\bar{t}/2) = 1$

$n = 1 \quad R_1 = S\left(\frac{\Delta\bar{t}}{2}\right) u\left(\frac{\Delta\bar{t}}{2}\right) \Delta\bar{t} = b_o u\left(\frac{\Delta\bar{t}}{2}\right) \Delta\bar{t}$

$n = 2 \quad R_2 = S\left(\frac{3\Delta\bar{t}}{2}\right) u\left(\frac{\Delta\bar{t}}{2}\right) \Delta\bar{t} = a_1 R_1 = a_1 S\left(\frac{\Delta\bar{t}}{2}\right) u\left(\frac{\Delta\bar{t}}{2}\right) \Delta\bar{t}$

$$b_o = S\left(\frac{\Delta\bar{t}}{2}\right) = K \frac{c_p^2}{4c_s^2} \frac{c_p}{a} \exp\left(-\frac{\Delta\bar{t}}{2}\right)$$

$$a_1 = \frac{S\left(\frac{3\Delta\bar{t}}{2}\right)}{S\left(\frac{\Delta\bar{t}}{2}\right)} = \exp(-\Delta\bar{t})$$

$$R_n = \exp(-\Delta\bar{t})\, R_{n-1} + K \frac{c_p^2}{4c_s^2} \frac{c_p}{a} \exp\left(-\frac{\Delta\bar{t}}{2}\right) U_n$$

$$R_n = \exp(-\Delta\bar{t})\, R_{n-1} + K \frac{c_p^2}{4c_s^2} \exp\left(-\frac{\Delta\bar{t}}{2}\right) u\left[\left(n - \frac{1}{2}\right)\Delta\bar{t}\right]\Delta\bar{t}$$

z-transform.

$$S(z) = \sum_{n=0}^{\infty} S(n) z^{-n}$$

$$= K \frac{c_p^2}{4c_s^2} \frac{c_p}{a} \sum_{n=0}^{\infty} \exp\left[-\frac{(2n+1)}{2} \overline{\Delta t}\right] z^{-n}$$

$$= K \frac{c_p^2}{4c_s^2} \frac{c_p}{a} \frac{\exp\left(-\frac{\overline{\Delta t}}{2}\right)}{1 - \exp(-\overline{\Delta t}) z^{-1}}$$

Input-output relationship in z-domain.

$$R(z) = S(z) U(z)$$

$$R_n = \exp(-\overline{\Delta t}) R_{n-1} + K \frac{c_p^2}{4c_s^2} \frac{c_p}{a} \exp\left(-\frac{\overline{\Delta t}}{2}\right) U_n$$

**6.15** Formulate the recursive equation for a one-degree-of-freedom system and for $M = L = 1$ based on a flexibility formulation of the impulse-invariant method. Introduce the unit-impulse forces in the form of a triangle (zero at $t = 0$, increasing linearly up to $t = \Delta t$ and then decreasing linearly up to $t = 2\Delta t$). Assume that the dynamic-flexibility coefficient $F(t)$ is piecewise linear over each time step.

*Solution*

Replacing $R$, $S$, and $u$ by $u$, $F$, and $R$, respectively, in Eq. 6.209 leads to

$$u_n = \frac{F(\Delta t) + 4F(2\Delta t) + F(3\Delta t)}{F(0) + 4F(\Delta t) + F(2\Delta t)} u_{n-1} + \frac{2F(0) + F(\Delta t)}{6} R_n \Delta t$$

$$+ \frac{1}{6}\left[F(0) + 4F(\Delta t) + F(2\Delta t) - \frac{F(\Delta t) + 4F(2\Delta t) + F(3\Delta t)}{F(0) + 4F(\Delta t) + F(2\Delta t)}(2F(0) + F(\Delta t))\right] R_{n-1} \Delta t$$

or rearranged

$$R_n = \left(-\frac{F(0) + 4F(\Delta t) + F(2\Delta t)}{2F(0) + F(\Delta t)} + \frac{F(\Delta t) + 4F(2\Delta t) + F(3\Delta t)}{F(0) + 4F(\Delta t) + F(2\Delta t)}\right) R_{n-1}$$

$$+ \frac{6}{\Delta t} \frac{1}{2F(0) + F(\Delta t)} u_n - \frac{6}{\Delta t} \frac{1}{2F(0) + F(\Delta t)} \frac{F(\Delta t) + 4F(2\Delta t) + F(3\Delta t)}{F(0) + 4F(\Delta t) + F(2\Delta t)} u_{n-1}$$

**6.16** The semi-infinite rod on an elastic foundation (Fig. 3-4a) with area $A$, modulus of elasticity $E$, mass density $\rho$ and spring coefficient per unit length $k_g$ is subjected to a prescribed support movement at point 0.

$$u_o(\bar{t}) = \frac{u_o}{2}\left[1 - \cos\left(2\pi \frac{\bar{t}}{\bar{t}_o}\right)\right] \quad 0 < \bar{t} < \bar{t}_o$$

$$u_o(\bar{t}) = 0 \quad \bar{t} > \bar{t}_o$$

The dimensionless time $\bar{t} = tc_l\kappa$ with $c_l = \sqrt{E/\rho}$ and $\kappa = \sqrt{k_g/(EA)}$ and $\bar{t}_o = 2$. Determine the dimensionless reaction force $\bar{P}_o(\bar{t}) = P_o(t)/(Ku_o)$ with a recursive equation based on the flexibility formulation of the impulse-invariant method with $M = L = 3$ ($K = 3$) and $\Delta \bar{t} = 0.1$. This problem is used as a benchmark in Chapter 3 (see Section 3.2.5). The exact solution can also be determined by evaluating the convolution integral directly using either the dynamic-stiffness coefficient or the flexibility coefficient in the time domain (Problem 6.9). The recursive solution based on the stiffness formulation of the impulse-invariant method is addressed in Section 6.9.6.

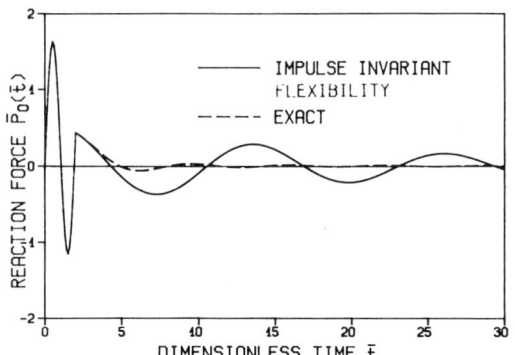

**Figure P6-16** Recursive calculation in time domain based on flexibility formulation of impulse-invariant method.

*Solution*

The flexibility formulation of the impulse-invariant method follows from the corresponding stiffness formulation (Section 6.9.2) by replacing $\{R\}$, $[S]$, $\{u\}$ by $\{u\}$, $[F]$, $\{R\}$, respectively. In particular, the unknown matrices $[a]_i$ and $[b]_i$ appearing in the recursive equation

$$\{u\}_n = \sum_{i=1}^{M} [a]_i \{u\}_{n-i} + \Delta t \sum_{i=0}^{L} [b]_i \{R\}_{n-i}$$

follow from Eq. 6.203 with the definition in Eq. 6.204, modified accordingly.

The reaction force is plotted in Fig. P6-16. The deviations from the exact solution are about the same as those for the corresponding stiffness formulation (Fig. 6-33).

**6.17** The convolution integral for $u_n = u(t = n\Delta t)$ in the flexibility formulation is specified as

$$u(t) = \int_o^t F(t - \tau) R(\tau) \, d\tau$$

Decompose this expression into two integrals. The first one from 0 to $t - \Delta t$ can be expanded in a Taylor series at $t - \Delta t$, which allows the integral to be expressed as a function of $u_{n-1}$, $R_{n-1}$ and $\dot{u}_{n-1}$. The second one from $t - \Delta t$ to $t$ will involve $R_{n-1}$ and $R_n$ assuming for the sake of conciseness a linear variation of $F(t)$ over the first time step. Derive the resulting recursive equation. Note that this approach is attractive, as $\ddot{u}_{n-1}$ is calculated in a numerical time integration scheme.

*Solution*

$$u(t) = \int_0^{t-\Delta t} F(t-\tau) R(\tau) d\tau + \int_{t-\Delta t}^{t} F(t-\tau) R(\tau) d\tau$$

$$\int_0^{t-\Delta t} F(t-\tau) R(\tau) d\tau = \int_0^{t-\Delta t} [F(t-\Delta t-\tau) + \Delta t \, \dot{F}(t-\Delta t-\tau)] R(\tau) d\tau$$

$$u(t-\Delta t) = \int_0^{t-\Delta t} F(t-\Delta t-\tau) R(\tau) d\tau$$

$$\dot{u}(t-\Delta t) = \int_0^{t-\Delta t} \dot{F}(t-\Delta t-\tau) R(\tau) d\tau + F(0) R(t-\Delta t)$$

$$\int_0^{t-\Delta t} F(t-\tau) R(\tau) d\tau = u(t-\Delta t) + \Delta t \, [\dot{u}(t-\Delta t) - F(0) R(t-\Delta t)]$$

$$\int_{t-\Delta t}^{t} F(t-\tau) R(\tau) d\tau = \frac{2F(0) + F(\Delta t)}{6} R_n \Delta t + \frac{F(0) + 2F(\Delta t)}{6} R_{n-1} \Delta t$$

for linear variation of $F$ and $R$

$$u_n = u_{n-1} + \Delta t \dot{u}_{n-1} + \frac{2F(0) + F(\Delta t)}{6} R_n \Delta t + \frac{-5F(0) + 2F(\Delta t)}{6} R_{n-1} \Delta t$$

**6.18** When $S(\omega)$ can be written as a ratio of two polynomials in $i\omega$

$$S(\omega) = \frac{p_o + p_1(i\omega) + p_2(i\omega)^2 + \cdots + p_L(i\omega)^L}{1 + q_1(i\omega) + q_2(i\omega)^2 + \cdots + q_M(i\omega)^M}$$

the following differential equation in the time domain relating the displacements $u$ to the forces $R$ applies for a system with causal behavior ($H$ = Heaviside function)

$$RH + q_1(RH)\dot{} + q_2(RH)\ddot{} + \cdots + q_M(RH)^{(M)}$$

$$= p_o uH + p_1(uH)\dot{} + p_2(uH)\ddot{} + \cdots + p_L(uH)^{(L)}$$

The superscript $i$ in parenthesis denotes the $i$th derivative with respect to time. The required initial conditions for $R, \dot{R}, \ldots, R^{(M-1)}$ follow from setting the coefficients of $\delta, \dot{\delta}, \ldots, \delta^{(M-1)}$ in the differential equation equal to zero.

On the wall of a spherical cavity of radius $a$ embedded in an infinite space the radial displacement $u(t)$ (which does not vary along the surface) is specified as

$$u(t) = \frac{q_o a}{4G} \left[1 - \exp\left(-\frac{c_p t}{a}\right)\right] H(t)$$

$G$ denotes the shear modulus and $c_p$ the dilatational-wave velocity. The pressure $q_o$ will turn out to be the value at $t = 0$ (for $c_p/c_s = 2$, that is $\nu = 1/3$).

Determine the wall pressure $p(t)$ by solving the differential equation subjected to the appropriate initial conditions.

## Solution

Spherical cavity with symmetric waves.
Total dynamic-stiffness coefficient (Eq. 2.24).

$$S(\omega) = \frac{4G}{a} \frac{1 + a/c_p(i\omega) + a^2/(4c_s^2)(i\omega)^2}{1 + a/c_p(i\omega)}$$

$$q_1 = \frac{a}{c_p}$$

$$p_o = \frac{4G}{a} \qquad p_1 = \frac{4G}{c_p} \qquad p_2 = \frac{Ga}{c_s^2}$$

Differential equation.

$$p(t)\,H(t) + q_1[p(t)\,H(t)]^\cdot = p_o u(t)\,H(t) + p_1[u(t)\,H(t)]^\cdot + p_2[u(t)H(t)]^{\cdot\cdot}$$

or

$$p(t)H(t) + q_1[\dot{p}(t)H(t) + p(t = 0^+)\,\delta(t)] = p_o u(t)H(t)$$
$$+ p_1[\dot{u}(t)H(t) + u(t = 0^+)\,\delta(t)] + p_2[\ddot{u}(t)H(t) + \dot{u}(t = 0^+)\,\delta(t) + u(t = 0^+)\,\dot{\delta}(t)]$$

Coefficient of $H(t)$.

$$p(t) + q_1 \dot{p}(t) = p_o u(t) + p_1 \dot{u}(t) + p_2 \ddot{u}(t)$$

Coefficient of $\delta(t)$ and $\dot{\delta}(t)$.

$$q_1 p(t = 0^+) = p_1 u(t = 0^+) + p_2 \dot{u}(t = 0^+)$$

$$p_2 u(t = 0^+) = 0$$

Differential equation.

$$p(t) + \frac{a}{c_p}\dot{p}(t) = q_o\left[1 - \frac{c_p^2}{4c_s^2}\exp\left(-\frac{c_p t}{a}\right)\right]$$

Initial condition.

$$p(t = 0^+) = q_o \frac{c_p^2}{4c_s^2}$$

Solution.

$$p(t) = q_o\left\{1 - \left[1 - \frac{1}{4}\frac{c_p^2}{c_s^2} + \frac{c_p t}{a}\frac{1}{4}\frac{c_p^2}{c_s^2}\right]\exp\left(-\frac{c_p t}{a}\right)\right\}$$

The same problem is addressed in Section 7.4.1 solving directly the differential equation of motion expressed in displacements.

**6.19** Recursive equations in the frequency domain are derived in Section 6.8. The interaction force-displacement relationship is formulated as follows. Based on Eqs. 6.173, 6.175, and 6.176, the displacement amplitude is specified as

$$\{u(\omega_j)\}_n = \{u(\omega_j)\}_{n-1} + c_1(\omega_j)\{u\}_{n-1} + c_2(\omega_j)\{u\}_n$$

whereby the left-hand side of Eq. 6.173 is denoted as $\{u(\omega_j)\}_n$ and the subscript $b$ and the tilde are omitted. The units of $\{u(\omega_j)\}$ and $\{u\}_n$ are length multiplied by time and length, respectively. Multiplying this equation by the regular part of the dynamic-stiffness matrix denoted as $[S(\omega_j)]$ and substituting Eq. 6.162 results in

$$\{R(\omega_j)\}_n = \{R(\omega_j)\}_{n-1} + c_1(\omega_j)[S(\omega_j)]\{u\}_{n-1} + c_2(\omega_j)[S(\omega_j)]\{u\}_n$$

with $\{R(\omega_j)\}_n$ denoting the amplitude of the regular part of the interaction forces. $\{R\}_n$ is formulated in Eq. 6.163 as

$$\{R\}_n = \frac{1}{N^T \Delta t} \sum_{j=0}^{N^T-1} \{R(\omega_j)\}_n \exp(i\omega_j n \Delta t)$$

with the period $T$ divided into $N^T$ time intervals of length $\Delta t$.

Similar recursive equations can be derived based on the $z$-transformation [01]. The right-sided $z$-transform is defined as

$$[S(z)] = \sum_{n=0}^{\infty} [S]_n z^{-n}$$

with $[S]_n$ specified as the discrete inverse Fourier transform

$$[S]_n = \frac{1}{N^T \Delta t} \sum_{j=0}^{N^T-1} [S(\omega_j)] \exp(i\omega_j n \Delta t)$$

After substitution and change of order of the sums

$$[S(z)] = \frac{1}{N^T \Delta t} \sum_{j=0}^{N^T-1} [S(\omega_j)] \sum_{n=0}^{\infty} \exp(i\omega_j n \Delta t) z^{-n}$$

$$= \frac{1}{N^T \Delta t} \sum_{j=0}^{N^T-1} [S(\omega_j)] \frac{1}{1 - \exp(i\omega_j \Delta t) z^{-1}}$$

results. Performing the inverse $z$-transformation based on the so-called parallel form (Section 6.9.4) leads to

$$\{R^j\}_n = \exp(i\omega_j \Delta t)\{R^j\}_{n-1} + \frac{1}{N^T \Delta t}[S(\omega_j)]\{U\}_n$$

$$\{R\}_n = \sum_{j=0}^{N^T-1} \{R^j\}_n$$

Substituting

$$\{U\}_n = \frac{\Delta t}{2}(\{u\}_{n-1} + \{u\}_n)$$

results in

$$\{R^j\}_n = \exp(i\omega_j \Delta t)\{R^j\}_{n-1} + \frac{1}{2N^T}[S(\omega_j)](\{u\}_{n-1} + \{u\}_n)$$

Prove that the recursive equation derived based on the $z$-transformation is from a practical point of view identical to that discussed in Section 6.8.

*Solution*

Result of Section 6.8, as specified in first paragraph of problem

$$\{R\}_n = \sum_{j=0}^{N^T-1} \{\overline{R^j}\}_n$$

with

$$\{\overline{R^j}\} = \frac{\exp(i\omega_j n\Delta t)}{N^T \Delta t} \{R(\omega_j)\}_{n-1}$$

$$+ \frac{\exp(i\omega_j n\Delta t)}{N^T \Delta t} c_1(\omega_j)[S(\omega_j)] \{u\}_{n-1}$$

$$+ \frac{\exp(i\omega_j n\Delta t)}{N^T \Delta t} c_2(\omega_j)[S(\omega_j)] \{u\}_n$$

First term on right-hand side

$$\exp(i\omega_j \Delta t) \frac{\exp[i\omega_j (n-1)\Delta t]}{N^T \Delta t} \{R(\omega_j)\}_{n-1} = \exp(i\omega_j \Delta t) \{\overline{R^j}\}_{n-1}$$

Second and third terms on right-hand side, using (slightly modified) definitions of $c_1(\omega_j)$, $c_2(\omega_j)$ (Eq. 6.175)

$$c_1(\omega_j) \cong c_2(\omega_j) = \exp(-i\omega_j n\Delta t) \frac{\Delta t}{2}$$

$$\frac{\exp(i\omega_j n\Delta t)}{N^T \Delta t} [S(\omega_j)](c_1(\omega_j)\{u\}_{n-1} + c_2(\omega_j)\{u\}_n) = \frac{1}{2N^T}[S(\omega_j)](\{u\}_{n-1} + \{u\}_n)$$

Complete agreement with result of derivation based on *z*-transformation.

# 7

# Substructure Method Using Green's Function of Soil

In the substructure method of analyzing soil–structure interaction, the interaction force–displacement relationship of the unbounded soil with respect to the degrees of freedom of the nodes located on the structure–soil interface in the time domain is determined. This contribution to the basic equation of motion is expressed in Chapter 6 in the form of convolution integrals involving either the dynamic-stiffness coefficients in the time domain and the corresponding motions or the dynamic-flexibility coefficients in the time domain and the corresponding interaction forces. To calculate the dynamic-stiffness or -flexibility coefficients in the frequency domain, which are then transformed to the time domain, use is made of the dynamic-flexibility coefficients in the frequency-wave number domain of the fictitious loads (Green's functions) when applying the boundary-element method in the frequency domain (Section 4.4.1). As many operations as possible are done in the frequency domain. The transformation back to the time domain is performed at the end of the method (on the level of the dynamic-stiffness or -flexibility coefficients).

In this chapter, an alternative procedure, based on the same assumptions as those discussed in Chapter 6, is described. However, the sequence of operations is different. The inverse transformation back to the time domain is already performed on the level of the dynamic-flexibility coefficients of the fictitious loads called Green's functions. The boundary-element method is thus applied directly in the time domain in this computational procedure. The unbounded soil's contribution to the basic equation of motion consists, in general, of convolution integrals of the displacement–force relationship in the time domain and of the interaction forces' history. As far as the variables appearing in the convolution integrals are concerned, this procedure based on Green's functions in the time domain is similar to the flexibility formulation of Chapter 6. However, many different techniques exist. In the direct boundary-element method, for instance, convolution integrals involving the time-histories of the surface tractions and the displacements will arise. The convolution integrals can also be evaluated recursively in the frequency and time domains as discussed in Sections 6.8 and 6.9.

For all the boundary-element methods discussed in the following, Green's functions for fictitious loads must be determined. These solutions in the time domain are equal to the displacements and surface tractions in a dynamic system, which can easily be analyzed. The procedures to calculate Green's functions for a layered half-space are discussed in Section 7.1. The spherical cavity with symmetric waves is again used for illustration. Also addressed is the full infinite homogeneous isotropic space, for which explicit expressions in the time domain exist. These can be used to develop the significant differences between the Green's functions for the two- and three-dimensional cases. The time-dependent boundary-integral equation, on which the boundary-element method is based—in particular, the elasto-dynamic reciprocity theorem—is examined in Section 7.2. The spatial and temporal discretizations of the different formulations of the boundary-integral equations are discussed in Section 7.3. The weighted-residual, the various forms of the indirect boundary-element, and the direct boundary-element methods are addressed and their numbers of operations are compared. The remainder of this chapter deals with applications: the cavity with a prescribed displacement is solved with all formulations as an illustrative example (Section 7.4); a rigid disk's dynamic flexibility coefficient is determined based on the indirect boundary-element method in the time domain and is compared to the Fourier transform of the corresponding value in the frequency domain (Section 7.5); and, finally, a structure with a partial basemat uplift (Section 7.6) and an embedded foundation with separation of the sidewall and uplift of the basemat (Section 7.7) are examined as examples of actual practice.

## 7.1 GREEN'S FUNCTION

### 7.1.1 Spherical Cavity with Symmetric Waves

**Calculation in time domain.** In all boundary-element methods, Green's functions—that is, displacements and surface tractions for prescribed fictitious loads—must be determined in a system that can easily be analyzed. These functions are calculated here for the system of a cavity of radius $a/2$ embedded in the same infinite space (Fig. 7-1) analyzed in Section 6.2.1. In a general soil–structure-interaction analysis, this system represents the free

## Sec. 7.1 Green's Function

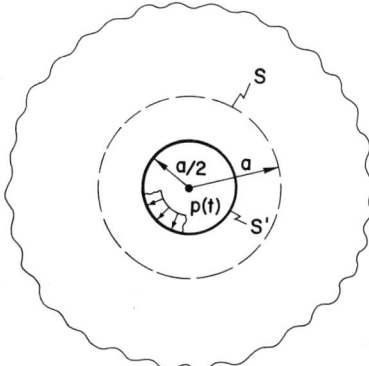

**Figure 7-1** Spherical cavity of radius a/2 (system free field).

field. Actually, the continuous infinite space without any cavity could have been used. As it turns out, however, the corresponding Green's functions are more complicated. Using the same nomenclature as for the calculations in the frequency domain (Section 4.4.3), $S'$ in Fig. 7-1 denotes the source surface, where the fictitious loads of the Green's functions are applied. $S$ represents the receiver surface (also called the observation surface)—that is, the structure–soil interface—where the Green's functions are evaluated (for the illustrative example of the cavity discussed in Section 7.4). Either an impulse pressure of unit value acting as a Dirac or a rectangular pulse pressure of unit value acting during $\Delta t$ (from time zero to $\Delta t$) will be used for the temporal discretization.

The displacements caused by a unit impulse pressure applied at the radius $a$ are calculated in Section 6.4.1. Replacing $a$ by $a/2$ (also for the derivation in the definition of $\alpha$, Eq. 6.101) and setting $r = a$ in Eq. 6.109 leads to the Green's function for the displacement $g_u(t)$ on $S$:

$$g_u(t) = \frac{a}{32G}\left[4\alpha \cos\beta\tau_1 - \frac{\alpha^2}{\beta}\left(4 - \frac{c_p^2}{c_s^2}\right)\sin\beta\tau_1\right]\exp(-\alpha\tau_1)H(\tau_1) \quad (7.1)$$

with

$$\tau_1 = t - \frac{a}{2c_p} \quad (7.2)$$

and where $\alpha$ and $\beta$ are still defined by Eqs. 6.101 and 6.108,

Substituting Eq. 6.109 with the same modifications into Eq. 6.24 leads to the radial stress $\sigma_r(r = a, t)$ and thus, with a change in sign, to the Green's function for the surface traction $g_t(t)$ on $S$:

$$g_t(t) = \frac{1}{2}\delta(\tau_1) + \frac{1}{8}\left[-4\alpha \cos\beta\tau_1 - \frac{\alpha^2}{\beta}\left(3\frac{c_p^2}{c_s^2} - 4\right)\sin\beta\tau_1\right]\exp(-\alpha\tau_1)H(\tau_1)$$

$$(7.3)$$

The Green's functions $g_u(t)$ and $g_t(t)$ for the impulse load are shown in Fig. 7-2, whereby the dimensionless time $\bar{t}$ is introduced (Eq. 6.26). The presence of the singular Dirac delta term for the surface traction in Eq. 7.3 is indicated by an arrow in Fig. 7-2b.

For the rectangular pulse, the differential equation of motion can again be solved directly in the time domain. It is appropriate to define the rectangular pulse as the difference of two Heaviside step functions—the first applied at $t = 0$ and the second at $t = \Delta t$. Proceeding as in Section 6.4.1 leads to the following Green's functions for the displacement $g_u(t)$ and surface traction $g_t(t)$ on the surface $S$ arising from the fictitious rectangular pulse applied on $S'$:

$$g_u(t) = \frac{a}{32G}\left\{1 + \left[3\frac{\alpha}{\beta}\sin\beta\tau_1 - \cos\beta\tau_1\right]\exp(-\alpha\tau_1)\right\}H(\tau_1)$$

$$-\frac{a}{32G}\left\{1 + \left[3\frac{\alpha}{\beta}\sin\beta\tau_2 - \cos\beta\tau_2\right]\exp(-\alpha\tau_2)\right\}H(\tau_2)$$
(7.4)

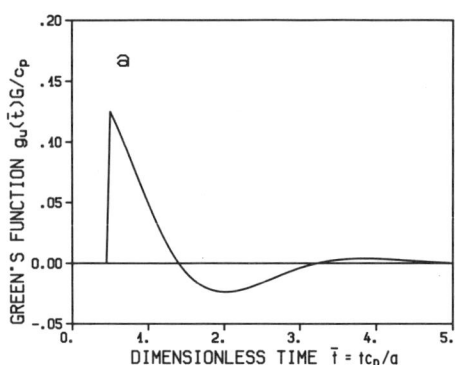

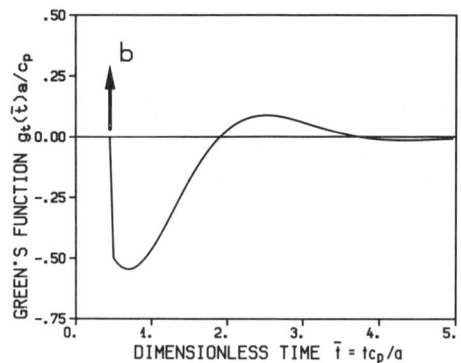

**Figure 7-2** Green's functions for Dirac impulse in time domain.
a. Displacement. b. Surface traction.

Sec. 6.1  Green's Function

$$g_t(t) = \frac{1}{8}\left\{1 + \left[-\frac{\alpha}{\beta}\sin\beta\tau_1 + 3\cos\beta\tau_1\right]\exp(-\alpha\tau_1)\right\}H(\tau_1) \qquad (7.5)$$

$$-\frac{1}{8}\left\{1 + \left[-\frac{\alpha}{\beta}\sin\beta\tau_2 + 3\cos\beta\tau_2\right]\exp(-\alpha\tau_2)\right\}H(\tau_2)$$

with

$$\tau_2 = t - \Delta t - \frac{a}{2c_p} \qquad (7.6)$$

The corresponding Green's functions are plotted in Fig. 7-3.

Alternatively, the Green's functions for a fictitious load having a general variation in time $f(t)$ can be determined from those of the Dirac impulse with the convolution integral $\int_0^\infty f(t - \tau)g(\tau)d\tau$. For the rectangular pulse of unit value and duration $\Delta t$, $f(t) = H(t) - H(t - \Delta t)$ applies.

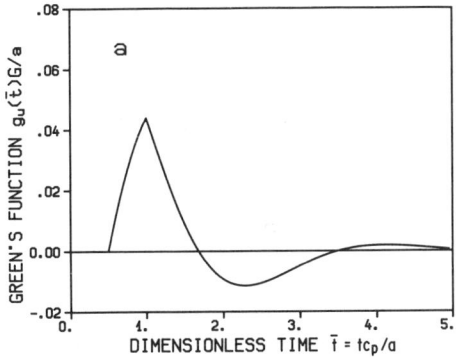

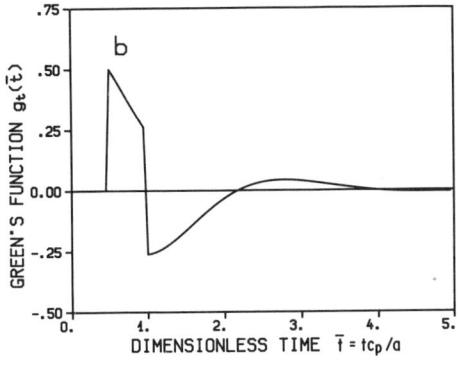

**Figure 7-3** Green's functions for rectangular pulse. a. Displacement. b. Surface traction.

**Transformation from frequency to time domain.** For a more general system—such as the one of the layered half-space—one cannot solve the equations of motion directly in the time domain, as one can for the spherical cavity (Section 7.1.1). Rather, the equations of motion must be transformed to the frequency domain and then solved. This then leads to inverse Fourier transforms to determine the Green's functions in the time domain. For harmonic motion with circular frequency $\omega$, the equation of motion for the cavity with radius $a/2$ embedded in an infinite space is specified as (Eq. 2.18)

$$u(r, \omega)_{,rr} + \frac{2}{r} u(r, \omega)_{,r} - \frac{2}{r^2} u(r, \omega) + \frac{\omega^2}{c_p^2} u(r, \omega) = 0 \qquad (7.7)$$

where $u(r, \omega)$ is the radial displacement's amplitude. For a pressure of unit amplitude applied on the wall at $r = a/2$, the solution of Eq. 7.7, after enforcing the radiation condition, is equal to (analogous to Eqs. 2.22 and 2.23)

$$u(r, a_o) = \frac{a}{8G} \frac{a}{r} \frac{a_o - i \frac{a}{r}}{2a_o + i \left( \frac{1}{4} \frac{c_p^2}{c_s^2} a_o^2 - 4 \right)} \exp \left[ -i \frac{a_o}{a} \left( r - \frac{a}{2} \right) \right] \qquad (7.8)$$

with the dimensionless frequency $a_o$ defined in Eq. 2.12.

Equation 7.8, formulated for $r = a$, is equal to the Green's function for the displacement $g_u(a_o)$

$$g_u(a_o) = \frac{a}{8G} \frac{a_o - i}{2a_o + i \left( \frac{1}{4} \frac{c_p^2}{c_s^2} a_o^2 - 4 \right)} \exp \left( -\frac{i}{2} a_o \right) \qquad (7.9)$$

Substituting Eq. 7.8 in Eq. 6.24 (formulated for the amplitudes) leads, for $r = a$, to the Green's function for the surface traction $g_t(a_o)$:

$$g_t(a_o) = \frac{1}{2} \left[ 1 + \frac{-a_o + 3i}{2a_o + i \left( \frac{1}{4} \frac{c_p^2}{c_s^2} a_o^2 - 4 \right)} \right] \exp \left( -\frac{i}{2} a_o \right) \qquad (7.10)$$

These Green's functions are plotted in Fig. 7-4.

Applying the inverse Fourier transform directly to the Green's functions in the frequency domain leads to the functions in the time domain for a Dirac impulse load. For instance,

$$g_u(t) = \frac{1}{2\pi} \int_{-\infty}^{+\infty} g_u(\omega) \exp(i\omega t) \, d\omega \qquad (7.11a)$$

Sec. 7.1   Green's Function

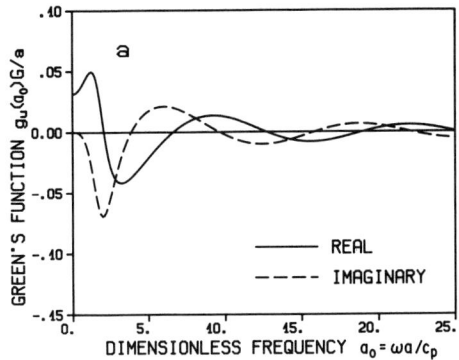

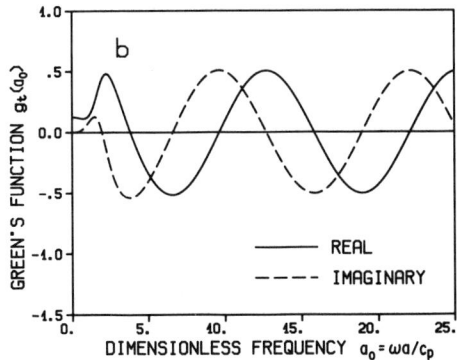

**Figure 7-4** Green's functions for unit pressure amplitude in frequency domain. a. Displacement. b. Surface traction.

holds, or, in dimensionless quantities,

$$g_u(\bar{t}) = \frac{c_p}{2\pi a}\int_{-\infty}^{+\infty} g_u(a_o)\exp(ia_o\bar{t})\,da_o \quad (7.11b)$$

which leads to Eq. 7.1. $g_t(t)$ of Eq. 7.3 is reobtained analogously. These two Green's functions in the time domain for a Dirac impulse are plotted in Fig. 7-2.

For a source load having a general variation in time $f(t)$, the Green's functions of Eqs. 7.9 and 7.10 must be multiplied by $f(\omega)$, which is the Fourier transform of $f(t)$. For instance, for a rectangular pulse of unit value and duration $\Delta t$,

$$f(\omega) = i\,\frac{\exp(-i\omega\Delta t) - 1}{\omega} \quad (7.12)$$

applies. The inverse transformations of $g_u(\omega) \cdot f(\omega)$ and $g_t(\omega) \cdot f(\omega)$ can, for this simple case, be performed analytically, resulting in the expressions of Eqs. 7.4 and 7.5. In more general cases, the transformations must be performed numerically. For the illustrative example, a numerical evaluation of the Green's functions associated with a rectangular pulse load leads to excellent agreement with those obtained analytically. Still better results are obtained for $g_t(t)$ if the first term of Eq. 7.10 is treated analytically. This is discussed when addressing the transformation of the flexibility coefficient (Section 6.4.2).

### 7.1.2 Scalar Wave in Full Infinite Space

**Three-dimensional case.** When discussing the properties of a Green's function and the propagation of waves in three dimensions in general, it is appropriate to address first the scalar-wave equation. For this case, details of directionality at the source and at the receiver do not—in contrast to the elasto-dynamic equation examined in Section 7.1.3—arise, which would tend to obscure the features. The scalar will, in general, be a potential. A clearer physical understanding is gained when the hydrodynamic pressure $p$ is selected as the scalar variable. Analogous to the derivation presented in Problem 6.12, the wave equation in three dimensions is formulated as

$$c^2(p_{,xx} + p_{,yy} + p_{,zz}) - \ddot{p} + f = 0 \qquad (7.13)$$

where the wave velocity $c$ equals

$$c = \sqrt{K/\rho} \qquad (7.14)$$

with the bulk modulus $K$ and the mass density $\rho$. $f$ denotes the source loading, which is spherically symmetric to avoid directionality at the source, located at the origin of the coordinate system. The concentrated source loading, consisting of a Dirac impulse applied at time zero, equals

$$f(r, t) = \delta(r)\,\delta(t) \qquad (7.15)$$

where $r$ and $t$ are the radial coordinate (of the receiver) and the time.

Formulating Eq. 7.13 in spherical coordinates leads to

$$c^2 \nabla^2 p(r, t) - \ddot{p}(r, t) + \delta(r)\,\delta(t) = 0 \qquad (7.16)$$

with the Laplace operator $\nabla^2$ of any scalar $p$ for spherical symmetry defined as

$$\nabla^2 p = \frac{1}{r^2}(r^2 p_{,r})_{,r} = \frac{1}{r}(rp)_{,rr} \qquad (7.17)$$

The solution for zero initial condition, which is equal to the Green's function (fundamental solution) for $p$, $g_p(r, t)$, is formulated as

$$g_p(r, t) = p(r, t) = \frac{1}{4\pi c^2}\frac{\delta\left(t - \dfrac{r}{c}\right)}{r} \qquad (7.18)$$

## Sec. 7.1  Green's Function

This remarkably simple result can be verified by substitution, leading to

$$c^2 \nabla^2 p(r, t) - \ddot{p}(r, t) = \frac{1}{4\pi} \delta\left(t - \frac{r}{c}\right) \nabla^2\left(\frac{1}{r}\right) \tag{7.19}$$

Using

$$\nabla^2\left(\frac{1}{r}\right) = -4\pi \, \delta(r) \tag{7.20}$$

the right-hand side equals $-\delta(t)\,\delta(r)$.

The Green's function (fundamental solution) for the normal derivative on a boundary $g_{p,n}(r, t)$ equals

$$g_{p,n}(r,t) = p_{,n}(r,t) = -\frac{1}{4\pi c^2 r}\left[\frac{\delta\left(t - \dfrac{r}{c}\right)}{r} - \frac{1}{c}\dot{\delta}\left(t - \frac{r}{c}\right)\right] r_{,n} \tag{7.21}$$

with $r_{,n} = \partial r / \partial n$.

If the source load acts at a point specified by a vector $\vec{x}'$ (symbolic coordinate $x'$) at time $\tau$, the Green's function at the receiver point with the vector $\vec{x}$ (coordinate $x$) at time $t$ equals

$$g_p(x, t, x', \tau) = \frac{1}{4\pi c^2} \frac{\delta\left(t - \tau - \dfrac{r}{c}\right)}{r} \tag{7.22}$$

with the distance between the two points $r = |\vec{x} - \vec{x}'|$. The quantity $t - \tau - r/c$ is called the retarded time—that is, the time it takes for the wave to propagate from the source to the receiver. If the source loading, which is still assumed to be concentrated spatially, consists of a function $f(t)$, then Eq. 7.22 still applies, with $f$ replacing $\delta$. If the source loading is extended in addition throughout a volume ($\vec{x}'$ varies), the resulting Green's function can be determined from superposition, whereby the solution depends on the retarded time, which is a function of $\vec{x}'$.

To discuss the features of three-dimensional waves, it is sufficient to address Eq. 7.18. For this three-dimensional case, the response at the receiver differs from zero only at the time $t = r/c$, and it has the same shape as that applied at the source, whereby the amplitude is proportional to the reciprocal of the distance $r$. In particular, the response returns to zero after the wave has passed. The elasto-dynamic wave propagation will exhibit similar properties, but the solution of the two-dimensional case is significantly different, as discussed next.

**Two-dimensional case.** The solution of Eq. 7.16 with Eq. 7.15 for the two-dimensional case in the plane x-z can be determined from that of the three-dimensional case, assuming sources on the y-axis and integrating accordingly (Fig. 7-5). The contribution of an impulse acting on an element $dy$ on the y-axis at a distance $r$ from the receiver which is located in the x-z plane at a distance $r_o$ from the origin is formulated, using Eq. 7.18, as

$$g_p(r_o, t) = \frac{1}{4\pi c^2} \int_{-\infty}^{+\infty} \frac{\delta\left(t - \frac{r}{c}\right)}{r} dy \tag{7.23}$$

with

$$r = \sqrt{r_o^2 + y^2} \tag{7.24}$$

Changing the integration variable to $r$ with $dy = r\, dr/y$ transforms Eq. 7.23 to

$$g_p(r_o, t) = \frac{1}{2\pi c^2} \int_{r_o}^{\infty} \frac{\delta\left(t - \frac{r}{c}\right)}{\sqrt{r^2 - r_o^2}} dr \tag{7.25}$$

or

$$g_p(r_o, t) = \frac{1}{2\pi c^2} \frac{1}{\sqrt{t^2 - \frac{r_o^2}{c^2}}} \qquad t > \frac{r_o}{c}$$

$$= 0 \qquad t < \frac{r_o}{c} \tag{7.26}$$

The Green's function (fundamental solution) for the normal derivative $g_{p,n}(r_o, t)$ is equal to

$$g_{p,n}(r_o, t) = \frac{1}{2\pi c^2}\left[\frac{r_o}{c^2\left(t^2 - \frac{r_o^2}{c^2}\right)^{3/2}} - \frac{\delta\left(t - \frac{r_o}{c}\right)}{c\sqrt{t^2 - \frac{r_o^2}{c^2}}}\right] r_{,n} \qquad t > \frac{r_o}{c}$$

$$= 0 \qquad t < \frac{r_o}{c} \tag{7.27}$$

## Sec. 7.1   Green's Function

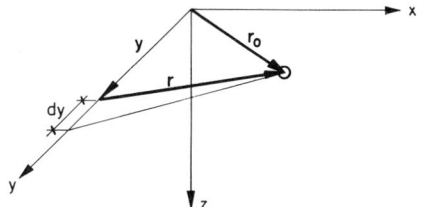

**Figure 7-5** Point-source solution for two-dimensional case, calculated from three-dimensional one through integration.

As another example of the two-dimensional scalar-wave equation, the out-of-plane motion with the displacement $v$ can be addressed. The wave equation (Eq. 3.118) is formulated in the presence of body loads per unit area $f(x, z, t)$:

$$v_{,xx} + v_{,zz} - \frac{\ddot{v}}{c_s^2} + \frac{f}{G} = 0 \tag{7.28}$$

In cylindrical coordinates with symmetry, Eq. 7.28 is transformed to (Eq. 6.83):

$$\nabla^2 v(r_o, t) - \frac{\ddot{v}}{c_s^2}(r_o, t) + \frac{\delta(r_o)\,\delta(t)}{G} = 0 \tag{7.29}$$

with the Laplace operator in symmetric cylindrical coordinates

$$\nabla^2 v = v_{,rr} + \frac{1}{r_o} v_{,r} \tag{7.30}$$

Based on Eq. 7.26, the Green's function (fundamental solution) for the out-of-plane displacement is thus equal to

$$g_v(r_o, t) = v(r_o, t) = \frac{1}{2\pi G} \frac{1}{\sqrt{t^2 - \frac{r_o^2}{c_s^2}}} \quad t > \frac{r_o}{c_s}$$

$$= 0 \quad\quad\quad\quad\quad\quad t < \frac{r_o}{c_s} \tag{7.31}$$

In Problem 7.2, the Green's function in the time domain (Eq. 7.31) is determined by the inverse Fourier transformation from the corresponding value in the frequency domain.

Comparing Eq. 7.26 with Eq. 7.18 demonstrates the significant difference between the solutions of the two- and three-dimensional scalar-wave equations. In both cases, a wave front exists—that is, the response remains zero until the wave has propagated with the velocity $c$ from the source to the receiver ($0 < t < r_o/c$ or $0 < t < r/c$). While, for the three-dimensional case, the response returns to zero immediately after the arrival of the pulse

($t > r/c$), for the two-dimensional case, it decays at the receiver with time but remains finite ($t > r_o/c$). In other words, the wave as a function of the position at a fixed time exhibits, not a sharply defined rear, but a tail, sometimes also called a wake. This, of course, is related to two-dimensional wave propagation's dispersive nature.

### 7.1.3 Elasto-Dynamic Wave in Full Infinite Space

**Three-dimensional case.** It is possible to reformulate the equations of motion of a homogeneous and isotropic medium introducing potentials. The displacements and surface tractions for a concentrated source loading, specified as a Dirac impulse in space and time, can then be determined working in the time domain. This classic derivation, which is quite lengthy, can be found in Refs. [E2, A1]. To discuss the important features of these Green's functions (fundamental solutions), it is sufficient to work with the result.

Without loss of generality, the concentrated load of unit impulse is applied in the $x_j$-direction at the origin of the infinite region at time zero. The receiver point is specified by the position vector $\tilde{x}$ (coordinate $x$) with the direction cosines $l_i$ ($i = 1, 2, 3$) and the magnitude $r = |\tilde{x}|$. The Green's function (fundamental solution) for the displacement $\{g_u(x, t)\}$ consists of the components of the displacement in the $x_i$-directions at time $t$ $u_{ij}(x, t)$

$$u_{ij}(x,t) = \frac{1}{4\pi\rho}(3l_i l_j - \delta_{ij})\frac{t}{r^3}\left[H\left(t - \frac{r}{c_p}\right) - H\left(t - \frac{r}{c_s}\right)\right]$$
$$+ \frac{1}{4\pi\rho c_p^2}l_i l_j \frac{1}{r}\delta\left(t - \frac{r}{c_p}\right) + \frac{1}{4\pi\rho c_s^2}(\delta_{ij} - l_i l_j)\frac{1}{r}\delta\left(t - \frac{r}{c_s}\right)$$
(7.32)

with the Heaviside function $H$. Equivalent formulations are possible [E2]. The Green's function for the surface tractions (or stresses) is specified in Ref. [E2].

If the source load acts in the $x_j$-direction at a location specified by $\tilde{x}'$ (coordinate $x'$) at time $\tau$, the displacement component in the $x_i$-direction at the receiver $\tilde{x}$ at time $t$ $u_{ij}(x, t, x', \tau)$ is specified as

$$u_{ij}(x,t,x',\tau) = \frac{1}{4\pi\rho}(3l_i l_j - \delta_{ij})\frac{t-\tau}{r^3}\left[H\left(t - \tau - \frac{r}{c_p}\right) - H\left(t - \tau - \frac{r}{c_s}\right)\right]$$
$$+ \frac{1}{4\pi\rho c_p^2}l_i l_j \frac{1}{r}\delta\left(t - \tau - \frac{r}{c_p}\right) + \frac{1}{4\pi\rho c_s^2}(\delta_{ij} - l_i l_j)\frac{1}{r}\delta\left(t - \tau - \frac{r}{c_s}\right)$$
(7.33)

Sec. 7.1  Green's Function

where, in this equation, $l_i$ denotes the components of the unit vector (direction cosines) from the source point $\vec{x}'$ to the receiver point $\vec{x}$, and $r = |\vec{x} - \vec{x}'|$ is the distance between these two points.

To discuss the properties of the three-dimensional wave propagation, Eq. 7.32 is examined. In contrast to the scalar-wave equation addressed in Section 7.1.2, the directionalities at the source, determined by $l_j$, and at the receiver, specified by $l_i$, appear. In addition, P- and S-waves occur, propagating with their corresponding velocities $c_p$ and $c_s$. The three terms on the right-hand side are discussed next [A1].

The first term is negligible in the far field—that is, for large $r$. It consists of P- and S-waves. For a fixed receiver ($r$ = constant), the first term vanishes for $t < r/c_p$ and again for $t > r/c_s$. (If the load consisted, not of a Dirac function in time, but of a function which returns to zero at time $t_o$, then the non-zero motion would be restricted to the time interval $r/c_p < t < r/c_s + t_o$—that is, the arrival time would equal $r/c_p$ and the duration would be equal to $r/c_s - r/c_p + t_o$; see Problem 7.3.)

The second term will be large in the far field. It describes a P-wave propagating with $c_p$ and has the same properties as discussed for the scalar-wave equation (Eq. 7.18). The displacement at the receiver appearing at $t = r/c_p$ has the same shape (the Dirac function) as the load applied at the source. In particular, it is zero before and after the retarded time $r/c_p$. The amplitude of the displacement decays as $1/r$. The $i$th component $u_{ij}$ is proportional to $l_i$, which means that the resulting displacement at the receiver is always radial—that is, has the same direction as that of the propagation. The displacement's magnitude is proportional to the cosine of the angle between the direction of the applied load and that of the displacement.

The third term will also be large in the far field, when compared to the first one. It corresponds to an S-wave propagating with $c_s$, exhibiting again the same properties as for the P-wave of the second term. The displacement's direction is, however, transverse—that is, in the direction perpendicular to that of the propagation. Its magnitude is again proportional to the cosine of the angle between the direction of the applied load and that of the displacement.

**Two-dimensional case.** For a two-dimensional homogeneous and isotropic infinite medium, the Green's function (fundamental solution) exhibits a different behavior. The Dirac-impulse source load in space and time is applied at the origin in the $x_j$-direction. The receiver point is specified by the position vector $\vec{x}$ (coordinate $x$), with the direction cosines $l_i$ ($i = 1, 2$) and the magnitude $r_o = |\vec{x}|$. The displacement component $u_{ij}(x, t)$ in the

$x_i$-direction at time $t$ equals [E2]

$$u_{ij}(x,t) = \frac{1}{2\pi\rho r_o^2}\left[-\sqrt{t^2 - \frac{r_o^2}{c_p^2}}\,\delta_{ij} + l_i l_j \frac{2t^2 - \frac{r_o^2}{c_p^2}}{\sqrt{t^2 - \frac{r_o^2}{c_p^2}}}\right]H\left(t - \frac{r_o}{c_p}\right)$$

$$-\frac{1}{2\pi\rho r_o^2}\left[-\left(\sqrt{t^2 - \frac{r_o^2}{c_s^2}} + \frac{r_o^2}{c_s^2\sqrt{t^2 - \frac{r_o^2}{c_s^2}}}\right)\delta_{ij} + l_i l_j \frac{2t^2 - \frac{r_o^2}{c_s^2}}{\sqrt{t^2 - \frac{r_o^2}{c_s^2}}}\right]H\left(t - \frac{r_o}{c_s}\right)$$

(7.34)

If the source load is applied at a location specified by $\tilde{x}'$ (coordinate $x'$) at time $\tau$, the corresponding equation for $u_{ij}(x, t, x', \tau)$ is easily formulated based on Eq. 7.34, replacing $t$ by $t - \tau$, with $l_i$ denoting the components of the unit vector from the source point to the receiver point and with $r_o$ being the distance between these two points.

The first term on the right-hand side of Eq. 7.34 describes $P$-waves; the second term, $S$-waves. A wave front exists; that is, the response remains zero until the wave has propagated, with the corresponding velocity, from the source to the receiver. Analogously to the solution of the two-dimensional scalar-wave equation (Eq. 7.26), a tail exists, which continues until time infinity. In contrast to the elasto-dynamic wave equation in three dimensions (Eq. 7.32), the motion thus does not return to zero after the arrival of the $S$-wave $(t > r_o/c_s)$.

### 7.1.4 Properties

An element of the Green's function (fundamental solution) is denoted as $g_{ij}(x, t, x', \tau)$. $\tilde{x}'$, $\tau$, and $j$ denote the position vector of the source point, the time of application of the Dirac impulse load, and its direction; $\tilde{x}$, $t$, and $i$ are the corresponding variables associated with the receiver point. The components of the displacement $u_{ij}$ (Eq. 7.33) and of the surface traction $t_{ij}$ are examples of $g_{ij}$.

In the boundary-element method discussed in this chapter, Green's functions are determined, not only for the full infinite space addressed so far, but

## Sec. 7.1  Green's Function

also for the layered half-space (Section 7.1.5). The conditions on this medium's exterior boundaries are assumed to be homogeneous (free surface or radiation condition at infinity). In particular, these boundary conditions do not depend on time. In addition, it is assumed that the system exhibits a so-called quiescent past—that is, before the source load is applied, the displacements and velocities vanish, which determines the initial conditions. For such a domain, the following properties exist for the Green's functions:

a. Causality—that is, the Green's function vanishes until the P-wave arrives at the receiver point.

$$g_{ij}(x, t, x', \tau) \equiv 0 \quad \text{for } t < \frac{|\vec{x} - \vec{x}'|}{c_p} \quad (7.35)$$

See also Section 6.5.1 for an example of non-causal behavior.

b. Translation in time and space. The Green's function depends only on the difference $t - \tau$; that is, the time origin can be shifted by $t_o$.

$$g_{ij}(x, t, x', \tau) = g_{ij}(x, t - t_o, x', \tau - t_o) \quad (7.36)$$

In particular, selecting the generally arbitrary $t_o$ equal to $t + \tau$ leads to a reciprocal relation for source and receiver times

$$g_{ij}(x, t, x', \tau) = g_{ij}(x, -\tau, x', -t) \quad (7.37)$$

As the Green's function depends only on the distance between the source and receiver points in a full infinite space, an arbitrary vector $\vec{x}_o$ (coordinate $x_o$) can be added.

$$g_{ij}(x, t, x' \tau) = g_{ij}(x + x_o, t, x' + x_o, \tau) \quad (7.38)$$

For $x_o = -x'$, this leads to

$$g_{ij}(x, t, x', \tau) = g_{ij}(x - x', t, 0, \tau) \quad (7.39)$$

This translation property in space applies only for a full infinite space consisting of a homogeneous and isotropic medium.

c. Reciprocity. Based on the dynamic reciprocity theorem of Maxwell-Betti, introduced in Section 7.2.2, it is easily shown that

$$g_{ij}(x, t, x', \tau) = g_{ji}(x', t, x, \tau) \quad (7.40)$$

applies (for homogeneous boundary conditions). This, of course, is the symmetry condition. It is possible to combine the reciprocity property in space (Eq. 7.40) and in time (Eq. 7.37), resulting in

$$g_{ij}(x, t, x', \tau) = g_{ji}(x', -\tau, x, -t) \quad (7.41)$$

### 7.1.5 Layered Half-Space

**General considerations.** The Green's functions specified in the previous sections apply to the full infinite space. They could be used to model a layered half-space, whereby the free surface and the interfaces between layers would also be discretized, truncating at some reasonable distance. However, it is conceptually preferable to analytically incorporate the conditions at the free surface and at the interfaces into the Green's functions, which leads to a finite discretization on the structure–soil interface only. This is also consistent with the philosophy of the boundary-element method, which reduces the dimensionality of the problem. This concept, of course, also applies to the analysis in the frequency domain (Section 4.4.3).

For a layered half-space, it is not possible to determine the Green's functions working directly in the time domain. A Fourier transformation at the start of the calculation from the time to the frequency domain and an inverse one at the end from the frequency to the time domain are added. All steps between these two transformations are the same as for the calculation of the Green's functions in the frequency domain. They are specified in Section 4.4.1 and summarized in the next paragraph.

**Transformation from time-space domain to frequency-wave number domain.** The source loading will in general consist of a distributed load $\{p(s', \tau)\}$ which acts at the source surface, characterized by the coordinate $s'$, at time $\tau$ (and not just of a Dirac function in space and time). The calculation of the Green's functions $[g(s, t)]$ (for the displacements and the surface tractions) at the structure–soil interface with the coordinate $s$ proceeds as follows. First applying the Fourier transformation to the temporal variation of $p(s', \tau)$, the corresponding amplitudes in the frequency domain $\{p(s', \omega)\}$ result. A transformation into the wave-number domain $k$ follows next. For the two-dimensional case, a Fourier transformation leads to $\{p(k, \omega)\}$; while for the three-dimensional case, expansions in a Fourier series in the circumferential direction (integer $n = 0, 1, 2, \ldots$) and in Bessel functions involving the wave number $k$ in the radial direction result in $\{p(n, k, \omega)\}$. The corresponding influence functions $[g(k, \omega)]$ in the frequency-wave number domain are then calculated. Applying the inverse Fourier or Bessel transformation determines the spatial functions $[g(s, \omega)]$, and another inverse Fourier transformation finally leads to the Green's functions in the time domain $[g(s, t)]$.

Specifically, the equations for the calculation of the displacement field as a function of time are summarized resulting from a vertical and a horizontal unit-impulse point load (Dirac function in time and space) applied within a horizontally layered half-space (Fig. 7-6). Introducing cylindrical coordinates $r$, $\theta$, $z$, a field variable with the vector $\{f(r, \theta, z, t)\}$ is transformed to $\{f(k, n, z, \omega)\}$ as follows. The transformation from the time to the frequency

## Sec. 7.1    Green's Function

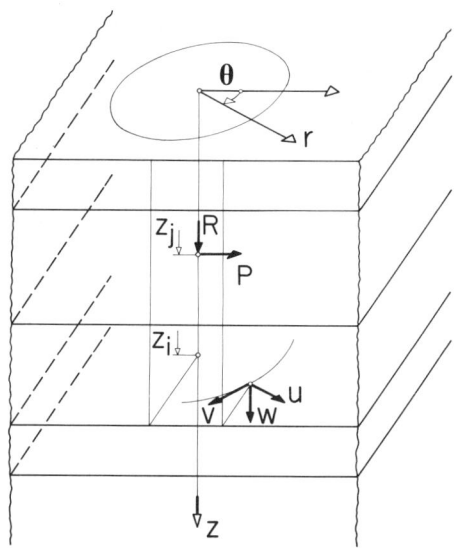

**Figure 7-6** Nomenclature of layered half-space.

domain is performed using the Fourier transformation. The field variable is expanded in a Fourier series in the circumferential direction $\theta$ (integer $n = 0, 1, 2, \ldots$) and into Bessel functions involving the wave number $k$ in the radial direction $r$. The following transform pair applies (Eq. 4.31):

$$\{f(k, n, \omega)\} = a_n \int_{r=0}^{\infty} r[C_n(kr)] \int_{\theta=0}^{2\pi} [D(n\theta)]\{f(r, \theta, \omega)\}\, d\theta\, dr \tag{7.42a}$$

$$\{f(r, \theta, \omega)\} = \sum_{n=0}^{\infty} [D(n\theta)] \int_{k=0}^{\infty} k[C_n(kr)]\{f(k, n, \omega)\}\, dk \tag{7.42b}$$

The argument $z$ has been omitted for the sake of conciseness. The scalar $a_n$ is the normalization factor, which is equal to $1/2\pi$ for $n = 0$ and to $1/\pi$ for $n \neq 0$. $[D(n\theta)]$ contains the sine and cosine functions of $n\theta$ on the diagonal. The matrix $[C_n(kr)]$ is defined as

$$[C_n(kr)] = \begin{bmatrix} \frac{1}{k} J_n(kr)_{,r} & \frac{n}{kr} J_n(kr) \\ \frac{n}{kr} J_n(kr) & \frac{1}{k} J_n(kr)_{,r} & -J_n(kr) \end{bmatrix} \tag{7.43}$$

where $J_n(kr)$ denotes the Bessel function of order $n$ of the first kind.

Assembling the dynamic-stiffness matrices of the individual layers and of the half-space results in the matrix of the site $[S(k, \omega)]$ which is independent of the Fourier index $n$. The site's discretized dynamic-equilibrium equation results in (Section 4.2.5)

$$[S(k, \omega)]\{u(k, n, \omega)\} = \{P(k, n, \omega)\} \tag{7.44}$$

where $\{u(k, n, \omega)\}$ and $\{P(k, n, \omega)\}$ denote the vectors of the amplitudes of the displacements and external loads at the interfaces between the layers. (See also Section 4.4.1.)

The vertical point load $R(t)$ whose Fourier transform is denoted as $R(\omega)$, which acts at the interface $j$ (depth $z_j$) is examined first (Fig. 7-6). Only the zeroth symmetric Fourier term ($n = 0$), resulting in constant values around the circumference, arises. Formulating Eq. 7.42a for the loads and making use of Eq. 7.43 leads to the load in the ($k, \omega$) domain

$$R(k, \omega) = -\frac{1}{2\pi} \int_{r=0}^{\infty} rJ_o(kr)\delta(r)dr \int_{\theta=0}^{2\pi} d\theta R(\omega) = -\frac{R(\omega)}{2\pi} \quad (7.45)$$

$R(k, \omega)$ represents the only non-zero term of $\{P(k, \omega)\}$ of Eq. 7.44. Solving the latter for the displacement at the interface $i$ (depth $z_i$) leads to

$$\begin{Bmatrix} u(k, \omega) \\ w(k, \omega) \end{Bmatrix} = \begin{Bmatrix} F_{uw}(k, \omega) \\ F_{ww}(k, \omega) \end{Bmatrix} R(k, \omega) \quad (7.46)$$

where $F(k, \omega)$ denotes the (in-plane) dynamic-flexibility coefficients. $u(k, \omega)$ and $w(k, \omega)$ represent the radial and vertical displacements in the ($k, \omega$)-domain.

The inverse transformation leading to $u(r, \omega)$ and $w(r, \omega)$ follows from Eq. 7.42b formulated for the displacements at interface $i$.

$$\begin{Bmatrix} u(r, \omega) \\ w(r, \omega) \end{Bmatrix} = \int_{k=0}^{\infty} \begin{bmatrix} J_o(kr)_{,r} \\ -kJ_o(kr) \end{bmatrix} \begin{Bmatrix} u(k, \omega) \\ w(k, \omega) \end{Bmatrix} dk \quad (7.47)$$

Substituting Eq. 7.46 in Eq. 7.47 results in

$$\begin{Bmatrix} u(r, \omega) \\ w(r, \omega) \end{Bmatrix} = \frac{1}{2\pi} \left( \int_{k=0}^{\infty} k \begin{Bmatrix} F_{uw}(k, \omega)J_1(kr) \\ F_{ww}(k, \omega)J_o(kr) \end{Bmatrix} dk \right) R(\omega) \quad (7.48)$$

Applying the inverse Fourier transformation to Eq. 7.48 leads to $u(r, t)$ and $w(r, t)$.

The horizontal point load $P(t)$, whose Fourier transform is denoted as $P(\omega)$, which acts at the interface $j$, is discussed next (Fig. 7-6). The first symmetric Fourier term is involved. The corresponding loads in the ($k,\omega$)-domain in the radial and circumferential directions are calculated using Eq. 7.42a:

$$P(k,\omega) = Q(k,\omega) = \frac{P(\omega)}{2\pi} \quad (7.49)$$

The corresponding displacements in the ($k,\omega$)-domain at interface $i$ follow from Eq. 7.44 as

$$\begin{Bmatrix} u(k,\omega) \\ v(k,\omega) \\ w(k,\omega) \end{Bmatrix} = \begin{bmatrix} F_{uu}(k,\omega) \\ & F_{vv}(k,\omega) \\ F_{wu}(k,\omega) \end{bmatrix} \begin{Bmatrix} P(k,\omega) \\ Q(k,\omega) \end{Bmatrix} \quad (7.50)$$

Sec. 7.1   Green's Function

Applying the inverse transformation of Eq. 7.42b to the displacements leads to the displacements $u(r,\theta,\omega)$, $v(r,\theta,\omega)$, and $w(r,\theta,\omega)$. Substituting Eq. 7.50 results in

$$\begin{Bmatrix} u(r,\theta,\omega) \\ v(r,\theta,\omega) \\ w(r,\theta,\omega) \end{Bmatrix} = \frac{1}{4\pi} \begin{bmatrix} \cos\theta & & \\ & -\sin\theta & \\ & & \cos\theta \end{bmatrix}$$
$$\times \left( \int_{k=0}^{\infty} k \begin{bmatrix} J_o(kr) - J_2(kr) & J_o(kr) + J_2(kr) & \\ J_o(kr) + J_2(kr) & J_o(kr) - J_2(kr) & \\ & & -2J_1(kr) \end{bmatrix} \begin{Bmatrix} F_{uu}(k,\omega) \\ F_{vv}(k,\omega) \\ F_{wu}(k,\omega) \end{Bmatrix} dk \right) P(\omega)$$

(7.51)

**Vertical load on free surface of half-space.** As an example, a unit-impulse point load is applied on the free surface of an undamped homogeneous half-space, and the corresponding displacements are evaluated at the free surface [W13]. The vertical displacement resulting from a vertical point load is addressed first (Lamb's problem). The temporal variation of the point load is represented by a Gaussian distribution, thus enabling the convenient computation of the Fourier transform.

$$R(t) = \frac{1}{2\sqrt{\pi}\sigma} \exp[-t^2/(4\sigma^2)] \qquad (7.52)$$

For the parameter $\sigma$ approaching zero, $R(t)$ converges to the Dirac delta function $\delta(t)$. The corresponding Fourier transform is equal to

$$R(\omega) = \int_{-\infty}^{+\infty} R(t) \exp(-i\omega t) dt = \exp(-\sigma^2\omega^2) \qquad (7.53)$$

The dynamic-flexibility coefficient $F_{ww}(k,\omega)$, appearing in Eq. 7.48, follows for the present case from the inversion of the in-plane dynamic-stiffness matrix of the half-space (Eq. 4.30)

$$F_{ww}(k,\omega) = -\frac{1}{G} \frac{k_s^2(k^2 - k_p^2)^{1/2}}{(2k^2 - k_s^2)^2 - 4k^2(k^2 - k_p^2)^{1/2}(k^2 - k_s^2)^{1/2}} \qquad (7.54)$$

where $G$ is the shear modulus. The wave numbers $k_s$ and $k_p$ are defined as

$$k_s = \frac{\omega}{c_s} \qquad (7.55a)$$

$$k_p = \frac{\omega}{c_p} \qquad (7.55b)$$

$c_s$ and $c_p$ denote the shear-wave and dilatational-wave velocities. $w(r,t)$ follows from the inverse Fourier transformation of $w(r,\omega)$:

$$w(r,t) = \frac{1}{2\pi} \int_{\omega=-\infty}^{+\infty} w(r,\omega)\exp(i\omega t)d\omega \tag{7.56}$$

Substituting the second row of Eq. 7.48 in Eq. 7.56 and using Eq. 7.53 leads to

$$w(r,t) = \frac{1}{4\pi^2} \int_{\omega=-\infty}^{+\infty} \exp(i\omega t - \sigma^2\omega^2) \left( \int_{k=0}^{\infty} kF_{ww}(k,\omega)J_o(kr)dk \right) d\omega \tag{7.57}$$

This expression can be simplified by introducing dimensionless variables:

$$\kappa = \frac{k}{k_s} \tag{7.58a}$$

$$a_o = \frac{\omega r}{c_s} \tag{7.58b}$$

$$\bar{t} = \frac{tc_s}{r} \tag{7.58c}$$

$$\bar{\sigma} = \frac{\sigma c_s}{r} \tag{7.58d}$$

Substituting Eq. 7.58 in Eq. 7.57 results in

$$w(r,t) = \frac{c_s}{Gr^2} \overline{w}(\bar{t}) \tag{7.59}$$

where

$$\overline{w}(\bar{t}) = -\frac{1}{4\pi^2} \int_{a_o=-\infty}^{+\infty} a_o \exp(ia_o\bar{t} - \bar{\sigma}^2 a_o^2)$$

$$\times \left( \int_{\kappa=0}^{\infty} \frac{\kappa(\kappa^2-\gamma^2)^{1/2}}{(2\kappa^2-1)^2 - 4\kappa^2(\kappa^2-\gamma^2)^{1/2}(\kappa^2-1)^{1/2}} J_o(\kappa a_o)d\kappa \right) da_o \tag{7.60}$$

with

$$\gamma = \frac{c_s}{c_p} \tag{7.61}$$

The evaluation of the integral over $\kappa$ in Eq. 7.60 is not straightforward. The denominator of the integrand represents the Rayleigh equation, which exhibits poles and branch points. Making use of the properties of Bessel functions, the integration bound can be made finite [P2].

### Sec. 7.1  Green's Function

The influence function $\bar{w}(\bar{t})$ (Green's function) is plotted for $\gamma = 0.5$, corresponding to Poisson's ratio $= 1/3$, in Fig. 7-7, using $\bar{\sigma} = 0.02$. No material damping is introduced. The arrival of the P-wave at $\bar{t} = 0.5$ is clearly visible; the arrivals of the S- and R-waves just above $\bar{t} = 1$ are close together (Rayleigh-wave velocity $c_R = 0.933\, c_s$).

The displacements associated with the arrivals of the P- and S-waves are opposite to the load application's direction. The Rayleigh wave contributes most, with the displacement's direction (the displacement is actually infinite for an exactly represented Dirac impulse) coinciding with the load's direction.

**Horizontal load on free surface of half-space.** Proceeding analogously for the horizontal point load using the first two rows of Eq. 7.51, the radial and circumferential displacements are expressed as

$$u(r,\theta,t) = \frac{c_s}{Gr^2} \cos\theta\, \bar{u}(\bar{t}) \tag{7.62a}$$

$$v(r,\theta,t) = -\frac{c_s}{Gr^2} \sin\theta\, \bar{v}(\bar{t}) \tag{7.62b}$$

where

$$\bar{u}(\bar{t}) =$$

$$-\frac{1}{4\pi^2}\int_{a_o=-\infty}^{\infty} a_o \exp(ia_o\bar{t} - \bar{\sigma}^2 a_o^2) \left( \int_{\kappa=0}^{\infty} \left\{ \frac{\kappa(\kappa^2-1)^{1/2}}{(2\kappa^2-1)^2 - 4\kappa^2(\kappa^2-\gamma^2)^{1/2}(\kappa^2-1)^{1/2}} \right.\right.$$

$$\left.\left. \cdot [J_o(\kappa a_o) - J_2(\kappa a_o)] + \frac{i\kappa}{(1-\kappa^2)^{1/2}} \cdot [J_o(\kappa a_o) + J_2(\kappa a_o)] \right\} d\kappa \right) da_o$$

$$\tag{7.63a}$$

$$\bar{v}(\bar{t}) =$$

$$-\frac{1}{4\pi^2}\int_{a_o=-\infty}^{\infty} a_o \exp(ia_o\bar{t} - \bar{\sigma}^2 a_o^2) \left( \int_{\kappa=0}^{\infty} \left\{ \frac{\kappa(\kappa^2-1)^{1/2}}{(2\kappa^2-1)^2 - 4\kappa^2(\kappa^2-\gamma^2)^{1/2}(\kappa^2-1)^{1/2}} \right.\right.$$

$$\left.\left. \cdot [J_o(\kappa a_o) + J_2(\kappa a_o)] + \frac{i\kappa}{(1-\kappa^2)^{1/2}} \cdot [J_o(\kappa a_o) - J_2(\kappa a_o)] \right\} d\kappa \right) da_o$$

$$\tag{7.63b}$$

For the same parameters, the radial and circumferential influence functions (Green's functions) $\bar{u}(\bar{t})$, $\bar{v}(\bar{t})$ are plotted in Fig. 7-8.

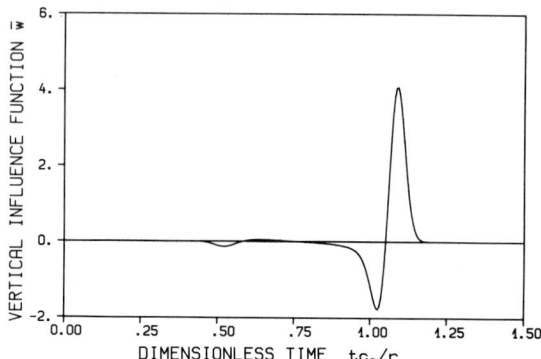

**Figure 7-7** Vertical displacement at surface of half-space from vertical surface point load.

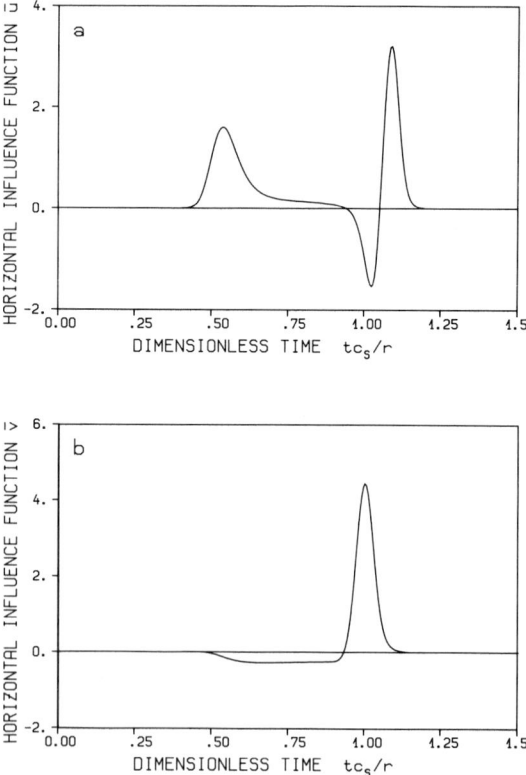

**Figure 7-8** Horizontal displacement at surface of half-space from horizontal surface point load. a. Radial. b. Circumferential.

**Comparison with two-dimensional case (half-plane).** As can be expected from the results of the full infinite space (Section 7.1.3), important differences exist between the solutions for a line impulse on the surface of a half-space (Dirac function on a half-plane) and those for an impulse on the surface of a half-space (Fig. 7-7).

The vertical displacement generated by a vertical impulse force is addressed. For the two-dimensional case [E2], the displacement also vanishes ahead of the $P$-wave front. The displacement (again opposite to the load application's direction) remains small until the arrival of the $S$-wave. The displacement then increases significantly, becoming unbounded at the Rayleigh wave's arrival time, when it also changes sign (two-sided singularity). For later times, the displacement, pointing in the same direction as the load, decreases, but does not decay to zero; that is, the two-dimensional case exhibits a tail.

For the point impulse load acting on the surface of a half-space, the singularity caused at the time of the Rayleigh-wave arrival is one-sided—that is, no change in sign. For later times, the displacement decays to zero.

## 7.2 TIME-DEPENDENT BOUNDARY-INTEGRAL EQUATION IN ELASTO-DYNAMICS

### 7.2.1 Fundamentals

A problem in elasto-dynamics is governed by a system of partial differential equations in space and time, and by the boundary and initial conditions. Instead of numerically solving this system directly by discretization in space and time, it is possible to convert it first to a boundary-integral–differential equation called boundary–integral equation for short. This is performed by formulating a weighted-residual statement in space and time, which will involve a convolution on the time variable. In general, first an expression for the displacement at a general point in space and time is derived, involving convolution integrals of the boundary values, the initial values, and the applied loading throughout the medium (the so-called representation theorem). The Green's functions discussed in Section 7.1 will appear in this equation. Through a limiting process, where the point of the displacement lies on the boundary, an integral equation is then derived which involves only the boundary values, the initial values, and the loading. As the initial values and the loading are specified, the only unknowns are located on the boundary. This boundary-integral equation is an equivalent statement of the partial differential equations and of the boundary and initial conditions, and can be discretized in space and time. In this boundary-integral–equation method, the number of spatial dimensions is reduced by one. It is also an advantage that the problem can be directly formulated in terms of the physically relevant boundary values. The motion at interior points need not be calculated. The medium's dynamic behavior is represented analytically without using elements and is entirely captured by the Green's functions (fundamental solutions). The radiation condition at infinity is incorporated into these functions and thus at the outset is contained in the formulation. The boundary-integral–equation method is thus well suited to model infinite domains. When the Green's functions also satisfy the conditions at the free surface and at the interfaces of the layers of a half-space (Section 7.1.5), the modeling for the numerical purpose is confined to the structure–soil interface—a surface of finite dimensions. This numerical procedure, based on the discretized boundary integral equation, is called the boundary-element method.

In the following, different forms of the reciprocity theorem are discussed, which describe a general relationship between a pair of elasto-dynamic states. By introducing Green's functions into the reciprocity theorem, the representation theorem is derived. These theorems form the basis of the so-called direct boundary-element method and can also be used to formulate the indirect one. The corresponding spatial and temporal discretizations are addressed in Section 7.3, where a standard weighted-residual formulation is also examined.

### 7.2.2 Reciprocity Theorem of Maxwell-Betti

The dynamic-reciprocity theorem specifies a relationship between a pair of solutions corresponding to two different loadings, boundary conditions, and initial conditions. The displacements of the first state $\{u(x,t)\}$ are caused by the loading $\{f(x,t)\}$ acting in the volume $V$ denoted by $x$, and by the boundary conditions on $S$ with the coordinate $s$, and by the initial conditions for the displacements $\{u(x,0)\}$ and velocities $\{\dot{u}(x,0)\}$. The second state is characterized analogously and is denoted by an asterisk. Denoting the corresponding surface tractions on $S$ as $\{t(s,t)\}$ (and $\{t^*(s,t)\}$), the basic dynamic-reciprocity theorem is formulated as [A1]

$$\int_V (\{f(x,t)\}^T - \rho\{\ddot{u}(x,t)\}^T) \{u^*(x,t)\} \, dv + \int_S \{t(s,t)\}^T \{u^*(s,t)\} \, ds$$
$$= \int_V (\{f^*(x,t)\}^T - \rho\{\ddot{u}^*(x,t)\}^T) \{u(x,t)\} \, dv + \int_S \{t^*(s,t)\}^T \{u(s,t)\} \, ds \quad (7.64)$$

Equation 7.64 expresses that the work done by the loads of the first state (including the inertia loads and the surface tractions) on the displacements of the second state is equal to that of the loads of the second state on the displacements of the first state. Including the inertia loads in the classic elasto-static relationship leads to the dynamic-reciprocity theorem.

The proof is also the same as for the static case. Using subscripts which

## Sec. 7.2  Time-Dependent Boundary-Integral Equation in Elasto-Dynamics

are equal to 1, 2, 3 and implying a summation over repeated subscripts, the equilibrium equations equal

$$\sigma_{ij,j} + f_i - \rho \ddot{u}_i = 0 \tag{7.65}$$

with the stress–displacement relationship

$$\sigma_{ij} = E_{ijkl} \frac{u_{k,l} + u_{l,k}}{2} \tag{7.66}$$

where $E_{ijkl}$ is the (symmetric) elasticity tensor. Substituting Eqs. 7.65 and 7.66 and applying the divergence theorem of Gauss (integration-by-parts formula)

$$-\int_V \sigma_{ij,j} u_i^* \, dv + \int_S \{t\}^T \{u^*\} \, ds = \int_V E_{ijkl} \frac{u_{k,l} + u_{l,k}}{2} \frac{u_{i,j}^* + u_{j,i}^*}{2} \, dv \tag{7.67}$$

to the left-hand side of Eq. 7.64 leads to

$$\int_V E_{ijkl} \frac{u_{k,l} + u_{l,k}}{2} \frac{u_{i,j}^* + u_{j,i}^*}{2} \, dv \tag{7.68}$$

Performing the same operations on the right-hand side of Eq. 7.64 results in an expression identical to Eq. 7.68, taking the symmetry of $E_{ijkl}$ into consideration.

In the form of the dynamic-reciprocity theorem that involves the accelerations (Eq. 7.64), the two states' initial values do not appear. The relationship still applies if the two states are evaluated at different times. Selecting the time $\tau$ for the first state and the time $t - \tau$ for the second one, and integrating from zero to $t$, the two terms with the accelerations, applying integration by parts, can be rewritten as

$$\int_0^t \{\ddot{u}(x,\tau)\}^T \{u^*(x,t-\tau)\} \, d\tau - \int_0^t \{u(x,\tau)\}^T \{\ddot{u}^*(x,t-\tau)\} \, d\tau$$

$$= \{\dot{u}(x,\tau)\}^T \{u^*(x,t-\tau)\} \Big|_0^t + \int_0^t \{\dot{u}(x,\tau)\}^T \{\dot{u}^*(x,t-\tau)\} \, d\tau$$

$$+ \{u(x,\tau)\}^T \{\dot{u}^*(x,t-\tau)\} \Big|_0^t - \int_0^t \{\dot{u}(x,\tau)\}^T \{\dot{u}^*(x,t-\tau)\} \, d\tau \tag{7.69}$$

$$= \{\dot{u}(x,t)\}^T \{u^*(x,0)\} - \{\dot{u}(x,0)\}^T \{u^*(x,t)\}$$
$$+ \{u(x,t)\}^T \{\dot{u}^*(x,0)\} - \{u(x,0)\}^T \{\dot{u}^*(x,t)\}$$

The acceleration terms are thus a function of the initial values. Substituting Eq. 7.69 into Eq. 7.64 leads to another formulation of the dynamic-reciprocity theorem:

$$\int_V \left( \int_o^t \{f(x,\tau)\}^T \{u^*(x,t - \tau)\} \, d\tau + \rho \{u(x,0)\}^T \{\ddot{u}^*(x,t)\} \right.$$

$$\left. + \rho \{\dot{u}(x,0)\}^T \{u^*(x,t)\} \right) dv + \int_S \int_o^t \{t(s,\tau)\}^T \{u^*(s,t - \tau)\} \, d\tau \, ds$$

$$= \int_V \left( \int_o^t \{f^*(x,t - \tau)\}^T \{u(x,\tau)\} \, d\tau + \rho \{u^*(x,0)\}^T \{\ddot{u}(x,t)\} \right.$$

$$\left. + \rho \{\dot{u}^*(x, 0)\}^T \{u(x, t)\} \right) dv + \int_S \int_o^t \{t^*(s, t - \tau)\}^T \{u(s, \tau)\} \, d\tau \, ds$$

(7.70)

If the initial conditions at $t = 0$ vanish—that is, a quiescent past exists—only the terms with convolution integrals in Eq. 7.70 will exist.

$$\int_V \int_o^t \{f(x, \tau)\}^T \{u^*(x, t - \tau)\} \, d\tau \, dv + \int_S \int_o^t \{t(s, \tau)\}^T \{u^*(s,t - \tau)\} \, d\tau \, ds$$

$$= \int_V \int_o^t \{f^*(x,t - \tau)\}^T \{u(x,\tau)\} \, d\tau \, dv + \int_S \int_o^t \{t^*(s,t - \tau)\}^T \{u(s,\tau)\} \, d\tau \, ds$$

(7.71)

The reciprocity property in space (Eq. 7.40) can be verified as follows. For homogeneous boundary conditions on $S$, the boundary integrals vanish in Eq. 7.71. Selecting $\{f\}$ as a unit impulse applied in the $i$-direction at $x$ and at time zero, and $\{f^*\}$ as a unit impulse applied in the $j$-direction at $x'$ and at time zero, leads to $u^*(x,t) = g_{ij}(x,t,x',\tau)$ and $u(x',t) = g_{ji}(x',t,x,\tau)$.

### 7.2.3 Representation Theorem

To express the displacement in space and time as a function of the applied loading, the boundary conditions, and the initial conditions, the Green's functions (fundamental solutions) are used as one of the two states in the dynamic-reciprocity theorem. Applying a unit impulse in space and time at the location $x'$ in the direction $j$ at $\tau$ (which corresponds to $\{f^*\}$) and denoting the Green's functions (fundamental solutions) for the displacements as $\{g_u(x,t,x',\tau)\} = \{g_u(x,t - \tau)\}$ in the volume and as $\{g_u(s,t - \tau)\}$ on the boundary and for the surface tractions as $\{g_t(s,t,x',\tau)\} = \{g_t(s,t - \tau)\}$, the dynamic-

## Sec. 7.2  Time-Dependent Boundary-Integral Equation in Elasto-Dynamics

reciprocity theorem (Eq. 7.71) for a quiescent past is formulated as

$$u_j(x',t) = \int_V \int_o^t \{f(x,\tau)\}^T \{g_u(x,t-\tau)\} d\tau\, dv$$

$$+ \int_S \int_o^t \{t(s,\tau)\}^T \{g_u(s,t-\tau)\} d\tau\, ds - \int_S \int_o^t \{u(s,\tau)\}^T \{g_t(s,t-\tau)\} d\tau\, ds \qquad (7.72)$$

where $u_j(x', t)$ is the displacement at the source point in the jth direction at time $t$.

This representation theorem specifies the contributions of the loading, of the surface tractions, and of the displacements on the boundary to the displacement at point $x'$. The weighting functions which appear in the convolution integrals are equal to the Green's functions (fundamental solutions)—for example, the displacement at $x$ arising from the Dirac impulse load in the jth direction at $x'$. It is also important to realize that the displacements $\{u(s,\tau)\}$ and the tractions $\{t(s,\tau)\}$ on the boundary cannot be prescribed independently.

If the point $x'$ is selected outside the volume $V$ (and also not on the boundary $S$), then the left-hand side of Eq. 7.72 will vanish. For later use, the equation for this case is stated, with minor rearrangements, assuming no body loads are acting:

$$\int_o^t \int_S \{g_u(s,t-\tau)\}^T \{t(s,\tau)\} ds\, d\tau = \int_o^t \int_S \{g_t(s,t-\tau)\}^T \{u(s,\tau)\} ds\, d\tau \qquad (7.73)$$

The representation theorem of Eq. 7.72 cannot be used directly to solve elasto-dynamic problems, since it requires the specification of all components of the displacements and surface tractions on the boundary $S$. For instance, if $S$ represents the structure–soil interface, then the displacements corresponding to the dynamic-stiffness coefficients will be specified, but the tractions will be unknown. Various possibilities exist to derive an integral-equation formulation that can be applied directly [E2].

Only one of the approaches is described next. Through a limiting process, which consists of choosing the point $x'$ on the boundary $S$ (denoted as $s'$), an integral representation is derived which involves only unknowns on the boundary. For a smooth boundary, it can be shown that Eq. 7.72 is transformed to

$$\frac{1}{2} u_j(s',t) = \int_V \int_o^t \{f(x,\tau)\}^T \{g_u(x,t-\tau)\} d\tau\, dv$$

$$+ \int_S \int_o^t \{t(s,\tau)\}^T \{g_u(s,t-\tau)\} d\tau\, ds - \int_S \int_o^t \{u(s,\tau)\}^T \{g_t(s,t-\tau)\} d\tau\, ds \qquad (7.74)$$

This singular integral equation relates the displacements and surface tractions on the boundary of the domain under consideration. It represents the starting point of the classical direct boundary-element method, which is discussed further in Section 7.3.4.

If the boundary is not smooth at $s'$ on $S$, such as at a corner, the factor 0.5 on the left-hand side of Eq. 7.74 is replaced by a value which depends only on the geometry at $s'$.

## 7.3 SPATIAL AND TEMPORAL DISCRETIZATIONS OF BOUNDARY-INTEGRAL EQUATION

### 7.3.1 Basic Procedure

Returning to the representation of the unbounded soil to be used in the basic equation of motion of soil–structure interaction in the time domain, the interaction force–displacement relationship, involving convolution integrals for the nodes located on the (generalized) structure–soil interface, must be calculated. This can be based on the reciprocity theorem discussed in Section 7.2.2 or on the limiting case of the representation theorem (Eq. 7.74). Analytical solutions will appear in these relations. These Green's functions are addressed in Section 7.1. If they do not correspond to unit-impulse loads in space and time but to (spatially and temporally) distributed loads, they can be calculated easily from the fundamental solutions for a Dirac impulse load, applying integration and convolution. These concepts of how to calculate the interaction force–displacement relationship in discretized form with boundary elements (so-called indirect and direct methods) are discussed in Sections 7.3.3 and 7.3.4.

To gain a better physical insight, it is appropriate to address the so-called weighted-residual method first (Section 7.3.2). This is based on the familiar principle of superposition. No use of the dynamic-reciprocity theorem of Section 7.2.2 need be made. The concept of the weighted-residual method to calculate the interaction force–displacement relationship can be summarized as follows. The Green's functions for Dirac impulse or (spatially and temporally) distributed loads will exactly satisfy the radiation condition, and the boundary conditions between two adjacent layers and at the free surface (Section 7.1.5); but, as they are constructed for the continuous layered half-space without excavation (soil system free field), they will not satisfy those on the structure–soil interface. On the exterior of the structure–soil interface (that is, on that part of the system free field that will subsequently be excavated), fictitious loads with unknown magnitudes are assumed to act. Using the Green's functions, the displacements on the line associated with the structure–soil interface can be calculated analytically. The fictitious loads' magnitudes are then determined so as to satisfy, in an average sense,

### Sec. 7.3   Spatial and Temporal Discretizations of Boundary-Integral Equation

the prescribed displacement conditions of the definition of the interaction force–displacement relationship on the structure–soil interface (stiffness matrix). The weighted-residual statement and thus the spatial and temporal discretizations are restricted to the (irregular) structure–soil interface. The surface tractions on the structure–soil interface arising from the fictitious loads are then determined using the corresponding Green's functions. The interaction forces are equal to the resultants of these surface tractions.

#### 7.3.2 Weighted-Residual Method

The structure–soil interface $S$ of the embedded foundation is discretized using boundary elements compatible with the structure's adjacent finite-element model. The nodes located on $S$ are denoted as $b$. In Fig. 7-9, the nomenclature used is illustrated for the in-plane motion—that is, for two dimensions. It is straightforwardly extended to three dimensions.

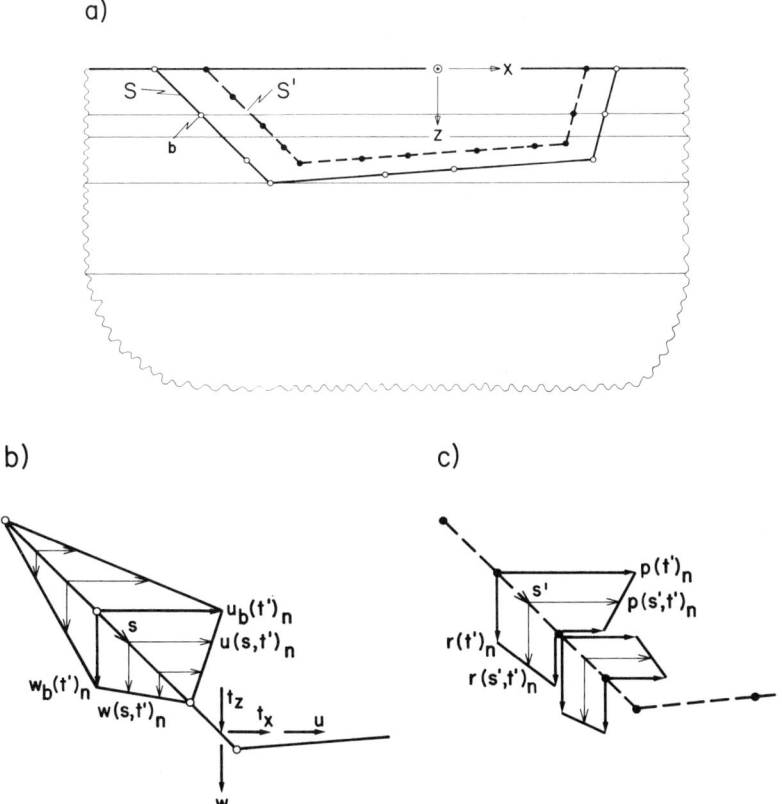

**Figure 7-9** Elements of discretization.   a. Source surface and structure–soil interface.   b. Specified displacements.   c. Selected load distribution.

The formulation of the $n$th step leading from time $(n - 1)\Delta t$, where all variables are known, to time $n\Delta t$ proceeds as follows [W14]. The displacements $\{u(s,t')\}_n$ on $S$, containing the three elements $u(s,t')_n$, $v(s,t')_n$, and $w(s,t')_n$ in the directions of the coordinate axes $x$, $y$, $z$ (Fig. 7-9b), are introduced first. In the foregoing expressions, the subscript $n$ denotes the $n$th time step, $t'$ indicates the time measured from the start of this time step ($0 \leq t' \leq \Delta t$), and $s$ symbolically represents a location on $S$. Introducing the shape functions $[N(s)]$, the total displacements (superscript $t$) $\{u^t(s,t')\}_n$ are formulated as

$$\{u^t(s,t')\}_n = [N(s)] \{u_b^t(t')\}_n \tag{7.75}$$

where $\{u_b^t(t')\}_n$ denotes the total displacements of the nodes $b$. The motion relative to the specified motion $\{u^g(s,t')\}_n$ equals

$$\{u(s,t')\}_n = \{u^t(s,t')\}_n - \{u^g(s,t')\}_n \tag{7.76}$$

The superscript $g$ refers to the generalized scattered motion. In an explicit algorithm, $\{u_b^t(t')\}_n$ can be predicted from the known motion at $(n - 1)\Delta t$ as specified in Eq. 6.135a.

In the weighted-residual formulation, a fictitious loading pattern, acting on that part of the free field which does not form the system ground, is introduced. These loads $\{p(s',t')\}_n$, with the three components $p(s',t')_n$, $q(s',t')_n$, and $r(s',t')_n$, are assumed to act on the source surface $S'$ (see Fig. 7-9c). $S'$ is always offset towards the soil region to be excavated—in the limiting case, by an infinitesimal amount. The loads act on the dynamic system of the free field consisting of the continuous soil—that is, on the layered half-space without excavation. These source loads, assumed to be constant over each time step (see Fig. 7-10a), are specified as

$$\{p(s',t')\}_n = [L(s')] \{p\}_n \tag{7.77}$$

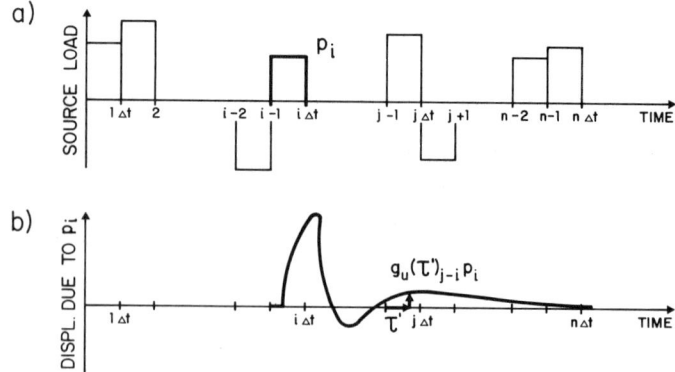

**Figure 7-10** Time variation of a. Source loads. b. Green's function.

### Sec. 7.3  Spatial and Temporal Discretizations of Boundary-Integral Equation    391

$\{p\}_n$ are the loads of the $n$th time step acting in nodes on $S'$, which are, in principle, independent of those on the structure–soil interface; and $[L(s')]$ represents the selected interpolation functions, with $s'$ denoting a location on $S'$. The number of parameters in $\{p\}_n$ must be larger than or equal to that in $\{u'_b(t')\}_n$. Discontinuities can be introduced, as shown in Fig. 7-9c, and time variations other than the selected constant one could be used in Eq. 7.77 (see Ref. [W13] for a linear time variation).

In the weighted-residual formulation, the displacements and tractions caused by the source loads must be calculated at any time $t$ on the surface $S$ that subsequently will form the structure–soil interface. For $t = (j - 1)\Delta t + \tau'$ $(j \leq n)$, the displacements $\{u_p(s,\tau')\}_j$ are written as

$$\{u_p(s,\tau')\}_j = \sum_{i=1}^{j} [g_u(s,\tau')]_{j-i} \{p\}_i \quad (7.78)$$

where $[g_u(s,\tau')]_{j-i}$ represents the Green's function for the displacements caused by unit nodal source loads acting during the $i$th time step. They can easily be calculated from the fundamental solution for a unit impulse load in space and time applying a convolution integral (Section 7.1.1). The time variation of one component of $[g_u(s,\tau')]_{j-i}$ is schematically shown in Fig. 7-10b. Analogously, the corresponding surface tractions $\{t_p(s,\tau')\}_j$ with the components $t_{px}(s,\tau')_j$, $t_{py}(s,\tau')_j$, and $t_{pz}(s,\tau')_j$ (Fig. 7-9b) are formulated as

$$\{t_p(s,\tau')\}_j = \sum_{i=1}^{j} [g_t(s,\tau')]_{j-i} \{p\}_i \quad (7.79)$$

Because only a finite number of load intensities can be introduced at distinct time steps, the displacement boundary conditions on the structure–soil interface $S$ cannot be exactly satisfied by the displacements caused by the source loads (that is, in every point on $S$ for every $t$), but only in an average sense as

$$\sum_{j=1}^{n} \int_{0}^{\Delta t} \int_{S} [W(s,\tau')]_{n-j}^{T} (\{u_p(s,\tau')\}_j - \{u(s,\tau')\}_j)\, ds\, d\tau' = 0 \quad (7.80)$$

The matrix $[W(s,\tau')]_{n-j}$ denotes the weighting functions applicable to the $j$th time step. Various choices are possible for $[W(s,\tau')]_{n-j}$. For instance, it can be selected as non-zero for the $n$th time step only (that is, for $j = n$). Then, omitting the subscript 0, Eq. 7.80 is transformed to

$$\int_{0}^{\Delta t} \int_{S} [W(s,t')]^{T} (\{u_p(s,t')\}_n - \{u(s,t')\}_n)\, ds\, dt' = 0 \quad (7.81)$$

which represents an integral over the $n$th time step only. Substituting Eq. 7.78 (formulated for $j = n$) and Eqs. 7.75 and 7.76 in Eq. 7.81 leads to

$$\sum_{i=1}^{n} [G]_{n-i} \{p\}_i = \{\bar{u}_b\}_n \quad (7.82)$$

where

$$[G]_{n-i} = \int_o^{\Delta t} \int_S [W(s,t')]^T [g_u(s,t')]_{n-i} \, ds \, dt' \qquad (7.83)$$

and

$$\{\bar{u}_b\}_n = \int_o^{\Delta t} \int_S [W(s,t')]^T [N(s)] \, ds \, \{u_b^t(t')\}_n \, dt' \\ - \int_o^{\Delta t} \int_S [W(s,t')]^T \{u^g(s,t')\}_n \, ds \, dt' \qquad (7.84)$$

$[G]_{n-i}$ is the generalized flexibility matrix. Moving the known values to the right-hand side, Eq. 7.82 is rewritten as

$$[G]_o \{p\}_n = \{\bar{u}_b\}_n - \{D\}_n \qquad (7.85)$$

with

$$\{D\}_n = \sum_{i=1}^{n-1} [G]_{n-i}\{p\}_i \qquad (7.86)$$

From virtual work considerations, the concentrated interaction forces at time $n\Delta t$ are obtained as

$$\{R_b\}_n = \int_S [N(s)]^T \{t_p(s,t' = \Delta t)\}_n \, ds \qquad (7.87)$$

Substituting Eq. 7.79 formulated for $j = n$ in Eq. 7.87 results in

$$\{R_b\}_n = \sum_{i=1}^{n} [T]_{n-i}^T \{p\}_i \qquad (7.88)$$

where

$$[T]_{n-i} = \int_S [g_t(s,t' = \Delta t)]_{n-i}^T [N(s)] \, ds \qquad (7.89)$$

For an embedded foundation, the source loads of all previous time steps contribute to the concentrated loads at time $n\Delta t$. Solving Eq. 7.85 for $\{p\}_n$ and substituting in Eq. 7.88 finally results in

$$\{R_b\}_n = [T]_o^T[G]_o^{-1}\{\bar{u}_b\}_n - [T]_o^T[G]_o^{-1}\{D\}_n + \sum_{i=1}^{n-1} [T]_{n-i}^T\{p\}_i \qquad (7.90)$$

To gain further insight, $\{u_b^t(t')\}_n$ is expressed, as an example, as

$$\{u_b^t(t')\}_n = f_o(t') \{u_b^t\}_n + f_1(t') \{\dot{u}_b^t\}_{n-1} + f_2(t') \{\ddot{u}_b^t\}_{n-1} \qquad (7.91)$$

with known functions $f_i(t')$ ($i = 0,1,2$). This relationship is somewhat different from that corresponding to the Newmark method (Eq. 6.134). Sub-

Sec. 7.3   Spatial and Temporal Discretizations of Boundary-Integral Equation

stituting Eq. 7.91 in Eq. 7.84 then leads to

$$\{\bar{u}_b\}_n = [U]_o\{u_b^t\}_n + [U]_1\{u_b^t\}_{n-1} + [U]_2\{u_b^t\}_{n-1} - \{U_g\}_n \tag{7.92}$$

where

$$[U]_i = \int_o^{\Delta t}\int_S [W(s,t')]^T[N(s)]ds f_i(t')dt' \tag{7.93a}$$

$$\{U_g\}_n = \int_o^{\Delta t}\int_S [W(s,t')]^T\{u^g(s,t')\}_n\, ds\, dt' \tag{7.93b}$$

Equation 7.92 transforms Eq. 7.90 to the interaction force–displacement relationship:

$$\{R_b\}_n = [S_{bb}]_o\{u_b^t\}_n + [T]_o^T[G]_o^{-1}[U]_1\{u_b^t\}_{n-1} + [T]_o^T[G]_o^{-1}[U]_2\{u_b^t\}_{n-1}$$
$$- [T]_o^T[G]_o^{-1}\{D\}_n + \sum_{i=1}^{n-1}[T]_{n-i}^T\{p\}_i - [T]_o^T[G]_o^{-1}\{U_g\}_n \tag{7.94}$$

with the instantaneous stiffness matrix $[S_{bb}]_o$ defined as

$$[S_{bb}]_o = [T]_o^T[G]_o^{-1}[U]_o \tag{7.95}$$

This weighted-residual technique, which is summarized for vanishing $\{u^g(s,t')\}_n$ on the left-hand side of Table 7-1, should be compared to the corresponding ones for the embedded foundation in the frequency domain (Table 4-2). The superscript $t$ can be dropped in this case.

### 7.3.3 Indirect Boundary-Element Method

For the indirect boundary-element method in the frequency domain, selecting the matrix of the weighting functions equal to that of the Green's functions for the surface tractions leads to a symmetric-stiffness matrix identical to the one obtained by application of the Maxwell-Betti reciprocity theorem (Section 4.4.3). Proceeding analogously in the time domain [W14], the weighting functions in the weighted-residual equation are selected as

$$[W(s,\tau')]_{n-j} = [g_t(s,\Delta t - \tau')]_{n-j} \tag{7.96}$$

Substituting Eq. 7.96 in Eq. 7.80 leads to

$$\sum_{j=1}^n \int_o^{\Delta t}\int_S [g_t(s,\Delta t - \tau')]_{n-j}^T (\{u_p(s,\tau')\}_j - \{u(s,\tau')\}_j)ds d\tau' = 0 \tag{7.97}$$

**TABLE 7-1  Various Formulations of Boundary-Element Method in Time Domain**

$$\{u(s, t')\}_n = [N(s)] \{u_b(t')\}_n$$

$$\{p(s', t')\}_n = [L(s')] \{p\}_n$$

$$\{u_p(s, \tau')\}_j = \sum_{i=1}^{j} [g_u(s, \tau')]_{j-i} \{p\}_i$$

$$\{t_p(s, \tau')\}_j = \sum_{i=1}^{j} [g_t(s, \tau')]_{j-i} \{p\}_i$$

**WEIGHTED RESIDUAL**

| Weighted-Residual Technique | Truncated Indirect Boundary-Element Method |
|---|---|
| $\{\overline{u_b}\}_n = \int_0^{\Delta t} \int_S [W(s, t')]^T [N(s)] \, ds \, \{u_b(t')\}_n \, dt'$ | $\{\overline{u_b}\}_n = \int_0^{\Delta t} \int_S [g_t(s, \Delta t - t')]_o^T [N(s)] \, ds \cdot \{u_b(t')\}_n \, dt'$ |
| $[G]_{n-i} = \int_0^{\Delta t} \int_S [W(s, t')]^T [g_u(s, t')]_{n-i} \, ds \, dt'$ | $[G]_{n-i} = \int_0^{\Delta t} \int_S [g_t(s, \Delta t - t')]_o^T [g_u(s, t')]_{n-i} \cdot ds \, dt'$ |
| $\{D\}_n = \sum_{i=1}^{n-1} [G]_{n-i} \{p\}_i$ | $\{D\}_n = \sum_{i=1}^{n-1} [G]_{n-i} \{p\}_i$ |
| $[G]_o \{p\}_n = \{\overline{u_b}\}_n - \{D\}_n$ | $[G]_o \{p\}_n = \{\overline{u_b}\}_n - \{D\}_n$ |
| $[T]_{n-i} = \int_S [g_t(s, t' = \Delta t)]_{n-i}^T [N(s)] \, ds$ | $[T]_{n-i} = \int_S [g_t(s, t' = \Delta t)]_{n-i}^T [N(s)] \, ds$ |
| $\{R_b\}_n = [T]_o^T [G]_o^{-1} \{\overline{u_b}\}_n - [T]_o^T [G]_o^{-1} \{D\}_n + \sum_{i=1}^{n-1} [T]_{n-i}^T \{p\}_i$ | $\{R_b\}_n = [T]_o^T [G]_o^{-1} \{\overline{u_b}\}_n - [T]_o^T [G]_o^{-1} \{D\}_n + \sum_{i=1}^{n-1} [T]_{n-i}^T \{p\}_i$ |

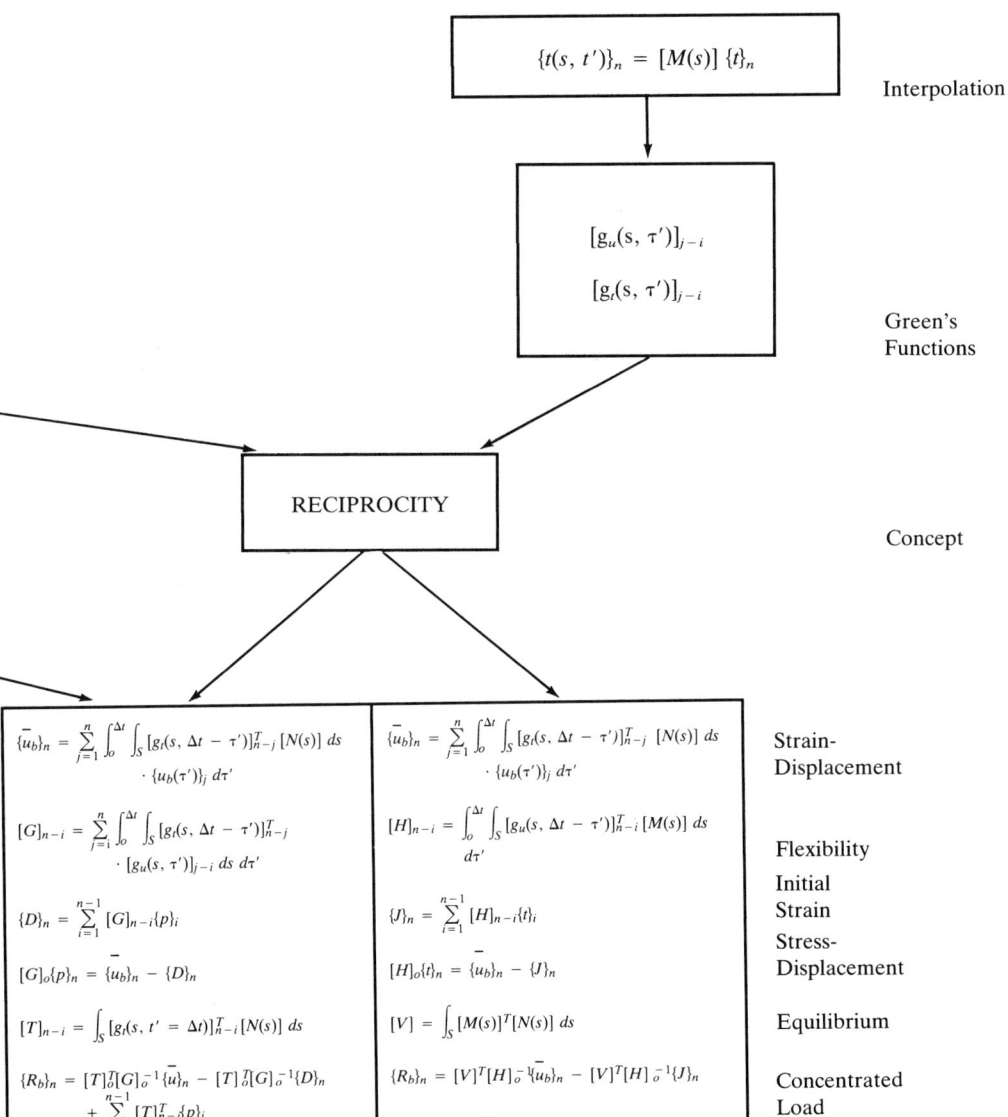

Equation 7.82 still applies, but with $[G]_{n-i}$ and $\{\bar{u}_b\}_n$ now defined as

$$[G]_{n-i} = \sum_{j=i}^{n} \int_o^{\Delta t} \int_S [g_t(s, \Delta t - \tau')]_{n-j}^T [g_u(s,\tau')]_{j-i} ds d\tau' \quad (7.98)$$

and

$$\{\bar{u}_b\}_n = \sum_{j=1}^{n} \int_o^{\Delta t} \int_S [g_t(s, \Delta t - \tau')]_{n-j}^T [N(s)] ds \{u_b^t(\tau')\}_j d\tau'$$

$$- \sum_{j=1}^{n} \int_o^{\Delta t} \int_S [g_t(s, \Delta t - \tau')]_{n-j}^T \{u^g(s,\tau')\}_j ds d\tau' \quad (7.99)$$

Substituting Eq. 7.91 in Eq. 7.99 results in

$$\{\bar{u}_b\}_n = \sum_{j=1}^{n} [T^*]_{n-j}^o \{u_b^t\}_j + \sum_{j=1}^{n} [T^*]_{n-j}^1 \{\dot{u}_b^t\}_{j-1}$$

$$+ \sum_{j=1}^{n} [T^*]_{n-j}^2 \{\ddot{u}_b^t\}_{j-1} - \sum_{j=1}^{n} \{U_g\}_j \quad (7.100)$$

where

$$[T^*]_{n-j}^o = \int_o^{\Delta t} \int_S [g_t(s, \Delta t - \tau')]_{n-j}^T [N(s)] ds f_o(\tau') d\tau' \quad (7.101a)$$

$$[T^*]_{n-j}^1 = \int_o^{\Delta t} \int_S [g_t(s, \Delta t - \tau')]_{n-j}^T [N(s)] ds f_1(\tau') d\tau' \quad (7.101b)$$

$$[T^*]_{n-j}^2 = \int_o^{\Delta t} \int_S [g_t(s, \Delta t - \tau')]_{n-j}^T [N(s)] ds f_2(\tau') d\tau' \quad (7.101c)$$

$$\{U_g\}_j = \int_o^{\Delta t} \int_S [g_t(s, \Delta t - \tau')]_{n-j}^T \{u^g(s,\tau')\}_j ds d\tau' \quad (7.101d)$$

This modifies Eqs. 7.94 and 7.95 as follows

$$\{R_b\}_n = [S_{bb}]_o \{u_b^t\}_n + \sum_{j=1}^{n-1} [T]_o^T [G]_o^{-1} [T^*]_{n-j}^o \{u_b^t\}_j$$

$$+ \sum_{j=1}^{n} [T]_o^T [G]_o^{-1} [T^*]_{n-j}^1 \{\dot{u}_b^t\}_{j-1} + \sum_{j=1}^{n} [T]_o^T [G]_o^{-1} [T^*]_{n-j}^2 \{\ddot{u}_b^t\}_{j-1} \quad (7.102)$$

$$- [T]_o^T [G]_o^{-1} \{D\}_n + \sum_{i=1}^{n-1} [T]_{n-i}^T \{p\}_i - \sum_{j=1}^{n} [T]_o^T [G]_o^{-1} \{U_g\}_j$$

with

$$[S_{bb}]_o = [T]_o^T [G]_o^{-1} [T^*]_o^o \quad (7.103)$$

Obviously, this procedure is computationally inefficient, as summations must be performed as indicated in Eqs. 7.98 and 7.99. This disappears if the weighting functions are selected as equal to zero for all time steps but the $n$th one (truncated indirect boundary-element method). All equations of Section 7.3.2 then apply, replacing $[W(s,t')]$ by $[g_t(s,\Delta t - t')]_o$. The matrix $[S_{bb}]_o$ corresponding to this truncated indirect method is identical to that of the complete indirect method. The formulations of the truncated indirect boundary-element method and of the indirect boundary-element method are summarized for $\{u^g(s,t')\}$ equal to zero in Table 7-1.

The indirect boundary-element formulation can also be derived with the reciprocity theorem of Maxwell-Betti, instead of using the weighted-residual approach. The dynamic-reciprocity theorem written in the form of the representation theorem, with the source point selected outside the unbounded soil, equals (Eq. 7.73, modified slightly)

$$\int_o^t \int_S \{u_{\delta p}(s,t-\tau)\}^T \{t(s,\tau)\}\,ds\,d\tau = \int_o^t \int_S \{t_{\delta p}(s,t-\tau)\}^T \{u(s,\tau)\}\,ds\,d\tau \qquad (7.104\mathrm{a})$$

or

$$\sum_{j=1}^n \int_o^{\Delta t} \int_S \{u_{\delta p}(s,\Delta t - \tau')\}_{n-j+1}^T \{t(s,\tau')\}_j\,ds\,d\tau'$$

$$= \sum_{j=1}^n \int_o^{\Delta t} \int_S \{t_{\delta p}(s,\Delta t - \tau')\}_{n-j+1}^T \{u(s,\tau')\}_j\,ds\,d\tau' \qquad (7.104\mathrm{b})$$

$\{t(s,\tau)\}$ represents the (unknown) surface tractions corresponding to $\{u(s,\tau)\}$, and $\{u_{\delta p}(s,\tau)\}$, $\{t_{\delta p}(s,\tau)\}$ describe the virtual state generated by the virtual source loads $\{\delta p(s',\tau')\}_1$. The latter exhibit the same spatial and temporal variation as the actual source loads, but they are selected to act only during the first time step. As in the weighted-residual formulation, the surface tractions arising from the specified displacements $\{t(s,\tau)\}$ are assumed to be equal to those of the source loads $\{t_p(s,\tau)\}$, Eq. 7.104b is thus rewritten as

$$\sum_{j=1}^n \int_o^{\Delta t} \int_S \{u_{\delta p}(s,\Delta t - \tau')\}_{n-j+1}^T \{t_p(s,\tau')\}_j\,ds\,d\tau'$$

$$= \sum_{j=1}^n \int_o^{\Delta t} \int_S \{t_{\delta p}(s,\Delta t - \tau')\}_{n-j+1}^T \{u(s,\tau')\}_j\,ds\,d\tau' \qquad (7.105)$$

The reciprocity law is then applied once again by using the state of the source loads and the virtual state

$$\sum_{j=1}^n \int_o^{\Delta t} \int_S \{u_{\delta p}(s,\Delta t - \tau')\}_{n-j+1}^T \{t_p(s,\tau')\}_j\,ds\,d\tau'$$

$$= \sum_{j=1}^n \int_o^{\Delta t} \int_S \{t_{\delta p}(s,\Delta t - \tau')\}_{n-j+1}^T \{u_p(s,\tau')\}_j\,ds\,d\tau' \qquad (7.106)$$

As the left-hand side of Eqs. 7.105 and 7.106 are identical,

$$\sum_{j=1}^{n} \int_{o}^{\Delta t} \int_{S} \{t_{\delta p}(s, \Delta t - \tau')\}_{n-j+1}^{T} \{u_p(s, \tau')\}_j \, ds \, d\tau'$$

$$= \sum_{j=1}^{n} \int_{o}^{\Delta t} \int_{S} \{t_{\delta p}(s, \Delta t - \tau')\}_{n-j+1}^{T} \{u(s, \tau')\}_j \, ds \, d\tau' \quad (7.107)$$

results. Substituting Eq. 7.79, formulated for the virtual state, leads, for arbitrary $\{\delta p\}_1$, to

$$\sum_{i=1}^{n} \int_{o}^{\Delta t} \int_{S} [g_t(s, \Delta t - \tau')]_{n-j}^{T} (\{u_p(s, \tau')\}_j - \{u(s, \tau')\}_j) \, ds \, d\tau' = 0 \quad (7.108)$$

which is identical to Eq. 7.97 (see Table 7-1).

In the indirect boundary-element method, the flexibility matrix $[G]_{n-i}$ defined in Eq. 7.98 is symmetric. For the proof of this property, it is appropriate to reformulate the definition of $[G]_{n-i}$ as follows:

$$[G]_{n-i} = \int_{t_{i-1}}^{t_n} \int_{S} [g_t(s, t_n - \tau)]^T [g_u(s, \tau - t_{i-1})] \, ds \, d\tau \quad (7.109)$$

where $t_n = n\Delta t$ and $t_{i-1} = (i-1)\Delta t$. The $(kl)$th element of $[G]_{n-i}$ equals

$$(G_{kl})_{n-i} = \int_{t_{i-1}}^{t_n} \int_{S} \{g_t(s, t_n - \tau)\}_k^T \{g_u(s, \tau - t_{i-1})\}_l \, ds \, d\tau \quad (7.110)$$

where $\{g_t(s,t_n - \tau)\}_k$ denotes the $k$th column of $[g_t(s,t_n - \tau)]$ and $\{g_u(s,\tau - t_{i-1})\}_l$ the $l$th one of $[g_u(s,\tau - t_{i-1})]$. Applying the reciprocity law to Eq. 7.110 leads to

$$(G_{kl})_{n-i} = \int_{t_{i-1}}^{t_n} \int_{S} \{g_u(s, t_n - \tau)\}_k^T \{g_t(s, \tau - t_{i-1})\}_l \, ds \, d\tau \quad (7.111)$$

which, using a property of convolution integrals, is equal to

$$(G_{kl})_{n-i} = \int_{t_{i-1}}^{t_n} \int_{S} \{g_u(s, \tau - t_{i-1})\}_k^T \{g_t(s, t_n - \tau)\}_l \, ds \, d\tau \quad (7.112)$$

or, as $(G_{kl})_{n-i}$ is a scalar quantity,

$$(G_{kl})_{n-i} = \int_{t_{i-1}}^{t_n} \int_{S} \{g_t(s, t_n - \tau)\}_l^T \{g_u(s, \tau - t_{i-1})\}_k \, ds \, d\tau \quad (7.113)$$

Comparing this equation with Eq. 7.110, it is concluded that the right-hand side of Eq. 7.113 equals $(G_{lk})_{n-i}$—that is,

$$(G_{kl})_{n-i} = (G_{lk})_{n-i} \quad (7.114)$$

In the weighted-residual formulation, $[G]_{n-i}$ of Eq. 7.83 is, in general, not

Sec. 7.3    Spatial and Temporal Discretizations of Boundary-Integral Equation

symmetric. In the truncated indirect boundary-element formulation, only $[G]_o$ is always symmetric.

### 7.3.4 Direct Boundary-Element Method

The starting point of this formulation is again the reciprocity theorem of Eq. 7.104b [W14]. In contrast to the indirect boundary-element method, the unknown surface tractions $\{t(s, t')\}_n$ are interpolated as a function of the nodal values as

$$\{t(s, t')\}_n = [M(s)] \{t\}_n \tag{7.115}$$

where $[M(s)]$ represents the selected interpolation functions. During the $n$th time step, the temporal variation of the surface tractions is assumed to be constant. Substituting Eqs. 7.78 and 7.79—both formulated for virtual loads acting during the first time step—in Eq. 7.104b and using Eqs. 7.75 and 7.115 leads to

$$\sum_{i=1}^{n} [H]_{n-i} \{t\}_i = \{\bar{u}_b\}_n \tag{7.116}$$

where

$$[H]_{n-i} = \int_0^{\Delta t} \int_S [g_u(s, \Delta t - \tau')]_{n-i}^T [M(s)] \, ds \, d\tau' \tag{7.117}$$

$$\{\bar{u}_b\}_n = \sum_{j=1}^{n} \int_0^{\Delta t} \int_S [g_t(s, \Delta t - \tau')]_{n-j}^T [N(s)] \, ds \, \{u_b^t(\tau')\}_j \, d\tau'$$

$$- \sum_{j=1}^{n} \int_0^{\Delta t} \int_S [g_t(s, \Delta t - \tau')]_{n-j}^T \{u^g(s, \tau')\}_j \, ds \, d\tau' \tag{7.118}$$

The number of virtual source loads must equal that of the surface tractions' nodal values to achieve square matrices $[H]_{n-i}$. Solving Eq. 7.116 for $\{t\}_n$ and substituting in the expression for the concentrated interaction forces analogously to Eq. 7.87

$$\{R_b\}_n = \int_S [N(s)]^T \{t(s, t' = \Delta t)\}_n \, ds \tag{7.119}$$

results in

$$\{R_b\}_n = [V]^T [H]_o^{-1} \{\bar{u}_b\}_n - [V]^T [H]_o^{-1} \{J\}_n \tag{7.120}$$

with

$$[V] = \int_S [M(s)]^T [N(s)] \, ds \qquad (7.121)$$

$$\{J\}_n = \sum_{i=1}^{n-1} [H]_{n-i} \{t\}_i \qquad (7.122)$$

Substituting Eq. 7.91 in Eq. 7.118 leads to the same equation as in the complete indirect boundary-element formulation (Eq. 7.100):

$$\{\bar{u}_b\}_n = \sum_{j=1}^{n} [T^*]_{n-j}^o \{u_b^t\}_j + \sum_{j=1}^{n} [T^*]_{n-j}^1 \{u_b^t\}_{j-1} + \sum_{j=1}^{n} [T^*]_{n-j}^2 \{\ddot{u}_b^t\}_{j-1} - \sum_{j=1}^{n} \{U_g\}_j$$

$$(7.123)$$

where $[T^*]_{n-j}$ and $\{U_g\}_j$ are still defined by Eq. 7.101. This modifies Eq. 7.120 as follows:

$$\{R_b\}_n = [S_{bb}]_o \{u_b^t\}_n + \sum_{j=1}^{n-1} [V]^T [H]_o^{-1} [T^*]_{n-j}^o \{u_b^t\}_j + \sum_{j=1}^{n} [V]^T [H]_o^{-1} [T^*]_{n-j}^1 \{u_b^t\}_{j-1}$$

$$+ \sum_{j=1}^{n} [V]^T [H]_o^{-1} [T^*]_{n-j}^2 \{\ddot{u}_b^t\}_{j-1} - [V]^T [H]_o^{-1} \{J\}_n - \sum_{j=1}^{n} [V]^T [H]_o^{-1} \{U_g\}_j$$

$$(7.124)$$

with

$$[S_{bb}]_o = [V]^T [H]_o^{-1} [T^*]_o^o \qquad (7.125)$$

In general, $[H]_o$ is not symmetric. The formulation of this direct boundary-element method is also summarized in Table 7-1 (omitting the terms associated with the earthquake excitation).

The standard direct boundary-element method, based on the singular integral equation (Eq. 7.74 with vanishing interior loading), is contained in this formulation. Assuming, for instance, that the displacements and surface tractions are constant over each boundary element and also over a time step, and formulating Eq. 7.74 for each boundary element, leads to Eq. 7.116 with Eqs. 7.117 and 7.118 applying. Singularities which arise when the source boundary-element becomes also a receiver element can be treated analytically.

The formulations for the three types of boundary-element methods in the time domain are thus analogous to those in the frequency domain, which are summarized in Table 4-2. In the latter, the equations are only specified for the calculation of the dynamic-stiffness matrix and thus it does not deal with the specified motion.

Sec. 7.3  Spatial and Temporal Discretizations of Boundary-Integral Equation    401

The convolution integrals appearing in the dynamic-reciprocity theorem are, of course, also present in the discretized form—for example, in the direct boundary-element method in the terms $\{\bar{u}_b\}_n$ (Eq. 7.118) and $\{J\}_n$ with $[H]_{n-i}$ (Eqs. 7.122 and 7.117). These terms consist of integrals, involving the Green's functions, of the displacements and surface tractions. The behavior of these influence functions for the full infinite space is quite different for the three-dimensional and two-dimensional cases. As in the former case, the Green's functions return to zero after the $S$-wave has passed (Eq. 7.32), the corresponding terms in the integrals will vanish. In the latter case, the Green's functions remain finite (Eq. 7.34), which means that all integrals must be calculated (after the arrival of the $P$-wave). For a layered half-space, in contrast to the full infinite space, the wave pattern is much more complicated. In general, in three dimensions, the Green's functions also do not return to zero, increasing the computational effort significantly for a layered system.

### 7.3.5 Number of Operations

Finally, the number of operations involved in the four methods summarized in Table 7-1 based on a nonrecursive evaluation of the convolution integrals is addressed. As all procedures require the calculation of Green's functions in the time domain, the corresponding computational effort can be disregarded in the comparison. The remaining numbers of operations are specified in Table 7-2. $N^s$ denotes the number of boundary elements introduced in the spatial discretization (with 3 degrees of freedom per element); and $N^s_G$ denotes the corresponding number of Gauss' points of integration. The temporal discretization is characterized by the number of time steps $N^t$ and Gauss' points $N^t_G$. The number of nodal values of the source loads is assumed to be equal to $3N^s$. As expected, the number of operations arising from the spatial discretization is essentially the same in all methods. The differences in the time-domain discretization arise from the various convolution integrals. The largest number of operations occurs in the indirect boundary-element method, as can be seen from the term

$$(N^t)^2(3N^s)^2[3N^s N^s_G N^t_G + \frac{1}{2} N^t_G + 1] \qquad (7.126)$$

followed by the direct boundary-element method. This is caused by the fact that in the indirect method the weighting functions, consisting of Green's functions, cause an additional convolution. The flexibility matrix $[G]_{n-i}$ thus contains a sum (Eq. 7.98), which is not the case in the direct method ($[H]_{n-i}$, Eq. 7.117). The weighted-residual and truncated indirect boundary-element methods require significantly fewer operations. The most efficient procedure is the point-collocation scheme, which follows from the weighted-residual method, setting $N^t_G = 1$. By way of illustration, the total number of operations is specified for the indicated spatial and temporal discretizations in

**TABLE 7-2  Number of Operations of Various Formulations of Boundary-Element Method in Time Domain**

| | Weighted-Residual Method Truncated Indirect BEM | Indirect BEM | Direct BEM |
|---|---|---|---|
| $\{\bar{u}_b\}_n$ | $(3N^s)^3 N_G^s N_G^t + N^t(3N^s)^2 N_G^t$ | $\dfrac{N^t(N^t+1)}{2}$ $\times [(3N^s)^3 N_G^s N_G^t + (3N^s)^2 N_G^t]$ | $\dfrac{N^t(N^t+1)}{2}$ $\times [(3N^s)^3 N_G^s N_G^t + (3N^s)^2 N_G^t]$ |
| $[G]_{n-i}$, resp. $[H]_{n-i}$ | $N^t(3N^s)^3 N_G^s N_G^t$ | $\dfrac{N^t(N^t+1)}{2}(3N^s)^3 N_G^s N_G^t$ | $N^t(3N^s)^3 N_G^s N_G^t$ |
| $\{D\}_n$, resp. $\{f\}_n$ | $\dfrac{(N^t-1)N^t}{2}(3N^s)^2$ | $\dfrac{(N^t-1)N^t}{2}(3N^s)^2$ | $\dfrac{(N^t-1)N^t}{2}(3N^s)^2$ |
| $\{p\}_n$, resp. $\{t\}_n$ | $N^t(3N^s)^2$ | $N^t(3N^s)^2$ | $N^t(3N^s)^2$ |
| $[T]_{n-i}$, resp. $[V]$ | $N^t(3N^s)^3 N_G^s$ | $N^t(3N^s)^3 N_G^s$ | $(3N^s)^3 N_G^s$ |
| $\{R_b\}_n$ | $\dfrac{(N^t-1)N^t}{2}(3N^s)^2 + N^t(3N^s)^2$ $+ (3N^s)^3$ | $\dfrac{(N^t-1)N^t}{2}(3N^s)^2$ $+ N^t(3N^s)^2 + (3N^s)^3$ | $N^t(3N^s)^2 + (3N^s)^3$ |
| TOTAL | $(N^t)^2(3N^s)^2 + N^t(3N^s)^2$ $\times [3N^s N_G^s(N_G^t+1) + N_G^t + 1]$ $+ (3N^s)^3(N_G^s N_G^t + 1)$ | $(N^t)^2(3N^s)^2 \left[ 3N^s N_G^s N_G^t + \dfrac{1}{2} N_G^t + 1 \right]$ $+ N^t(3N^s)^2$ $\times \left[ 3N^s N_G^s(N_G^t+1) + \dfrac{1}{2} N_G^t + 1 \right]$ $+ (3N^s)^3$ | $(N^t)^2(3N^s)^2 \left[ 3N^s \dfrac{1}{2} N_G^s N_G^t + \dfrac{1}{2} N_G^t + \dfrac{1}{2} \right]$ $+ N^t(3N^s)^2 \left[ 3N^s \dfrac{3}{2} N_G^s N_G^t + \dfrac{1}{2} N_G^t + \dfrac{3}{2} \right]$ $+ (3N^s)^3(N_G^s+1)$ |

TABLE 7-3  Example of Number of Operations of Various Methods

| $N^s$ | $N^t$ | Indirect BEM | Direct BEM | Weighted-Residual Truncated Indirect BEM | Point Collocation | Time-Domain Stiffness Formulation | Frequency-Domain Stiffness Formulation |
|---|---|---|---|---|---|---|---|
| 10 | 1'024 | $1.15 \cdot 10^{11}$ | $5.82 \cdot 10^{10}$ | $1.11 \cdot 10^9$ | $1.06 \cdot 10^9$ | $9.45 \cdot 10^8$ | $4.67 \cdot 10^6$ |
|  | 2'048 | $4.61 \cdot 10^{11}$ | $2.32 \cdot 10^{11}$ | $4.11 \cdot 10^9$ | $4.00 \cdot 10^9$ | $3.78 \cdot 10^9$ | $9.34 \cdot 10^6$ |
| 20 | 1'024 | $9.15 \cdot 10^{11}$ | $4.60 \cdot 10^{11}$ | $5.11 \cdot 10^9$ | $4.67 \cdot 10^9$ | $3.78 \cdot 10^9$ | $1.67 \cdot 10^7$ |
|  | 2'048 | $3.66 \cdot 10^{12}$ | $1.84 \cdot 10^{12}$ | $1.78 \cdot 10^{10}$ | $1.69 \cdot 10^{10}$ | $1.51 \cdot 10^{10}$ | $3.34 \cdot 10^7$ |

Table 7-3. The number of Gauss' points for the spatial discretization $N_G^s$ is equal to 2; and that of the temporal discretization $N_G^t$ is equal to 2 (with the exception of point collocation). The first four columns confirm the conclusions just reached. Differences of two orders of magnitude do arise.

Instead of working with the Green's functions in the time domain, the dynamic-stiffness or -flexibility coefficients of the discretized structure–soil interface S are initially determined in the frequency domain and then transformed to the time domain, as described in Chapter 6. The latter then appear directly in a convolution integral (Eqs. 6.2 or 6.10), whose evaluation requires

$$(N^t)^2(3N^s)^2 \frac{1}{2} N_G^t + N^t(3N^s)^2 N_G^t \frac{1}{2} N_G^t \tag{7.127}$$

operations. In contrast to the methods discussed in Table 7-2, the factor of the term $(N^t)^2$ equals $(3N^s)^2 \frac{1}{2} N_G^t$. For $N_G^t = 2$, this stiffness formulation in the time domain leads essentially to the same number of operations (fifth column of Table 7-3) as point collocation. For $N_G^t = 1$, a significant reduction occurs.

Finally, the standard procedure, which works in the frequency domain and which is applicable to a linear system, is briefly discussed. Again considering the contribution of the soil only, the Fourier transformation from $\{u_b(t)\}$ to $\{u_b(\omega)\}$, the evaluation of $\{R_b(\omega)\}$, and the inverse transformation resulting in $\{R_b(t)\}$ lead to

$$N^t \ln(2N^t) \, 4 \, (3N^s) + N^t \, 4(3N^s)^2 \tag{7.128}$$

operations. It is assumed that the Fast Fourier Transform is used, and that these operations are performed for $N^t$ frequencies. As is visible from the last column of Table 7-3, the computational effort is reduced by at least two orders of magnitude! This is confirmed in a practical example (Section 7.7).

## 7.4 LOADED SPHERICAL CAVITY WITH SYMMETRIC WAVES

### 7.4.1 Analytical Solution

The various boundary-element methods in the time domain described in Section 7.3 are illustrated and compared by using the one-dimensional dynamic problem of the spherical cavity, subjected to a prescribed displacement time-history on its wall, in an infinite space [W17]. The corresponding pressure on the cavity's wall is to be determined—a calculation analogous to that of establishing a dynamic-stiffness coefficient in the time domain. The emphasis is placed on the temporal discretization, as the spatial one is also addressed in a standard frequency-domain calculation, which is quite well known.

## Sec. 7.4 Loaded Spherical Cavity with Symmetric Waves

On the wall $S$ of the spherical cavity of radius $a$ embedded in an infinite space (Fig. 7-11), the radial displacement $u_o(t)$ (which is constant in the longitudinal and colatitudinal directions) is prescribed as

$$u_o(t) = \frac{q_o a}{4G}\left[1 - \exp\left(-\frac{c_p t}{a}\right)\right] \qquad t \geq 0 \qquad (7.129)$$

$G$ denotes the shear modulus and $c_p$ the dilatational-wave velocity. The pressure $q_o$ will turn out to be the value applied at $t = 0$. $u_o(t)$ is shown in Fig. 7-12a.

In the nomenclature of a general soil–structure interaction problem, $S$ represents the structure–soil interface, and the infinite space with the spherical cavity of radius $a$ represents the system ground—that is, the substructure soil.

Based on the results of the spherical cavity with symmetric waves discussed in Sections 3.1.3 and 6.2.1, the radial displacement is specified as (Eq. 6.13)

$$u(r, t) = -\frac{f\left(t - \frac{r-a}{c_p}\right)}{r^2} - \frac{f'\left(t - \frac{r-a}{c_p}\right)}{rc_p} \qquad (7.130)$$

with $f' = df[t - (r-a)/c_p]/d[t - (r-a)/c_p]$.

The initial conditions are formulated as

$$u(r, t = 0) = 0 \qquad (7.131a)$$

$$\dot{u}(r, t = 0) = 0 \qquad (7.131b)$$

and the boundary condition as

$$u(r = a, t) = u_o(t) \qquad (7.132)$$

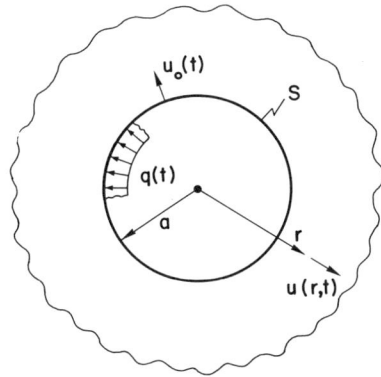

**Figure 7-11** Spherical cavity of radius $a$ (system ground).

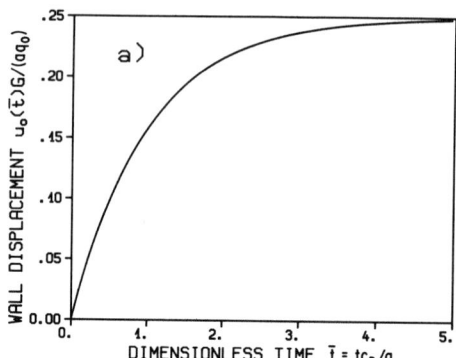
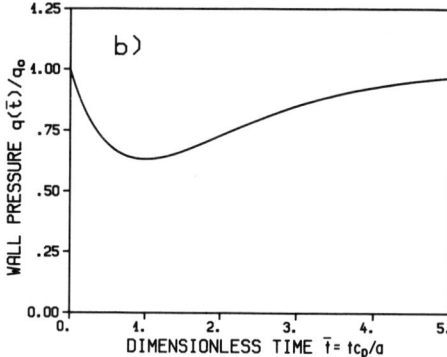

**Figure 7-12** Response of spherical cavity of radius $a$.  a. Prescribed displacement time-history.  b. Resulting pressure time-history.

Substituting Eq. 7.130 in Eq. 7.132 results in the ordinary differential equation of first order

$$\frac{\dot{f}(t)}{ac_p} + \frac{f(t)}{a^2} = -u_o(t) \tag{7.133}$$

Proceeding analogously to Section 6.2.1, the general solution equals

$$u(r, t) = \frac{q_o a}{4G} \left\{ \frac{a^2}{r^2} - \left[ \frac{a^2}{r^2} + \frac{c_p \tau}{a} \left( \frac{a^2}{r^2} - \frac{a}{r} \right) \right] \exp\left( -\frac{c_p \tau}{a} \right) \right\} \tag{7.134}$$

with

$$\tau = t - \frac{r - a}{c_p} \tag{7.135}$$

Substituting Eq. 7.134 in the radial stress $\sigma_r$–displacement relation (Eq. 6.24)

$$\sigma_r(r, t) = G \left[ \frac{c_p^2}{c_s^2} u(r, t)_{,r} + \left( \frac{c_p^2}{c_s^2} - 2 \right) \frac{2}{r} u(r, t) \right] \tag{7.136}$$

Sec. 7.4  Loaded Spherical Cavity with Symmetric Waves                407

leads to

$\sigma_r(r, t)$

$$= q_o \left\{ -\frac{a^3}{r^3} + \left[\frac{a^3}{r^3} - \frac{1}{4}\frac{c_p^2 a}{c_s^2 r} + \frac{c_p \tau}{a}\left(\frac{a^3}{r^3} - \frac{a^2}{r^2} + \frac{1}{4}\frac{c_p^2 a}{c_s^2 r}\right)\right] \exp\left(-\frac{c_p \tau}{a}\right) \right\}$$

(7.137)

where $c_s$ is the shear-wave velocity. The corresponding pressure $q(t)$ on the wall follows as

$$q(t) = -\sigma_r(a, t) = q_o \left\{ 1 - \left[1 - \frac{1}{4}\frac{c_p^2}{c_s^2} + \frac{c_p t}{a}\frac{1}{4}\frac{c_p^2}{c_s^2}\right] \exp\left(-\frac{c_p t}{a}\right) \right\} \quad (7.138)$$

$q(t)$ is plotted in Fig. 7-12b for a Poisson's ratio equal to 1/3.

### 7.4.2 Discretization

The Green's functions, used in the various boundary-element methods, are calculated for the system of a cavity of radius $a/2$ which can easily be analyzed (Fig 7-1). These influence functions are derived in Section 7.1.1. For a unit impulse, the Green's functions $g_u(t)$ and $g_t(t)$ are specified in Eqs. 7.1 and 7.3 and for a rectangular pulse in Eqs. 7.4 and 7.5.

In the illustrative example of the cavity embedded in an infinite space, the problem is one-dimensional, and the spatial discretization is thus straightforward. Unless otherwise indicated, all forthcoming calculations are performed by using the crude time step $\Delta t = 0.5\ a/c_p$, and the source loads (or the surface tractions, in the direct method) are assumed to be constant over each time step (rectangular pulse). Also, to avoid numerical integration errors, closed-form analytical expressions are used in the various formulations.

### 7.4.3 Result

The results of the weighted-residual technique (Section 7.3.2) are addressed first. Examining the case of a constant-weighting function $W(t')$ (applicable to the last time step only), the wall pressure $q(t)$ of the cavity (which is equal to the surface traction $t_p(t)$) is plotted in Fig. 7-13a versus the dimensionless time $\bar{t} = tc_p/a$. The exact solution of Eq. 7.138 is shown as a dashed line. The zigzagging of the pressure curve arises from the jumps in the Green's function $g_t$, shown in Fig. 7-3b. In Fig. 7-13b, the wall displacement $u_p(t)$ (determined from $g_u$ and the known source loads) is compared to the prescribed displacement. The corresponding source loads $p(t)$ are

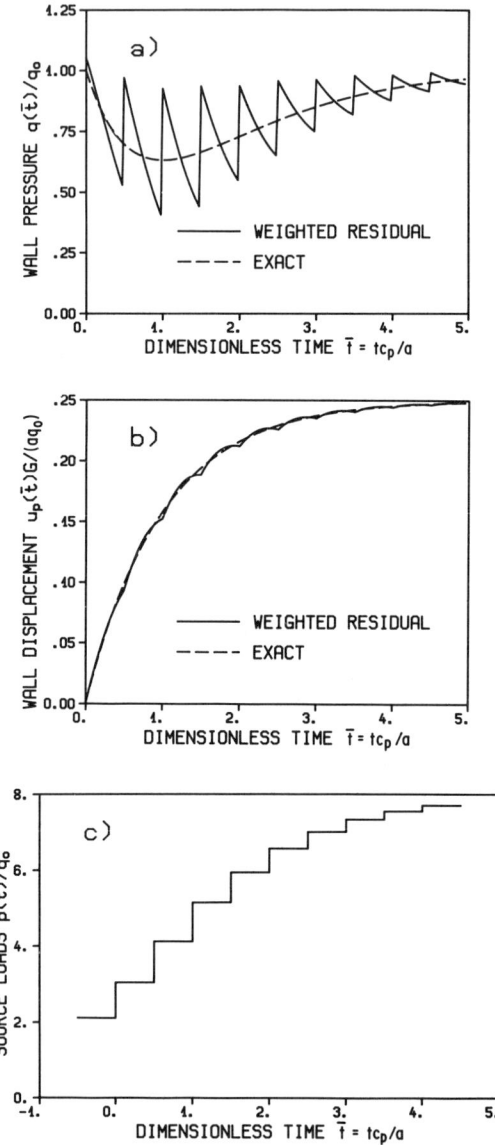

**Figure 7-13** Weighted-residual technique with constant-weighting function. a. Wall pressure. b. Wall displacement. c. Source loads.

shown in Fig. 7-13c. It is worth mentioning that the first source load is applied at $t = -0.5\ a/c_p$ to allow for the front of the wave to reach the cavity's wall at $t = 0$. Selecting the time step 10 times smaller leads to a better agreement of the wall pressure with the exact solution (Fig. 7-14). The weighted-residual technique can also be applied using a Dirac function at the

Sec. 7.4 Loaded Spherical Cavity with Symmetric Waves

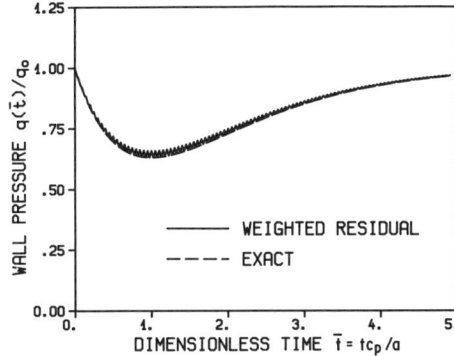

**Figure 7-14** Weighted-residual technique with constant-weighting function, $\triangle t = 0.05\, a/c_p$.

end of the time step as a weighting function—that is, performing point collocation. This leads to complete agreement of the calculated displacement $u_p(t)$ with the exact value at the end of each time step, as is visible in Fig. 7-15. The wall pressure and the source loads, which are not shown, hardly differ from those obtained by using a constant-weighting function.

Applying the truncated indirect boundary-element and the indirect boundary-element method (Section 7.3.3) leads to results which are very similar to those using the weighted-residual technique (not presented).

Finally, the direct boundary-element method (Section 7.3.4) is addressed. Selecting a rectangular pulse as virtual load (which affects $g_u(\tau')$ and $g_t(\tau')$) leads to the wall pressure shown in Fig. 7-16. This virtual load is applied on the source surface $S'$ at time $t = -0.5\, a/c_p$. As the surface traction (pressure) is assumed to be piecewise constant, the agreement with the exact solution is somewhat better than in the other methods. The wall displacement corresponding to the discretization cannot be determined in the direct method. In Fig. 7-17, the wall pressure obtained by application of the direct method using a Dirac impulse as virtual load is presented. The agreement is slightly better.

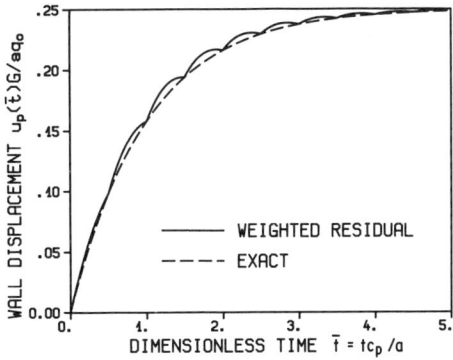

**Figure 7-15** Weighted-residual technique with point collocation.

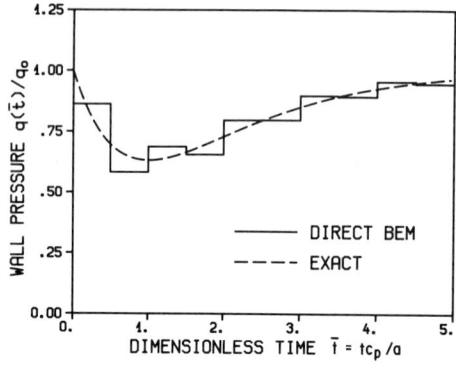

**Figure 7-16** Direct boundary-element method with rectangular pulse as virtual load.

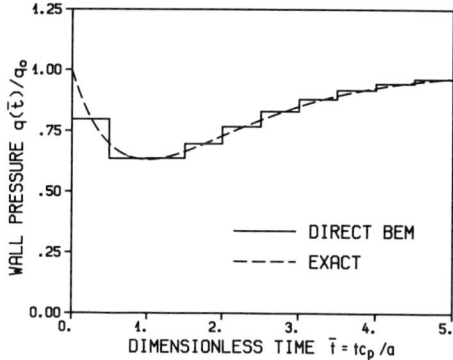

**Figure 7-17** Direct boundary-element method with Dirac impulse as virtual load.

## 7.5 FLEXIBILITY OF RIGID CIRCULAR DISK

As an application of the truncated indirect boundary-element method, the dynamic-flexibility coefficients in the time domain, in the three translational and rotational directions at the center of a rigid massless circular disk of radius $a$ resting on the surface of an undamped elastic half-space with Poisson's ratio 1/3, are calculated [W13]. Relaxed contact is assumed—that is, the vertical displacement due to the horizontal load and the horizontal displacement due to the vertical load are omitted.

The contact area is discretized into 112 square boundary elements with constant distributed source loads over each element (see Fig. 7-21). At first, the rigid-body constraint defined by

$$\{u_b(t)\} = [A] \{u_o(t)\} \tag{7.139}$$

is introduced in the force–displacement relationship (Eq. 7.90). The superscript $t$ is omitted. $\{u_o\}$ contains the rigid disk's 6 degrees of freedom. The matrix $[A]$ represents the kinematic transformation with geometric quantities only. Selecting $f_1(t')$ and $f_2(t')$ in Eq. 7.91 as zero and solving for $\{u_o\}_n$ leads to (see also Table 7-1)

$$\{u_o\}_n = ([A]^T[S_{bb}]_o[A])^{-1}(\{R_o\}_n + [A]^T[T]_o^T[G]_o^{-1}\{D\}_n - [A]^T \sum_{i=1}^{n-1} [T]_{n-i}^T\{p\}_i)$$

(7.140)

$\{R_o\}_n$ contains the corresponding prescribed forces—that is, a Dirac delta function (distributed over the first time step). For the first time step, the last two terms in Eq. 7.140 vanish. For all other time steps, the loads $\{p\}_i$ are self-equilibrating.

As an example, the nondimensional vertical flexibility coefficient in the time domain $a/c_s \, \overline{F}(t)$ is plotted versus the dimensionless time $\bar{t} = t\,c_s/a$ in Fig. 7-18. The flexibility coefficient is divided by the static value $(1-\nu)/(4Ga)$ [denoted as $\overline{F}(\bar{t})$]. This solution, using boundary elements in the time domain, agrees well with the result determined by the inverse transformation of the flexibility coefficient in the frequency domain. The latter, shown as a dashed line, is discussed in Section 6.4.2 (Fig. 6-12b).

## 7.6 STRUCTURE WITH PARTIAL BASEMAT UPLIFT

For severe earthquakes, large overturning moments arise which may lead to tension in part of the area of contact of the basemat of the structure and of the soil, according to a calculation based on a linear theory. As tension is incompatible with the constitutive model of soil, the basemat will become partially separated from the underlying soil. A thorough discussion of this partial-uplift phenomenon lies outside the scope of this book. Approximate procedures, using springs and dampers with frequency-independent coefficients, are described in Sections 2.14 and 3.11. For the purpose of illustration, it is sufficient to examine the behavior of the simple structure with a rigid basemat resting on the surface of a half-space, shown in Fig. 7-19.

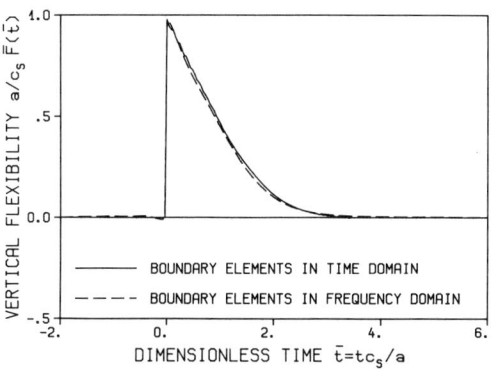

Figure 7-18 Dynamic-flexibility coefficient in time domain, comparison of calculation procedures.

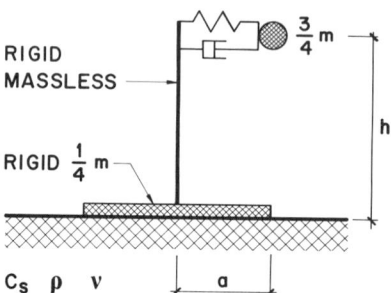

**Figure 7-19** Model of investigated structure.

Three-quarters of the mass $m$ is concentrated at the height $h$. The mass moment of inertia associated with the rocking degree of freedom at the circular basemat of radius $a$ equals $1/4\, a^2 m$. The fixed-base frequency and the damping ratio of the structure equal 4 Hz and 0.07, respectively. The following parameters apply: $a/c_s = 0.06$ s, $h/a = 1.5$, $m/(\rho a^3) = 3$ ($\rho$ = mass density of soil), and Poisson's ratio $\nu$ of the undamped soil = $1/3$. Horizontal and vertical artificial time-histories which follow the U.S. NRC response spectra (Regulatory Guide 1.60) [U2], normalized to 0.4g and 0.267g, respectively, are used. Vertically incident waves are assumed [W13].

The computational method proceeds as follows. For the $n$th step, the total displacements $\{u_o^t(t')\}_n$, and thus $\{u_o^t\}_n$ at the center of the basemat at time $n\Delta t$, are calculated as a prediction from the motion at time $(n-1)\Delta t$ using Eq. 6.135a ($\beta = 0$). In analogy to Eq. 7.139, the total displacements $\{u_b^t\}_n$ of all boundary elements follow as

$$\{u_b^t\}_n = [A]\{u_o^t\}_n \qquad (7.141)$$

$[A]$ represents the rigid body constraints. In the actual calculation, 112 square boundary elements with a constant spatial variation of the loads are used (Fig. 7-21). The force–displacement relationship of the truncated indirect boundary-element method (Eqs. 7.94 and 7.102) is formulated as

$$\begin{aligned}
\{R_b\}_n = {}& [S_{bb}]_o \{u_b^t\}_n \\
& + [T]_o^T [G]_o^{-1} [T^*]_o^1 \{u_b^t\}_{n-1} \\
& + [T]_o^T [G]_o^{-1} [T^*]_o^2 \{\ddot{u}_b^t\}_{n-1} \qquad (7.142) \\
& - [T]_o^T [G]_o^{-1} \{D\}_n + \sum_{i=1}^{n-1} [T]_{n-i}^T \{p\}_i \\
& - [T]_o^T [G]_o^{-1} \{U_g\}_n
\end{aligned}$$

In the definition of $\{U_g\}_n$ (Eq. 7.101d), the free-field motion $\{u^f(s, t')\}_n$ replaces $\{u^g(s, t')\}_n$.

Equation 7.142 is evaluated for all boundary elements. If tension arises, the corresponding components are set equal to zero. In addition, if the

Sec. 7.6   Strucutre with Partial Basemat Uplift                               413

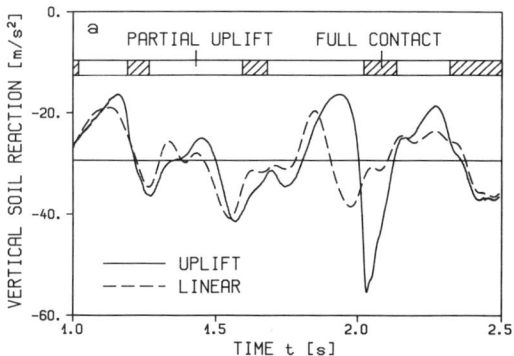

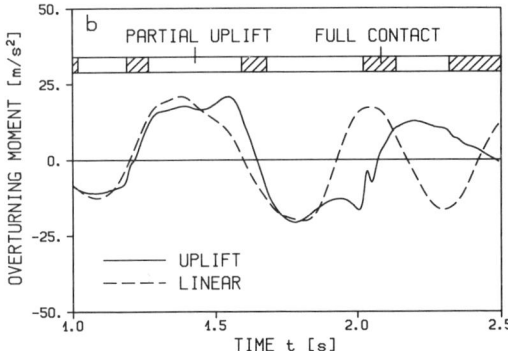

**Figure 7-20** Time-history of total soil reaction. a. Vertical force (factor $\rho a^3$). b. Overturning moment (factor $\rho a^4$).

resultant shear force exceeds the allowable value (= normal force multiplied by the friction coefficient), the corresponding modification is performed. Because, for a surface structure, the $[T]_o$-matrix is diagonal (Eq. 7.89 with $[g_t(s)]_o^T = [L(s)]^T$), those elements of the loads $\{p\}_n$ which are affected by this modification are easily calculated (see also Eq. 7.87). To be able to calculate the right-hand side of Eq. 7.142, the history of the updated loads on the boundary elements $\{p\}_i$ ($i = 1, 2, \ldots, n-1$) is needed. Formulating equilibrium leads to the resultant soil reactions at the center of the basemat.

$$\{R_o\}_n = [A]^T \{R_b\}_n \tag{7.143}$$

This algorithm is remarkably simple. However, the computational effort in evaluating the convolution integrals that affect Eq. 7.142's right-hand side is substantial.

For the simple structure shown in Fig. 7-19, slipping is disregarded. No uplift occurs during the first second of the time-history. The time-histories of the vertical soil reaction and of the overturning moment in the following 1.5 s are plotted in Fig. 7-20. For instance, the values of the vertical soil

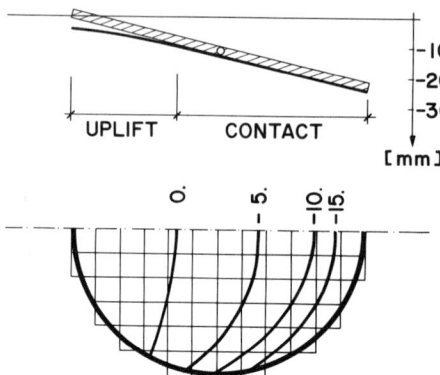

**Figure 7-21** Elevation of basemat and vertical soil pressure ([m/s $^2$], factor $\rho a$) at $t = 1.10$ s.

reaction shown in Fig. 7-20a, specified in m/s², must be multiplied by the factor $\rho a^3$ given in the caption to obtain the actual results. Partial uplift occurs most of the time. Compared to the linear analysis, a significant increase of the vertical reaction occurs, leading to high-frequency oscillations. The elevation of the basemat and the vertical soil pressure shortly after uplift starts are plotted in Fig. 7-21.

## 7.7 EMBEDDED FOUNDATION WITH SEPARATION OF SIDEWALL AND UPLIFT OF BASEMAT

### 7.7.1 Illustrative Example

As another example, the simple structure shown in Fig. 7-22, having a cylindrical foundation embedded in an elastic half-space, is examined for a vertical earthquake [W14]. In the area of contact between the structure and the soil, it is assumed that no tension can arise. This leads to a local nonlinearity consisting of the partial separation of the sidewall and partial uplift of the basemat.

The structure, which represents a typical nuclear-reactor building, is modeled by a single degree of freedom in the vertical direction with a fixed-base frequency of 10 Hz. The mass $m_s$ of the structure equals $50 \cdot 10^6$ kg and the base's mass $m_o = 25 \cdot 10^6$ kg. The ratio of the structure's viscous damping equals 0.05. The rigid cylindrical base of depth $e = 10$ m and radius $a = 20$ m is embedded in the viscoelastic half-space having a shear-wave velocity $c_s = 500$ m/s, and a mass density $\rho = 2.4 \cdot 10^3$ kg/m³ (which results in a shear modulus $G = 0.6 \cdot 10^9$ N/m²). Poisson's ratio $\nu = 0.33$ and a hysteretic damping ratio $\zeta = 0.05$. The vertically propagating waves' control motion is defined at the free surface and consists of an artificial time-history of 10 s

Sec. 7.7    Embedded Foundation with Separation of Sidewall and Uplift    415

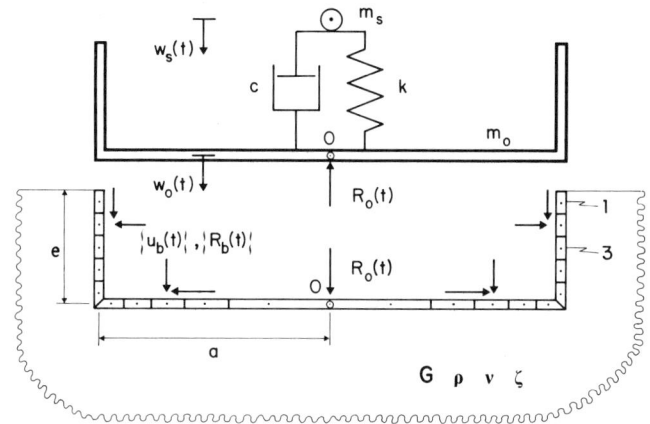

**Figure 7-22** Discretized structure–soil system.

duration with a response spectrum which closely follows that of the U.S. NRC Regulatory Guide: 1.60 normalized to 0.27g [U2].

### 7.7.2 Computational Procedure

The equations of motion in the time domain (Eq. 6.3) are formulated as (Fig. 7.22).

$$\begin{bmatrix} m_s & 0 \\ 0 & m_o \end{bmatrix} \begin{Bmatrix} \ddot{w}_s^t(t) \\ \ddot{w}_o^t(t) \end{Bmatrix} + \begin{bmatrix} c & -c \\ -c & c \end{bmatrix} \begin{Bmatrix} \dot{w}_s^t(t) \\ \dot{w}_o^t(t) \end{Bmatrix} + \begin{bmatrix} k & -k \\ -k & k \end{bmatrix} \begin{Bmatrix} w_s^t(t) \\ w_o^t(t) \end{Bmatrix} = \begin{Bmatrix} 0 \\ -R_o(t) \end{Bmatrix}$$

(7.144)

where $w_s^t(t)$ and $w_o^t(t)$ are the total displacements of the mass point and the base, respectively. $R_o(t)$ is the soil's interaction force. The structure's spring and damper coefficients are denoted as $k$ and $c$.

The structure–soil interface is discretized with 10 boundary elements: the sidewall with 5 cylinders of equal height and the basemat with 5 annular rings of equal area. The nodes $b$ are selected in the centers of the boundary elements, where the radial and vertical components of the displacements $\{u_b(t)\}$ are introduced. The shape functions $[N(s)]$ applicable to the boundary elements are selected as piecewise constant over each element.

The truncated indirect boundary-element method is used. To calculate $R_o(t)$ from the interaction forces $\{R_b(t)\}$ equilibrium is formulated as

$$R_o(t) = \{A\}^T \{R_b(t)\} \tag{7.145}$$

where the vector of kinematic transformation $\{A\}$ contains only zeroes and ones. The source line $S'$ is offset from the structure–soil interface $S$ by an infinitesimal amount, and the nodes associated with the loading coincide with those of the boundary elements. Over each element with the same dimension as the adjacent boundary element, the two components of the loads are constant, and thereby $[L(s')]$ is defined.

As the soil is assumed to remain elastic, the generalized scattered motion $\{u_b^g(t)\}$ can be calculated from that of the free field $\{u_b^f(t)\}$ working in the frequency domain (Eq. 6.12). While only vertical components will arise in $\{u_b^f(t)\}$, $\{u_b^g(t)\}$ contains, in general, radial and vertical components.

The computational procedure for the $n$th time step (from $(n-1)\Delta t$ to $n\Delta t$) proceeds as follows. The discussion can be restricted to the analysis of the base and the adjacent soil. All variables are known up to time $(n-1)\Delta t$. From the total motion at the center of the basemat at time $(n-1)\Delta t$, the vertical displacement in the same point at time $n\Delta t$ $(w_o^t)_n$ is predicted ($\beta = 0$). The interaction forces, formulated according to the truncated indirect boundary-element method, follow from Eq. 7.90, but with $\{\bar{u}_b\}_n$ specified as (Eqs. 7.84 and 7.96)

$$\{\bar{u}_b\}_n = \int_o^{\Delta t}\int_S [g_t(s,\Delta t - t')]_o^T [N(s)]\,ds\,\{u_b^t(t')\}_n\,dt' \\ - \int_o^{\Delta t}\int_S [g_t(s,\Delta t - t')]_o^T \{u^g(s,t')\}_n\,ds\,dt' \quad (7.146)$$

If tension arises for the components of $\{R_b\}_n$ in the direction normal to the structure–soil interface, the soil's corresponding boundary elements lose contact with the adjacent base of the structure. For the corresponding nodes, the forces' normal and tangential components are then set equal to zero, which modifies $\{R_b\}_n$. This also affects the source-load parameters $\{p\}_n$. Solving Eq. 7.88 for $\{p\}_n$ leads to

$$\{p\}_n = ([T]_o^T)^{-1}\left(-\sum_{i=1}^{n-1}[T]_{n-i}^T\{p\}_i + \{R_b\}_n\right) \quad (7.147)$$

This modified $\{p\}_n$ is used in all subsequent time steps and the modified $\{R_b\}_n$ is used to calculate $\{R_o\}_n$ (Eq. 7.145), which represents the resulting contribution of the soil's interaction forces to the equilibrium equation at the center of the basemat. This completes the calculation of the $n$th time step when an explicit integration scheme is used. The use of an implicit scheme requires iterations before proceeding to the next time step.

In all calculations using the boundary-element approach, an implicit scheme with a time step $\Delta t = 0.001$ s is chosen.

### 7.7.3 Linear Analysis

So comparisons can be made, a linear analysis in the frequency domain based on the indirect boundary-element method is performed using the same spatial discretization. As an example of an intermediate result, the dynamic-

Sec. 7.7  Embedded Foundation with Separation of Sidewall and Uplift    417

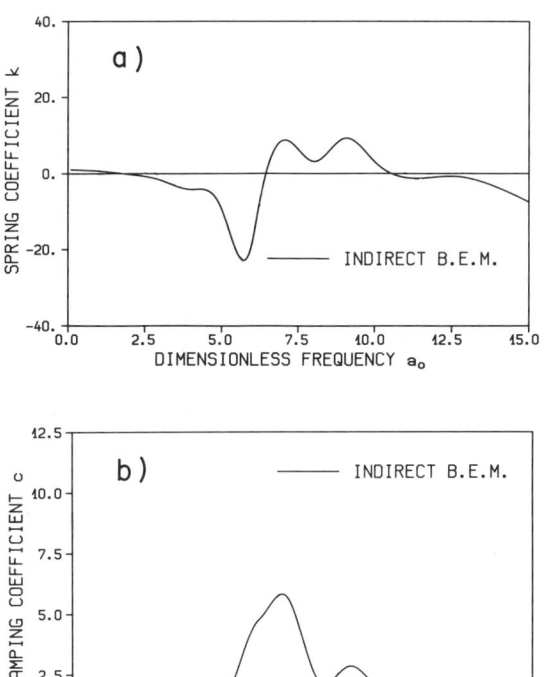

**Figure 7-23** Vertical dynamic-stiffness coefficient of free field.

stiffness coefficient in the vertical direction of the rigid foundation is presented in Fig. 7-23. This coefficient $S_{oo}^f$ refers to the system free field—that is, to the continuous soil. It is equal to the sum of the coefficients of the system ground $S_{oo}^g$ and of the excavated part of the soil $S_{oo}^e$

$$S_{oo}^f = S_{oo}^g + S_{oo}^e \qquad (7.148)$$

$S_{oo}^f$ is nondimensionalized as

$$S_{oo}^f = K(k + ia_o c) \qquad (7.149)$$

where the static-stiffness coefficient $K$ is equal to 7.64 $Ga$, and $a_o$ denotes the dimensionless frequency, defined as

$$a_o = \omega \frac{a}{c_s} \qquad (7.150)$$

$k$ and $c$ are the spring and damping coefficients.

For the purpose of checking, a linear analysis is also performed working exclusively in the time domain using the truncated indirect boundary-element formulation. The results agree extremely well with those obtained in the frequency domain (within 3%). The computational effort is, however, increased by two orders of magnitude.

### 7.7.4 Nonlinear Analysis

As an example of the temporal variation of Green's functions, the vertical displacement $g_u(\bar{t})$ and the vertical surface traction $g_t(\bar{t})$ in the upper of the two Gauss points of integration of the third boundary element indicated in Fig. 7-22 are plotted in Fig. 7-24. They are due to a vertical unit source load acting on the same source element. The dimensionless time $\bar{t}$ is introduced as

$$\bar{t} = \frac{tc_s}{a} \tag{7.151}$$

The initial value and the initial slope in the time domain ($\bar{t} = 0^+$) can be checked by using the corresponding asymptotic values in the frequency

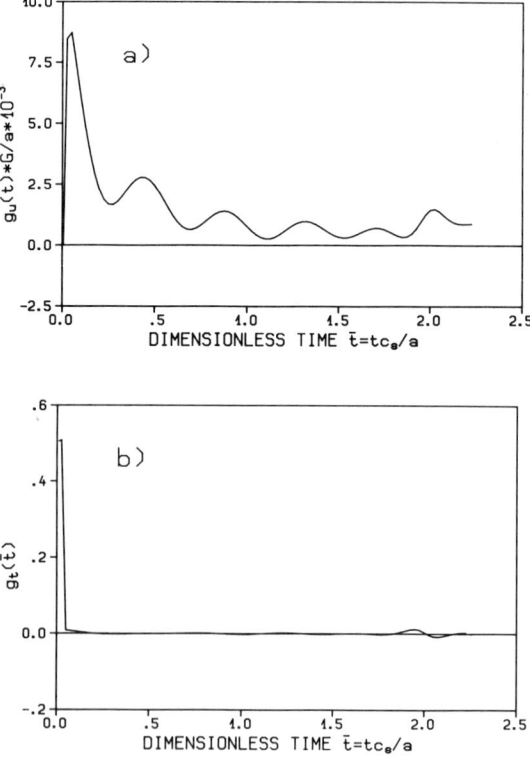

**Figure 7-24** Green's functions in time domain. a. Displacement. b. Surface traction.

Sec. 7.7   Embedded Foundation with Separation of Sidewall and Uplift        419

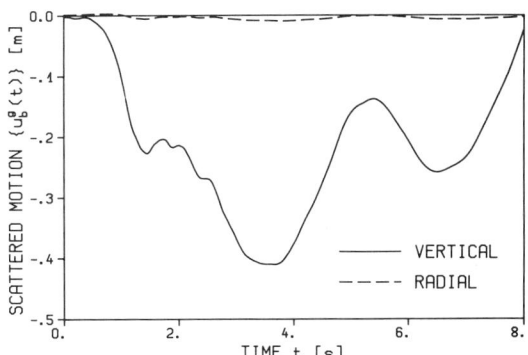

**Figure 7-25**   Vertical and radial displacements of generalized scattered motion.

domain ($ia_o \to \infty$, Eq. 6.118). The influence of the arrival of the shear wave traveling across the foundation is clearly visible at $\bar{t} = 2$.

The displacements of the generalized scattered motion $\{u_b^g(t)\}$ in the vertical and radial directions in the center of element 1 shown in Fig. 7-22 are presented in Fig. 7-25. The radial component is, as expected, small.

The final results of the nonlinear analysis are addressed next. The first 2.5 s of the time-histories of the total acceleration $\ddot{w}_s^t(t)$ of the mass point representing the structure and of the spring force of the structure are shown in Figs. 7-26 and 7-27. Compared to the results of the linear analysis, the calculation taking uplift and separation into account results in a slightly larger response and exhibits somewhat higher frequencies. As can be seen from the time-history of the number of boundary elements which lose contact as presented in Fig. 7-28, a strong nonlinear behavior occurs.

Finally, the stability of the algorithm is briefly discussed. To reduce the computational effort, it is tempting, in the truncated indirect boundary-element method, to retain only a limited number of terms in the two discre-

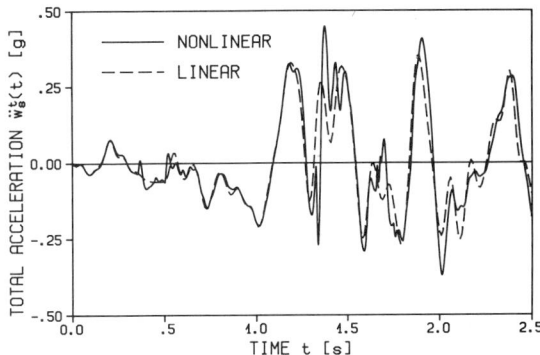

**Figure 7-26**   Total acceleration of mass.

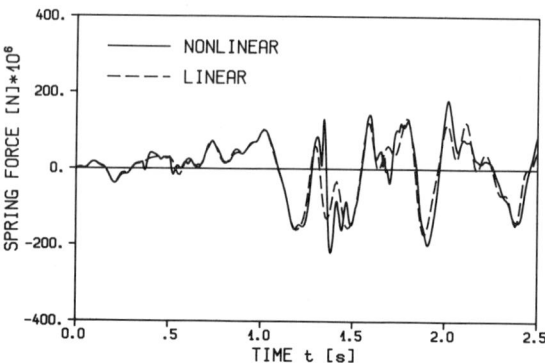

**Figure 7-27** Spring force.

tized convolution integrals on the right-hand side of Eq. 7.94. Besides decreasing the accuracy, this also affects the stability behavior. This is verified in Fig. 7-29, where the time-history of the interaction force $R_o(t)$ is examined. The smaller the number of retained terms, the earlier the algorithm becomes unstable.

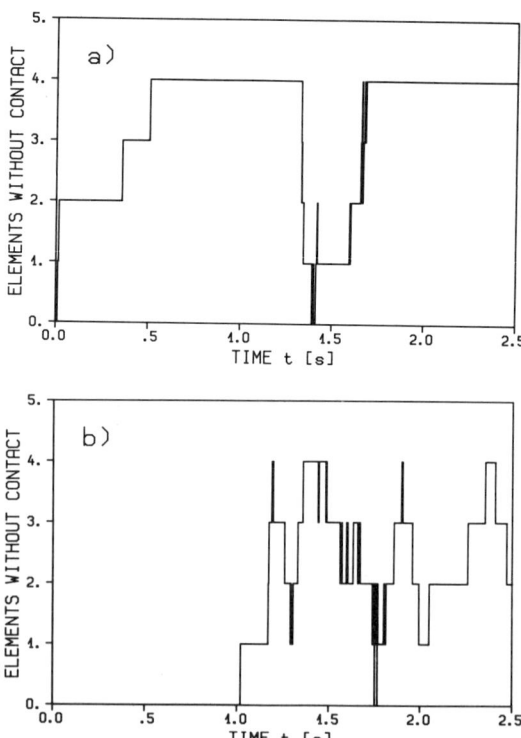

**Figure 7-28** Number of boundary elements with loss of contact. a. Vertical elements on sidewall. b. Horizontal elements on basemat.

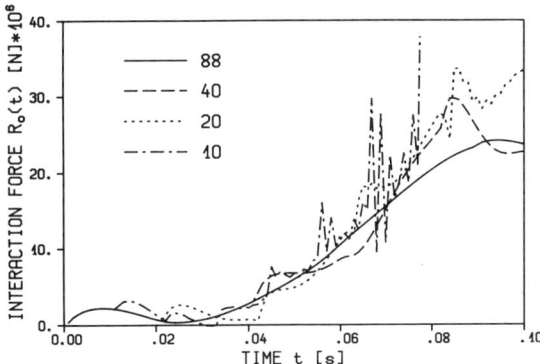

**Figure 7-29** Stability behavior as a function of number of retained terms in convolution integrals.

## SUMMARY

1. In all boundary-element methods in the time domain, Green's functions—that is, displacements and surface tractions for prescribed fictitious loads, such as a Dirac impulse in space and time—are calculated in a system which can easily be analyzed.

2. Closed-form expressions, which can be derived directly in the time domain, exist for the fundamental solution arising from a Dirac impulse source load applied to a full infinite space. The two- and three-dimensional solutions show different behaviors. While both exhibit a wave front (that is, the response vanishes until the P-wave has propagated from the source to the receiver), after the S-wave's arrival, the three-dimensional solution returns to zero, but the two-dimensional result has a tail—that is, the solution decays with time, but remains finite.

3. The Green's functions exhibit causality (that is, they vanish ahead of the P-wave front), translation in time (that is, the time origin can be shifted), and, for homogeneous boundary conditions, reciprocity (that is, symmetry).

4. It is conceptually preferable to analytically incorporate the conditions at the free surface and at the interfaces of the layered half-space into the Green's functions. Transformations from the time-space domain to the frequency-wave number domain and vice versa are applied. To determine the interaction force–displacement relationship, only a finite discretization on the structure–soil interface has then to be performed.

5. A problem in elasto-dynamics is governed by partial differential equations and by boundary and initial conditions or, alternatively, by an integral equation with time convolution, involving the boundary values, the loading, and the initial values. In this boundary-integral–equation

method, the number of the spatial dimensions is reduced by one; the problem can be formulated directly in the physically relevant boundary values, whereby the dynamic behavior of the unbounded domain, including the radiation condition, is captured by the Green's functions.

6. The direct boundary-element method can be based on the dynamic-reciprocity theorem which specifies a relationship between a pair of elasto-dynamic states. The first one will consist of the (specified) displacements and the (unknown) surface tractions corresponding to the interaction force–displacement relationship, and the second one will be made up of (known) Green's functions.

7. The various boundary-element methods are developed in the time domain for a foundation embedded in a layered half-space. They are the weighted-residual technique and the indirect boundary-element method, based on a weighted-residual equation, and the direct boundary-element method, based on a reciprocity equation; both equations involve time and space. In the indirect approach, formulating the weighted-residual equation over the last time step only, results in the truncated indirect boundary-element formulation. In all cases, discretized convolution integrals occur. In the weighted-residual technique and the truncated indirect boundary-element method, convolution integrals involving the source load's time-history arise; in the direct boundary-element method, convolution integrals arise with the time-histories of the surface tractions and the displacements. The displacements and the surface tractions (Green's functions) along the structure–soil interface are calculated for pulses of constant magnitude over one time step.

8. All boundary-element methods presented in the time domain are highly reliable, as is verified when the results of a linear analysis in the time domain are compared to those of the corresponding one in the frequency domain.

9. The evaluation of the convolution integrals is computationally expensive. The largest number of operations occurs in the indirect boundary-element method. This is reduced by a factor of two in the direct boundary-element method for typical problems. The weighted-residual and the truncated indirect boundary-element methods require significantly fewer operations. The most efficient procedure is the point-collocation scheme.

10. The procedure, based on Green's functions in the time domain, described in this chapter is based on the same assumptions as that based on the dynamic stiffness in the time domain developed in Chapter 6. Both make use of the Green's functions in the frequency-wave number domain and apply the boundary-element method. However, the sequence of the operations is different. In the method based on the dynamic stiffness, the transformation back to the time domain is per-

formed at the end of the method—that is, as many operations as possible are done in the frequency domain. It is thus efficient in those cases where most of the problem can be solved in this way. The calculation of the rigid disk's dynamic-flexibility coefficient is such an example, where the inverse transformation is restricted to the final result. On the other hand, the inverse transformation back to the time domain is performed in the method working with Green's functions (fundamental solutions) in the time domain already on the level of the individual boundary elements. It is thus well suited for problems where the corresponding degrees of freedom cannot be eliminated—as, for example, in the case of the partial basemat uplift.
11. As an example, the nonlinear soil–structure-interaction analysis of a structure embedded in a half-space with partial uplift of the basemat and separation of the sidewall is investigated.

## PROBLEMS

**7.1** Determine the Green's functions of the axial displacement $g_u(\bar{t})$ and of the normal force $g_t(\bar{t})$ for the semi-infinite rod on elastic foundation specified in Section 3.2.1 (Fig. 3-4). The unit impulse load is applied at $x = 0$ and the Green's functions are calculated at $x = 1/\kappa$, with $\kappa$ specified in Eq. 3.56a. Use a transformation into the frequency domain. Plot $g_u(a_o)$, $g_t(a_o)$, $g_u(\bar{t})$, and $g_t(\bar{t})$ in dimensionless form.

*Solution*

Frequency domain.
Dynamic-flexibility coefficient (Eq. 3.80 or Problem 6.9):

$$F(a_o) = \frac{1}{S(a_o)} = \frac{1}{K\sqrt{1-a_o^2}}$$

Displacement (Eqs. 3.59, 3.60):

$$u(x,a_o) = F(a_o) \exp(-\kappa x \sqrt{1-a_o^2})$$

Green's function of displacement:

$$g_u(a_o) = u(x = 1/\kappa, a_o) = \frac{1}{K} \frac{\exp(-\sqrt{1-a_o^2})}{\sqrt{1-a_o^2}}$$

Plot Fig. P 7-1a
Normal force (Eq. 3.54):

$$N(x,a_o) = EAu(a_o)_{,x}$$

Green's function of normal force:

$$g_t(a_o) = -N(x = 1/\kappa, a_o) = \exp(-\sqrt{1-a_o^2})$$

Plot Fig. P 7-1b.

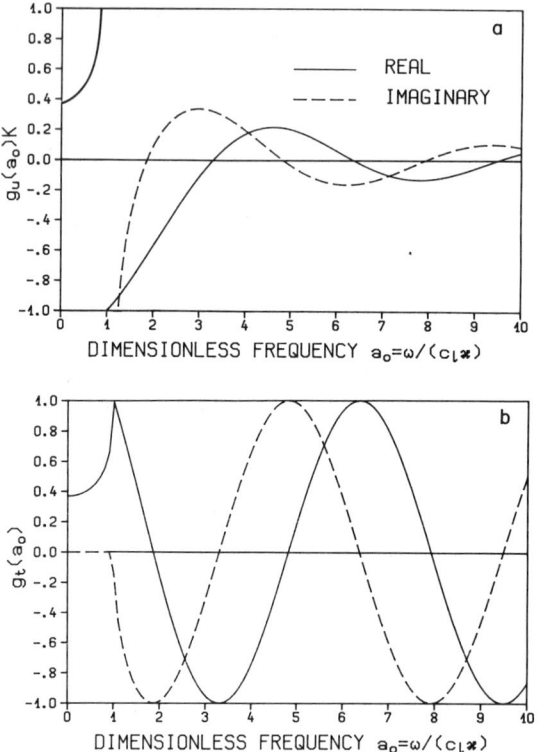

**Figure P7-1** Green's functions of semi-infinite rod on elastic foundation for Dirac impulse. a. Displacement in frequency domain. b. Normal force in frequency domain. c. Displacement in time domain. d. Normal force in time domain.

Transformation to time domain (Eq. 7.11b):

$$g(\bar{t}) = \frac{c_l \kappa}{2\pi} \int_{-\infty}^{+\infty} g(a_o) \exp(ia_o \bar{t}) \, da_o$$

Green's function of displacement:

$$g_u(\bar{t}) = \frac{c_l \kappa}{2\pi K} \int_{-\infty}^{+\infty} \frac{\exp(-\sqrt{1 - a_o^2})}{\sqrt{1 - a_o^2}} \exp(ia_o \bar{t}) \, da_o$$

Ref. [C1, No. 866]:

$$g_u(\bar{t}) = \frac{1}{\rho c_l A} J_o(\sqrt{\bar{t}^2 - 1}) \quad \bar{t} > 1$$

$$= 0 \quad \bar{t} < 1$$

Plot Fig. P 7-1c.

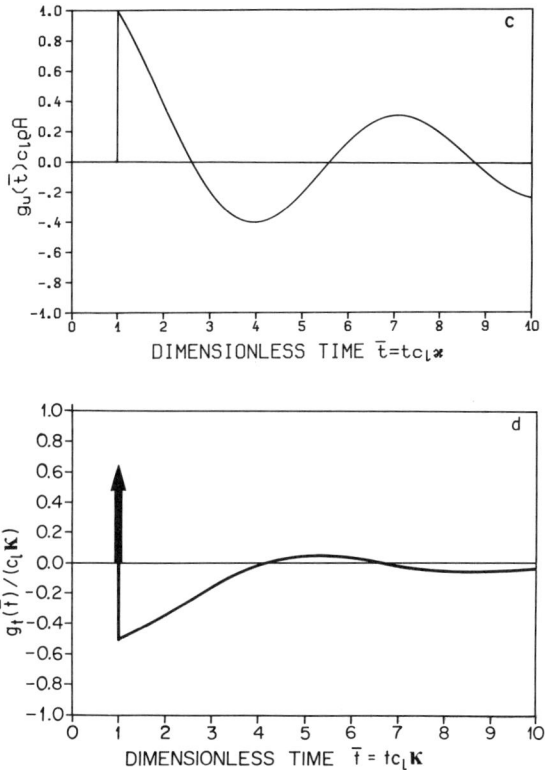

**Figure P7-1** Continued.

Green's function of normal force:

$$g_t(\bar{t}) = +\frac{c_l\kappa}{2\pi}\int_{-\infty}^{+\infty} \exp(-\sqrt{1-a_o^2})\exp(ia_o\bar{t})\, da_o$$

Ref. [C1, No. 865.1 and 601]:

$$g_t(\bar{t}) = -c_l\kappa\frac{J_1(\sqrt{\bar{t}^2-1})}{\sqrt{\bar{t}^2-1}} + c_l\kappa\delta(\bar{t}-1) \qquad \bar{t} > 1$$

$$= 0 \qquad\qquad\qquad\qquad\qquad\qquad \bar{t} < 1$$

Plot Fig. P 7-1d.

**7.2** The wave equation for out-of-plane motion in symmetric cylindrical coordinates $r$ for harmonic motion in the presence of a concentrated body load with unit amplitude acting at the origin is formulated as (Eq. 6.83)

$$v_{,rr}(r,\omega) + \frac{1}{r}v_{,r}(r,\omega) + \frac{\omega^2}{c_s^2}v(r,\omega) + \frac{1}{G} = 0$$

The Green's function (fundamental solution) in the frequency domain equals

$$g_v(r,\omega) = v(r,\omega) = \frac{-i}{4G} H_o^{(2)}\left(\frac{\omega}{c_s} r\right)$$

Determine the Green's function in the time domain by applying the inverse Fourier transformation.

*Solution*

Transformation of Hankel function to modified Bessel function of the second kind with imaginary argument (Ref. [C1]):

$$H_o^{(2)}(z) = \frac{2}{\pi} i K_o(iz)$$

With $z = \frac{\omega}{c_s} r$:

$$g_v(r,\omega) = \frac{1}{2\pi G} K_o\left(\frac{i\omega r}{c_s}\right)$$

Fourier transform:

$$g_v(r,t) = \frac{1}{2\pi} \int_{-\infty}^{+\infty} g_v(r,\omega) \exp(i\omega t)\, d\omega$$

Ref. [C1, No. 912.3]:

$$g_v(r,t) = \frac{1}{2\pi G} \frac{1}{\sqrt{t^2 - \frac{r^2}{c_s^2}}} \qquad t > \frac{r}{c_s}$$

which agrees with the result determined directly in the time domain from the three-dimensional scalar equation (Eq. 7.31).

**7.3** A rectangular pulse with magnitude $P_o$ and duration $\Delta t$ acts at $\tau = 0$ as a concentrated force in the x-direction at the origin of the full infinite space (Fig. P 7-3a). Determine and plot the displacement $u$ (Green's function) in the x-direction at a point with polar coordinates $(r, 45°)$ located in the x-y plane as a function of the dimensionless time $tc_p/r$, assuming a. the three-dimensional case and b. the two-dimensional case. Select $r/(c_p \Delta t) = 1$ and $c_p/c_s = \sqrt{2}$ (Poisson's ratio $= 0$).

*Solution*
a. Three-dimensional case.
Eq. 7.32, unit impulse:

$$u_{ii}(t) = \frac{1}{4\pi\rho}\left(\frac{3}{2} - 1\right)\frac{t}{r^3}\left[H\left(t - \frac{r}{c_p}\right) - H\left(t - \frac{r}{c_s}\right)\right]$$

$$+ \frac{1}{4\pi\rho c_p^2}\frac{1}{2r}\delta\left(t - \frac{r}{c_p}\right) + \frac{1}{4\pi\rho c_s^2}\left(1 - \frac{1}{2}\right)\frac{1}{r}\delta\left(t - \frac{r}{c_s}\right)$$

$$u(t) = \int_o^{\Delta t} u_{ii}(t - \tau) P_o [H(\tau) - H(\tau - \Delta t)]\, d\tau$$

$$u(t) = \frac{P_o}{8\pi\rho r^3}\left[f_p(t)\left\{H\left(t-\frac{r}{c_p}\right) - H\left(t-\Delta t-\frac{r}{c_p}\right)\right\} - f_s(t)\left\{H\left(t-\frac{r}{c_s}\right)\right.\right.$$
$$\left.\left. - H\left(t-\Delta t-\frac{r}{c_s}\right)\right\}\right] + \frac{P_o}{8\pi\rho r c_p^2}\left[H\left(t-\frac{r}{c_p}\right) - H\left(t-\Delta t-\frac{r}{c_p}\right)\right]$$
$$+ \frac{P_o}{8\pi\rho r c_s^2}\left[H\left(t-\frac{r}{c_s}\right) - H\left(t-\Delta t-\frac{r}{c_s}\right)\right]$$

with

$$f_p(t) = \begin{cases} \frac{1}{2}\left[t^2 - \left(\frac{r}{c_p}\right)^2\right] & \text{for } \frac{r}{c_p} < t < \Delta t + \frac{r}{c_p} \\ \frac{1}{2}[2t\Delta t - \Delta t^2] & t > \Delta t + \frac{r}{c_p} \end{cases}$$

$$f_s(t) = \begin{cases} \frac{1}{2}\left[t^2 - \left(\frac{r}{c_s}\right)^2\right] & \text{for } \frac{r}{c_s} < t < \Delta t + \frac{r}{c_s} \\ \frac{1}{2}[2t\Delta t - \Delta t^2] & t > \Delta t + \frac{r}{c_s} \end{cases}$$

For $\frac{r}{c_p} + \Delta t > \frac{r}{c_s}$, the first term equals

$$\frac{P_o}{8\pi\rho r^3}\begin{cases} \frac{1}{2}\left[t^2 - \left(\frac{r}{c_p}\right)^2\right] & \text{for } \frac{r}{c_p} < t < \frac{r}{c_s} \\ \frac{1}{2}\left[\left(\frac{r}{c_s}\right)^2 - \left(\frac{r}{c_p}\right)^2\right] & \frac{r}{c_s} < t < \frac{r}{c_p} + \Delta t \\ \frac{1}{2}\left[2t\Delta t - \Delta t^2 - t^2 + \left(\frac{r}{c_s}\right)^2\right] & \frac{r}{c_p} + \Delta t < t < \frac{r}{c_s} + \Delta t \end{cases}$$

$\bar{u}(t)$ is expressed as

$$\bar{u}(t) = \frac{P_o}{8\pi\rho r c_p^2}\bar{\bar{u}}(t)$$

Plot of dimensionless function $\bar{\bar{u}}(t = tc_p/r)$ in Fig. P 7-3b.

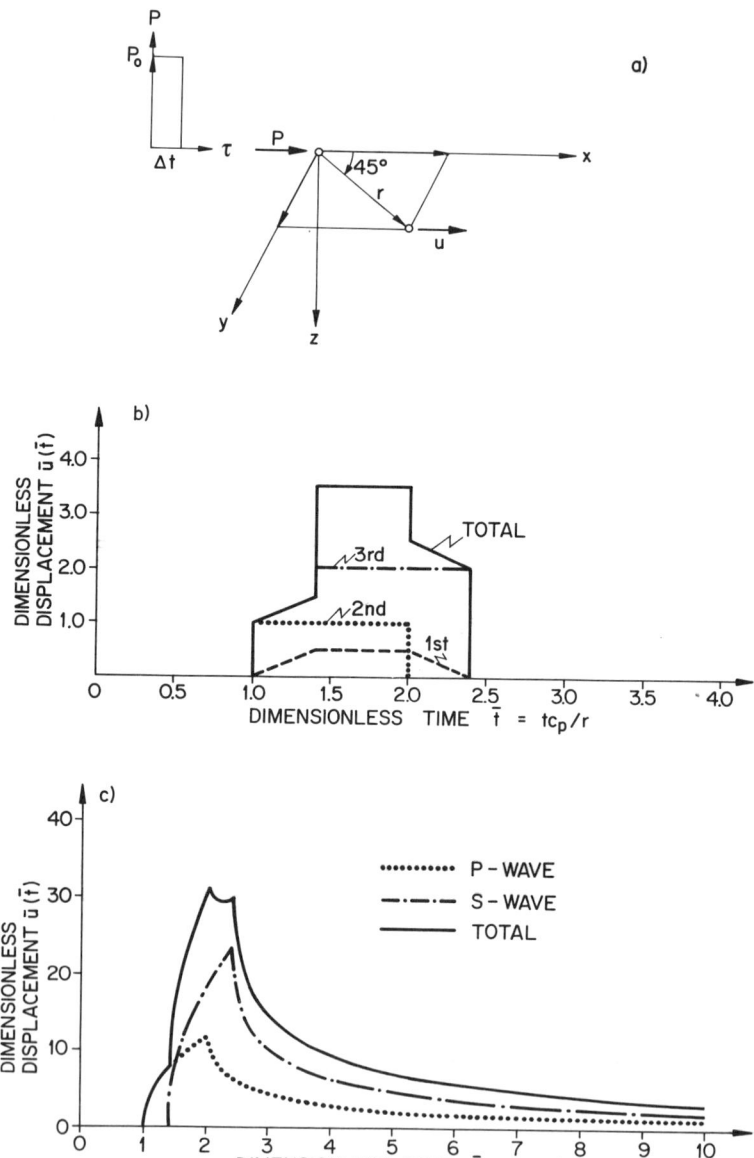

**Figure P7-3** Rectangular pulse acting in three- and two-dimensional full infinite spaces. a. Problem investigated. b. Green's function for three-dimensional case with first, second, and third terms. c. Green's function for two-dimensional case.

**Chap. 7   Problems**                                                   **429**

b. Two-dimensional case.
Eq. 7.34, unit impulse:

$$u_{ii}(t) = \frac{1}{2\pi\rho r^2}\left[-\sqrt{t^2 - \frac{r^2}{c_p^2}} + \frac{t^2 - \frac{r^2}{2c_p^2}}{\sqrt{t^2 - \frac{r^2}{c_p^2}}}\right]H\left(t - \frac{r}{c_p}\right)$$

$$-\frac{1}{2\pi\rho r^2}\left[-\sqrt{t^2 - \frac{r^2}{c_s^2}} + \frac{t^2 - \frac{3r^2}{2c_s^2}}{\sqrt{t^2 - \frac{r^2}{c_s^2}}}\right]H\left(t - \frac{r}{c_s}\right)$$

$$u(t) = \int_0^{\Delta t} u_{ii}(t-\tau)P_o[H(\tau) - H(\tau - \Delta t)]\,d\tau$$

$u(t)$ is expressed as

$$u(\bar{t}) = \frac{P_o}{2\pi\rho c_p^2}\bar{u}(\bar{t})$$

Plot of dimensionless function $\bar{u}(\bar{t} = tc_p/r)$ in Fig. P7-3c.

**7.4** Specify the Green's function for a unit impulse in space and time (fundamental solution) in the case of the out-of-plane motion for a homogeneous half-plane. Make use of the corresponding Green's function for the full infinite plane (Fig. P7-4).

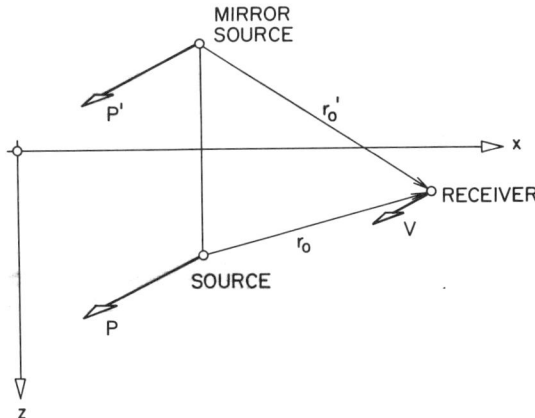

**Figure P7-4**   Green's function for half-plane, out-of-plane motion.

*Solution*

Green's function for full infinite plane (Eq. 7.31):

$$v(r_o, t) = \frac{1}{2\pi G} \frac{1}{\sqrt{t^2 - \frac{r_o^2}{c_s^2}}}$$

Mirror source symmetric with respect to free surface of half-plane with load $P'$ acting in same direction as $P$:

$$v(r'_o, t) = \frac{1}{2\pi G} \frac{1}{\sqrt{t^2 - \frac{r'^2_o}{c_s^2}}}$$

Superposition of two source loads leads to $\tau_{yz}(x) = 0$ on $z = 0$. Green's function for half-plane:

$$g_v(r_o, t) = \frac{1}{2\pi G} \left[ \frac{1}{\sqrt{t^2 - \frac{r_o^2}{c_s^2}}} + \frac{1}{\sqrt{t^2 - \frac{r'^2_o}{c_s^2}}} \right]$$

# References

[A1]  K. Aki and P.G. Richards, *Quantitative Seismology*, Vol. I (San Francisco, CA: W.H. Freeman and Company, 1980).

[A2]  R.J. Apsel, "Dynamic Green's Functions for Layered Media and Applications to Boundary-Value Problems" (Ph.D. Thesis, University of California, San Diego, CA 1979).

[B1]  E.P. Bayo and E.L. Wilson, "Use of Ritz Vectors in Wave Propagation and Foundation Response," *Earthquake Engineering and Structural Dynamics*, 12 (1984), 499–505.

[B2]  G.N. Bycroft, "Soil–Structure Interaction at Higher Frequency Factors," *Earthquake Engineering and Structural Dynamics*, 5 (1977), 235–248.

[B3]  R. Bracewell, *The Fourier Transform and its Application* (New York, NY: McGraw-Hill, 1965)

[C1]  G.A. Campbell and R.M. Foster, *Fourier Integrals for Practical Applications* (Princeton, NJ: Van Nostrand, 1967).

[C2]  A.K. Chopra, "Hydrodynamic Pressures on Dams During Earthquakes," *Journal of the Engineering Mechanics Division*, ASCE, 93 (1967), 205–223.

[C3]  R. Clayton and B. Engquist, "Absorbing Boundary Conditions for Acoustic and Elastic Wave Equations," *Bulletin of the Seismological Society of America*, 67 (1977), 1529–1540.

[C4]  M. Cohen and P.C. Jennings, "Silent Boundary Methods for Transient Analysis," in *Computational Methods for Transient Analysis*, T. Belytschko and T.J.R. Hughes, eds. (Amsterdam: Elsevier Science Publishers, 1983), pp. 301–360.

[C5]  S.H. Crandall, *Dynamic Response of Systems with Structural Damping* (Cambridge, MA: Massachusetts Institute of Technolgoy, Air Force Office of Scientific Research, 1961), AFOSR 1561.

[C6] P.A. Cundall, R.R. Kunar, P.C. Carpenter, and J. Marti, "Solution of Infinite Dynamic Problems by Finite Modelling in the Time Domain," *Proceedings of the 2nd International Conference on Applied Numerical Modelling*, Madrid, London: Pentech Press, 1974, 339–351.

[D1] G.R. Darbre and J.P. Wolf, "Criterion of Stability and Implementation Issues of Hybrid Frequency-Time Domain Procedure for Nonlinear Dynamic Analysis," *Earthquake Engineering and Structural Dynamics* (submitted).

[D2] Z.H. Duron, "A Non-Reflecting Boundary for Finite Elements" (Masters Thesis, Department of Civil Engineering, Massachusetts Institute of Technology, Cambridge, MA, 1983).

[E1] B. Engquist and A. Majda, "Absorbing Boundary Conditions for the Numerical Simulation of Waves," *Mathematics of Computation*, 31 (1977), 629–651.

[E2] A.C. Eringen and E.S. Suhubi, "Elastodynamics," vol. II, *Linear Theory*, (New York, NY: Academic Press, 1975).

[G1] G. Gazetas, "Analysis of Machine Foundations: State of the Art," *Soil Dynamics and Earthquake Engineering*, 2 (1983), 2–42.

[G2] G. Gazetas, "Rocking of Strip and Circular Footings," *Proceedings of the International Symposium on Dynamic Soil–Structure Interaction*, D.E. Beskos, T. Krauthammer, and I. Vardoulakis, eds. (Minneapolis and Rotterdam: A.A. Balkema, 1984), pp. 3–10.

[G3] K.F. Graff, *Wave Motion in Elastic Solids* (Oxford: Clarendon Press, 1975).

[H1] M. Harwood and M. Novak, "Uplift in Hammer Foundations," *Soil Dynamics and Earthquake Engineering*, 5 (1986), 102–117.

[H2] W.C. Hurty and M.F. Rubinstein, *Dynamics of Structures* (Englewood Cliffs, NJ: Prentice-Hall, 1964).

[K1] E. Kausel and J.M. Roesset, "Stiffness Matrices for Layered Soils," *Bulletin of the Seismological Society of America*, 71 (1981), 1743–1761.

[K2] E. Kausel and Z.H. Duron, "Theoretical Considerations on the Smith-Cundall-et al. Boundary," *Proceedings Fourth Engineering Mechanics Division Specialty Conference, ASCE*, vol. I (Lafayette, IN: Purdue University, 1983), pp. 154–157.

[K3] E. Kausel and J.M. Roesset, "Soil Amplification: Some Refinements," *Soil Dynamics and Earthquake Engineering*, 3 (1984), 116–123.

[K4] J.D. Kawamoto, "Solution of Nonlinear Dynamic Structural Systems by a Hybrid Frequency–Time-Domain Approach," Research Report R 83-5, Department of Civil Engineering, Massachusetts Institute of Technology, Cambridge, MA, 1983.

[K5] H. Kawase, T. Sato, S. Sato, and G. Tanaka, "Analytical and Observational Study on Seismic Scattering Property of a Site with Geologic Irregularity," *Transactions of the 8th International Conference on Structural Mechanics in Reactor Technology*, Brussels, 1985, vol. K(a), Paper K1/1, 1–6.

[K6] A.M. Kaynia, "Dynamic Stiffness and Seismic and Response of Pile Groups," Research Report R 82-03, Department of Civil Engineering, Massachusetts Institute of Technology, Cambridge, MA, 1982.

[K7] R.R. Kunar and J. Marti, "A Nonreflecting Boundary for Explicit Calculations," *Computational Methods for Infinite Domain Media-Structure Interaction*, Applied Mechanics Division, Vol. 46, ASME, 1981, 183–204.

[L1] Z.P. Liao and H.L. Wong, "A Transmitting Boundary for the Numerical Simulation of Elastic Wave Propagation," *Soil Dynamics and Earthquake Engineering*, 3 (1984), 174–183.

[L2] J.E. Luco and R.A. Westmann, "Dynamic Response of Circular Footings," *Journal of the Engineering Mechanics Division*, ASCE, 97 (1971), 1381–1395.

[L3] J.E. Luco and R.A. Westmann, "Dynamic Response of a Rigid Footing Bonded to an Elastic Halfspace," *Journal of Applied Mechanics*, ASME, 39 E (1972), 527–534.

[L4] J.E. Luco, "Linear Soil–Structure Interaction: A Review," *Applied Mechanics Division*, Vol. 53, ASME, 1982, 41–57

[L5] J.E. Luco, "On the Relation Between Radiation and Scattering Problems for Foundations Embedded in an Elastic Half-Space," *Soil Dynamics and Earthquake Engineering*, 5 (1986), 97–101.

[L6] J. Lysmer and R.L. Kuhlemeyer, "Finite Dynamic Model for Infinite Media," *Journal of the Engineering Mechanics Division*, ASCE, 95 (1969), 859–877.

[L7] J. Lysmer, "Analytical Procedures in Soil Dynamics," *Proceedings Specialty Conference on Earthquake Engineering and Soil Dynamics*, Geotechnical Engineering Division, ASCE, Pasadena, CA, 1978, Vol. III, 1267–1316.

[L8] J. Lysmer, M. Tabatabaie, F. Tajirian, S. Vahdani, and F. Ostadan, "SASSI, A System for Analysis of Soil-Structure Interaction," *UCB/GT*, 81–02, University of California, Berkeley, CA, 1981.

[L9] J. E. Luco, "CLASSI, A Program for Analysis of Soil-Structure Interaction," *Personal Communication*, University of California, San Diego, CA.

[M1] J.W. Meek and A.S. Veletsos, "Simple Models for Foundations in Lateral and Rocking Motion," *Proceedings of the 5th World Conference on Earthquake Engineering*, Rome, Italy, 1974, Vol. 2, 2610–2613.

[M2] G.F. Miller and H. Pursey, "The Field and Radiation Impedance of Mechanical Radiators on the Free Surface of a Semi-Infinite Isotropic Solid," *Proceedings Royal Society*, A 223 (1954), 521–541.

[M3] G.F. Miller and H. Pursey, "On the Partition of Energy between Elastic Waves in a Semi-Infinite Solid," *Proceedings Royal Society*, A 233 (1955), 55–69.

[M4] S.K. Mohasseb and J.P. Wolf, "Recursive Evaluation of Interaction Forces of Unbounded Soil in Frequency Domain," *Soil Dynamics and Earthquake Engineering*, (submitted).

[M5] S.K. Mohasseb, "Nonlinear Seismic Analysis of Fully Base-Isolated Structures on Flexible Soils" (Doctoral Dissertation, Institute of Structural Engineering, Swiss Federal Institute of Technology, Zurich, 1987).

[M6] M. Motosaka and J.P. Wolf, "Recursive Evaluation of Interaction Forces of

Unbounded Soil in Time Domain," *Transactions of the 9th International Conference on Structural Mechanics In Reactor Technology*, Lausanne, 1987, Vol. K1, Paper K 6/11, 355–362.

[N1]  *National Earthquake Hazards Reduction Program (NEHRP) Recommended Provision for the Development of Seismic Regulations for New Buildings*, Building Seismic Safety Council, Washington, D.C., 1986.

[N2]  B. Nour-Omid and R.W. Clough, "Dynamic Analysis of Structures Using Lanczos Co-ordinates," *Earthquake Engineering and Structural Dynamics*, 12 (1984), 565–577

[O1]  A.V. Oppenheim and R.W. Schaefer, *Digital Signal Processing* (Englewood Cliffs, NJ: Prentice-Hall, 1975).

[P1]  A. Pais and E. Kausel, "Stochastic Response of Foundations," Research Report R 85-6, Department of Civil Engineering, Massachusetts Institute of Technology, Cambridge, MA, 1985.

[P2]  C.L. Pekeris, "The Seismic Surface Pulse," *Proceedings National Academy of Science, Geophysics*, 41 (1955), 469–480.

[R1]  F.E. Richart, R.D. Woods, and J.R. Hall, *Vibrations of Soils and Foundations* (Englewood Cliffs, NJ: Prentice-Hall, 1970).

[R2]  F.J. Rizzo, D.J. Shippy, and M. Rezayat, "A Boundary Integral Equation Method for Radiation and Scattering of Elastic Waves in Three-Dimensions," *International Journal for Numerical Methods in Engineering*, 21 (1985), 115–129.

[R3]  J.M. Roesset and M.M. Ettouney, "Transmitting Boundaries: A Comparison," *International Journal for Numerical and Analytical Methods in Geomechanics*, 1 (1977), 151–176.

[R4]  J.M. Roesset, "Stiffness and Damping Coefficients of Foundations," *Dynamic Response of Pile Foundations: Analytical Aspects*, M.W. O'Neill and R. Dobry, eds., *Proceedings of the Geotechnical Engineering Division*, ASCE, October 1980, 1–30.

[R5]  Du Ruiming and E.L. Wilson, "An Effective Modified Ritz Vector Direct Superposition Method," *Transactions of the 8th International Conference on Structural Mechanics in Reactor Technology*, Brussels, 1985, Vol. B. Paper B 10/2, 375–380.

[S1]  I.S. Sandler, "A Method of Successive Approximations for Structure Interaction Problems," *Computational Methods for Infinite Domain Media-Structure Interaction*, Applied Mechanics Division, Vol. 46, ASME, 1981, 67–82.

[S2]  H.B.Seed and J. Lysmer, "The Significance of Site Response in Soil–Structure Interaction Analysis for Nuclear Facilities," *Proceedings Second ASCE Conference on Civil Engineering and Nuclear Power*, Knoxville, TN, 1980, Vol. II, Paper 14-1.

[S3]  W.D. Smith, "A Nonreflecting Plane Boundary for Wave Propagation Problems," *Journal of Computational Physics*, 15 (1974), 492–503.

[S4]  J.A. Stricklin and W.E. Hasler, "Formulation and Solution Procedures for Nonlinear Structural Analysis," *Computers and Structures*, 7 (1977), 125–136.

# References

[U1] P. Underwood and T.L. Geers, "Doubly Asymptotic Boundary-Element Analysis of Dynamic Soil–Structure Interaction," *International Journal of Solids and Structures*, 17 (1981), 687–697.

[U2] U.S. Nuclear Regulatory Commission, Washington, D.C., *Design Response Spectra for Seismic Design of Nuclear-Power Plants*, Regulatory Guide 1.60 (Washington, D.C.: U.S. Nuclear Regulatory Commission, October 1973).

[V1] D.K. Vaughan, G.L. Wojcik, and J. Isenberg, "Influence of Boundary Approximations on Soil–Structure Interaction Response," *Transactions of the 8th International Conference on Structural Mechanics in Reactor Technology*, Brussels, 1985, Vol. K(a), Paper K6/1, 209–214.

[V2] A.S. Veletsos and Y.T. Wei, "Lateral and Rocking Vibration of Footings," *Journal of the Soil Mechanics and Foundation Division*, ASCE, 97 (1971), 1227–1248.

[V3] A.S. Veletsos and B. Verbic, "Vibration of Viscoelastic Foundations," *Earthquake Engineering and Structural Dynamics*, 2 (1973), 87–102.

[V4] A.S. Veletsos and V.D. Nair, "Torsional Vibration of Viscoelastic Foundations," *Journal of the Geotechnical Engineering Division*, ASCE, 100 (1974), 225–245.

[V5] A.S. Veletsos and C.E. Ventura, "Modal Analysis of Non-Classically Damped Linear Systems," *Earthquake Engineering and Structural Dynamics*, 14 (1986), 217–243.

[V6] B. Verbic, "Analysis of Certain Structure–Foundation Interaction Systems" (Doctoral Thesis, Department of Civil Engineering, Rice University, Houston, TX, 1972)

[W1] M. Watabe, H. Yamanouchi, I. Ohkawa, O. Chiba, and M. Tohdo, "Building Effects of Irregular Site Condition on Seismic Response of Nuclear Power Plant," *Transactions of the 8th International Conference on Structural Mechanics in Reactor Technology*, Brussels, 1985, Vol. K(a), Paper K 5/8, 195–201.

[W2] D. Wepf, "Time-Domain Dam–Reservoir Interaction Analysis Based on Boundary Elements" (Doctoral Dissertation, Institute of Structural Engineering, Swiss Federal Institute of Technology, Zurich, 1987) (in German).

[W3] R.V. Whitman, "Soil–Platform Interaction," *Proceedings of Conference on Behaviour of Offshore Structures*, Norwegian Geotechnical Institute, Oslo, Vol. 1, 1976, 817–829.

[W4] E.L. Wilson, M.W. Yuan, and J.M. Dickens, "Dynamic Analysis by Direct Superposition of Ritz Vectors," *Earthquake Engineering and Structural Dynamics*, 10 (1982), 813–823.

[W5] J.P. Wolf, "Soil–Structure Interaction with Separation of Base Mat from Soil (Lifting-Off)" *Nuclear Engineering and Design*, 38 (1976), 357–384.

[W6] J.P. Wolf and P.E. Skrikerud, "Seismic Excitation with Large Overturning Moments: Tensile Capacity, Projecting Base Mat or Lifting-Off?," *Nuclear Engineering and Design*, 50 (1978), 305–321.

[W7] J.P. Wolf and P. Obernhuber, "Response of Structures Permitting Lift-Off to Rotational Input Motion from Horizontally Propagating Waves," *Proceedings*

*Second ASCE Conference on Civil Engineering and Nuclear Power*, Knoxville, TN, 1980, Vol. VI, Paper 4-2.

[W8] J.P. Wolf and P. Obernhuber, "Effects of Horizontally Propagating Waves on the Response of Structures with a Soft First Storey," *Earthquake Engineering and Structural Dynamics*, 9 (1981), 1–21.

[W9] J.P. Wolf and G.R. Darbre, "Dynamic-Stiffness Matrix of Soil by the Boundary Element Method: Conceptual Aspects," *Earthquake Engineering and Structural Dynamics*, 12 (1984), 385–400.

[W10] J.P. Wolf and G.R. Darbre, "Dynamic-Stiffness Matrix of Soil by the Boundary-Element Method: Embedded Foundation," *Earthquake Engineering and Structural Dynamics*, 12 (1984), 401–416.

[W11] J.P. Wolf, *Dynamic Soil–Structure Interaction* (Englewood Cliffs, NJ: Prentice-Hall, 1985).

[W12] J.P. Wolf and P. Obernhuber, "Nonlinear Soil–Structure Interaction Analysis Using Dynamic Stiffness or Flexibility of Soil in the Time Domain," *Earthquake Engineering and Structural Dynamics*, 13 (1985), 195–212.

[W13] J.P. Wolf and P. Obernhuber, "Nonlinear Soil–Structure-Interaction Analysis Using Green's Function of Soil in the Time Domain," *Earthquake Engineering and Structural Dynamics*, 13 (1985), 213–223.

[W14] J.P. Wolf and G.R. Darbre, "Nonlinear Soil–Structure-Interaction Analysis Based on the Boundary-Element Method in Time Domain with Application to Embedded Foundation," *Earthquake Engineering and Structural Dynamics*, 14 (1986), 83–101.

[W15] J.P. Wolf, "A Comparison of Time-Domain Transmitting Boundaries," *Earthquake Engineering and Structural Dynamics*, 14 (1986), 655–673.

[W16] J.P. Wolf and D.R. Somaini, "Approximate Dynamic Model of Embedded Foundation in Time Domain," *Earthquake Engineering and Structural Dynamics*, 14 (1986), 683–703.

[W17] J.P. Wolf and G.R. Darbre, "Time-Domain Boundary-Element Method in Visco-Elasticity with Application to a Spherical Cavity," *Soil Dynamics and Earthquake Engineering*, 5 (1986), 138–148.

[W18] J.P. Wolf, "Nonlinear Soil–Structure Interaction Analysis Based on Hybrid Frequency–Time-Domain Formulation," *Proceedings of 8th European Conference on Earthquake Engineering*, Lisbon, 1986, Vol. 2, 5.5/25-5.5/32.

[W19] J.P. Wolf and B. Weber, "Approximate Dynamic Stiffness of Embedded Foundation Based on Independent Thin Layers with Separation of Soil," *Proceedings of 8th European Conference on Earthquake Engineering*, Lisbon, 1986, Vol. 2, 5.6/33-5.6/40.

[W20] J.P. Wolf and M. Motosaka, "Recursive Evaluation of Interaction Forces of Unbounded Soil in Time Domain," and "Recursive Evaluation of Interaction Forces of Unbounded Soil in Time Domain from Dynamic-Stiffness Coefficients in Frequency Domain," *Earthquake Engineering and Structural Dynamics* (submitted).

[W21] H.L. Wong and J.E. Luco, "Tables of Impedance Functions and Input Motions for Rectangular Foundations," Report CE 78-15, Department of Civil Engineering, University of Southern California, Los Angeles, CA, 1978.

[W22] F.S. Wong and P. Weidlinger, "Dynamic Soil–Structure Interaction and the Design of Underground Shelters," *Proceedings, Design of Protective Structures, Bundesakademie für Wehrverwaltung und Wehrtechnik*, Germany, 1982.

[W23] H.L. Wong and J.E. Luco, "Tables of Impedance Functions for Square Foundations on Layered Media," *Soil Dynamics and Earthquake Engineering*, 4 (1985), 64–81.

# Index

Absorbing boundary, 76
Angle of incidence:
    in-plane motion, 183
    out-of-plane motion, 74, 113, 170, 181, 188
Anvil, 50
Artificial boundary, 5, 6, 97, 100, 125–147

Base isolation, 3, 153–157, 284–286
Basic equation of motion:
    frequency domain, 176, 202–203, 208
    time domain
        flexibility formulation, 249, 335
        stiffness formulation, 247, 248
Benchmark problem:
    definition, 96–97
    doubly asymptotic approximation, 110
    extrapolation algorithm, 120–123
    fictitious material damping, 107
    paraxial boundary, 114–118
    superposition boundary, 99–100, 106
    viscous damper, 109
Bessel transformation, 186, 376–377

Binomial coefficient, 120
Boundary condition:
    frequency-independent, 7, 76
    global, 6, 76, 244
    local, 7, 76
Boundary-element method:
    frequency domain, 195–201, 203–208
    static, 110
    time domain, 388–402
Boundary-integral-differential equation (*see* Boundary-integral equation)
Boundary-integral equation:
    frequency domain
        dynamic-stiffness coefficient, 194–195
        scattered motion, 202
    time domain, 383–388
Boundary zone, 101

Carrier, 93
Cauchy principle, 199
Causality, 276–277, 375
Characteristic equation, 134, 139
Coefficients for dampers and masses:
    conical rod in shear, 84
    conical rod in torsion, 88
    cylindrical cavity, 69

# Index

definition
  discrete model, 21
  lumped-parameter model, 15
disk on half-space
  discrete model, 24
  lumped-parameter model, 16
embedded cylinder, 29
embedded prism, 36
influence of damping, 47
rectangle on half-space, 36
spherical cavity, 22
square on layered half-space, 43
strip
  embedded, 70
  surface, 40
Complex eigenvalue approach, 54–57
Complex modal analysis, 54–57
Complex response, 175
Computational procedure:
  explicit integration
    flexibility formulation, 281–282
    stiffness formulation, 280–281
  implicit integration
    flexibility formulation, 283–284
    stiffness formulation, 282–283
Conical rod in shear, 81–84, 335
Conical rod in torsion, 86–88, 160
Conservatism, 3
Control motion, 8, 187
Control point, 8, 187
Convergence criterion, 216
Convergence properties, 227–228
Convolution integral:
  flexibility formulation, 249
  recursive evaluation
    frequency domain, 293–301
    time domain, 302–321
  stiffness formulation, 247
Corrector, 53, 279
Correspondence principle, 44, 47, 142, 180, 276–278, 341, 342
Curve fitting:
  discrete model, 22
  lumped-parameter model, 15
Cutoff frequency:
  fluid, 352
  layer, 23, 90, 132, 135, 141
  rod on elastic foundation, 89, 91, 126
  rod with exponentially increasing area, 166
Cylindrical cavity:
  discrete model, 69
  dynamic-stiffness coefficient, 69
  equation of motion, 171

Damper, frequency independent (*see also* Discrete model and Lumped-parameter model), 109, 148, 275–276, 285, 345
Damping coefficient:
  circular cavity
    horizontal, 63
    rocking, 64
    vertical, 64
  cylindrical cavity, 70
  discrete model, 22
  disk on half-space
    rocking, 25
    torsion, 66
    vertical, 26
  embedded cylinder
    coupling, 33
    horizontal, 30
    rocking, 32
    torsion, 34
    vertical, 31
  fluid, 351
  free field, 417
  layer, 136
  one-degree-of-freedom system
    hysteretic, 14
    viscous, 13
  rectangle on half-space
    rocking, 37, 38
    torsion, 39
  rod on elastic foundation, 95
  rod with exponentially increasing area, 338
  spherical cavity, 19, 20
  square on layered half-space
    horizontal, 44
    rocking, 46
    vertical, 45
  strip
    horizontal, 41
    rocking, 42
    rocking, embedded, 70
Damping ratio:
  hysteretic, 14
  modal, 54
  viscous, 13, 44, 257

Dam-reservoir interaction, 348
Decay function, 292
Diffraction (*see also* Scattered motion), 201
Direct boundary-element method:
 frequency domain
  dynamic-stiffness coefficient, 199, 200
  scattered motion, 204, 206–208
 time domain, 400–401, 409–410
Directionality of waves, 126, 258–260
Direct method:
 definition, 6–7
 doubly asymptotic approximation, 109–111
 extrapolation algorithm, 119–124
 free field, 7, 147–148
 fictitious material damping, 106–107
 location of artificial boundary, 125–147
 paraxial boundary, 111–119
 superposition boundary, 98–106
 transmitting boundary, 76
 viscous damper, 107–109
Discrete model:
 conical rod in torsion, 87–88, 160
 damping, 41, 44–48
 definition, 12, 20
 disk with mass on half-space, 49, 56
 dynamic-stiffness coefficient
  frequency domain, 22
 free field, 74
 rigid block on half-space, 73
 spherical cavity, 22, 85
 table
  disk on half-space, 24
  embedded cylinder, 29
  embedded prism, 36
  rectangle on half-space, 36
  square on layered half-space, 43
  strip, 40
 time domain, 336
Disk on half-space:
 conical rod in shear, 83–84
 conical rod in torsion, 88, 160
 discrete model, 24
 dynamic-flexibility coefficient in time domain, 274, 411

dynamic-stiffness coefficient in time domain, 258
 equivalent, 58
 lumped-parameter model, 16
Disk on layer, dynamic-flexibility coefficient in time domain, 275
Disk with mass on half-space, 48–50, 55–57
Dispersion:
 layer, 90
 rod on elastic foundation, 89, 91
 rod with exponentially increasing area, 166
Doubly asymptotic approximation, 109–111, 148, 166
Duhamel integral, 247
Dynamic-flexibility coefficient, frequency domain:
 free field, 192, 194
 rod on elastic foundation, 344
Dynamic-flexibility coefficient, time domain:
 calculation
  time domain, 267–269
  transformation from frequency domain, 270–276
 definition, 249
 disk on half-space, 274, 411
 disk on layer, 275
 hysteretic damping, 278
 initial characteristics, 271
 initial value, 272
 rod on elastic foundation, 345
 spherical cavity, 269
Dynamic-reciprocity theorem (*see* Reciprocity theorem)
Dynamic-stiffness coefficient:
 conical rod in shear, 84
 influence of damping, 45
 sidewall of cylinder, 63–64
 uplift, 59
Dynamic-stiffness coefficient, frequency and wave-number domains:
 halfspace
  in-plane motion, 184
  out-of-plane motion, 182
 layer
  in-plane motion, 185
  out-of-plane motion, 182
 site, 187
Dynamic-stiffness coefficient, frequency domain:
 circular cavity
  horizontal, 63

# Index

rocking, 64
vertical, 64
conical rod in torsion, 87
cylindrical cavity, 69
cylindrical segment, 266
definition, 6, 13
discrete model, 21, 22
disk on half-space
  rocking, 25
  torsion, 66
  vertical, 66
embedded cylinder
  coupling, 33
  horizontal, 30
  rocking, 32
  torsion, 34
  vertical, 31, 266
embedded foundation, 189–201
excavation, 177, 201
fluid, 352
free field, 177, 201, 417
ground, 177, 200
infinite frequency, 260–267
layer
  exact, 135
  extrapolation algorithm, 143, 174
  paraxial boundary, 141–142
  viscous damper, 140, 144
rectangle on half-space
  rocking, 37, 38
  torsion, 39
regular part, 254, 256, 292
rod on elastic foundation, 95
rod with exponentially increasing
  area
    Kelvin model, 341
    undamped, 336
    Voigt model, 342
singular part, 254, 256, 292
spherical cavity, 19, 20, 267
square on layered half-space
  horizontal, 44
  rocking, 46
  vertical, 45
strip
  horizontal, 41
  rocking, 42
  rocking, embedded, 70
structure, 176
Dynamic-stiffness coefficient, time
  domain:
  calculation
    time domain, 249–253
    transformation from frequency
      domain, 253–255
  definition, 246
  discrete model, 336
  disk on half-space, 257–258
  embedded foundation, 256,
    388–401
  excavation, 248, 256
  fluid, 352
  half-space, 256
  hysteretic damping, 277
  initial characteristics, 255
  layer, 340
  rod on elastic foundation, 344
  rod with exponentially increasing
    area
      Kelvin model, 341
      undamped, 337
      Voigt model, 342
  spherical cavity, 251, 252, 255

Eigenvalue, 55–57, 189, 226, 238
Effective density, 74
Elementary boundary condition, 98
Embedded cylinder:
  discrete model, 29
  hammer foundation, 52
Embedded prism, discrete model,
  32–34, 36
Embedment:
  cylinder, 25
  prism, 30
Equation of motion (see Wave
  equation)
Equivalent:
  disk, 58, 63
  radius, 59, 62
Error function, 105, 164
Euclidean norm, 216, 317
Excavation, dynamic-stiffness
  coefficient:
    frequency domain, 177, 201
    time domain, 248, 256
Explicit integration:
  extrapolation algorithm, 119
  flexibility formulation, 281–282
  formulation, 52–53, 279
  recursive evaluation
    flexibility formulation, 297–298
    stiffness formulation, 293–297
  superposition boundary, 106
  stiffness formulation, 280–281

Extrapolation algorithm, 119–124, 128, 139, 148, 170, 172, 173–174

Factor of safety, 4
Far field, 129, 173, 373
Fast Fourier Transform, 6, 214, 271, 291, 309
Flexibility coefficient, dynamic (*see* Dynamic-flexibility coefficient)
Fluid:
    dynamic-stiffness coefficient
        frequency domain, 352
        time domain, 352
    wave equation, 349
Fluid-structure interaction, 348
Fourier integral theorem, 254
Fourier series, 6, 134, 137, 175, 184, 377
Fourier transformation, 91, 92, 175, 182, 184, 237, 253, 254, 289–293, 376
Free field:
    analysis, 3
    boundary-integral equation, 202
    calculation, 187–189
    definition, 7
    dynamic-stiffness coefficient, 201, 417
    Green's function
        frequency domain, 191–194, 196
        time domain, 376–379
    loading, 147–148, 177, 247–248
    nonlinearity, 3, 333–334
Frequency:
    damped, 54, 56
    natural, 155
    undamped, 56
Frequency, cutoff (*see* Cutoff frequency)
Frequency-time-domain analysis (*see* Hybrid frequency-time-domain analysis)
Friction coefficient, 151, 155, 284
Friction plate, 3, 155, 284
Fundamental solution (*see* Green's function)

Gaussian distribution, 379
Generalized coordinate, 76
Geometrical spreading, 2
Green's function:
    frequency domain
        half-plane, 193–194
        half-space, 191–193, 196, 206
        infinite plane, 191, 426
        infinite space, 190
        rod on elastic foundation, 423
        spherical cavity, 366
    time domain
        embedded foundation, 418
        half-plane, 376, 383, 430
        half-space, 376–382
        horizontal load on free surface of half-space, 381
        infinite plane, 370, 371, 374
        infinite space, 368, 369, 372
        properties, 375
        rod on elastic foundation, 424–425
        spherical cavity, 363, 364–365
        vertical load on free surface of half-space, 379–381
Ground:
    dynamic-stiffness coefficient
        frequency domain, 198, 199
        time domain, 247, 256
    interaction force
        frequency domain, 177
        time domain, 247
    motion (*see* Scattered motion)
Group velocity, 90–91, 93–94

Half-plane:
    discrete model (*see* Discrete model)
    Green's function
        frequency domain, 193–194
        time domain, 376
Half-space:
    discrete model (*see* Discrete model)
    Green's function
        frequency domain, 191–193
        time domain, 376–379
Hammer foundation, 50–54
Head, 50
High-frequency behavior:
    dynamic-flexibility coefficient, 271
    increased directionality, 258–260
    regular part of dynamic-stiffness coefficient, 255
    singular part of dynamic-stiffness coefficient, 256
Hilbert transform, 278
Hybrid frequency-time-domain analysis:
    convergence criterion, 216
    formulation, 213–216
    segments, 227–228

Index 443

stability criterion
 harmonic excitation, 220–227
 transient excitation, 237–238
Hysteretic damping, 14, 41, 142, 177, 180, 276–278

Impedance, 80
Implicit integration:
 flexibility formulation, 283–284
 recursive evaluation
  flexibility formulation, 300–301
  stiffness formulation, 299–300
 stiffness formulation, 282–283
Impulse-invariant method, 304–309
Incoming wave, 19, 87, 112, 133, 166
Incompressible:
 fluid, 351, 353
 spherical cavity, 20
Indirect boundary-element method:
 frequency domain
  dynamic-stiffness coefficient, 198, 200, 208
  scattered motion, 204, 205–206, 208
 time domain, 393–399
 truncated, 394, 398
Infinite plane, Green's function:
 frequency domain, 191, 426
 time domain, 370, 371, 374
Infinite space, Green's function:
 frequency domain, 190
 time domain, 368, 369, 372
Initial characteristics:
 dynamic-flexibility coefficient, 271
 regular part of dynamic-stiffness coefficient, 255
Initial-stress approach, 213, 227
Initial value, 272
Initial-value theorem, 237, 255
Instability, 309
Interaction horizon, 4–6, 7
Interpolation:
 frequency domain, 301
 time domain, 310, 315
Irregular site, 205–208
Isolation (see Base isolation)

Kelvin model, 257, 266, 274, 275, 341, 346
Kinematic transformation matrix, 178, 410, 415

Lamb's problem, 379
Lanczos coordinate, 77
Law of momentum, 52, 272
Least-square method, 317–318
Loading environment, 3, 7, 147
Location of artificial boundary, 125–147, 172, 173
Love wave, 132, 152
Lumped-parameter model:
 definition, 12
 disk with mass on half-space, 49, 56
 generalized, 66–69
 rigid block on half-space, 72
 table, 16
 uplift, 59

Material damping, fictitious, 106–107
Maxwell-Betti (see Reciprocity theorem)
Mirror image, 100
Mirror source, 430
Modal analysis, complex, 54–57
Mode shape, 54, 55, 77, 189, 226
Modulation, 93
Multiple reflections, 100

Narrow-banded pulse, 92–94
Natural frequency, undamped, 13, 54
Newmark, 52–53, 106, 278–279, 412, 416
Non-causal behavior, 276–277, 375
Nonlinearity:
 base isolation, 153–157, 284
 free field, 3, 333
 hammer foundation, 50
 structure, 3, 246
 unbounded soil, 333
 uplift (see Uplift)
Nonreflecting boundary, 76
Normal mode, 54
Nuclear Regulatory Commission (see U.S. Nuclear Regulatory Commission)
Number of operations:
 direct boundary-element method, 397, 403
 frequency domain, stiffness formulation, 403
 indirect boundary-element method, 403
 point collocation, 403
 recursive evaluation, frequency domain, 301

Number of operations (continued)
  recursive evaluation, time domain, 321
  time domain, stiffness formulation, 403
  truncated indirect boundary-element method, 403
  weighted residual method, 403
Nyquist frequency, 237–238, 240

Outcrop, 8, 187
Outgoing wave, 4, 19, 85, 87, 100, 112, 119, 166
Overdamped system, 55

Paraxial boundary, 111–119, 138, 146, 148, 166–170, 171
Partial-fraction expansion, 316
Point collocation, 307, 403, 409
Potential, 18, 84, 368
Predictor, 52, 279, 412, 416
Prismatic rod, 78–81
Pseudo-force, 215
Pseudo-load, 215

Quiescent past, 375, 386, 387

Radiation condition:
  boundary-integral equation, 384
  conical rod in shear, 83
  cylindrical cavity, 265
  free field, 196
  prismatic rod, 80, 132
  spherical cavity, 85
Radiation damping, 4, 20, 95, 106, 254
Radiation energy, 13, 90, 182
Radiation problem, 202
Ratio of polynomials, 315, 357
Rayleigh wave, 142, 152
Receiver, 190, 363
Reciprocity theorem:
  frequency domain, 195, 198, 199
  time domain, 375, 384–386, 398
Rectangle on half-space, discrete model, 36
Recursive equation:
  exact, 309
  frequency domain, 294–295, 298, 300
  time domain
    direct form, 303, 316
    parallel form, 317, 359
Recursive evaluation, frequency domain:
  explicit integration
    flexibility formulation, 297–298
    stiffness formulation, 293–297, 353–354, 358–360
  implicit integration
    flexibility formulation, 300–301
    stiffness formulation, 299–300
Recursive evaluation, time domain:
  directly from frequency domain, 315–321
  impulse-invariant method, 304–309, 355, 356
  segment approach, 310–315
Regular part (*see* Dynamic-stiffness coefficient)
Representation theorem:
  frequency domain, 195, 202
  time domain, 386–388
Restitution coefficient, 50
Retarded time, 369
Retrograde motion, 152
Ritz vector, 77
Rod on elastic foundation:
  definition of benchmark problem, 96–97
  dynamic-flexibility coefficient
    frequency domain, 334
    time domain, 335
  dynamic-stiffness coefficient
    frequency domain, 95
    time domain, 334
  equation of motion, 89
  Green's function
    frequency domain, 423
    time domain, 424–425
  location of artificial boundary, 126
  transmitting boundary, 99–100, 106, 107, 109, 110, 114–118, 120–123
  wave motion, 88–94
Rod with exponentially increasing area:
  dynamic-flexibility coefficient
    frequency domain, 346
    time domain, 346
  dynamic-stiffness coefficient
    frequency domain, 169, 341, 342
    time domain, 337, 341, 342
  transmitting boundary, 164–169

San Fernando earthquake, 151–153
Scattered motion, generalized, 178, 201–209, 249, 419
Segment approach, 310–315
Separation of sidewall:
  approximate analysis, 62–65
  rigorous analysis, 415–416
Silent boundary, 76
Singular part (*see* Dynamic-stiffness coefficient)
Slipping, 64, 151, 155
Source surface, 196, 201, 203, 205–206, 208, 390
Spectral radius, 226
Spherical cavity:
  boundary element, 407–410
  discrete model, 22, 85
  dynamic-flexibility coefficient
    frequency domain, 270
    time domain, 269, 278
  dynamic-stiffness coefficient
    frequency domain, 19, 277
    time domain, 251–252, 277
  equation of motion
    frequency domain, 18
    time domain, 84
  Green's function
    frequency domain, 366
    time domain, 363, 364–365
Spherical cavity with mass, 287–289, 292–293, 296–297, 298, 300, 301
Spring coefficient:
  circular cavity
    horizontal, 63
    rocking, 64
    vertical, 64
  cylindrical cavity, 70
  discrete model, 22
  disk on half-space
    rocking, 25
    torsion, 66
    vertical, 26
  embedded cylinder
    coupling, 33
    horizontal, 30
    rocking, 32
    torsion, 34
    vertical, 31
  fluid, 351
  free field, 417
  layer, 136

one-degree-of-freedom system
  hysteretic, 14
  viscous, 13
rectangle on half-space
  rocking, 37, 38
  torsion, 39
rod on elastic foundation, 95
rod with exponentially increasing area, 338
spherical cavity, 19, 20
square on layered half-space
  horizontal, 44
  rocking, 46
  vertical, 45
strip
  horizontal, 41
  rocking, 42
  rocking, embedded, 70
Spring, frequency independent (*see also* Discrete model and Lumped-parameter model), 109, 148, 275–276, 285, 345
Square on layered half-space, discrete model, 43
Stability, 420
Stability criterion:
  harmonic excitation
    multiple-degree-of-freedom system, 226
    one-degree-of-freedom system, 223, 227
  transient excitation, 238
Static-stiffness coefficient:
  conical rod in shear, 84
  conical rod in torsion, 88
  cylindrical cavity, 69
  disk on half-space, 16, 24
  embedded cylinder, 29
  embedded prism, 32–34
  layer, 135
  rod on elastic foundation, 95
  rod with exponentially increasing area, 336
  spherical cavity, 19
  square on layered half-space, 43
  strip, embedded, rocking, 70
Stiffness coefficient, dynamic (*see* Dynamic-stiffness coefficient)
Storage requirement:
  recursive evaluation, frequency domain, 286
  recursive evaluation, time domain, 321

Strip, discrete model
  embedded, 70
  surface, 40
Substructure method:
  definition, 4–6
  free field, 7, 187
  frequency domain, 175–209
  time domain
    dynamic-stiffness coefficient, 244–326
    Green's function, 361–421
Successive Fourier transformations, 289–293
Superposition boundary, 98–106, 125, 127, 138, 148, 161–162, 162, 162–164, 166, 172

Three-parameter model (see Kelvin model)
Through-soil coupling, 2
Transformation matrix, kinematic (see Kinematic transformation matrix)
Transmitting boundary:
  doubly asymptotic approximation, 109–111
  extrapolation algorithm, 119–124
  fictitious material damping, 106–107
  location, 125–147
  paraxial boundary, 111–119
  superposition boundary, 98–106
  viscous damper, 107–109
Truncated indirect boundary-element method, 394, 398, 410–411, 412–413, 415–416

Underdamped system, 55
Uplift:
  analysis
    equivalent radii, 58–62
    frequency-independent springs and dampers, 148–151
    iterative, 232–236
    rigorous, 414, 415–416
  anvil, 50
  criterion, 58, 151
  parametric study, 61–62
  reactor building, 61–62, 151–153, 414–421
  rigid block, 232–236
U.S. Nuclear Regulatory Commission, 61, 284, 287, 412, 415

Viscous damper, 107–109, 126, 128, 138, 146, 147, 166, 171, 172
Viscous damping (see also Kelvin and Voigt models), 13, 41, 257
Voigt model, 41, 257, 267, 274, 275, 342, 346

Water (see Fluid)
Wave equation:
  conical rod in shear, 82
  conical rod in torsion, 86
  cylindrical cavity, 171
  fluid, 348
  out-of-plane motion, 111
  prismatic rod, 79
  rod on elastic foundation, 89
  rod with exponentially increasing area, 165
  scalar, 368
  spherical cavity
    frequency domain, 18
    time domain, 85
Wave pattern, 8, 187
Weighted-residual method:
  frequency domain
    dynamic-stiffness coefficient, 196–198, 200
    scattered motion, 203–204
  time domain
    boundary-integral equation, 383
    dynamic-stiffness coefficient, 389–393
    spherical cavity, 407–409

Z-Transform, 312, 316, 359